Lecture Notes in Physics

Volume 828

For further volumes:
http://www.springer.com/series/5304

The Lecture Notes in Physics

The series Lecture Notes in Physics (LNP), founded in 1969, reports new developments in physics research and teaching—quickly and informally, but with a high quality and the explicit aim to summarize and communicate current knowledge in an accessible way. Books published in this series are conceived as bridging material between advanced graduate textbooks and the forefront of research and to serve three purposes:

- to be a compact and modern up-to-date source of reference on a well-defined topic
- to serve as an accessible introduction to the field to postgraduate students and nonspecialist researchers from related areas
- to be a source of advanced teaching material for specialized seminars, courses and schools

Both monographs and multi-author volumes will be considered for publication. Edited volumes should, however, consist of a very limited number of contributions only. Proceedings will not be considered for LNP.

Volumes published in LNP are disseminated both in print and in electronic formats, the electronic archive being available at springerlink.com. The series content is indexed, abstracted and referenced by many abstracting and information services, bibliographic networks, subscription agencies, library networks, and consortia.

Proposals should be sent to a member of the Editorial Board, or directly to the managing editor at Springer:

Christian Caron
Springer Heidelberg
Physics Editorial Department I
Tiergartenstrasse 17
69121 Heidelberg/Germany
christian.caron@springer.com

Eleftherios Papantonopoulos
Editor

From Gravity to Thermal Gauge Theories: The Ads/CFT Correspondence

Springer

Editor
Eleftherios Papantonopoulos
Department of Physics
National Technical University
Zografou Campus
157 80 Athens
Greece
e-mail: lpapa@central.ntua.gr

ISSN 0075-8450 e-ISSN 1616-6361

ISBN 978-3-642-04863-0 e-ISBN 978-3-642-04864-7

DOI 10.1007/978-3-642-04864-7

Springer Heidelberg Dordrecht London New York

Cover design: eStudio Calamar, Berlin/Figueres

Printed on acid-free paper

Springer is part of Springer Science+Business Media (www.springer.com)

Preface

This book is an edited version of the review talks given in the Fifth Aegean School on the AdS/CFT correspondence, held in Adamas on Milos Island, Greece, from 21st to 26th of September 2009. The aim of this book is to present an advanced multiauthored textbook which meets the needs of both the postgraduate students and the young researchers, in the fields of Field Theory, String Theory, Gravity and Condensed Matter Physics.

The AdS/CFT correspondence is a powerful tool in studying strongly coupled phenomena in field theory using results from a weakly coupled gravity background. The principle of the AdS/CFT was developed within the string theory and it was proved to be very useful in describing strongly coupled phenomena in gauge theories like quark-gluon plasma and heavy ions collisions. Soon it was realized that its applicability can be extended, in a more phenomenological approach, to condensed matter systems and to systems described by fluid dynamics.

The selected contributions to this volume are aimed to describe the principle of the AdS/CFT correspondence in its field theoretic formulation in string theory, its applicability to holographic QCD and to heavy ions collisions and to give an account of processes in fluid dynamics and of phenomena in condensed matter physics, which can be studied with the use of this principle.

In the introductory part of the book the article by Christos Charmousis after introducing the basic properties of the anti de Sitter spacetime, it discusses the static black holes in this space, their basic properties and the novel topological effects due to the presence of a negative cosmological constant. The anti-de Sitter spacetime is a necessary ingredient to build up the gravity sector of the dual conformal field theory. To extract information about the transport coefficients of the boundary dual theory the properties of the background gravity sector should be known, best described by perturbation theory. The article by George Siopsis reviews the perturbations of black holes in asymptotically anti-de Sitter space and it shows how the quasi-normal modes governing these perturbations can be calculated analytically and it discusses the implications on the hydrodynamics of gauge theory fluids per the AdS/CFT correspondence.

The contribution by Philip Argyres introduces the basic concepts of the correspondence and its foundation in string theory. He first reviews the properties of d-dimensional conformal field theories and describes their relation to quantum gravitational theories on $(d + 1)$-dimensional anti-de Sitter spacetimes. The 't Hooft limit of $U(N)$ Yang-Mills theory is reviewed and then it is described how an appropriate limit of type IIB superstring theory with D3-branes can be used to motivate a precise and computable correspondence between a 4-dimensional conformal field theory and a quantum gravitational theory on $AdS_5 \times S^5$. He finally discusses an extension of this construction in which probe branes on the AdS spacetime are included.

The second part of the book deals with the holographic realization of the AdS/CFT correspondence. The first article presented by Elias Kiritsis reviews the applications of the correspondence to QCD. In particular it provides a detail review of holographic models based on Einstein-dilaton gravity with a potential in 5 dimensions. Such theories, for a judicious choice of potential are very close to the physics of large-N Yang-Mills theory both at zero and finite temperature. The model can be used to calculate transport coefficients, like bulk viscosity, the drag force jet quenching parameters, relevant for the physics of the quark-gluon plasma.

The next article by Romuald Janik in a pedagogical way presents the techniques of the AdS/CFT correspondence which can be applied to the study of real time dynamics of a strongly coupled plasma system. These methods are based on solving gravitational Einstein's equations on the string/gravity side of the AdS/CFT correspondence. These AdS/CFT methods provide a fascinating arena interrelating General Relativity phenomena with strongly coupled gauge theory physics.

The contribution by Veronika Hubeny discusses the fluid/gravity correspondence which originates from the AdS/CFT correspondence. This correspondence constitutes a one-to-one map between configurations of a conformal fluid dynamics in d dimensions and solutions to Einsteins equations in $d + 1$ dimensions. In particular, the bulk solutions describe a regular generic, non-uniform and dynamical black hole which at late times settles down to a stationary planar black hole. The iterative construction of such solutions is indicated and the key physical properties are extracted.

The article by Amos Yarom is on heavy ions collisions. This contribution provides a review of two particular applications of the gauge-gravity duality to heavy ion collisions. The first involves a study of the wake of a quark as it travels through the quark gluon plasma and its possible connection to measurements of jet correlations carried out at the relativistic heavy ion collider at Brookhaven. The second section provides, via the gauge/gravity duality, a lower bound on the entropy produced in a collision of two energetic distributions. This is then compared to particle multiplicity in gold–gold collisions.

This part of the book ends with the article by Jiro Soda. In his article, he explains how the AdS/CFT correspondence is related to the Randall–Sundrum braneworld models. There are two different versions of these braneworlds models, namely, the single-brane model and the two-brane model. In the case of the single-brane model, the relation between the geometrical and the AdS/CFT correspondence approach using the gradient expansion method is revealed. In the case of two-brane system, he shows that the AdS/CFT correspondence play an important role in the sense that the low energy effective field theory can be described by the conformally coupled scalar-tensor theory where the radion plays the role of the scalar field.

The last part of the book consists of four articles and they discuss various aspects of the application of the AdS/CFT correspondence to condensed matter physics. In the first article Subir Sachdev reviews two classes of strong coupling problems in condensed matter physics, and describes insights gained by application of the AdS/CFT correspondence. The first class concerns non-zero temperature dynamics and transport in the vicinity of quantum critical points described by relativistic field theories. The second class concerns symmetry breaking transitions of two-dimensional systems in the presence of gapless electronic excitations at isolated points or along lines (i.e. Fermi surfaces) in the Brillouin zone.

The next article by Gary Horowitz gives an introduction to the theory of holographic superconductors. These are superconductors that have a dual gravitational description using gauge/gravity duality. After introducing a suitable gravitational theory, he discusses its properties in various regimes: the probe limit, the effects of backreaction, the zero temperature limit, and the addition of magnetic fields. Using the gauge/gravity dictionary, these properties reproduce many of the standard features of superconductors.

The string theory realization of holographic superconductors is given in the next contribution by Matthias Kaminski. After reviewing the basic D-brane physics and gauge/gravity methods at finite temperature he constructs the gravity dual of a D3/D7-brane system yielding a superconducting or superfluid vector-condensate. The corresponding gauge theory is $3 + 1$-dimensional $N = 2$ supersymmetric Yang-Mills theory with $SU(N_c)$ color and $SU(2)$ favor symmetry and it shows a second order phase transition typical to superconductivity. Condensates of this nature are comparable to those recently found experimentally in p-wave superconductors such as a ruthenate compound.

The final article is by Tassos Petkou who discusses some attempts to construct a Kalb-Ramond superconductor. The article starts by explaining theholographic implications of torsional degrees of freedom in the context of AdS_4/CFT_3, emphasizing in particular the physical interpretation of the latter as carriers of the non-trivial gravitational magnetic field. It presents a new exact 4-dimensional gravitational background with torsion and argue that it corresponds to the holographic dual of a 3d system undergoing parity symmetry breaking. Finally, it compares the new gravitational background with known wormhole solutions—with and without cosmological constant—and argue that they can all be unified under an intriguing Kalb-Ramond superconductivity framework.

The Fifth Aegean School and consequently this book, became possible with the kind support of many people and organizations. The School was organized by the Physics Department of the National Technical University of Athens, and supported by the Physics Department of the University of Tennessee. We received financial support from the following sources and organizations and this is gratefully acknowledged: Ministry of National Education and Religious Affairs, Alexander Onassis Foundation, the "S&B Industrial Minerals S.A.", General Secretariat of Aegean and Islands Policy, Springer Lecture Notes in Physics. We specially thank the staff of the Milos Conference Center "George Iliopoulos" for making available to us all the excellent facilities of the Center and for helping us to run smoothly the school. The administrative support of the Fifth Aegean School was taken up with great care by Fani and Christina Siatra. We acknowledge the help of Vassilis Zamarias who designed and maintained the website of the School. We also thank Kostas Anagnostopoulos for helping us in editing this book.

Last, but not least, we are grateful to the staff of Springer-Verlag, responsible for the Lecture Notes in Physics, whose abilities and help contributed greatly to the appearance of this book.

Athens, July 2010 Lefteris Papantonopoulos

Contents

List of Contributors

Amos Yarom, Joseph Henry Laboratories, Princeton University, 08540, Princeton, NJ, USA; e-mail: ayarom@princeton.edu

Anastasios C. Petkou, Department of Physics, University of Crete, 71003, Heraklion, Greece; e-mail: petkou@physics.uoc.gr

Christos Charmousis, LPT, Université de Paris-Sud, Bât 210, 91405, Orsay CEDEX, France; and LMPT, Parc de Grandmont, Université Francois Rabelais, 37200, Tours, France; e-mail: Christos.Charmousis@th.u-psud.fr

Elias Kiritsis, Crete Center for Theoretical Physics, Department of Physics, University of Crete, 71003, Heraklion, Greece; and APC, Université Paris, 7, (UMR du CNRS 7164) Bâtiment Condorcet, F-75205, Paris Cedex 13, France; e-mail: kiritsis@physics.uoc.gr

Francesco Nitti, APC, Université Paris, 7, Bâtiment Condorcet, F-75205, Paris Cedex 13, France; e-mail: nitti@cpht.polytechnique.fr

Gary T. Horowitz, Physics Department, UCSB, 93106, Santa Barbara, CA, USA; e-mail: gary@physics.ucsb.edu

George Michalogiorgakis, Department of Physics, Purdue University, 525 Northwestern Avenue, 47907-2036, West Lafayette, IN, USA; e-mail: michalogiorgakis@gmail.com

George Siopsis, Department of Physics and Astronomy, The University of Tennessee, 37996-1200, Knoxville, TN, USA; e-mail: siopsis@tennessee.edu

Jiro Soda, Department of Physics, Kyoto University, 606-8501, Kyoto, Japan; e-mail: jiro@tap.scphys.kyoto-u.ac.jp

Liuba Mazzanti, Department of Particle Physics and IGFAE, University of Santiago de Compostela, E-15782, Santiago de Compostela, Spain; e-mail: mazzanti@fpaxp1.usc.es

Matthias Kaminski, Department of Physics, Princeton University, 08544, Princeton, NJ, USA; e-mail: mkaminsk@princeton.edu

Philip C. Argyres, U. Cincinnati, P.O. Box 210011, 45221-0011, Cincinnati, OH, USA; e-mail: argyres@physics.uc.edu

Romuald A. Janik, Institute of Physics, Jagiellonian University, ul. Reymonta 4, 30-059, Kraków, Poland; e-mail: romuald@th.if.uj.edu.pl

Subir Sachdev, Department of Physics, Harvard University, 02138, Cambridge, MA, USA; e-mail: sachdev@physics.harvard.edu

Veronika E. Hubeny, Centre for Particle Theory & Department of Mathematical Sciences, Science Laboratories, South Road, DH1 3LE, Durham, UK; e-mail: veronika.hubeny@durham.ac.uk

Umut Gursoy, Institute for Theoretical Physics, Utrecht University, Leuvenlaan 4, 3584, Utrecht, CE, The Netherlands; e-mail: u.gursoy@uu.nl

Part I
Introduction to the AdS/CFT Correspondence

Chapter 1
Introduction to Anti de Sitter Black Holes

Christos Charmousis

Abstract These introductory notes concern basic properties of negative constant curvature spacetimes and their black holes. For comparison purposes we will begin by reviewing flat spacetime, the spacetime diagram and two particular patches, Milne and Rindler. We will then discuss anti de Sitter, its symmetries, basic properties and the construction of the spacetime diagram. We then look into the properties of anti de Sitter spacetime giving some global and local parametrisations. We will study the static black holes and then discuss their basic properties and novel topological effects due to the presence of a negative cosmological constant. We show using the classical Euclidean path integral approach their thermodynamic properties in the canonical ensemble with a heat bath of constant temperature. Finally we discuss, rather briefly, stationary and axially symmetric spacetimes and some properties of the rotating black holes.

1.1 Spacetimes of Constant Curvature

1.1.1 Spaces of Maximal Symmetry and Constant Curvature

Let us consider in d dimensions Einstein's equations with cosmological constant Λ,

$$G_{AB} + \Lambda g_{AB} = 0. \tag{1.1}$$

C. Charmousis (✉)
LPT, Université de Paris-Sud, Bât 210, 91405, Orsay CEDEX, France
e-mail: Christos.Charmousis@th.u-psud.fr

C. Charmousis
LMPT, Parc de Grandmont, Université Francois Rabelais, Tours, France

E. Papantonopoulos (ed.), *From Gravity to Thermal Gauge Theories: The Ads/CFT Correspondence,* Lecture Notes in Physics, 828,
DOI: 10.1007/978-3-642-04864-7_1,

Here, G_{AB} is the Einstein tensor and capital Latin letters denote d dimensional spacetime indices. The simplest of solutions to (1.1) we can consider are spaces locally characterised by the geometric condition,

$$R_{ABCD} = \frac{R}{(d-1)d}(g_{AC}\,g_{BD} - g_{AD}\,g_{BC}), \tag{1.2}$$

where d is the spacetime dimension and R, Ricci scalar curvature. Using (1.1) we see that,

$$G_{AB} = -R\,g_{AB}\frac{d-2}{2d} = -\Lambda\,g_{AB}.$$

In other words $R = \frac{2d}{d-2}\Lambda$ are the, locally, constant curvature solutions of the Einstein equations with cosmological constant. They are Einstein spaces which are locally homogeneous. In particular for $\Lambda = 0$ we have flat-Minkowski spacetime, for $\Lambda > 0$ positively curved, de Sitter (dS) spacetime and for $\Lambda < 0$ anti de Sitter spacetime (adS).

The above three spacetimes are of maximal symmetry and therefore by definition admit the maximal number of Killing vectors. Flat spacetime for example is isometric under the group of Poincaré transformations, in other words d-dimensional Lorentz coordinate transformations plus d translations. This is obvious in the inertial Minkowskian system of coordinates. Therefore we have $d + d(d-1)/2 = d(d+1)/2$ Killing vectors as generators of these symmetries all together. Since this is the maximal number of Killing vectors (see *Exercise* 1) this spacetime is by definition maximally symmetric. Flat spacetime is defined as the unique space of zero Riemann curvature (and of trivial topology).

Before moving on to non zero constant curvature spacetimes it is useful to classify the maximally symmetric $n = d-1$-spacelike sections. They are respectively representing locally Euclidean, spherical and hyperbolic sections and their line element can be written in a compact fashion as

$$dK_n^2 = \frac{d\chi^2}{1-\kappa\chi^2} + \chi^2 d\Omega_{n-1}^2, \quad \chi \geq 0, \tag{1.3}$$

for normalised curvature $\kappa = 0, 1, -1$ respectively. The spherical line element $d\Omega_{n-1}^2$ $(n > 1)$ is given by the iterative relation,

$$\begin{aligned} d\Omega_k^2 &= d\theta_k^2 + \sin^2(\theta_k)d\Omega_{k-1}^2, \ldots, \quad d\Omega_1 = d\theta_1, \\ \theta_k &\in [0,\pi[, \ldots, \theta_1 \in [0, 2\pi[, \quad k = 2, \ldots, n-1. \end{aligned} \tag{1.4}$$

Setting $\chi = \sin\phi$ or $\chi = \sinh\phi$ in (1.3) for $\kappa = 1, -1$ respectively gives us the usual line element for the unit sphere and hyperboloid,

$$d\Omega_n^2 = d\phi^2 + \sin^2\phi\, d\Omega_{n-1}^2, \quad \phi \in [0, \pi] \tag{1.5}$$

$$dH_n^2 = d^2 + \sinh^2 d\Omega_{n-1}^2, \quad \in [0, +\infty[. \tag{1.6}$$

We will be making extensive use of these parametrisations later on.

1.1.2 Flat Spacetime

1.1.2.1 Conformal Spacetime Diagram

To understand some of the basic properties of anti de Sitter space it is useful to first make a rapid go through the case of flat space. This way we can introduce relevant definitions in a more intuitive setting and compare properties. We turn our attention to flat spacetime and in particular to its conformal spacetime diagram (see for example [1]). One can write d dimensional Minkowski spacetime (1.3) as

$$ds^2 = -dt^2 + dr^2 + r^2 d\Omega_{n-1}^2, \tag{1.7}$$

in a spherical coordinate system (compare with (1.3)). To obtain the structure of flat spacetime at infinity we will go to a spacetime metric which is conformally equivalent,

$$\tilde{g}_{AB}(x) = \Omega^2(x) g_{AB},$$

to the initial one (1.7). The important point to realise is that the structure we obtain at asymptotic infinity will be common to all asymptotically flat spacetimes. This is due to the fact that a conformal transformation will shrink or stretch spacetime lengths but will not alter the null cones and therefore asymptotic properties. The main idea therefore is to use conformal transformations in order to bring asymptotic infinities to finite values for the conformally transformed metric. This is the central idea of Carter–Penrose diagrams.

To achieve this for flat spacetime we first pass to null retarded and advanced coordinates, $u = t - r, v = t + r$, where u or $v =$ constant correspond to null radial geodesics of (1.7). Now we can bring coordinate infinity to finite values via the coordinate transformation, $\tan \tilde{U} = u, \tan \tilde{V} = v$ with $-\frac{1}{2}\pi < \tilde{U}, \tilde{V} < \frac{1}{2}\pi$. Go once again to time-space like coordinates $\tilde{t} = \tilde{U} + \tilde{V}, \tilde{r} = \tilde{V} - \tilde{U}$. Then (1.7) is conformally (and not coordinate) equivalent to

$$d\tilde{s}^2 = -d\tilde{t}^2 + d\tilde{r}^2 + \sin^2 \tilde{r} d\Omega_{n-1}^2$$

with conformal factor, $\Omega = 2\cos(\tilde{U})\cos(\tilde{V})$, *i.e.*, $ds^2 = \Omega^2 d\tilde{s}^2$ (compare with (1.3)). Its important to keep track of the coordinate ranges,

$$-\pi < \tilde{t} + \tilde{r} < \pi, \quad -\pi < \tilde{t} - \tilde{r} < \pi, \quad \tilde{r} \geq 0. \tag{1.8}$$

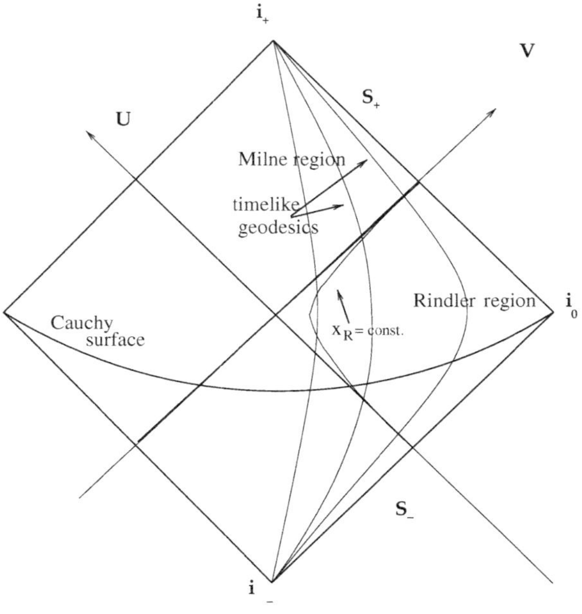

Fig. 1.1 Three dimensional diagram and spacetime diagram (*right*) of Minkowski spacetime. The quarter diamonds on the *left* are the regions of Rindler and Milne spacetimes

We can now draw the Carter–Penrose spacetime diagram for flat spacetime (see Fig. 1.1). The domain defined by (1.8) is a triangle with its boundary defining asymptotic infinity of flat spacetime. It will correspond to a diamond shaped region of the Einstein cylinder (see [1]). In the diagram $\Im^-$ and $\Im^+$ stand for null past infinity and future null infinity, respectively, whereas i^-, i^+ are the endpoints of timelike geodesics $r = constant$. i^0 is that of spacelike geodesics $t = constant$.

A Cauchy surface is a spacelike surface intersecting all null and timelike inextensible geodesics. It will inevitably touch i^0 asymptotically. As its name indicates a Cauchy surface is therefore a valid set of initial data (for our second order Einstein field equations). A non-geodesic curve, such as $x_R = constant$, can reach null infinity $\Im^+$ if it is uniformly accelerated approaching the speed of light as $t \to +\infty$. The tangent curve emanating from $x_R = constant$ at $\Im^+$ is the acceleration horizon of the worldline $x_R = constant$. No events above this line can be witnessed by the observer whose worldline is $x_R = constant$. These effects concern Rindler and Milne spacetimes which we briefly turn to now.

1.1.2.2 Rindler and Milne Spacetimes

The line element for (spherical) Rindler spacetime is given by

$$ds^2 = -x_R^2 dt_R^2 + dx_R^2 + x_R^2 \cosh^2 t_R d\Omega_{n-1}^2, \quad x_R > 0, \tag{1.9}$$

and we immediately note that for $x_R = 0$ we have a singularity. Direct calculation of the Riemann tensor for this spacetime gives us identically zero i.e., we have a coordinate (and not curvature) singularity and furthermore Rindler spacetime is just a coordinate patch of flat spacetime. Indeed the coordinate transformation relating it to (1.7) is given by

$$\tanh t_R = \frac{t}{r} = \frac{\sin(\tilde{V}+\tilde{U})}{\sin(\tilde{V}-\tilde{U})}, \quad x_R = \sqrt{r^2 - t^2} = \sqrt{-\tan\tilde{U}\tan\tilde{V}}. \tag{1.10}$$

From the above we see that Rindler spacetime covers only part of the global spacetime diagram since $\tilde{U}\tilde{V} < 0$. The lines $\tilde{V} = 0$ and $\tilde{U} = 0$ corresponding to, $x_R = 0$ and $t_R \to \pm\infty$, are event horizons for the observer with coordinate time t_R and worldline $x_R = constant$. An observer with worldline $x_R = constant$ is in uniform acceleration $a = \frac{1}{x_R}$ since his trajectory in the original inertial Minkowski coordinates are hyperbolas (1.10). The inverse transformation reads,

$$t = x_R \sinh t_R, \quad r = x_R \cosh t_R.$$

To get Milne spacetime we can consider,

$$x_R \to it_M, \quad t_R \to x_M + \frac{i\pi}{2}.$$

This transformation, is one involving complex coordinates, however, it maps us to a real metric which is a different portion of flat spacetime ($\tilde{U}\tilde{V} > 0$). It is Milne spacetime,

$$ds^2 = -dt_M^2 + t_M^2 dH_n^2, \quad t_M > 0, \tag{1.11}$$

which is now a cosmological or time dependent type of metric. Notice it has no initial big bang singularity. An observer of Milne spacetime has rather a cosmological horizon at $t_M = 0$ and has Hubble expansion rate $H = \frac{1}{t_M}$ in proper time. Note that (just like in de Sitter spacetime) there is no horizon problem in this cosmological metric due to the fact that Milne spacetime is free of curvature singularities.

1.1.3 Anti de Sitter Spacetime

1.1.3.1 Definition, Boundary and Isometries

Useful and elegant representations of de Sitter and anti de Sitter spacetimes are obtained by embedding d dimensional hypersurfaces in $d+1$ dimensional flat spacetime. Then de Sitter spacetime of curvature scale a is defined as the hyperboloid,

$$-X_0^2 + \sum_{i=1}^{d} X_i^2 = a^2, \tag{1.12}$$

embedded in $d+1$ dimensional Minkowski spacetime. De Sitter space is topologically $\Re \times S^{d-1}$ and thus its spatial sections are compact (see also [1 or 2–4]). We shall focus on adS space from now on.

AdS spacetime in turn is obtained by considering the hyperboloid,

$$X_0^2 + X_d^2 - \sum_{i=1}^{d-1} X_i^2 = l^2 \tag{1.13}$$

of radius of curvature $l > 0$ embedded in the $d+1$ dimensional spacetime,

$$ds^2 = -(dX_0^2 + dX_d^2) + \sum_{i=1}^{d-1} dX_i^2, \tag{1.14}$$

where note the double time coordinates. Clearly, any element of the Lorentz group $SO(2, d-1)$ will leave (1.14) and (1.13) unchanged (by construction). Also we see that translation invariance is broken by (1.13). Since $SO(2, d-1)$ has $d(d+1)/2$ Killing generators just like flat spacetime, which is of maximal symmetry, $SO(2, d-1)$ is the precise isometry group of adS.

A second important point is that adS spacetime has a conformal boundary[1] at the hyperboloid infinity. To see this we rescale all coordinates by $X_A \to X_A\Lambda$ and take $\Lambda \to \infty$. This limit defines the boundary as

$$X_0^2 + X_d^2 - \sum_{i=1}^{d-1} X_i^2 = 0, \tag{1.15}$$

$$X_A = X_A\Lambda, \Lambda \in \Re. \tag{1.16}$$

Suppose first that $X_0 \neq 0$: we can then divide by X_0 and then rescale. We therefore have that the boundary verifies,

$$-X_d^2 + \sum_{i=1}^{d-1} X_i^2 = 1,$$

which is a hyperboloid in $d-1$ dimensions. This is just $d-1$ dimensional de Sitter space according to (1.12). The topology is that of $\Re \times S^{d-2}$ If $X_0 = 0$ on the other hand, from (1.15) we have a sphere in $d-2$ dimensions (times a point). Adding the two spaces together the boundary is a maximally symmetric space $S^1 \times S^{d-2}$.

[1] The boundary is called conformal for it admits not one but an equivalence class of metrics which are related via a conformal transformation, $g_{\mu\nu}^{boundary} = \Omega\Gamma_{\mu\nu}^{boundary}$. This equivalence class is manifest in the arbitrariness of Λ in (1.16).

Since the adS isometry acts on this space the boundary preserves $SO(1,d-1)$ symmetry. Note, however, that additionally we still have d extra transformations (1.16) for the boundary metric, 1 dilatation and $d-1$ special conformal transformations. The symmetry group is the conformal group $\mathcal{C}(1,d-2)$ that possesses, Lorentz and angle preserving transformations. Due to this last property, which has to do with the fact that we are dealing with an equivalence class of metrics, the boundary of adS admits as many symmetries as the bulk adS space. In other words the symmetry group of the boundary $\mathcal{C}(1,d-2)$, is isomorphic to the symmetry group of the bulk adS space, $SO(2,d-1)$.

1.1.3.2 Parametrisations

Let us now construct line elements for adS space in d dimensions. A global parametrisation is constructed as follows. Given the form of (1.13) we consider two spheres $X_0^2+X_d^2=r_1^2$, $\sum_{i=1}^{d-1}X_i^2=r_2^2$ of radii r_1,r_2 such that

$$r_1^2-r_2^2=l^2. \tag{1.17}$$

This equation is solved setting $r_1=l\cosh(u/l), r_2=l\sinh(u/l)$ where $u\in[0,+\infty[$. Now we just take the relevant parametrisations in polar spherical coordinates, (1.3) for $\kappa=0$, and replace them in the line element (1.14),

$$\begin{aligned} ds^2 &= -(dr_1^2+r_1^2d^2)+dr_2^2+r_2^2d\Omega_{d-2}^2 \\ &= -l^2\cosh^2(u/l)d^2+du^2+l^2\sinh^2(u/l)d\Omega_{d-2}^2. \end{aligned} \tag{1.18}$$

The embedding in flat space is thus defined in coordinates, $(t,u,\theta_1,...,\theta_{d-2})$ as,

$$\begin{aligned} X_0 = l\cosh(u/l)\sin, \quad X_d = l\cosh(u/l)\cos, \\ X_1 = l\sinh(u/l)\cos\theta_1,...,X_{d-1} = l\sinh(u/l)\sin\theta_1...\cos\theta_{d-2}. \end{aligned} \tag{1.19}$$

The boundary lies at $u\to\infty$. This is the global parametrisation of adS since all points of the hyperboloid are taken into account exactly once. This metric is solution to the Einstein equations with cosmological constant,

$$2\Lambda=-\frac{(d-1)(d-2)}{l^2}. \tag{1.20}$$

Note that the timelike coordinate is an angular coordinate $\in[-\pi,\pi[$. This signifies that adS is a spacetime with closed timelike curves! We can, however, get around this; since the space is not simply connected (i.e. the time circle cannot be topologically reduced to a point) we can unwrap the circle of the time coordinate and take a new coordinate $t\in[-\infty,+\infty]$ with t $\equiv$ in each 2π-interval. This means that we are effectively taking infinite copies of the hyperboloid. This is the universal covering of adS space,

$$ds^2=-l^2\cosh^2(u/l)dt^2+du^2+l^2\sinh^2(u/l)d\Omega_{d-2}^2. \tag{1.21}$$

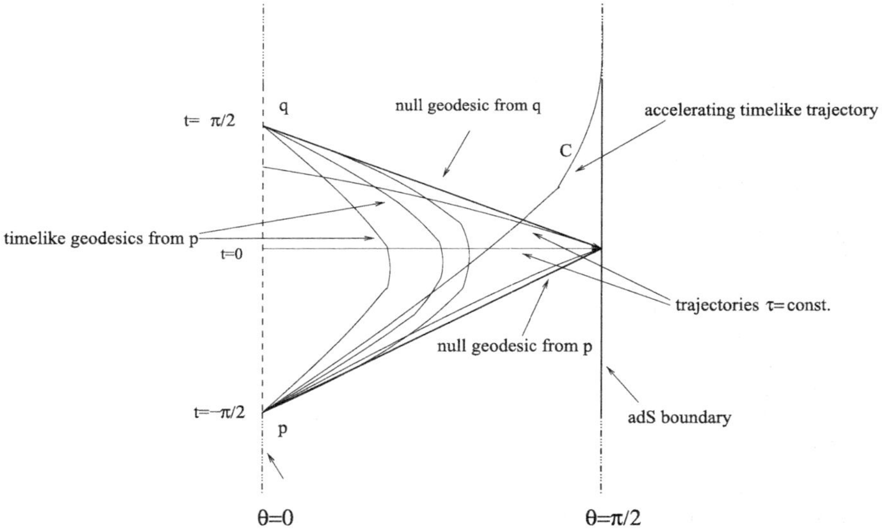

Fig. 1.2 Spacetime diagram with timelike geodesics of adS spacetime (1.29). All past and future timelike geodesics from $\tau = 0$ focus in the past at point p and and in the future at point q. The resulting "geodesic" triangular region covers only part of the timelike future of p. Similarly, the Cauchy surface at $t = 0$ covers only the triangular region in between p and q and their null past and future, respectively. Any event beyond this triangle is not causally connected to $\tau = 0$ unless suitable reflective boundary conditions are imposed at the adS boundary. The accelerating timelike curve C "escapes" from this bounded region and gets to see beyond, ending up at the boundary of adS

This space is, however, not a Cauchy space (or in other words globally hyperbolic). We can of course define a Cauchy surface but it represents no longer sufficient initial data to describe the entire space of adS. Data on the boundary of adS have to be also specified (see Fig. 1.2). We will come back to this briefly when studying the spacetime diagram.

Also note that for the normal adapted coordinate u, ∂_u is not a Killing vector for adS. In adS space we miss a global spacelike Killing vector generating translational invariance in u (in de Sitter space we do not have a global timelike Killing vector). Taking $u = u_0$ constant with u_0 large we see that,

$$ds^2 \sim l^2 e^{\frac{2u_0}{l}} \left(-dt^2 + d\Omega_{d-2}^2\right),$$

that the geometry of the boundary is indeed topologically $\Re \times S^{d-2}$ for the universal covering of adS. Had we considered we would have got, $S^1 \times S^{d-2}$.

Let us now consider a local parametrisation defined by

$$y = l \ln \frac{X_d + X_{d-1}}{l}, \quad \tilde{t} = \frac{X_0}{X_d + X_{d-1}},$$
$$x_i = \frac{X_i}{X_d + X_{d-1}}, \quad i = 1, ..., d-2. \tag{1.22}$$

This parametrisation covers only half of the hyperboloid since we must have $X_d + X_{d-1} > 0$ in order for (1.22) to be well defined. In order to find the relevant line element we must invert (1.22). Using (1.13) it is easy to show that,

$$X_d + X_{d-1} = le^{\frac{y}{l}}, \quad X_d - X_{d-1} = le^{\frac{-y}{l}}\, 1 + e^{\frac{2y}{l}} \frac{\sum_{j=1}^{d-2} x_j^2 - \tilde{t}}{l^2}\Bigg).$$

Taking the sum and the difference we then obtain the inverse transformation,

$$\begin{aligned} X_{d-1} &= l \sinh\left(\frac{y}{l}\right) - \frac{l}{2} e^{\frac{y}{l}} (\mathrm{x}^2 - \tilde{t}^2), \\ X_0 &= l\tilde{t}e^{\frac{y}{l}}, \quad X_i = l x_i e^{\frac{y}{l}}, \\ X_d &= l \cosh\left(\frac{y}{l}\right) + \frac{l}{2} e^{\frac{y}{l}} (\mathrm{x}^2 - \tilde{t}^2), \end{aligned} \tag{1.23}$$

where $\mathrm{x}^2 = \sum_{j=1}^{d-2} x_j^2$ (compare with the global parametrisation (1.19)). Inserting into (1.14) we now get the desired line element,

$$ds^2 = l^2 e^{\frac{2y}{l}} (-d\tilde{t}^2 + d\mathrm{x}^2) + dy^2, \tag{1.24}$$

with $y \in [-\infty, +\infty]$ measuring proper distance. Note that by rescaling the $(\tilde{t}, \mathrm{x})$ coordinates we can get rid of the l^2 factor. Planar coordinates are obtained by setting $r = le^{\frac{y}{l}}$. We obtain,

$$ds^2 = r^2(-d\tilde{t}^2 + d\mathrm{x}^2) + \frac{l^2 dr^2}{r^2}, \tag{1.25}$$

with $r > 0$. We will use this chart for planar black holes. Again note from (1.22) that planar coordinates cover only half of the hyperboloid. To cover all of adS we take two portions $r > 0$ and $r < 0$. Note that the boundary is attained at $r \to \infty$ whereas we have a horizon at $r = 0$ due to the fact that we cannot cover all of adS space. This is a degenerate (i.e. of zero temperature) Killing horizon associated to the Poincaré flat slicing of adS space (we will come back to this in a moment). Finally the Poincaré chart is obtained as the conformally flat version of adS by setting $z = \frac{l^2}{r}$,

$$ds^2 = \frac{l^2}{z^2}(-d\tilde{t}^2 + d\mathrm{x}^2 + dz^2), \tag{1.26}$$

where now the boundary is at $z = 0$ for this chart.

Let us now find a proper time parametrisation of adS. We proceed as for the global parametrisation modulo we now start by considering, $X_0^2 = \rho_1^2, X_d^2 - \sum_{i=1}^{d-1} X_i^2 = \rho_2^2$ of radii ρ_1, ρ_2 such that

$$\rho_1^2 + \rho_2^2 = l^2, \tag{1.27}$$

which is solved setting $\rho_1 = l \sin\tau, \rho_2 = l\cos\tau$. We repeat the same procedure for the hyperboloid $X_d^2 - \sum_{i=1}^{d-1} X_i^2 = \rho_2^2$ with $X_d = \rho_2 \cosh$ and then we parametrise $\sum_{i=1}^{d-1} X_i^2$ using (1.3) for $\kappa = 0$. All in all we have,

$$\begin{aligned} &X_0 = \sin\tau, \quad X_d = l\cos\tau\cosh, \\ &X_1 = l\cos\tau\sinh \quad \cos\theta_1, ..., X_{d-1} = l\cos\tau\sinh \quad \sin\theta_1 ... \cos\theta_{d-1} \end{aligned} \tag{1.28}$$

and replacing in the line element (1.14) we get,

$$ds^2 = -l^2 d\tau^2 + l^2 \cos^2\tau dH_{d-1}^2. \tag{1.29}$$

This coordinate system is only defined for $\tau \in]-\frac{1}{2}\pi, \frac{1}{2}\pi[$. Therefore in this patch $-l \le X_0 \le l$ (see Fig. 1.2).

1.1.3.3 Spacetime Diagram

Let us turn our attention now to the spacetime diagram for adS. Our starting point is the global coordinate system (1.21). As before we write the metric in a conformally flat form and define novel coordinates so as to bring infinity of the radial coordinate to a new finite coordinate value. Time, however, is either periodic in or infinite in t. This amounts to solving $d\theta = dul \cosh\frac{u}{l}$ and therefore considering the coordinate transformation,

$$\tan\left(\frac{\theta}{2} + \frac{\pi}{4}\right) = e^{\frac{u}{l}}, \tag{1.30}$$

with $\theta \in [0, \frac{\pi}{2}[$. The line element is

$$ds^2 = \frac{l^2}{\cos^2\theta}\left(-dt^2 + d\theta^2 + \sin^2\theta d\Omega_{n-1}^2\right) \tag{1.31}$$

and it is conformally equivalent to a quarter of the Einstein cylinder (1.7). The difference here is that $\theta \in [0, \frac{\pi}{2}]$ rather than $\theta \in [0, \pi[$ for flat space. Also note that we cannot make a conformal transformation which brings time infinities to finite values for then the conformal factor explodes. It is useful for comparison between different charts to note that,

$$\begin{aligned} &X_0 = l\frac{\sin t}{\cos\theta}, X_d = l\frac{\cos t}{\cos\rho}, \\ &X_1 = l\tan\rho\cos\theta_1, ..., X_{d-1} = l\tan\rho\sin\theta_1 .. \cos\theta_{n-1}. \end{aligned} \tag{1.32}$$

Therefore the spacetime diagram consists of an infinite strip of length $\pi/2$ (for the universal covering). The boundary resides at $\theta = \frac{\pi}{2}$, the endpoint of both future and past null geodesics. For timelike geodesics we use the proper time slicing, (1.29). Indeed comparing X_0, X_d for (1.32) and (1.28) we get

$$\sin t = \sin\tau\cos\theta, \qquad \cos t = \frac{1}{\tanh}\sin\theta \tag{1.33}$$

for τ and constant respectively (see Fig. 1.2).

Note that timelike infinity in the past i^- or future i^+ is infinite. As a consequence there is no finite Cauchy spacelike surface at all in this space. This is because although any constant t_0 surface ($t = 0$ or $\tau = 0$ in Fig. 1.2), X say, covers the whole of adS at instant $t = t_0$, however, we can find null surfaces that never intersect X. This means that universal adS is not globally hyperbolic: Cauchy data on arbitrary spacelike surface X, determines the system's evolution only in a region bounded by a null hypersurface called a Cauchy horizon for X (the triangular region in Fig. 1.2). Physics on adS depends also on the boundary conditions imposed at the boundary. Secondly all timelike geodesics emanating from $t = 0$ focus at point q and diverge at p never reaching the boundary of adS. Unlike de Sitter space which inflates spacetime events away from each other, anti-de-Sitter never allows free-falling particles to escape to the boundary. There are actually regions in the future of p which can never be reached by any geodesic as is shown in the figure but only by accelerated observers as the curve C in (1.2). This is a sign that the gravitational potential in adS will be a bounding potential for gravitons or free particles.

Planar (or Poincaré) coordinates cover the region $r \geq 0$ which gives us the triangular region in the spacetime diagram. Indeed we have (1.23) $r = X_d + X_{d-1} = \frac{l}{\cos\theta}(\cos t + \sin\theta\cos\Omega)$ given the global patch (1.19) in coordinates (1.31) (see Fig. 1.3) Note that the $r = const$ worldlines are now accelerating trajectories and as a result we have the presence of an adS horizon at $r \sim \infty$. The Poincaré patch is in this sense similar to the Rindler patch in flat spacetime.

Exercise 1 Show that for a Killing vector we have

$$\nabla_A \nabla_B \xi_C = -R^D_{BCA}\xi_D.$$

Deduce that any Killing vector can be obtained by its values ξ^A and $\nabla_A \xi_B$ at any point $P \in \mathcal{M}$. From Killing's relation deduce therefore that there are at most $\frac{d(d+1)}{2}$ linearly independent Killing vectors in $\mathcal{M}$ and hence at most $\frac{d(d+1)}{2}$ isometries of the metric (for more details see [5]).

Exercise 2 The constant curvature slicings of adS are given by the following line element,

$$ds^2 = -(\kappa + \frac{r^2}{l^2})dt^2 + \frac{dr^2}{\kappa + \frac{r^2}{l^2}} + r^2 dK^2_{n-1}. \tag{1.34}$$

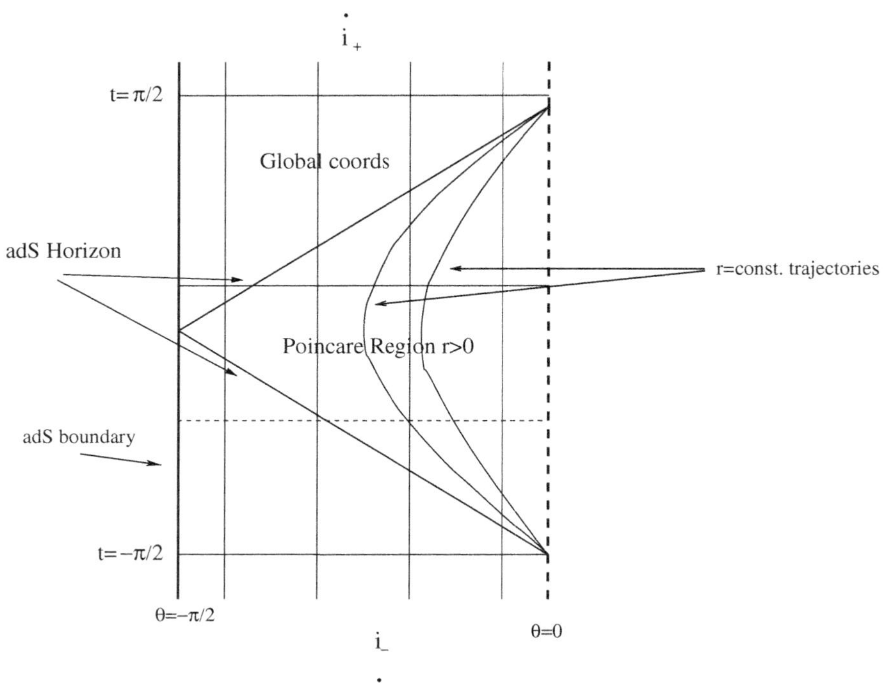

Fig. 1.3 On the right a 3 dimensional representation of (universal) adS space with all but one (symbolically Ω in the figure) of the angular coordinates fixed. The boundary of adS is represented by the infinite cylinder, the bulk of adS space by its interior. The fundamental domain is the cylinder bounded the horizontal cups $t = \pm\pi/2$. The Poincaré region is bounded by the tilted adS horizon caps $r = 0$ and covers half of the fundamental domain for $t \in [-\pi/2, \pi/2]$. On the left a projected 2 dimensional view of the Poincaré region [3]

For $\kappa = 0$ we obtain the flat slicing already encountered (1.25). Show that the spherical slicing is nothing but that of global coordinates (1.21). Find the coordinate transformation relating the flat and spherical parametrisations. Find then the region of validity of the Poincaré coordinates. Show that the hyperbolic slicing gives a time dependent version of adS. Find its domain of validity.

Exercise 3 Show that the global coordinate system for de Sitter space is,

$$ds^2 = -dt^2 + a^2 \cosh^2\left(\frac{t}{a}\right) d\Omega_n^2. \tag{1.35}$$

Therefore there is no global timelike Killing vector in de Sitter space. Construct the Poincaré slicing and the Carter–Penrose diagram for de-Sitter space (see for example [1]).

1.2 Static Black Holes

1.2.1 Basic Properties

Assume a $d = n + 1$ dimensional spacetime such that it has $n - 1$ dimensional sections of constant curvature given by the line element (1.3). Then one can show, for metrics of such symmetry, that the general solution of Einstein's equations with cosmological constant admits a locally timelike Killing vector. This is a slightly generalised version of Birkhoff' s theorem for vacuum spacetime (see for example [6]). The solution reads (see [7, 8])

$$ds^2 = -V(r)dt^2 + \frac{dr^2}{V(r)} + \frac{r^2}{l^2} dK_{n-1}^2, \quad n \geq 2, \tag{1.36}$$

where

$$V(r) = \kappa - \frac{\mu}{r^{n-2}} + \frac{r^2}{l^2}, \tag{1.37}$$

where the $(n-1)$ dimensional metric dK_{n-1}^2 is given by (1.3) and the adS curvature length l is related to the cosmological constant (1.20). Parameter μ is an integration constant which, as we will see, is associated to the black hole mass [9, 10]. For the charged version of these black holes and their thermodynamics see [11].

Unlike de Sitter or flat spacetime, black holes in adS [12] will exist for all three values of $\kappa = 0, 1, -1$. In other words black hole horizons do not have only spherical topology in adS, we can have toroidal or even hyperbolic black holes. The horizons have to undergo topological identifications so as to make the horizon surface compact. For flat horizons, $\kappa = 0$, one can simply identify all flat directions in order to get a torus of $n - 1$ dimensions. One can identify flat directions non-trivially, creating non orientable horizon surfaces such as the Klein bottle in 2 dimensions and so forth. For $\kappa = -1$ the construction is more complicated. Here we give a rapid overview but the interested reader should consult [13–16]. Take a locally H^2 horizon (in a 4 dimensional hyperbolic black hole) and consider a quotient space $\Sigma = H^2/\Gamma$. Here, H^2 is modded out by its discrete subgroup Γ made of discrete boosts of $SO(2,1)$ (which as we saw earlier belongs to the symmetry group of H^2). Then Σ turns out to be a compact space of genus g. The compact space has $4g$ sides and the sum of its angles has to give an overall angle of 2π in order to avoid conical singularities (see Fig. 1.4). The fundamental domain of Σ is a polygon whose edges are geodesics of H^2. The action of Γ will identify opposing edges. The simplest case free of conical singularities is a regular hyperbolic octagon with opposite edges identified [13]. Often these black holes are referred to as topological black holes [7, 14–17] due to the identifications one has to undergo in order to compactify the horizon geometries. In fact, any Einstein space of dimension $n - 1$ can form a horizon for an adS black hole [7].

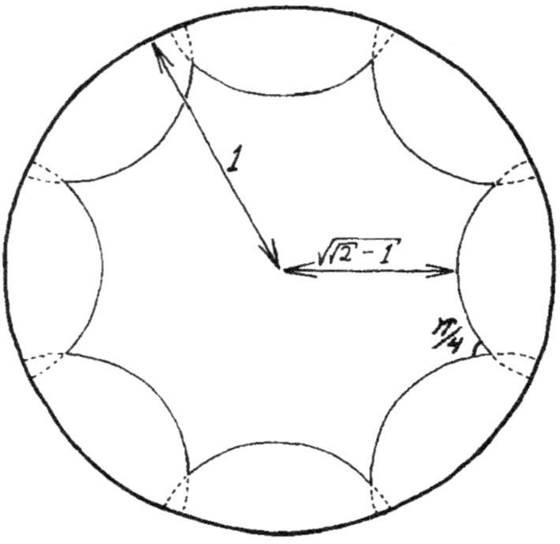

Fig. 1.4 The fundamental region depicting a regular hyperbolic octagon in the Poincaré hyperbolic unit disk, $D=\left\{\mathrm{x}\in\Re^2 : |\mathrm{x}|<1\right\}$ with metric $ds^2=\frac{dx^2+dy^2}{(1-x^2-y^2)^2}$. Notice how distances get warped as we approach the boundary. Each depicted arc is orthogonal to the disks boundary. Thus each polygon side is a geodesic segment in this space. The sum of angles gives 2π ensuring the absence of conical singularities. The opposite edges of the polygon are identified via the action of Γ. The thus constructed horizon surface is a compact space which is locally hyperbolic. This figure is taken from [6] where the detailed construction can be found

The solutions for $\kappa=+1$ are sometimes called "Schwarzschild-adS" solutions because they reduce to the standard Schwarzschild solution when the cosmological constant vanishes, $l\to\infty$, and to adS in global coordinates when $\mu=0$. In fact, these are the genuine asymptotically adS solutions precisely because asymptotically we recover the global patch of adS space. Moreover, their topology is $\mathrm{I\!R}^2\times S^{n-1}$, and the horizon is the sphere S^{n-1}, like that of the Schwarzschild solution.

Given that these black holes are static their horizons, in this coordinate system, are the zeros of the potential. We will denote their (largest root) event horizon by $r=r_h, V(r_h)=0$. This event horizon has typically a non-zero temperature which can be calculated the standard Euclidean way: Indeed consider $t\to i\tau$. We have

$$ds^2=V(r)d\tau^2+\frac{dr^2}{V(r)}+r^2dK^2_{n-1} \tag{1.38}$$

and the metric is then of Euclidean signature for $r>r_h$. This can be seen by expanding around $r=r_h$,

$$ds^2\sim\left(\frac{1}{4}V'^2_{r_h}\right)\rho^2d\theta^2+d\rho^2+\cdots, \tag{1.39}$$

with radial isotropic (or cylindrical) coordinate $\rho=\sqrt{\frac{2(r-r_h)}{|V'_{r_h}|}}$. Clearly in order to evade a conical singularity at the origin of the axis, $r=r_h$, we must impose the periodicity,

$$\beta = \frac{4\pi}{|V'(r_h)|}. \tag{1.40}$$

As we will see, the Euclidean quantum field propagator, with the above periodic boundary conditions, describes a canonical ensemble of states in thermal equilibrium with a heat bath of temperature $T = \beta^{-1}$ [18] where

$$\beta = \frac{4\pi l^2 r_h}{n r_h^2 + \kappa(n-2)l^2}. \tag{1.41}$$

This relation can be inverted to find

$$r_h = \frac{2\pi l^2}{n\beta}\left[1 \pm \sqrt{1 - \kappa\frac{n(n-2)\beta^2}{4\pi^2 l^2}}\right], \tag{1.42}$$

which allows us to take β as the parameter that determines the solution. For $\kappa = +1$ the minus branch for r_h exists and corresponds to small black holes in adS. Notice that in the limit where $r_h \gg l$ the $\kappa = \pm 1$ classes of solutions approach the planar black hole class $\kappa = 0$. This admits an interpretation in terms of an "infinite volume" limit, in which the curvature radius of S^{n-1} or H^{n-1} is much larger than the thermal wavelength of the system [19].

Setting $\mu = 0$ we recover differing patches of adS space. For $\kappa = 0$ we obtain the Poincaré patch which covers only part of adS. In this case $r \to \infty$ is the boundary of adS whereas $r = 0$ is a horizon. This is a degenerate Killing horizon meaning that there is no temperature associated with it (This is also true for the $\kappa = 1$ case). These are then the ground states for their respective class of solutions parametrised by μ. In other words adS space is not associated with some temperature or inversely we can construct any temperature adS state by suitably fixing the period of Euclidean time. This is similar to flat spacetime and in contrast to de Sitter spacetime. For $\kappa = -1$ however, we have a bifurcate Killing horizon at $r_h = l$ with $r > l$. The hyperbolic slicing covers a yet smaller triangular portion of adS but the horizon in question has now temperature, $\beta = 2\pi l$. This patch is very similar to the Rindler patch of flat spacetime. Therefore in this case it is not clear what is the ground state of the system. We stress, however, that the $\mu = 0$ case is the only one which has no curvature singularity. The issue of the ground state is an important question since when calculating the partition function it is important to specify the background solution with which to annihilate divergences. AdS/CFT is capital in resolving this and we will come back to this in a moment [9, 10].

Furthermore, for the $\kappa = -1$ class of black holes [8], and in contrast to the $\kappa = +1, 0$ classes, the zero temperature solution exists and is different from the one that is isometric to adS. In fact, for $\kappa = -1$ there is a range of negative values for μ such that the solutions still possess regular horizons. When we are in such a configuration we also have an inner horizon much like for a RN black hole. The minimum values of μ and r_h that are compatible with cosmic censorship and for which the horizon is degenerate are [7],

$$\mu_{\text{ext}} = -\frac{2}{n-2}\left(\frac{n-2}{n}\right)^{n/2} l^{n-2}, \quad r_{\text{ext}} = \sqrt{\frac{n-2}{n}}l, \tag{1.43}$$

and,

$$\mu_{\text{ext}} = -\frac{l^2}{4}, \quad r_{\text{ext}} = \frac{l}{\sqrt{2}}, \quad \text{for } n = 4. \tag{1.44}$$

For these values of the parameters, the black hole is extremal with coinciding inner and event horizons.

1.2.2 Thermodynamics

The thermodynamics of adS black holes have many interesting properties which are rather different from their asymptotically flat or de Sitter cousins. We will use here the path integral method developed by Hartle and Hawking [20] in order to calculate the partition function and then, using standard thermodynamic formulas, the basic thermodynamic quantities for adS black holes. Already as we mentioned above, the $\kappa = -1$ case, presents subtleties due to the fact that the background adS solution possesses a temperature at $r_h = l$ whereas at the same time there is an extremal black hole of zero temperature. Therefore it is not clear, at least naively, which is the right background one should use. The adS/CFT correspondence cures this ambiguity by providing through the boundary CFT the correct geometric counter-terms. These are used to cancel out the infinities and provide a background independent way to calculate the partition function [9, 10, 21, 22]. For this presentation we will concentrate on pre-adS/CFT method as a motivation and comparison with adS/CFT which will be studied later on. We therefore concentrate here solely on the $\kappa = 1$ case (for the conserved charges see [23, 24]) giving in particular an overview of the Hawking–Page phase transition.

We shall consider a canonical ensemble with a constant temperature heat bath. The system has thus fixed temperature but is allowed to exchange energy with the heat bath. In asymptotically flat space although a black hole can be in equilibrium with thermal radiation at some common constant temperature T_0, this equilibrium is unstable once gravitational corrections are taken into account. In other words, if the mass of the black hole were to increase due to some infalling matter then the temperature would decrease and the black hole would continue to grow. This means that the canonical ensemble, where the black hole is in thermal equilibrium with a heat bath of constant temperature, is ill defined if gravitational energy flow is not held fixed. In adS space, however, the thermal radiation remains confined close to the black hole since the gravitational potential, $V \sim r^2/l^2$ increases for large r. This is true for any choice of origin since adS is a homogeneous space. Non-zero rest mass particles are confined and prevented from escaping to infinity and one can consider a canonical ensemble description for any given temperature T. Effectively, though volume of space is infinite, adS provides a confining gravitational box. The canonical ensemble partition function is defined via a path integral [25] that takes us from a given configuration S, $(S, g, \phi) \longrightarrow (S', g', \phi')$ to S' via all possible paths. These paths represent arbitrary matter fields and arbitrary metrics which are asymptotically flowing to zero and adS, respectively, in periodic time τ with period β (1.40). Here ϕ represents collectively matter fields and g Euclidean metrics. Symbolically,

$$\langle g', \phi', S' | g, \phi, S \rangle = \int D(\phi, g) \exp\left(-I[\phi, g]\right). \tag{1.45}$$

The path integral is taken in Euclidean signature in order to optimise regularity. Indeed, this permits the amplitude to be an elliptic operator with an exponential *damping* factor rather than an oscillating one. All defined above possible configurations are allowed but one can argue that regular classical solutions, i.e. saddle points of the action, are going to give the dominant contributions to this integral and in particular to the partition function $Z = e^{-I}$ where,

$$I_{sol} = -\frac{1}{16\pi G}\int d^d x\sqrt{g}(R - 2\Lambda) + BT = \frac{d-1}{8\pi G}\int d^d x\sqrt{g} = \frac{d-1}{8\pi G} Vol \tag{1.46}$$

Here BT is the Gibbons–Hawking boundary term yielding a zero contribution for adS (unlike flat space where it is the sole term giving a contribution). We have immediately run into a problem: the volume integral resulting from this calculation is clearly infinite, both for adS space and the black hole in question. We will therefore here, have to regularise by some vacuum (background) solution behaving similarly at asymptotic infinity to the black hole in question, so as to get rid of the infinities. We choose the background as the ground state since it is a regular solution satisfying the same boundary conditions *for any temperature of the heat bath*. We resort to [26] considering an upper cut-off $r < R$, subtracting the two volume integrals and then taking the limit $R \rightarrow \infty$.

Let us take a closer look [19]. Our background in question is pure adS in global coordinates. For adS and the black hole we have, respectively,

$$Vol_1(R) = \int_0^{\beta_0} d\tau \int_0^R dr \int_{S^{d-2}} d\Omega r^{d-2}, \tag{1.47}$$

$$Vol_2(R) = \int_0^{\beta} d\tau \int_{r_+}^R dr \int_{S^{d-2}} d\Omega r^{d-2}. \tag{1.48}$$

and note the differences in the bounds of integration for the two integrals. Although the period β for the black hole is fixed by (1.41), for adS space it is arbitrary, β_0. We can therefore fix β_0 so that *the temperature of both configurations is the same at* $r = R$. This is consistent with the fact that we are assuming a common heat bath for the canonical ensemble at temperature T. This sets,

$$\beta_0 = \frac{\beta\sqrt{1 + \frac{R^2}{l^2} - \frac{\mu}{R^{d-3}}}}{\sqrt{1 + \frac{R^2}{l^2}}}. \tag{1.49}$$

After evaluating the action difference and taking the limit we finally obtain,

$$-log(Z) = I = \frac{Vol(S^{d-2})r_+^{d-2}(l^2 - r_+^2)}{4G[(d-1)r_+^2 + (d-3)l^2]}. \tag{1.50}$$

The action I we have calculated is essentially the difference in between the saddle points present in the partition fuction. We see that I is positive for *small* r_h compared to l which means that the tunnelling probability from adS to a black hole is exponentially *suppressed*. We have thus semi-classical stability. On the contrary the sign is inverted for larger r_h than l, which points to an instability, physically favouring tunnelling to black holes. In between we can expect to find a phase transition. Let us now pursue to find in a standard way the thermodynamic quantities. The mean energy of thermal radiation is $E = \sum_{states} p_i E_i$, where $p_i = \frac{1}{Z} e^{-\beta E_i}$ and therefore,

$$E = -\frac{\partial}{\partial \beta} logZ = \frac{(d-2)Vol(S^{d-2})(l^{-2} r_+^{d-1} - r_+^{d-3})}{16\pi G} = M, \tag{1.51}$$

where $M = \frac{(d-2)Vol(S^{d-2})\mu}{16\pi G}$ is the gravitational mass of the black hole ([23, 24] for $\kappa = 1$, with the general conserved charges for locally asymptotically adS space-times given in [21]). The entropy $S = \sum_{states} p_i \log p_i$ is given by

$$S = \beta E - I = \frac{1}{4G} r_h^{d-2} Vol(S^{d-2}) = \frac{A}{4G} \tag{1.52}$$

where A is the volume of the horizon. Here we see how in adS the entropy-area relation is verified. The heat capacity i.e., the amount of heat energy required to increase the temperature by a unit quantity, is given by

$$\begin{aligned} C &= \frac{\partial E}{\partial T} = \frac{\partial E}{\partial r_h}\frac{\partial r_h}{\partial T} \\ &= \frac{(n-1)Vol(S^{d-2})r_h^{n-3}}{8n}(nr_h^2 \\ &+ (n-2)l^2)\left[1 \pm \sqrt{1 - \kappa\frac{n(n-2)\beta^2}{4\pi^2 l^2}}\right], \end{aligned} \tag{1.53}$$

with C, the second derivative of the action with respect to temperature, gives thermodynamic stability. The free energy is given by $F = E - TS = IT$

We can finally check the possible phases: first we note that r_h is an increasing function of the mass M. Now it is easy to check that $0 \leq \beta < \beta_{max}$, where $\beta_{max} = \frac{2\pi l}{\sqrt{(d-3)(d-1)}}$. Hence for temperatures $T < T_{min} = \frac{1}{\beta_{max}}$ there is no black hole and we have a pure (adS) phase of thermal radiation with negative free energy. For $T > T_{min}$ there are two black holes with $M_- < M_+$ (1.42). The black hole with the smaller mass has negative specific heat. It is therefore unstable to decay either to a larger black hole or to pure thermal adS. The bigger black hole has on the contrary positive specific heat and is thus thermodynamically stable. Furthermore for $T_{min} < T < T_{HP}$ the free energy of the plus branch is less than pure thermal adS which is energetically favoured. However, for $T > T_{HP}$ the situation is inversed and the large black hole state has less free energy than thermal adS space. AdS space can therefore tunnel towards this large black hole which becomes the energetically preferred state.

1.3 Beyond Static Black Holes

Suppose we now reduce the symmetry of spacetime, requiring a stationary and axially symmetric spacetime. As is often the case once symmetry is reduced the field equations are no longer integrable and knowledge of the full spectrum of solutions is lost. The presence of the cosmological constant presents a genuine complication with respect to the vacuum case which remains integrable for the stationary axially symmetric case (see [27–33] and references within).

Our aim in this section is to describe the difficulty associated to the presence of Λ and to discuss briefly this open problem. Let us focus here mostly on the case of $d = 4$ dimensions to simplify the problem as much as possible. Consider a spacetime with two killing vectors one which is locally timelike and one axial which are no longer orthogonal to each other. In other words we have lost $d - 2$ maximal symmetry and we have traded stationarity for staticity. The general metric anzatz reads,

$$ds^2 = e^{2\nu}\alpha^{-1/2}(dr^2 + dz^2) + \alpha e^{-\frac{\Omega}{2}}d\varphi^2 - \alpha e^{\frac{\Omega}{2}}(dt + Ad\varphi)^2, \tag{1.54}$$

where $\partial_t, \partial\varphi$ are the Killing vectors in this adapted coordinate system. The metric functions, ν, α and Ω depend on r and z. Note that we can still undertake 2 dimensional conformal transformations in the (r, z) plane without loss of generality. The field equations read,

$$\Delta\alpha = -2\Lambda\alpha^{\frac{1}{2}}e^{2\nu}, \tag{1.55}$$

$$\vec{\nabla} \cdot \left(e^{\Omega}\alpha\vec{\nabla}A\right) = 0, \tag{1.56}$$

$$\frac{1}{\alpha}\vec{\nabla}\cdot\left(\alpha\vec{\nabla}\Omega\right)+2e^{\Omega}(\vec{\nabla}A)^2=0, \tag{1.57}$$

$$\Delta v+\frac{1}{4}(\vec{\nabla}\Omega)^2-e^{\Omega}(\vec{\nabla}A)^2=\frac{\Delta\alpha}{\alpha}, \tag{1.58}$$

$$2v_{,u}\frac{\alpha_{,u}}{\alpha}-\frac{\alpha_{,\zeta\zeta}}{\alpha}=\frac{1}{8}(\Omega_{,\zeta})^2-\frac{1}{2}e^{\Omega}A^2_{,\zeta}\quad(\zeta\leftrightarrow\bar{\zeta}), \tag{1.59}$$

and correspond to a 4 dimensional spacetime with cosmological constant [16]. The derivative operators are the usual flat 2 dimensional operators with respect to r and z and $\zeta=r+iz$.

This system is a slight generalisation of the Lewis–Papapetrou system for $\Lambda=0$ (see for example [27–33]). In particular static and axially symmetric metrics (usually called Weyl metrics) with $\Lambda=0$ are obtained setting $A=0$. The vacuum case is long known to be integrable [27–33]. This is no longer true in the presence of the cosmological constant and also in higher dimensions [34]. In higher dimensions this is due to the appearance of novel curvature scales, much like the cosmological constant, associated to the additional Killing vectors. However, when all curvature scales are set to zero (all spatial Killing vectors have $U(1)$ symmetry) the system is integrable for $D>4$ and has been studied extensively in [35]. To picture how this works for vacuum and fails for $\Lambda\neq 0$ let us switch off in (1.55–1.59) the curvature scale Λ and A (for simplicity). Note then from (1.55) that α is a 2 dimensional harmonic function and we can use the 2 dimensional conformal transformations of the metric (1.54) to fix, without any loss of generality $\alpha=r$. Then (1.57), is just Laplace's equation, written in 3 dimensional cylindrical coordinates, a linear ODE! Solving for Ω is in fact the same as solving for a Newtonnian axial source! Given any Newtonian source we can find (and even superpose) Ω, the corresponding Weyl potential for the static and axisymmetric metric (1.54). In fact using this Newtonian analogy we can construct multi black hole solutions (see for example [34, 36–38]). In essence once we have determined the source corresponding to a black hole (a linear massive rod lying on the $r=0$ axis) we just superpose such sources on the symmetry axis. For a non-linear theory as GR, however, this is not all. Using the non-linear (1.59) we solve for v obtaining the full metric solution. Typically v picks up conical singularities on the axis $r=0$ which are interpreted as keeping the static equilibrium between attractive gravitational sources. For a double black hole we have either two infinite cosmic strings lying on $r=0$ pulling apart the black holes or a rigid strut in between the rods keeping them apart [36–38]. The former corresponds to a conical defect, $v(0,z)>0$, whereas the latter to a conical excess, $v(0,z)<0$.

Firstly go back to $\Lambda\neq 0$. Note now that the curvature scale Λ spoils this integrability at step 1 given that we can no longer adequately fix the coordinate system as before (1.55). Furthermore secondly, switching back on A takes us from the static to the stationary case. Integrability for $\Lambda=0$ is still maintained largely due to the fact that (1.56–1.57) combine together to form a complex equation, the Ernst equation with respect to a complex potential $\mathcal{E}$ that has as its real part the

Weyl potential and as its imaginary part a dual of the rotation A [39, 40]. It is interesting that this mathematical property remains true even when $\Lambda \neq 0$. [41], [42]. The Ernst equation replacing (1.56), (1.57) reads,

$$\frac{1}{\alpha}\vec{\nabla}\cdot\left(\alpha\vec{\nabla}\mathcal{E}_-\right) = \frac{(\vec{\nabla}\mathcal{E}_-)^2}{Re(\mathcal{E}_-)} + Re(\mathcal{E}_-)\frac{\Delta\alpha}{\alpha}, \tag{1.60}$$

with $\mathcal{E} = e^{\frac{\Omega}{2}}\alpha + i\omega$.

Let us now look at a specific example solution first found by Carter, which generalises the Kerr metric in the presence of a cosmological constant. The solution reads,

$$\begin{aligned} ds_4^2 = &-\frac{\Delta}{\rho^2}\left(dt - \frac{a\sin^2\theta}{\Xi_a}d\varphi\right)^2 + \frac{\Delta_\theta\sin^2\theta}{\rho^2}\left(adt - \frac{r^2+a^2}{\Xi_a}d\varphi\right)^2 \\ &+ \rho^2\left(\frac{dr^2}{\Delta} + \frac{d\theta^2}{\Delta_\theta}\right), \end{aligned} \tag{1.61}$$

where k is the adS curvature scale, M is the black hole mass, a the angular momentum parameter and

$$\Delta = (r^2 + a^2)(1 + k^2r^2) - 2Mr, \tag{1.62}$$

$$\Delta_\theta = 1 - a^2k^2\cos^2\theta, \quad \Xi_a = 1 - a^2k^2, \tag{1.63}$$

$$\rho^2 = r^2 + a^2\cos^2\theta, \quad \Lambda = -3k^2. \tag{1.64}$$

Rotation is bounded since the solution remains valid for $a^2 < k^2$ and is singular at $a^2 = k^2$. The exterior event horizon is the largest zero of $\Delta(r_+) = 0$ and the extremal limit is obtained when the inner and and outer horizons coincide. This also sets the lower bound for the mass parameter, $m \geq m_{ext}$ which is extremised for $a^2 = k^2$. Unlike the static black holes here the zeros of Δ do not correspond to the position of the Killing horizon of ∂_t. In fact the geometric region defined by the null Killing generator of the horizon and the Killing horizon of ∂_t defines the ergoregion as in the case of Kerr spacetime. In this region we have superradiance effects for certain wave modes due to the possible energy extraction of the black hole with the reflected wave mode being of greater energy than the initial one ($E_{scattered} > E_{initial}$). Now an intriguing possibility arises [43]. AdS, with its confining gravitational potential, can act as an amplifier to this effect. This is true for massive modes due to the confining potential not allowing them to escape to infinity but shooting them back into the hole, and massless modes alike, due to the reflecting boundary conditions imposed at the adS boundary. One expects through this effect that the black hole will gradually loose its angular momentum [43]. Such an instability was found for small black holes in [44].

In 5 dimensions an interesting difference occurs [43] (for the generalization of Carter' s solution to higher dimensions see [45–48]) and this effect, can in some cases, be washed away. First there is an additional angular momentum parameter due to the extra Killing vector. The null generator of the horizon reads, $\chi = \partial_t + \Omega_\phi \partial_\phi + \Omega \partial$ where Ω_i is the rotational velocity associated to each angular momentum parameter. If the hole is sufficiently slow rotating $\Omega_i < k$ with respect to the curvature scale of adS, the vector χ remains timelike all the way up to infinity. In other words in adS matter can be in a co-rotating frame with the black hole, something impossible for a Kerr spacetime where χ is spacelike and an external observer would have to move at a speed greater than light to keep up with the black hole rotation. This effect also agrees with the rotating black hole thermodynamics in adS [43] where a similar phase transition occurs as for the static case we studied earlier (see also [49] for a more general discussion including the case of adS black rings and other assorted higher dimensional black objects).

Let us finally give the Ernst potential for the rotating black hole solution found by Carter. We can transit in between the coordinate system of (1.61) and (1.54) by setting

$$\frac{dr^2}{\Delta} = d\mathrm{r}^2, \quad \frac{d\theta^2}{\Delta_\theta} = dz^2, \tag{1.65}$$

meaning that z and r are implicitly given as functions or θ and r, respectively. Using (1.54), this is all we need to know in order to identify the different components:

$$\alpha = \frac{\sin\theta}{\Xi_a}\sqrt{\Delta\Delta_\theta}, \tag{1.66}$$

$$A = \frac{a\sin^2\theta(\Delta - \Delta_\theta(r^2 + a^2))}{\Xi_a(a^2\Delta_\theta\sin^2\theta - \Delta)}, \tag{1.67}$$

$$e^\Omega = \frac{\Xi_a^2(\Delta - a^2\Delta_\theta\sin^2\theta)^2}{\Delta\Delta_\theta\rho^4\sin^2\theta}, \tag{1.68}$$

$$e^{2\nu} = \rho^2\alpha^{1/2}. \tag{1.69}$$

The electric Ernst potential for Carter's solution is given by

$$\mathcal{E} = \frac{1}{\rho^2}\left(\Delta - a^2\sin^2\theta\Delta_\theta - 2ia\cos\theta(k^2\rho^2 r + M)\right). \tag{1.70}$$

If there is no rotation, $a = 0$, the Ernst potential is real and corresponds to the $\kappa = 1$ black holes we studied in the previous section. For $M = 0$ we have pure adS but the potential is still complex since the metric has non-zero angular momentum.[2] If $\Lambda = 0, \mathcal{E}$ is the usual Ernst potential in the coordinates of (1.61).

[2] This is quite unlike the situation for Kerr's solution at asymptotic infinity.

Acknowledgements It is a pleasure to thank Bruno Boisseau, Jeff Dufaux, Renaud Parentani and Kostas Skenderis for helpful comments and discussion. I am especially grateful to Roberto Emparan and Bernard Linet for reading through the manuscript and making numerous helpful and critical comments and corrections. I last but not least thank Jihad Mourad for discussion and explanations regarding properties of adS space.

References

1. Hawking, S.W., Ellis, G.F.R.: The The Large Scale Structure of Space-Time. Cambridge University press, Cambridge (1973)
2. Birell, N.D., Davies, D.C.W.: Quantum Fields in Curved Space, Cambridge Monographs in Mathematical Physics. Cambridge University press, New York (1982)
3. Balasubramanian, V., Kraus, P., Lawrence, A.E.: Phys. Rev. D **59**, 046003 (1999) [arXiv:hep-th/9805171]
4. Dufaux, J.-F.: Modèles branaires en théories de gravité généralisés. PhD thesis, Orsay (2004)
5. Wald, R.M.: General Relativity. The University of Chicago Press, Chicago (1984)
6. Bowcock, P., Charmousis, C., Gregory, R.: Class. Quant. Grav. **17**, 4745 (2000) [arXiv:hep-th/0007177]
7. Birmingham, D.: Class. Quant. Grav. **16**, 1197 (1999) [hep-th/9808032]
8. Emparan, R.: Phys. Lett. B **432**, 74 (1998), [arXiv:hep-th/9804031]
9. Papadimitriou, I., Skenderis, K.: JHEP **0508**, 004 (2005) [arXiv:hep-th/0505190]
10. Skenderis, K.: Class. Quant. Grav. **19**, 5849 (2002) [arXiv:hep-th/0209067]
11. Chamblin, A., Emparan, R., Johnson, C.V., Myers, R.C.: Phys. Rev. D **60**, 064018 (1999) [arXiv:hep-th/9902170]
12. Emparan, R., Reall, H.S.: Living Rev. Rel. **11**, 6 (2008) [arXiv:0801.3471 [hep-th]]
13. Aminneborg, S., Bengtsson, I., Holst, S., Peldán, P.: Class. Quant. Grav. **13**, 2707 (1996) [gr-qc/9604005]
14. Lemos, J.P.S.: Phys. Lett. B **353**, 46 (1995) [arXiv:gr-qc/9404041]
15. Lemos, J.P.S.: Class. Quant. Grav. **12**, 1081 (1995) [arXiv:gr-qc/9407024]
16. Brill, D., Louko, J., Peldán, P.: Phys. Rev. D **56**, 3600 (1997) gr-qc/9705012
17. Emparan, R.: JHEP **9906**, 036 (1999) [arXiv:hep-th/9906040]
18. Gibbons, G.W., Hawking, S.W.: Phys. Rev. D **15**, 2752 (1977)
19. Witten, E.: Adv. Theor. Math. Phys. **2**, 505 (1998) [hep-th/9803131]
20. Hartle, J.B., Hawking, S.W.: Phys. Rev. D **13**, 2188 (1976)
21. Balasubramanian, V., Kraus, P.: Commun. Math. Phys. **208**, 413 (1999) [arXiv:hep-th/9902121]
22. Emparan, R., Johnson, C.V., Myers, R.C.: Phys. Rev. D **60**, 104001 (1999) [arXiv:hep-th/9903238] [hep-th/9903238]
23. Ashtekar, A., Magnon, A.: Class. Quant. Grav. **1**, L39 (1984)
24. Ashtekar, A., Das, S.: Class. Quant. Grav. **17**, L17 (2000) [arXiv:hep-th/9911230]
25. Hawking, S.W., Israel, W. (eds.): General Relativity, An Einstein Centenary Survey. Cambridge University Press, Cambridge (1979)
26. Hawking, S.W., Page, D.N.: Commun. Math. Phys. **87**, 577 (1983)
27. Stephani, H., Kramer, D., MacCallum, M., Hoenselaers, C., Herlt, E.: Exact Solutions to Einstein's Equations (2nd edn.) Cambridge University Press, Cambridge (1984)
28. Weyl, H.: Annalen Physik **54**, 117 (1917)
29. Lewis, T.: In: Proceedings of the Royal Society of London A, **136**, 176 (1932)
30. Papapetrou, A.: Ann. Phys. **12**, 309 (1953)
31. Papapetrou, A.: Ann. Inst. H. Poincaré **A4**, 83 (1966)
32. Belinskii, V.A., Zakharov, V.E.: Sov. Phys. JETP **77**, 3
33. Hoenselaers, C., Kinnersley, W., Xanthopoulos, B.C.: J. Math. Phys. **20**, 2530 (1979)

34. Charmousis, C., Gregory, R.: Class. Quant. Grav. **21**, 527 (2004) [arXiv:gr-qc/0306069]
35. Emparan, R., Reall, H.S.: Phys. Rev. D **65**, 084025 (2002) [arXiv:hep-th/0110258]
36. Israel, W., Khan, K.A.: Nuovo Cim. **33**, 331 (1964)
37. Myers, R.C.: Phys. Rev. D **35**, 455 (1987)
38. Dowker, H.F., Thambyahpillai, S.N.: Class. Quant. Grav. **20**, 127 (2003) [arXiv:gr-qc/0105044]
39. Ernst, F.J.: Phys. Rev. **167**, 1175 (1968)
40. Ernst, F.J.: Phys. Rev. **168**, 1415 (1968)
41. Charmousis, C.: J. Phys. Conf. Ser. **68**, 012007 (2007) [arXiv:0710.0472 [gr-qc]]
42. Hawking, S.W., Reall, H.S.: Phys. Rev. D **61**, 024014 (2000) [arXiv:hep-th/9908109]
43. Cardoso, V., Dias, O.J.C.: Phys. Rev. D **70**, 084011 (2004) [arXiv:hep-th/0405006]
44. Myers, R.C., Perry, M.J.: Annals Phys. **172**, 304 (1986)
45. Hawking, S.W., Hunter, C.J., Taylor-Robinson, M.M.: Phys. Rev. D **59**, 064005 (1999) [hep-th/9811056]
46. Gibbons, G.W., Lu, H., Page, D.N., Pope, C.N.: Phys. Rev. Lett. **93**, 171102 (2004), [arXiv:hep-th/0409155]
47. Gibbons, G.W., Lu, H., Page, D.N., Pope, C.N.: J. Geom. Phys. **53**, 49 (2005) [arXiv:hep-th/0404008]
48. Caldarelli, M.M., Emparan, R., Rodriguez, M.J.: JHEP **0811**, 011 (2008) [arXiv:0806.1954 [hep-th]]
49. Carter, B.: Commun. Math. Phys. **10**, 280 (1968)

Chapter 2
Perturbations of Anti de Sitter Black Holes

George Siopsis

Abstract I review perturbations of black holes in asymptotically anti de Sitter space. I show how the quasi-normal modes governing these perturbations can be calculated analytically and discuss the implications on the hydrodynamics of gauge theory fluids per the AdS/CFT correspondence. I also discuss phase transitions of hairy black holes with hyperbolic horizons and the dual superconductors emphasizing the analytical calculation of their properties.

2.1 Introduction

The perturbations of a black hole are governed by quasi-normal modes (QNMs). The latter are typically obtained by solving a wave equation for small fluctuations in the black hole background subject to the conditions that the flux be ingoing at the horizon and outgoing at asymptotic infinity. These boundary conditions in general lead to a discrete spectrum of complex frequencies whose imaginary part determines the decay time of the small fluctuations

$$\Im\omega = \frac{1}{\tau}. \tag{2.1}$$

There is a vast literature on quasi-normal modes and I make no attempt to review it here. Instead, I concentrate on obtaining analytic expressions for QNMs of black hole perturbations in asymptotically AdS space. One can rarely obtain analytic expressions in closed form. Instead, I discuss techniques which allow one to

G. Siopsis (✉)
Department of Physics and Astronomy, The University of Tennessee, Knoxville, TN 37996-1200, USA
e-mail: Siopsis@tennessee.edu

E. Papantonopoulos (ed.), *From Gravity to Thermal Gauge Theories: The Ads/CFT Correspondence*, Lecture Notes in Physics, 828,
DOI: 10.1007/978-3-642-04864-7_2,

calculate the spectrum perturbatively in an asymptotic regime (high or low overtones). For high overtones, the frequencies at leading order are proportional to the radius of the horizon. For low overtones, one in general obtains an additional frequency which is inversely proportional to the horizon radius. Thus for large black holes there is a gap between the lowest frequency and the rest of the spectrum of QNMs. I pay special attention to the lowest frequencies because they govern the behavior of the gauge theory fluid on the boundary according to the AdS/CFT correspondence. The latter may have experimental consequences pertaining to the formation of the quark-gluon plasma in heavy ion collisions. Moreover, I discuss phase transitions to hairy black holes which correspond to a dual superconducting phase. I concentrate on the case of black holes with a hyperbolic horizon because their properties can be understood analytically.

In Sect. 2.2 I discuss scalar, gravitational and electromagnetic perturbations of an AdS Schwarzschild black hole analytically calculating the QNM spectrum in the high frequency regime. In Sect. 2.3 I calculate the QNM spectrum analytically in the low frequency regime and discuss its relevance to the hyrdodynamic behavior of the dual gauge theory fluid on the boundary. In Sect. 2.4 I introduce hairy black holes and discuss their phase transition in the case of a hyperbolic horizon which can be understood analytically. The dual gauge theory corresponds to a superconductor whose properties can be calculated via electromagnetic perturbations. Finally, I conclude with Sect. 2.5.

2.2 Perturbations

In this section I discuss scalar, gravitational and electromagnetic perturbations of an AdS Schwarzschild black hole in d dimensions analytically calculating the QNM spectrum in the high frequency regime. Low overtones will be discussed in the next section.

The metric of an AdS Schwarzschild black hole is

$$ds^2 = -\left(\frac{r^2}{l^2} + K - \frac{2\mu}{r^{d-3}}\right)dt^2 + \frac{dr^2}{\frac{r^2}{l^2} + K - \frac{2\mu}{r^{d-3}}} + r^2 d\Sigma^2_{K,d-2}. \tag{2.2}$$

I shall choose units so that the AdS radius $l = 1$. The horizon radius and Hawking temperature are, respectively,

$$2\mu = r_+^{d-1}\left(1 + \frac{K}{r_+^2}\right), \quad T_H = \frac{(d-1)r_+^2 + K(d-3)}{4\pi r_+}. \tag{2.3}$$

The mass and entropy of the hole are, respectively,

$$M = (d-2)(K + r_+^2)\frac{r_+^{d-3}}{16\pi G} Vol(\Sigma_{K,d-2}), \quad S = \frac{r_+^{d-2}}{4G} Vol(\Sigma_{K,d-2}). \tag{2.4}$$

The parameter K determines the curvature of the horizon and the boundary of AdS space. For $K = 0, +1, -1$ we have, respectively, a flat ($\mathbb{R}^{d-2}$), spherical ($\mathbb{S}^{d-2}$) and hyperbolic ($\mathbb{H}^{d-2}/\Gamma$, topological black hole, where Γ is a discrete group of isometries) horizon (boundary).

The harmonics on $\Sigma_{K,d-2}$ satisfy

$$(\nabla^2 + k^2)\mathbb{T} = 0. \tag{2.5}$$

For $K = 0$, k is the momentum; for $K = +1$, the eigenvalues are quantized,

$$k^2 = l(l + d - 3) - \delta, \tag{2.6}$$

whereas for $K = -1$,

$$k^2 = \xi^2 + \left(\frac{d-3}{2}\right)^2 + \delta, \tag{2.7}$$

where ξ is discrete for non-trivial Γ. $\delta = 0, 1, 2$ for scalar, vector, or tensor perturbations, respectively.

According to the AdS/CFT correspondence, QNMs of AdS black holes are expected to correspond to perturbations of the dual Conformal Field Theory (CFT) on the boundary. The establishment of such a correspondence is hindered by difficulties in solving the wave equation governing the various types of perturbation. In three dimensions one obtains a hypergeometric equation which leads to explicit analytic expressions for the QNMs [1, 2]. In five dimensions one obtains a Heun equation and a derivation of analytic expressions for QNMs is no longer possible. On the other hand, numerical results exist in four, five and seven dimensions [3–5].

2.2.1 Scalar Perturbations

To find the asymptotic form of QNMs, we need to find an approximation to the wave equation valid in the high frequency regime. In three dimensions the resulting wave equation will be an exact equation (hypergeometric equation). In five dimensions, I shall turn the Heun equation into a hypergeometric equation which will lead to an analytic expression for the asymptotic form of QNM frequencies in agreement with numerical results.

2.2.1.1 AdS$_3$

In three dimensions the wave equation for a massless scalar field is

$$\frac{1}{r}\partial_r\left(r^3\left(1 - \frac{r_+^2}{r^2}\right)\partial_r\Phi\right) - \frac{1}{r^2 - r_+^2}\partial_t^2\Phi + \frac{1}{r^2}\partial_x^2\Phi = 0. \tag{2.8}$$

Writing the wavefunction in the form

$$\Phi = e^{i(\omega t - px)}\Psi(y), \quad y = \frac{r_+^2}{r^2}, \tag{2.9}$$

the wave function becomes

$$y^2(y-1)((y-1)\Psi')' + \hat{\omega}^2 y\Psi + \hat{p}^2 y(y-1)\Psi = 0 \tag{2.10}$$

to be solved in the interval $0 < y < 1$, where

$$\hat{\omega} = \frac{\omega}{2r_+} = \frac{\omega}{4\pi T_H}, \quad \hat{p} = \frac{p}{2r_+} = \frac{p}{4\pi T_H} . \tag{2.11}$$

For QNMs, we are interested in the solution

$$\Psi(y) = y(1-y)^{i\hat{\omega}}{}_2F_1(1 + i(\hat{\omega} + \hat{p}), 1 + i(\hat{\omega} - \hat{p}); 2; y), \tag{2.12}$$

which vanishes at the boundary ($y \to 0$). Near the horizon ($y \to 1$), we obtain a mixture of ingoing and outgoing waves,

$$\Psi \sim A_+(1-y)^{-i\hat{\omega}} + A_-(1-y)^{+i\hat{\omega}}, \quad A_\pm = \frac{\Gamma(\pm 2i\hat{\omega})}{\Gamma(1 \pm i(\hat{\omega} + \hat{p}))\Gamma(1 \pm i(\hat{\omega} - \hat{p}))}.$$

Setting $A_- = 0$, we deduce the quasi-normal frequencies

$$\hat{\omega} = \pm\hat{p} - in, \quad n = 1, 2, \ldots \tag{2.13}$$

which form a discrete spectrum of complex frequencies with $\Im\hat{\omega} < 0$.

2.2.1.2 AdS$_5$

Restricting attention to the case of a large black hole, the massless scalar wave equation reads

$$\frac{1}{r^3}\partial_r(r^5 f(r)\,\partial_r\Phi) - \frac{1}{r^2 f(r)}\partial_t^2\Phi - \frac{1}{r^2}\,\nabla^2\Phi = 0, \quad f(r) = 1 - \frac{r_+^4}{r^4}. \tag{2.14}$$

Writing the solution in the form

$$\Phi = e^{i(\omega t - p\cdot x)}\Psi(y), \quad y = \frac{r^2}{r_+^2} \tag{2.15}$$

the radial wave equation becomes

$$(y^2-1)\big(y(y^2-1)\Psi'\big)' + \left(\frac{\hat{\omega}^2}{4}y^2 - \frac{\hat{p}^2}{4}(y^2-1)\right)\Psi = 0. \tag{2.16}$$

For QNMs, we are interested in the analytic solution which vanishes at the boundary and behaves as an ingoing wave at the horizon. The wave equation contains an additional (unphysical) singularity at $y = -1$, at which the wavefunction behaves as $\Psi \sim (y+1)^{\pm\hat{\omega}/4}$. Isolating the behavior of the wavefunction near the singularities $y = \pm 1$,

$$\Psi(y) = (y-1)^{-i\hat{\omega}/4}(y+1)^{\pm\hat{\omega}/4}F_{\pm}(y), \tag{2.17}$$

we shall obtain two sets of modes with the same $\Im\hat{\omega}$, but opposite $\Re\hat{\omega}$.

$F_{\pm}(y)$ satisfies the Heun equation

$$y(y^2-1)F\pm'' + \left\{\left(3 - \frac{i\pm 1}{2}\hat{\omega}\right)y^2 - \frac{i\pm 1}{2}\hat{\omega}y - 1\right\}F'_{\pm} + \left\{\frac{\hat{\omega}}{2}\left(\pm\frac{i\hat{\omega}}{4} \mp 1 - i\right)y - (i\mp 1)\frac{\hat{\omega}}{4} - \frac{\hat{p}^2}{4}\right\}F_{\pm} = 0 \tag{2.18}$$

to be solved in a region in the complex y-plane containing $|y| \geq 1$ which includes the physical regime $r > r_+$.

For large $\hat{\omega}$, the constant terms in the polynomial coefficients of F' and F are small compared with the other terms, therefore they may be dropped. The wave equation may then be approximated by a hypergeometric equation

$$(y^2-1)F''_{\pm} + \left\{\left(3 - \frac{i\pm 1}{2}\hat{\omega}\right)y - \frac{i\pm 1}{2}\hat{\omega}\right\}F'_{\pm} + \frac{\hat{\omega}}{2}\left(\pm\frac{i\hat{\omega}}{4} \mp 1 - i\right)F_{\pm} = 0, \tag{2.19}$$

in the asymptotic limit of large frequencies $\hat{\omega}$. The acceptable solution is

$$F_0(x) = {}_2F_1(a_+, a_-; c; (y+1)/2), \quad a_{\pm} = 1 - \tfrac{i\pm 1}{4}\hat{\omega} \pm 1, \quad c = \tfrac{3}{2} \pm \tfrac{1}{2}\hat{\omega}. \tag{2.20}$$

For proper behavior at the boundary ($y \to \infty$), we demand that F be a *polynomial*, which leads to the condition

$$a_+ = -n, \quad n = 1, 2, \ldots \tag{2.21}$$

Indeed, it implies that F is a polynomial of order n, so as $y \to \infty$, $F \sim y^n \sim y^{-a_+}$ and $\Psi \sim y^{-i\hat{\omega}/4}y^{\pm\hat{\omega}/4}y^{-a_+} \sim y^{-2}$, as expected.

We deduce the quasi-normal frequencies [6]

$$\hat{\omega} = \frac{\omega}{4\pi T_H} = 2n(\pm 1 - i) \tag{2.22}$$

in agreement with numerical results.

It is perhaps worth mentioning that these frequencies may also be deduced by a simple monodromy argument [6]. Considering the monodromies around the singularities, if the wavefunction has no singularities other than $y = \pm 1$, the contour around $y = +1$ may be unobstructedly deformed into the contour around $y = -1$, which yields

$$\mathcal{M}(1)\mathcal{M}(-1) = 1. \tag{2.23}$$

Since the respective monodromies are $\mathcal{M}(1) = e^{\pi\hat{\omega}/2}$ and $\mathcal{M}(-1) = e^{\mp i\pi\hat{\omega}/2}$, using $\Im\hat{\omega} < 0$, we deduce $\hat{\omega} = 2n(\pm 1 - i)$, in agreement with our result above.

2.2.2 Gravitational Perturbations

Next I consider gravitational perturbations. For definiteness, I concentrate on the case of spherical black holes ($K = +1$). I shall derive analytic expressions for QNMs [7] including first-order corrections [8]. The results are in good agreement with results of numerical analysis [9]. Extension to other forms of the horizon is straightforward [10].

The radial wave equation for gravitational perturbations in the black-hole background (2.2) can be cast into a Schrödinger-like form,

$$-\frac{d^2\Psi}{dr_*^2} + V[r(r_*)]\Psi = \omega^2\Psi \,, \tag{2.24}$$

in terms of the tortoise coordinate defined by

$$\frac{dr_*}{dr} = \frac{1}{f(r)} \,. \tag{2.25}$$

The potential V for the various types of perturbation has been found by Ishibashi and Kodama [11]. For tensor, vector and scalar perturbations, one obtains, respectively,

$$V_{\mathsf{T}}(r) = f(r)\left\{\frac{\ell(\ell+d-3)}{r^2} + \frac{(d-2)(d-4)f(r)}{4r^2} + \frac{(d-2)f'(r)}{2r}\right\} \tag{2.26}$$

$$V_{\mathsf{V}}(r) = f(r)\left\{\frac{\ell(\ell+d-3)}{r^2} + \frac{(d-2)(d-4)f(r)}{4r^2} - \frac{rf'''(r)}{2(d-3)}\right\} \tag{2.27}$$

$$\begin{aligned}
V_{\mathsf{S}}(r) = {} & \frac{f(r)}{4r^2}\left[\ell(\ell+d-3)-(d-2)+\frac{(d-1)(d-2)\mu}{r^{d-3}}\right]^{-2} \\
& \times\left\{\frac{d(d-1)^2(d-2)^3\mu^2}{R^2r^{2d-8}} - \frac{6(d-1)(d-2)^2(d-4)[\ell(\ell+d-3)-(d-2)]\mu}{R^2r^{d-5}}\right. \\
& + \frac{(d-4)(d-6)[\ell(\ell+d-3)-(d-2)]^2r^2}{R^2} + \frac{2(d-1)^2(d-2)^4\mu^3}{r^{3d-9}} \\
& + \frac{4(d-1)(d-2)(2d^2-11d+18)[\ell(\ell+d-3)-(d-2)]\mu^2}{r^{2d-6}} \\
& + \frac{(d-1)^2(d-2)^2(d-4)(d-6)\mu^2}{r^{2d-6}} \\
& - \frac{6(d-2)(d-6)[\ell(\ell+d-3)-(d-2)]^2\mu}{r^{d-3}} \\
& - \frac{6(d-1)(d-2)^2(d-4)[\ell(\ell+d-3)-(d-2)]\mu}{r^{d-3}} \\
& \left. + 4[\ell(\ell+d-3)-(d-2)]^3 + d(d-2)[\ell(\ell+d-3)-(d-2)]^2\right\},
\end{aligned}$$

Near the black hole singularity ($r \sim 0$),

$$V_{\mathsf{T}} = -\frac{1}{4r_*^2} + \frac{\mathcal{A}_{\mathsf{T}}}{[-2(d-2)\mu]^{\frac{1}{d-2}}} r_*^{-\frac{d-1}{d-2}} + \cdots, \quad \mathcal{A}_{\mathsf{T}} = \frac{(d-3)^2}{2(2d-5)} + \frac{\ell(\ell+d-3)}{d-2}, \tag{2.28}$$

$$V_{\mathsf{V}} = \frac{3}{4r_*^2} + \frac{\mathcal{A}_{\mathsf{V}}}{[-2(d-2)\mu]^{\frac{1}{d-2}}} r_*^{-\frac{d-1}{d-2}} + \cdots, \quad \mathcal{A}_{\mathsf{V}} = \frac{d^2-8d+13}{2(2d-15)} + \frac{\ell(\ell+d-3)}{d-2} \tag{2.29}$$

and

$$V_{\mathsf{S}} = -\frac{1}{4r_*^2} + \frac{\mathcal{A}_{\mathsf{S}}}{[-2(d-2)\mu]^{\frac{1}{d-2}}} r_*^{-\frac{d-1}{d-2}} + \cdots, \tag{2.30}$$

where

$$\mathcal{A}_{\mathsf{S}} = \frac{(2d^3-24d^2+94d-116)}{4(2d-5)(d-2)} + \frac{(d^2-7d+14)[\ell(\ell+d-3)-(d-2)]}{(d-1)(d-2)^2}. \tag{2.31}$$

I have included only the terms which contribute to the order I am interested in. The behavior of the potential near the origin may be summarized by

$$V = \frac{j^2-1}{4r_*^2} + \mathcal{A}\, r_*^{-\frac{d-1}{d-2}} + \cdots \tag{2.32}$$

where $j = 0$ (2) for scalar and tensor (vector) perturbations.

On the other hand, near the boundary (large r),

$$V = \frac{j_\infty^2 - 1}{4(r_* - \bar{r}_*)^2} + \cdots, \quad \bar{r}_* = \int_0^\infty \frac{dr}{f(r)}, \tag{2.33}$$

where $j_\infty = d-1$, $d-3$ and $d-5$ for tensor, vector and scalar perturbations, respectively.

After rescaling the tortoise coordinate ($z = \omega r_*$), the wave equation to first order becomes

$$\left(\mathcal{H}_0 + \omega^{-\frac{d-3}{d-2}} \mathcal{H}_1\right)\Psi = 0, \tag{2.34}$$

where

$$\mathcal{H}_0 = \frac{d^2}{dz^2} - \left[\frac{j^2-1}{4z^2} - 1\right], \quad \mathcal{H}_1 = -\mathcal{A}\, z^{-\frac{d-1}{d-2}}. \tag{2.35}$$

By treating $\mathcal{H}_1$ as a perturbation, one may expand the wave function

$$\Psi(z) = \Psi_0(z) + \omega^{-\frac{d-3}{d-2}} \Psi_1(z) + \cdots \tag{2.36}$$

and solve the wave equation perturbatively.

The zeroth-order wave equation,

$$\mathcal{H}_0 \Psi_0(z) = 0, \tag{2.37}$$

may be solved in terms of Bessel functions,

$$\Psi_0(z) = A_1 \sqrt{z} J_{\frac{j}{2}}(z) + A_2 \sqrt{z} N_{\frac{j}{2}}(z). \tag{2.38}$$

For large z, it behaves as

$$\begin{aligned}\Psi_0(z) &\sim \sqrt{\frac{2}{\pi}} [A_1 \cos(z - \alpha_+) + A_2 \sin(z - \alpha_+)] \\ &= \frac{1}{\sqrt{2\pi}} (A_1 - iA_2) e^{-i\alpha_+} e^{iz} + \frac{1}{\sqrt{2\pi}} (A_1 + iA_2) e^{+i\alpha_+} e^{-iz},\end{aligned}$$

where $\alpha_\pm = \frac{\pi}{4}(1 \pm j)$.

At the boundary ($r \to \infty$), the wavefunction ought to vanish, therefore the acceptable solution is

$$\Psi_0(r_*) = B \sqrt{\omega(r_* - \bar{r}_*)}\, J_{\frac{j_\infty}{2}}(\omega(r_* - \bar{r}_*)). \tag{2.39}$$

Indeed, $\Psi \to 0$ as $r_* \to \bar{r}_*$, as desired.

Asymptotically (large z), it behaves as

$$\Psi(r_*) \sim \sqrt{\frac{2}{\pi}} B \cos[\omega(r_* - \bar{r}_*) + \beta], \quad \beta = \frac{\pi}{4}(1 + j_\infty). \tag{2.40}$$

This ought to be matched to the asymptotic form of the wavefunction in the vicinity of the black-hole singularity along the Stokes line $\Im z = \Im(\omega r_*) = 0$. This leads to a constraint on the coefficients A_1, A_2,

$$A_1 \tan(\omega \bar{r}_* - \beta - \alpha_+) - A_2 = 0. \tag{2.41}$$

By imposing the boundary condition at the horizon

$$\Psi(z) \sim e^{iz}, \quad z \to -\infty, \tag{2.42}$$

one obtains a second constraint. To find it, one needs to analytically continue the wavefunction near the black hole singularity ($z = 0$) to negative values of z. A rotation of z by $-\pi$ corresponds to a rotation by $-\frac{\pi}{d-2}$ near the origin in the complex r-plane. Using the known behavior of Bessel functions

$$J_\nu(e^{-i\pi} z) = e^{-i\pi\nu} J_\nu(z), \quad N_\nu(e^{-i\pi} z) = e^{i\pi\nu} N_\nu(z) - 2i \cos \pi\nu\, J_\nu(z), \tag{2.43}$$

for $z < 0$ the wavefunction changes to

$$\Psi_0(z) = e^{-i\pi(j+1)/2}\sqrt{-z}\left\{\left[A_1 - i(1+e^{i\pi j})A_2\right] J_{\frac{j}{2}}(-z) + A_2 e^{i\pi j} N_{\frac{j}{2}}(-z)\right\} \tag{2.44}$$

whose asymptotic behavior is given by

$$\Psi \sim \frac{e^{-i\pi(j+1)/2}}{\sqrt{2\pi}}\left[A_1 - i(1+2e^{j\pi i})A_2\right] e^{-iz} + \frac{e^{-i\pi(j+1)/2}}{\sqrt{2\pi}}\left[A_1 - iA_2\right] e^{iz}. \tag{2.45}$$

Therefore one obtains a second constraint

$$A_1 - i(1+2e^{j\pi i})A_2 = 0. \tag{2.46}$$

The two constraints are compatible provided

$$\begin{vmatrix} 1 & -i(1+2e^{j\pi i}) \\ \tan(\omega\bar{r}_* - \beta - \alpha_+) & -1 \end{vmatrix} = 0, \tag{2.47}$$

which yields the quasi-normal frequencies [7]

$$\omega\bar{r}_* = \frac{\pi}{4}(2 + j + j_\infty) - \tan^{-1}\frac{i}{1+2e^{j\pi i}} + n\pi. \tag{2.48}$$

The first-order correction to the above asymptotic expression may be found by standard perturbation theory [8]. To first order, the wave equation becomes

$$\mathcal{H}_0\Psi_1 + \mathcal{H}_1\Psi_0 = 0. \tag{2.49}$$

The solution is

$$\begin{aligned}\Psi_1(z) = {} & \sqrt{z}N_{\frac{j}{2}}(z)\int_0^z dz' \frac{\sqrt{z'}J_{\frac{j}{2}}(z')\mathcal{H}_1\Psi_0(z')}{\mathcal{W}} \\ & - \sqrt{z}J_{\frac{j}{2}}(z)\int_0^z dz' \frac{\sqrt{z'}N_{\frac{j}{2}}(z')\mathcal{H}_1\Psi_0(z')}{\mathcal{W}}.\end{aligned} \tag{2.50}$$

where $\mathcal{W} = 2/\pi$ is the Wronskian.

The wavefunction to first order reads

$$\Psi(z) = \{A_1[1-b(z)] - A_2a_2(z)\}\sqrt{z}J_{\frac{j}{2}}(z) + \{A_2[1+b(z)] + A_1a_1(z)\}\sqrt{z}N_{\frac{j}{2}}(z), \tag{2.51}$$

where

$$a_1(z) = \frac{\pi\mathcal{A}}{2}\,\omega^{-\frac{d-3}{d-2}}\int_0^z dz'\, z'^{-\frac{1}{d-2}}J_{\frac{j}{2}}(z')J_{\frac{j}{2}}(z'),$$

$$a_2(z) = \frac{\pi\mathcal{A}}{2}\,\omega^{-\frac{d-3}{d-2}}\int_0^z dz'\, z'^{-\frac{1}{d-2}}N_{\frac{j}{2}}(z')N_{\frac{j}{2}}(z'),$$

$$b(z) = \frac{\pi\mathcal{A}}{2}\,\omega^{-\frac{d-3}{d-2}}\int_0^z dz'\, z'^{-\frac{1}{d-2}}J_{\frac{j}{2}}(z')N_{\frac{j}{2}}(z'),$$

and $\mathcal{A}$ depends on the type of perturbation.

Asymptotically, it behaves as

$$\Psi(z) \sim \sqrt{\frac{2}{\pi}}[A_1' \cos(z - \alpha_+) + A_2' \sin(z - \alpha_+)], \tag{2.52}$$

where

$$A_1' = [1 - \bar{b}]A_1 - \bar{a}_2 A_2, \quad A_2' = [1 + \bar{b}]A_2 + \bar{a}_1 A_1 \tag{2.53}$$

and I introduced the notation

$$\bar{a}_1 = a_1(\infty), \quad \bar{a}_2 = a_2(\infty), \quad \bar{b} = b(\infty). \tag{2.54}$$

The first constraint is modified to

$$A_1' \tan(\omega \bar{r}_* - \beta - \alpha_+) - A_2' = 0. \tag{2.55}$$

Explicitly,

$$[(1 - \bar{b}) \tan(\omega \bar{r}_* - \beta - \alpha_+) - \bar{a}_1]A_1 - [1 + \bar{b} + \bar{a}_2 \tan(\omega \bar{r}_* - \beta - \alpha_+)]A_2 = 0. \tag{2.56}$$

To find the second constraint to first order, one needs to approach the horizon. This entails a rotation by $-\pi$ in the z-plane. Using

$$a_1(e^{-i\pi}z) = e^{-i\pi\frac{d-3}{d-2}} e^{-i\pi j} a_1(z),$$
$$a_2(e^{-i\pi}z) = e^{-i\pi\frac{d-3}{d-2}} \left[e^{i\pi j} a_2(z) - 4\cos^2\frac{\pi j}{2} a_1(z) - 2i(1 + e^{i\pi j}) b(z) \right],$$
$$b(e^{-i\pi}z) = e^{-i\pi\frac{d-3}{d-2}} \left[b(z) - i(1 + e^{-i\pi j}) a_1(z) \right],$$

in the limit $z \to -\infty$ one obtains

$$\Psi(z) \sim -ie^{-ij\pi/2} B_1 \cos(-z - \alpha_+) - ie^{ij\pi/2} B_2 \sin(-z - \alpha_+), \tag{2.57}$$

where

$$\begin{aligned} B_1 = A_1 &- A_1 e^{-i\pi\frac{d-3}{d-2}} [\bar{b} - i(1 + e^{-i\pi j})\bar{a}_1] \\ &- A_2 e^{-i\pi\frac{d-3}{d-2}} \left[e^{+i\pi j}\bar{a}_2 - 4\cos^2\frac{\pi j}{2}\bar{a}_1 - 2i(1 + e^{+i\pi j})\bar{b} \right] \\ &- i(1 + e^{i\pi j}) \left[A_2 + A_2 e^{-i\pi\frac{d-3}{d-2}} [\bar{b} - i(1 + e^{-i\pi j})\bar{a}_1] + A_1 e^{-i\pi\frac{d-3}{d-2}} e^{-i\pi j}\bar{a}_1 \right] \\ B_2 = A_2 &+ A_2 e^{-i\pi\frac{d-3}{d-2}} [\bar{b} - i(1 + e^{-i\pi j})\bar{a}_1] + A_1 e^{-i\pi\frac{d-3}{d-2}} e^{-i\pi j}\bar{a}_1. \end{aligned}$$

Therefore the second constraint to first order reads

$$[1 - e^{-i\pi\frac{d-3}{d-2}}(i\bar{a}_1 + \bar{b})]A_1 - [i(1 + 2e^{i\pi j}) + e^{-i\pi\frac{d-3}{d-2}}((1 + e^{i\pi j})\bar{a}_1 + e^{i\pi j}\bar{a}_2 - i\bar{b})]A_2 = 0 \tag{2.58}$$

Compatibility of the two first-order constraints yields

$$\begin{vmatrix} 1+\bar{b}+\bar{a}_2\tan(\omega\bar{r}_* - \beta - \alpha_+) & i(1+2e^{i\pi j}) + e^{-i\pi\frac{d-3}{d-2}}((1+e^{i\pi j})\bar{a}_1 + e^{i\pi j}\bar{a}_2 - i\bar{b}) \\ (1-\bar{b})\tan(\omega\bar{r}_* - \beta - \alpha_+) - \bar{a}_1 & 1 - e^{-i\pi\frac{d-3}{d-2}}(i\bar{a}_1 + \bar{b}) \end{vmatrix} = 0, \tag{2.59}$$

leading to the first-order expression for quasi-normal frequencies,

$$\begin{aligned} \omega\bar{r}_* &= \frac{\pi}{4}(2 + j + j_\infty) + \frac{1}{2i}\ln 2 + n\pi \\ &\quad - \frac{1}{8}\left\{6i\bar{b} - 2ie^{-i\pi\frac{d-3}{d-2}}\bar{b} - 9\bar{a}_1 + e^{-i\pi\frac{d-3}{d-2}}\bar{a}_1 + \bar{a}_2 - e^{-i\pi\frac{d-3}{d-2}}\bar{a}_2\right\}, \end{aligned}$$

where

$$\begin{aligned} \bar{a}_1 &= \frac{\pi\mathcal{A}}{4}\left(\frac{n\pi}{2\bar{r}_*}\right)^{-\frac{d-3}{d-2}} \frac{\Gamma(\frac{1}{d-2})\Gamma(\frac{j}{2} + \frac{d-3}{2(d-2)})}{\Gamma^2(\frac{d-1}{2(d-2)})\Gamma(\frac{j}{2} + \frac{d-1}{2(d-2)})} \\ \bar{a}_2 &= \left[1 + 2\cot\frac{\pi(d-3)}{2(d-2)}\cot\frac{\pi}{2}\left(-j + \frac{d-3}{d-2}\right)\right]\bar{a}_1 \\ \bar{b} &= -\cot\frac{\pi(d-3)}{2(d-2)}\bar{a}_1. \end{aligned}$$

Thus the first-order correction is $\sim \mathcal{O}(n^{-\frac{d-3}{d-2}})$.

The above analytic results are in good agreement with numerical results [9] (see Ref. [8] for a detailed comparison).

2.2.3 Electromagnetic Perturbations

The electromagnetic potential in four dimensions is

$$V_{\mathsf{EM}} = \frac{\ell(\ell+1)}{r^2}f(r). \tag{2.60}$$

Near the origin,

$$V_{\mathsf{EM}} = \frac{j^2 - 1}{4r_*^2} + \frac{\ell(\ell+1)r_*^{-3/2}}{2\sqrt{-4\mu}} + \cdots, \tag{2.61}$$

where $j = 1$. Therefore a vanishing potential to zeroth order is obtained. To calculate the QNM spectrum one needs to include first-order corrections from the outset. Working as with gravitational perturbations, one obtains the QNMs

$$\omega \bar{r}_* = n\pi - \frac{i}{4} \ln n + \frac{1}{2i} \ln\left(2(1+i)\mathcal{A}\sqrt{\bar{r}_*}\right), \quad \mathcal{A} = \frac{\ell(\ell+1)}{2\sqrt{-4\mu}}. \tag{2.62}$$

Notice that the first-order correction behaves as $\ln n$, a fact which may be associated with gauge invariance.

As with gravitational perturbations, the above analytic results are in good agreement with numerical results [9] (see Ref. [8] for a detailed comparison).

2.3 Hydrodynamics

There is a correspondence between $\mathcal{N} = 4$ Super Yang–Mills (SYM) theory in the large N limit and type-IIB string theory in $AdS_5 \times S^5$ (AdS/CFT correspondence). In the low energy limit, string theory is reduced to classical supergravity and the AdS/CFT correspondence allows one to calculate all gauge field-theory correlation functions in the strong coupling limit leading to non-trivial predictions on the behavior of gauge theory fluids. For example, the entropy of $\mathcal{N} = 4$ SYM theory in the limit of large 't Hooft coupling is precisely 3/4 its value in the zero coupling limit.

The long-distance (low-frequency) behavior of any interacting theory at finite temperature must be described by fluid mechanics (hydrodynamics). This leads to a universality in physical properties because hydrodynamics implies very precise constraints on correlation functions of conserved currents and the stress-energy tensor. Their correlators are fixed once a few transport coefficients are known.

2.3.1 Vector Perturbations

I start with vector perturbations and work in the d-dimensional Schwarzschild background (2.2) with $K = +1$ (spherical horizon and boundary). It is convenient to introduce the coordinate [12]

$$u = \left(\frac{r_+}{r}\right)^{d-3}. \tag{2.63}$$

The wave equation becomes

$$-(d-3)^2 u^{\frac{d-4}{d-3}} \hat{f}(u) \left(u^{\frac{d-4}{d-3}} \hat{f}(u) \Psi'\right)' + \hat{V}_{\mathrm{V}}(u)\Psi = \hat{\omega}^2 \Psi, \quad \hat{\omega} = \frac{\omega}{r_+}, \tag{2.64}$$

where prime denotes differentiation with respect to u and I have defined

$$\hat{f}(u) \equiv \frac{f(r)}{r^2} = 1 - u^{\frac{2}{d-3}}\left(u - \frac{1-u}{r_+^2}\right) \tag{2.65}$$

$$\hat{V}_{\mathsf{V}}(u) \equiv \frac{V_{\mathsf{V}}}{r_+^2} = \hat{f}(u)\left\{\hat{L}^2 + \frac{(d-2)(d-4)}{4}u^{-\frac{2}{d-3}}\hat{f}(u) - \frac{(d-1)(d-2)\left(1+\frac{1}{r_+^2}\right)}{2}u\right\}, \tag{2.66}$$

where $\hat{L}^2 = \frac{\ell(\ell+d-3)}{r_+^2}$.

First I consider the large black hole limit $r_+ \to \infty$ keeping $\hat{\omega}$ and $\hat{L}$ fixed (small). Factoring out the behavior at the horizon ($u = 1$)

$$\Psi = (1-u)^{-i\frac{\hat{\omega}}{d-1}}F(u), \tag{2.67}$$

the wave equation simplifies to

$$\mathcal{A}F'' + \mathcal{B}_{\hat{\omega}}F' + \mathcal{C}_{\hat{\omega},\hat{L}}F = 0, \tag{2.68}$$

where

$$\begin{aligned}
\mathcal{A} &= -(d-3)^2 u^{\frac{2d-8}{d-3}}\left(1 - u^{\frac{d-1}{d-3}}\right)\\
\mathcal{B}_{\hat{\omega}} &= -(d-3)\left[d-4-(2d-5)u^{\frac{d-1}{d-3}}\right]u^{\frac{d-5}{d-3}} - 2(d-3)^2\frac{i\hat{\omega}}{d-1}\frac{u^{\frac{2d-8}{d-3}}\left(1-u^{\frac{d-1}{d-3}}\right)}{1-u}\\
\mathcal{C}_{\hat{\omega},\hat{L}} &= \hat{L}^2 + \frac{(d-2)\left[d-4-3(d-2)u^{\frac{d-1}{d-3}}\right]}{4}u^{-\frac{2}{d-3}}\\
&\quad - \frac{\hat{\omega}^2}{1-u^{\frac{d-1}{d-3}}} + (d-3)^2\frac{\hat{\omega}^2}{(d-1)^2}\frac{u^{\frac{2d-8}{d-3}}\left(1-u^{\frac{d-1}{d-3}}\right)}{(1-u)^2}\\
&\quad - (d-3)\frac{i\hat{\omega}}{d-1}\frac{\left[d-4-(2d-5)u^{\frac{d-1}{d-3}}\right]u^{\frac{d-5}{d-3}}}{1-u} - (d-3)^2\frac{i\hat{\omega}}{d-1}\frac{u^{\frac{2d-8}{d-3}}\left(1-u^{\frac{d-1}{d-3}}\right)}{(1-u)^2}.
\end{aligned}$$

One may solve this equation perturbatively by separating

$$(\mathcal{H}_0 + \mathcal{H}_1)F = 0, \tag{2.69}$$

where

$$\begin{aligned}
\mathcal{H}_0 F &\equiv \mathcal{A}F'' + \mathcal{B}_0 F' + \mathcal{C}_{0,0}F\\
\mathcal{H}_1 F &\equiv (\mathcal{B}_{\hat{\omega}} - \mathcal{B}_0)F' + (\mathcal{C}_{\hat{\omega},\hat{L}} - \mathcal{C}_{0,0})F.
\end{aligned}$$

Expanding the wavefunction perturbatively,

$$F = F_0 + F_1 + \cdots \tag{2.70}$$

at zeroth order the wave equation reads

$$\mathcal{H}_0 F_0 = 0 \tag{2.71}$$

whose acceptable solution is

$$F_0 = u^{\frac{d-2}{2(d-3)}}, \tag{2.72}$$

being regular at both the horizon ($u = 1$) and the boundary ($u = 0$, or $\Psi \sim r^{-\frac{d-2}{2}} \to 0$ as $r \to \infty$). The Wronskian is

$$\mathcal{W} = \frac{1}{u^{\frac{d-4}{d-3}}(1 - u^{\frac{d-1}{d-3}})} \tag{2.73}$$

and another linearly independent solution is

$$\check{F}_0 = F_0 \int \frac{\mathcal{W}}{F_0^2}, \tag{2.74}$$

which is unacceptable because it diverges at both the horizon ($\check{F}_0 \sim \ln(1-u)$ for $u \approx 1$) and the boundary ($\check{F}_0 \sim u^{-\frac{d-4}{2(d-3)}}$ for $u \approx 0$, or $\Psi \sim r^{\frac{d-4}{2}} \to \infty$ as $r \to \infty$).

At first order the wave equation reads

$$\mathcal{H}_0 F_1 = -\mathcal{H}_1 F_0 \tag{2.75}$$

whose solution may be written as

$$F_1 = F_0 \int \frac{\mathcal{W}}{F_0^2} \int \frac{F_0 \mathcal{H}_1 F_0}{\mathcal{A}\mathcal{W}}. \tag{2.76}$$

The limits of the inner integral may be adjusted at will because this amounts to adding an arbitrary amount of the unacceptable solution. To ensure regularity at the horizon, choose one of the limits of integration at $u = 1$ rendering the integrand regular at the horizon. Then at the boundary ($u = 0$),

$$F_1 = \check{F}_0 \int_0^1 \frac{F_0 \mathcal{H}_1 F_0}{\mathcal{A}\mathcal{W}} + \text{regular terms.} \tag{2.77}$$

The coefficient of the singularity ought to vanish,

$$\int_0^1 \frac{F_0 \mathcal{H}_1 F_0}{\mathcal{A}\mathcal{W}} = 0, \tag{2.78}$$

which yields a constraint on the parameters (dispersion relation)

$$\mathbf{a}_0 \hat{L}^2 - i\mathbf{a}_1 \hat{\omega} - \mathbf{a}_2 \hat{\omega}^2 = 0. \tag{2.79}$$

After some algebra, one arrives at

$$\mathbf{a}_0 = \frac{d-3}{d-1}, \quad \mathbf{a}_1 = d-3. \tag{2.80}$$

The coefficient $\mathbf{a}_2$ may also be found explicitly for each dimension d, but it cannot be written as a function of d in closed form. It does not contribute to the dispersion relation at lowest order. E.g., for $d = 4, 5$, one obtains, respectively

$$\mathbf{a}_2 = \frac{65}{108} - \frac{1}{3}\ln 3, \quad \frac{5}{6} - \frac{1}{2}\ln 2. \tag{2.81}$$

Equation (2.79) is quadratic in $\hat{\omega}$ and has two solutions,

$$\hat{\omega}_0 \approx -i\frac{\hat{L}^2}{d-1}, \quad \hat{\omega}_1 \approx -i\frac{d-3}{\mathbf{a}_2} + i\frac{\hat{L}^2}{d-1}. \tag{2.82}$$

In terms of the frequency ω and the quantum number ℓ,

$$\omega_0 \approx -i\frac{\ell(\ell+d-3)}{(d-1)r_+}, \quad \frac{\omega_1}{r_+} \approx -i\frac{d-3}{\mathbf{a}_2} + i\frac{\ell(\ell+d-3)}{(d-1)r_+^2}. \tag{2.83}$$

The smaller of the two, ω_0, is inversely proportional to the radius of the horizon and is not included in the asymptotic spectrum. The other solution, ω_1, is a crude estimate of the first overtone in the asymptotic spectrum, nevertheless it shares two important features with the asymptotic spectrum: it is proportional to r_+ and its dependence on ℓ is $\mathcal{O}(1/r_+^2)$. The approximation may be improved by including higher-order terms. This increases the degree of the polynomial in the dispersion relation (2.79) whose roots then yield approximate values of more QNMs. This method reproduces the asymptotic spectrum derived earlier albeit not in an efficient way.

To include finite size effects, I shall use perturbation theory (assuming $1/r_+$ is small) and replace $\mathcal{H}_1$ by

$$\mathcal{H}_1' = \mathcal{H}_1 + \frac{1}{r_+^2}\mathcal{H}_+ \tag{2.84}$$

where

$$\mathcal{H}_+ F \equiv \mathcal{A}_+ F'' + \mathcal{B}_+ F' + \mathcal{C}_+ F. \tag{2.85}$$

The coefficients may be easily deduced by collecting $\mathcal{O}(1/r_+^2)$ terms in the exact wave equation. One obtains

$$\begin{aligned}
\mathcal{A}_+ &= -2(d-3)^2 u^2(1-u) \\
\mathcal{B}_+ &= -(d-3)u\left[(d-3)(2-3u) - (d-1)\frac{1-u}{1-u^{\frac{d-1}{d-3}}}u^{\frac{d-1}{d-3}}\right] \\
\mathcal{C}_+ &= \frac{d-2}{2}\left[d-4-(2d-5)u-(d-1)\frac{1-u}{1-u^{\frac{d-1}{d-3}}}u^{\frac{d-1}{d-3}}\right].
\end{aligned}$$

Interestingly, the zeroth order wavefunction F_0 is an eigenfunction of $\mathcal{H}_+$,

$$\mathcal{H}_+ F_0 = -(d-2)F_0, \tag{2.86}$$

therefore the first-order finite-size effect is a simple shift of the angular momentum operator

$$\hat{L}^2 \to \hat{L}^2 - \frac{d-2}{r_+^2}. \tag{2.87}$$

The QNMs of lowest frequency are modified to

$$\omega_0 = -i\frac{\ell(\ell+d-3)-(d-2)}{(d-1)r_+} + \mathcal{O}(1/r_+^2). \tag{2.88}$$

For $d = 4, 5$, we have respectively,

$$\omega_0 = -i\frac{(\ell-1)(\ell+2)}{3r_+}, \quad -i\frac{(\ell+1)^2-4}{4r_+} \tag{2.89}$$

in agreement with numerical results [9, 13].

One deduces from (2.88) the maximum lifetime of the vector modes,

$$\tau_{\max} = \frac{4\pi}{d}T_H. \tag{2.90}$$

In the case of a flat horizon ($K = 0$),

$$\omega_0 = -i\frac{k^2}{(d-1)r_+}, \tag{2.91}$$

which leads to the diffusion constant

$$D = \frac{1}{4\pi T_H}. \tag{2.92}$$

In the case of a hyperbolic horizon ($K = -1$), a similar calculation yields [10]

$$\omega_0 = -i\frac{\xi^2 + \frac{(d-1)^2}{4}}{(d-1)r_+}, \quad \tau = \frac{1}{|\omega_0|} < \frac{16\pi}{(d-1)^2}T_H. \tag{2.93}$$

It follows that for $d = 5$, these modes live longer than their spherical counterparts which is important for plasma behavior.

2.3.2 Scalar Perturbations

Next I consider scalar perturbations which are calculationally more involved but phenomenologically more important because their spectrum contains the lowest frequencies and therefore the longest living modes. For a scalar perturbation we ought to replace the potential $\hat{V}_{\mathrm{V}}$ by

$$\begin{aligned}\hat{V}_{\mathrm{S}}(u) = {} & \frac{\hat{f}(u)}{4}\left[\hat{m} + \left(1+\frac{1}{r_+^2}\right)u\right]^{-2} \\ & \times \left\{ d(d-2)\left(1+\frac{1}{r_+^2}\right)^2 u^{\frac{2d-8}{d-3}} - 6(d-2)(d-4)\hat{m}\left(1+\frac{1}{r_+^2}\right)u^{\frac{d-5}{d-3}} \right. \\ & + (d-4)(d-6)\hat{m}^2 u^{-\frac{2}{d-3}} + (d-2)^2\left(1+\frac{1}{r_+^2}\right)^3 u^3 \\ & + 2(2d^2-11d+18)\hat{m}\left(1+\frac{1}{r_+^2}\right)^2 u^2 \\ & + \frac{(d-4)(d-6)\left(1+\frac{1}{r_+^2}\right)^2}{r_+^2}u^2 - 3(d-2)(d-6)\hat{m}^2\left(1+\frac{1}{r_+^2}\right)u \\ & \left. - \frac{6(d-2)(d-4)\hat{m}\left(1+\frac{1}{r_+^2}\right)}{r_+^2}u + 2(d-1)(d-2)\hat{m}^3 + d(d-2)\frac{\hat{m}^2}{r_+^2} \right\},\end{aligned} \tag{2.94}$$

where $\hat{m} = 2\frac{\ell(\ell+d-3)-(d-2)}{(d-1)(d-2)r_+^2} = \frac{2(\ell+d-2)(\ell-1)}{(d-1)(d-2)r_+^2}$.

In the large black hole limit $r_+ \to \infty$ with $\hat{m}$ fixed (small), the potential simplifies to

$$\begin{aligned}\hat{V}_{\mathrm{S}}^{(0)}(u) = {} & \frac{1-u^{\frac{d-1}{d-3}}}{4(\hat{m}+u)^2}\left\{ d(d-2)u^{\frac{2d-8}{d-3}} - 6(d-2)(d-4)\hat{m}u^{\frac{d-5}{d-3}}\right. \\ & + (d-4)(d-6)\hat{m}^2 u^{-\frac{2}{d-3}} + (d-2)^2 u^3 \\ & \left. +2(2d^2-11d+18)\hat{m}u^2 - 3(d-2)(d-6)\hat{m}^2 u + 2(d-1)(d-2)\hat{m}^3 \right\}.\end{aligned} \tag{2.95}$$

The wave equation has an additional singularity due to the double pole of the scalar potential at $u = -\hat{m}$. It is desirable to factor out the behavior not only at the horizon, but also at the boundary and the pole of the scalar potential,

$$\Psi = (1-u)^{-i\frac{\hat{\omega}}{d-1}} \frac{u^{\frac{d-4}{2(d-3)}}}{\hat{m}+u} F(u). \tag{2.96}$$

Then the wave equation reads

$$\mathcal{A}F'' + \mathcal{B}_{\hat{\omega}}F' + \mathcal{C}_{\hat{\omega}}F = 0, \tag{2.97}$$

where

$$\begin{aligned}
\mathcal{A} =& -(d-3)^2 u^{\frac{2d-8}{d-3}}(1-u^{\frac{d-1}{d-3}}) \\
\mathcal{B}_{\hat{\omega}} =& -(d-3)u^{\frac{2d-8}{d-3}}(1-u^{\frac{d-1}{d-3}})\left[\frac{d-4}{u} - \frac{2(d-3)}{\hat{m}+u}\right] \\
& -(d-3)[d-4-(2d-5)u^{\frac{d-1}{d-3}}]u^{\frac{d-5}{d-3}} - 2(d-3)^2\frac{i\hat{\omega}}{d-1}\frac{u^{\frac{2d-8}{d-3}}(1-u^{\frac{d-1}{d-3}})}{1-u} \\
\mathcal{C}_{\hat{\omega}} =& -u^{\frac{2d-8}{d-3}}(1-u^{\frac{d-1}{d-3}})\left[-\frac{(d-2)(d-4)}{4u^2} - \frac{(d-3)(d-4)}{u(\hat{m}+u)} + \frac{2(d-3)^2}{(\hat{m}+u)^2}\right] \\
& -\left[\left\{d-4-(2d-5)u^{\frac{d-1}{d-3}}\right\}u^{\frac{d-5}{d-3}} + 2(d-3)\frac{i\hat{\omega}}{d-1}\frac{u^{\frac{2d-8}{d-3}}(1-u^{\frac{d-1}{d-3}})}{1-u}\right] \\
& \times\left[\frac{d-4}{2u} - \frac{d-3}{\hat{m}+u}\right] - (d-3)\frac{i\hat{\omega}}{d-1}\frac{[d-4-(2d-5)u^{\frac{d-1}{d-3}}]u^{\frac{d-5}{d-3}}}{1-u} \\
& -(d-3)^2\frac{i\hat{\omega}}{d-1}\frac{u^{\frac{2d-8}{d-3}}(1-u^{\frac{d-1}{d-3}})}{(1-u)^2} + \frac{\hat{V}_{\mathsf{S}}^{(0)}(u) - \hat{\omega}^2}{1-u^{\frac{d-1}{d-3}}} \\
& +(d-3)^2\frac{\hat{\omega}^2}{(d-1)^2}\frac{u^{\frac{2d-8}{d-3}}(1-u^{\frac{d-1}{d-3}})}{(1-u)^2}
\end{aligned}$$

I shall define the zeroth-order wave equation as $\mathcal{H}_0 F_0 = 0$, where

$$\mathcal{H}_0 F \equiv \mathcal{A}F'' + \mathcal{B}_0 F'. \tag{2.98}$$

The acceptable zeroth-order solution is

$$F_0(u) = 1, \tag{2.99}$$

which is plainly regular at all singular points ($u = 0, 1, -\hat{m}$). It corresponds to a wavefunction vanishing at the boundary ($\Psi \sim r^{-\frac{d-4}{2}}$ as $r \to \infty$).

The Wronskian is

$$\mathcal{W} = \frac{(\hat{m}+u)^2}{u^{\frac{2d-8}{d-3}}(1-u^{\frac{d-1}{d-3}})} \tag{2.100}$$

and an unacceptable solution is $\check{F}_0 = \int \mathcal{W}$. It can be written in terms of hypergeometric functions. For $d \geq 6$, it has a singularity at the boundary, $\check{F}_0 \sim u^{-\frac{d-5}{d-3}}$ for $u \approx 0$, or $\Psi \sim r^{\frac{d-6}{2}} \to \infty$ as $r \to \infty$. For $d = 5$, the acceptable wavefunction behaves as $r^{-1/2}$ whereas the unacceptable one behaves as $r^{-1/2}\ln r$. For $d = 4$, the roles of F_0 and $\check{F}_0$ are reversed, however the results still valid because the correct boundary condition at the boundary is a Robin boundary condition [12, 14]. Finally, note that $\check{F}_0$ is also singular (logarithmically) at the horizon ($u = 1$).

Working as in the case of vector modes, one arrives at the first-order constraint

$$\int_0^1 \frac{\mathcal{C}_{\hat{\omega}}}{\mathcal{A}\mathcal{W}} = 0, \tag{2.101}$$

because $\mathcal{H}_1 F_0 \equiv (\mathcal{B}_{\hat{\omega}} - \mathcal{B}_0)F_0' + \mathcal{C}_{\hat{\omega}} F_0 = \mathcal{C}_{\hat{\omega}}$. This leads to the dispersion relation

$$\mathbf{a}_0 - \mathbf{a}_1 i\hat{\omega} - \mathbf{a}_2\hat{\omega}^2 = 0, \tag{2.102}$$

After some algebra, one obtains

$$\mathbf{a}_0 = \frac{d-1}{2}\frac{1+(d-2)\hat{m}}{(1+\hat{m})^2}, \quad \mathbf{a}_1 = \frac{d-3}{(1+\hat{m})^2}, \quad \mathbf{a}_2 = \frac{1}{\hat{m}}\{1+O(\hat{m})\}. \tag{2.103}$$

For small $\hat{m}$, the quadratic equation has solutions

$$\hat{\omega}_0^{\pm} \approx -i\frac{d-3}{2}\hat{m} \pm \sqrt{\frac{d-1}{2}}\hat{m} \tag{2.104}$$

related to each other by $\hat{\omega}_0^{+} = -\hat{\omega}_0^{-*}$, which is a general symmetry of the spectrum.

Finite size effects at first order amount to a shift of the coefficient $\mathbf{a}_0$ in the dispersion relation

$$\mathbf{a}_0 \to \mathbf{a}_0 + \frac{1}{r_+^2}\mathbf{a}_+ \tag{2.105}$$

After some tedious but straightforward algebra, we obtain

$$\mathbf{a}_+ = \frac{1}{\hat{m}}\{1+O(\hat{m})\}. \tag{2.106}$$

The modified dispersion relation yields the modes

$$\hat{\omega}_0^{\pm} \approx -i\frac{d-3}{2}\hat{m} \pm \sqrt{\frac{d-1}{2}\hat{m}+1}. \tag{2.107}$$

In terms of the quantum number ℓ,

$$\omega_0^{\pm} \approx -i(d-3)\frac{\ell(\ell+d-3)-(d-2)}{(d-1)(d-2)r_+} \pm \sqrt{\frac{\ell(\ell+d-3)}{d-2}}, \tag{2.108}$$

in agreement with numerical results [13].

Notice that the imaginary part is inversely proportional to r_+, as in vector case. In the scalar case, we also obtained a finite real part independent of r_+.

The maximum lifetime of a gravitational scalar mode is found from (2.108) to be

$$\tau_{\max} = \frac{d-2}{(d-3)d} 4\pi T_H. \tag{2.109}$$

In the case of a flat horizon ($K = 0$), one obtains

$$\omega = \pm \frac{k}{\sqrt{d-2}} - i \frac{d-3}{(d-1)(d-2)r_+} k^2, \tag{2.110}$$

showing that the speed of sound is

$$v = \frac{1}{\sqrt{d-2}} \tag{2.111}$$

as expected for a CFT and the diffusion constant is

$$D = \frac{d-3}{d-2} \frac{1}{4\pi T_H}. \tag{2.112}$$

For a hyperbolic horizon ($K = -1$), a similar calculation yields [10]

$$\omega = \pm \sqrt{\frac{\xi^2 + (\frac{d-3}{2})^2}{d-2}} - i \frac{(d-3)[\xi^2 + \frac{(d-1)^2}{4}]}{(d-1)(d-2)r_+}, \quad \tau < \frac{4(d-2)}{(d-3)(d-1)^2} 4\pi T_H. \tag{2.113}$$

In the physically relevant case $d = 5$, evidently the $K = -1$ scalar modes live longer than any other modes, which is important for plasma behavior.

2.3.3 Tensor Perturbations

Finally, for completeness I discuss the case of tensor perturbations. Unlike the other two cases of gravitational perturbations, the asymptotic spectrum of tensor perturbations is the entire spectrum. To see this, note that in the large black hole limit, the wave equation reads

$$- (d-3)^2 (u^{\frac{2d-8}{d-3}} - u^3) \Psi'' - (d-3)[(d-4)u^{\frac{d-5}{d-3}} - (2d-5)u^2]\Psi'$$
$$+ \left\{ \hat{L}^2 + \frac{d(d-2)}{4} u^{-\frac{2}{d-3}} + \frac{(d-2)^2}{4} u - \frac{\hat{\omega}^2}{1 - u^{\frac{d-1}{d-3}}} \right\} \Psi = 0.$$

For the zeroth-order equation, we may set $\hat{L} = 0 = \hat{\omega}$. The resulting equation may be solved exactly. Two linearly independent solutions are ($\Psi = F_0$ at zeroth order)

$$F_0(u) = u^{\frac{d-2}{2(d-3)}}, \quad \check{F}_0(u) = u^{-\frac{d-2}{2(d-3)}} \ln\left(1 - u^{\frac{d-1}{d-3}}\right). \tag{2.114}$$

Neither behaves nicely at both ends ($u = 0, 1$). Therefore both are unacceptable which makes it impossible to build a perturbation theory to calculate small frequencies which are inversely proportional to r_0. This negative result is in agreement with numerical results [9, 13] and in accordance with the AdS/CFT correspondence. Indeed, there is no ansatz that can be built from tensor spherical harmonics $\mathbb{T}_{ij}$ satisfying the linearized hydrodynamic equations, because of the conservation and tracelessness properties of $\mathbb{T}_{ij}$.

2.3.4 Hydrodynamics on the AdS Boundary

The above results in the bulk dictate the hydrodynamic behavior of the dual gauge theory fluid on the conformal boundary. To see the correspondence, one needs to understand the hydrodynamics in the linearized regime of a $d-1$ dimensional fluid with dissipative effects. The fluid lives on a space with metric

$$ds_{\partial}^2 = -dt^2 + d\Sigma_{K,d-2}^2. \tag{2.115}$$

The hydrodynamic equations are simply the requirement that the stress-energy momentum tensor be conserved,

$$\nabla_{\mu} T^{\mu\nu} = 0. \tag{2.116}$$

As the duality corresponds to a conformal field theory one must also demand scale invariance which implies

$$T_{\mu}^{\mu} = 0, \quad \epsilon = (d-2)p, \quad \zeta = 0, \tag{2.117}$$

where ϵ, p and ζ are the energy density, pressure and bulk viscosity of the fluid. In the rest frame of the fluid, the velocity field is $u^{\mu} = (1, 0, 0, 0)$ and the pressure p_0 is constant. Consider a perturbation

$$u^{\mu} = (1, u^i), \quad p = p_0 + \delta p, \tag{2.118}$$

Applying the hydrodynamic equations, one obtains

$$\begin{aligned} &(d-2)\partial_t \delta p + (d-1)p_0 \nabla_i u^i = 0 \\ &(d-1)p_0 \partial_t u^i + \partial^i \delta p - \eta \left[\nabla^j \nabla_j u^i + K(d-3)u^i + \frac{d-4}{d-2} \partial^i (\nabla_j u^j) \right] = 0, \end{aligned} \tag{2.119}$$

where I used the curvature tensor $R_{ij} = K(d-3)g_{ij}$.

For *vector perturbations*, consider the *ansatz*

$$\delta p = 0, \quad u^i = \mathcal{C}_V e^{-i\omega t} \mathbb{V}^i, \tag{2.120}$$

where $\mathbb{V}^i$ is a vector harmonic.

The hydrodynamic equations imply

$$-i\omega(d-1)p_0 + \eta\left[k_V^2 - K(d-3)\right] = 0. \tag{2.121}$$

Using

$$\frac{\eta}{p_0} = (d-2)\frac{\eta}{s}\frac{S}{M} = \frac{4\pi\eta}{s}\frac{r_+}{K+r_+^2}, \tag{2.122}$$

with ω from the gravity dual, one obtains for large r_+,

$$\frac{\eta}{s} = \frac{1}{4\pi} \tag{2.123}$$

which is the standard value of the ratio in gauge theory fluids with a gravity dual [15].

For *scalar perturbations*, consider the *ansatz*

$$u^i = \mathcal{A}_S e^{-i\omega t}\partial^i \mathbb{S}, \quad \delta p = \mathcal{B}_S e^{-i\omega t}\mathbb{S}, \tag{2.124}$$

where $\mathbb{S}$ is a scalar harmonic.

The hydrodynamic equations imply the system of equations

$$\begin{aligned}
&(d-2)i\omega\mathcal{B}_S + (d-1)p_0 k_S^2 \mathcal{A}_S = 0 \\
&\mathcal{B}_S + \mathcal{A}_S\left[-i\omega(d-1)p_0 - 2(d-3)K\eta + 2\eta k_S^2\frac{d-3}{d-2}\right] = 0.
\end{aligned} \tag{2.125}$$

The determinant must vanish,

$$\begin{vmatrix} (d-2)i\omega & (d-1)p_0 k_S^2 \\ 1 & -i\omega(d-1)p_0 - 2(d-3)K\eta + 2\eta k_S^2\,\frac{d-3}{d-2} \end{vmatrix} = 0. \tag{2.126}$$

Arguing along the same lines as for vector perturbations, we arrive at

$$\frac{\eta}{s} = \frac{1}{4\pi} \tag{2.127}$$

which is the same result as the one obtained with vector QNMs.

2.3.5 Conformal Soliton Flow

The above results have been applied to the study of the quark-gluon plasma which forms in heavy ion collisions (at the Relativistic Heavy Ion Collider (RHIC) and elsewhere). In the case of a spherical horizon ($K = +1$), the boundary of space-time is $S^3 \times \mathbb{R}$. This may be conformally mapped onto a flat Minkowski space. Then by holographic renormalization, the AdS$_5$-Schwarzschild black hole is dual to a spherical shell of plasma on the four-dimensional Minkowski space which first contracts and then expands (conformal soliton flow) [13].

Quasi-normal modes govern the properties of this plasma with long-lived modes (i.e., of small $\Im\omega$) having the most influence. For example, one obtains the ratio

$$\frac{v_2}{\delta} = \frac{1}{6\pi}\Re\frac{\omega^4 - 40\omega^2 + 72}{\omega^3 - 4\omega}\sin\frac{\pi\omega}{2}, \tag{2.128}$$

where $v_2 = \langle\cos 2\phi\rangle$ evaluated at $\theta = \frac{\pi}{2}$ (mid-rapidity) and averaged with respect to the energy density at late times; $\delta = \frac{\langle y^2 - x^2\rangle}{\langle y^2 + x^2\rangle}$ is the eccentricity at time $t = 0$. Numerically, $\frac{v_2}{\delta} = 0.37$, which compares well with the result from RHIC data, $\frac{v_2}{\delta} \approx 0.323$ [16].

Another observable is the thermalization time which is found to be

$$\tau = \frac{1}{2|\Im\omega|} \approx \frac{1}{8.6T_{\text{peak}}} \approx 0.08\ \text{fm/c}, \quad T_{\text{peak}} = 300\,\text{MeV} \tag{2.129}$$

not in agreement with the RHIC result $\tau \sim 0.6$ fm/c [17], but still encouragingly small. For comparison, the corresponding result from perturbative QCD is $\tau \gtrsim 2.5$ fm/c [18, 19].

In the case of a hyperbolic horizon (topological black hole; $K = -1$), one needs to work with a conformal map from $\mathbb{H}^{d-2}/\Gamma \times \mathbb{R}$ to a $(d-1)$-dimensional Minkowski space. Finding an explicit form of this map for $d = 5$ involves a considerable amount of numerical work. However, it is important that one consider this case because the modes of hyperbolic black holes live the longest [10].

2.4 Phase Transitions

In this section I discuss hairy black holes in asymptotically AdS space and their duals. At low temperatures, an instability leads to symmetry breaking and the formation of a dual superconductor. Electromagnetic perturbations of the black hole determine the conductivity in the bulk. First I review the case of a flat horizon ($K = 0$) [20] and then I discuss the case of hyperbolic horizon ($K = -1$) where exact analytical results are obtained [21].

2.4.1 $K = 0$

Consider a scalar Ψ of mass $m^2 = -2$, which is above the Breitenlohner-Freedman (BF) bound coupled to an electromagnetic potential A_μ in 3 + 1 dimensions. The Lagrangian density is

$$\mathcal{L} = -\frac{1}{2}\partial_\mu \Psi \partial^\mu \Psi + \Psi^2 - \frac{1}{4} F_{\mu\nu} F^{\mu\nu} - \frac{q^2}{2} \Psi^2 (\partial_\mu \theta - A_\mu)(\partial^\mu \theta - A^\mu), \qquad (2.130)$$

where θ is a Stückelberg field. q is an arbitrary parameter which can be thought of as the electric charge of the scalar field Ψ (one may instead turn Ψ into a complex scalar field of charge q coupled to an electromagnetic potential in a standard fashion).

The Lagrangian density (2.130) is invariant under the $U(1)$ gauge transformation

$$A_\mu \to A_\mu + \partial_\mu \omega, \quad \theta \to \theta + \omega. \qquad (2.131)$$

To fix the gauge, set

$$\theta = 0. \qquad (2.132)$$

Working in the probe limit ($q \to \infty$) in which there is no back reaction to the metric, assume that the fields propagate in the black hole background (2.2) with $d = 4$ and $K = 0$. The radius of the horizon and Hawking temperature are, respectively,

$$r_+ = (2\mu)^{1/3}, \quad T = \frac{3r_+}{4\pi}. \qquad (2.133)$$

Assuming spherical symmetry and an electrostatic potential $A_0 = \Phi(r)$, the field equations yield two coupled non-linear differential equations [20]

$$\begin{aligned} \Psi'' + \left(\frac{f'}{f} + \frac{2}{r}\right)\Psi' + \left(\frac{\Phi}{f}\right)^2 \Psi + \frac{2}{f}\Psi &= 0 \\ \Phi'' + \frac{2}{r}\Phi' - \frac{2\Psi^2}{f}\Phi &= 0, \end{aligned} \qquad (2.134)$$

where I set $q = 1$ and

$$f(r) = r^2 - \frac{2\mu}{r} \qquad (2.135)$$

As $r \to \infty$, one obtains the boundary behavior

$$\Psi = \frac{\Psi^{(1)}}{r} + \frac{\Psi^{(2)}}{r^2} + \cdots, \quad \Phi = \Phi^{(0)} + \frac{\Phi^{(1)}}{r} + \cdots \qquad (2.136)$$

where one of the $\Psi^{(i)} = 0$ ($i = 1, 2$) for stability, $\Phi^{(0)}$ is the chemical potential and $\Phi^{(1)} = -\rho$ (charge density).

Below a critical temperature T_0 a condensate forms,

$$\langle \mathcal{O}_i \rangle = \sqrt{2}\Psi^{(i)} \qquad (2.137)$$

of an operator of dimension $\Delta = i$.

At $T = T_0$, one may set $\Psi = 0$ in the equation for Φ and deduce (using $\Phi(r_+) = 0$)

$$\Phi = \rho\left(\frac{1}{r_+} - \frac{1}{r}\right). \tag{2.138}$$

Then the equation for Ψ turns into an eigenvalue problem which yields

$$T_0 \approx 0.226\sqrt{\rho}, \quad 0.118\sqrt{\rho}$$

depending on the boundary conditions.

To study the properties of the dual CFT, apply an electromagnetic perturbation. It obeys the wave equation

$$A'' + \frac{f'}{f}A' + \left(\frac{\omega^2}{f^2} - \frac{2\Psi^2}{f}\right)A = 0, \tag{2.139}$$

to be solved subject to the boundary conditions that it be ingoing at the horizon, $A \sim f^{-i\omega/(4\pi T)}$, and at the boundary ($r \to \infty$),

$$A = A^{(0)} + \frac{A^{(1)}}{r} + \cdots \tag{2.140}$$

Ohm's law yields the conductivity

$$\sigma(\omega) = \frac{A^{(1)}}{i\omega A^{(0)}}. \tag{2.141}$$

For $T \geq T_0$, $\Psi = 0$, therefore $A \sim e^{i\omega r_*}$ where $r_* = \int dr/f(r)$ is the tortoise coordinate. It follows that

$$\sigma(\omega) = 1. \tag{2.142}$$

At low T, for $\langle \mathcal{O}_1 \rangle \neq 0$, we have

$$\Psi \approx \frac{\langle \mathcal{O}_1 \rangle}{\sqrt{2}\, r}$$

Since $r_+ \to 0$, we obtain $A \sim e^{i\omega' r_*}$, where $\omega' = \sqrt{\omega^2 - \langle \mathcal{O}_1 \rangle^2}$. Therefore, for $\omega < \langle \mathcal{O}_1 \rangle$, $\Re\sigma = 0$, i.e., we obtain a superconductor with a gap.

2.4.2 $K = -1$

Turning to the case of a hyperbolic horizon [21], choose a scalar Ψ of mass $m^2 = -2$, as before, but conformally coupled with potential

$$V(\Psi) = \frac{8\pi G}{3}\Psi^4$$

The system has an exact solution (MTZ black hole [22])

$$ds^2 = -f_{MTZ}(r)dt^2 + \frac{dr^2}{f_{MTZ}(r)} + r^2 d\sigma^2, \quad f_{MTZ} = r^2 - \left(1 + \frac{r_0}{r}\right)^2, \tag{2.143}$$

with

$$\Psi(r) = -\sqrt{\frac{3}{4\pi G}\frac{r_0}{r + r_0}}, \quad \Phi = 0. \tag{2.144}$$

The temperature, entropy and mass are, respectively,

$$T = \frac{1}{\pi}\left(r_+ - \frac{1}{2}\right), \quad S_{MTZ} = \frac{\sigma}{4G}(2r_+ - 1), \quad M_{MTZ} = \frac{\sigma r_+}{4\pi G}(r_+ - 1). \tag{2.145}$$

and the law of thermodynamics $dM = TdS$ holds.

At $M = 0$, the MTZ black hole coincides with the topological black hole with no hair (Eq. 2.2) with $d = 4$, $K = -1$),

$$ds^2_{\text{AdS}} = -(r^2 - 1)dt^2 + \frac{dr^2}{r^2 - 1} + r^2 d\Sigma^2 \tag{2.146}$$

and an enhanced scaling symmetry (pure AdS space) emerges at the critical temperature

$$T_0 = \frac{1}{2\pi} \tag{2.147}$$

At this point there is a phase transition which can be seen by calculating the difference in free energies,

$$\Delta F = F_{TBH} - F_{MTZ} = -\frac{\sigma}{8\pi G}\pi^3 l^3 (T - T_0)^3 + \cdots, \tag{2.148}$$

showing that there is a third-order phase transition at T_0.

Perturbative stability of the MTZ black hole has also been demonstrated for $T < T_0$ ($M < 0$) [21]. Comparing with the flat case, note that here both $\Psi^{(1)}$ and $\Psi^{(2)}$ are non-vanishing, yet the MTZ black hole is stable. However this is true only if the mass is negative which is never the case with a flat horizon. Also, here the condensation of the scalar field has a geometrical origin and is due entirely to its coupling to gravity.

Moreover, the heat capacities in the normal and superconducting (corresponding to the MTZ black hole) phases, respectively, as $T \to 0$ exhibit a power-law behavior

$$C_n \approx \frac{\pi\sigma}{3\sqrt{3}G} T, \quad C_s \approx \frac{\pi\sigma}{2G} T, \tag{2.149}$$

Since both $\Psi^{(1)}$ and $\Psi^{(2)}$ are non-vanishing, we have a multi-trace deformation of the CFT [23] with a condensate

$$\langle \mathcal{O}_1 \rangle = \sqrt{\frac{3\pi^3}{2G}} (T_0^2 - T^2) \,. \tag{2.150}$$

It should be noted that the deformation does not break the global $U(1)$ symmetry because Ψ is a real field (see Eq. 2.130).

To study the conductivity, apply an electromagnetic perturbation. It obeys the wave equation (2.139) which may be solved using first-order perturbation theory in q^2,

$$A = e^{-i\omega r_*} + \frac{q^2}{2i\omega} e^{i_*} \int_{r_+}^{r} dr' \Psi^2(r') e^{-2i_*} - \frac{q^2}{2i\omega} e^{-i_*} \int_{r_+}^{r} dr' \Psi^2(r'). \tag{2.151}$$

The conductivity to first order in q^2 is

$$\sigma(\omega) = \frac{A^{(1)}}{i\omega A^{(0)}} = 1 - \frac{q^2}{i\omega} \int_{r_+}^{\infty} dr \Psi^2(r) e^{-2i\omega r_*}. \tag{2.152}$$

The superfluid density is found from

$$\Re[\sigma(\omega)] \sim \pi n_s \delta(\omega), \quad \Im[\sigma(\omega)] \sim \frac{n_s}{\omega}, \quad \omega \to 0. \tag{2.153}$$

One obtains

$$n_s = q^2 \int_{r_+}^{\infty} dr \Psi^2(r) = \frac{3q^2}{4\pi G} \frac{r_0^2}{r_+ + r_0} = \alpha (T_0 - T)^2, \quad \alpha = \frac{3\pi q^2}{4G}. \tag{2.154}$$

Near $T = 0$,

$$n_s(0) - n_s(T) \approx \frac{\alpha}{\pi} T^\delta, \quad \delta = 1 \tag{2.155}$$

In Table 2.1, this analytic prediction is compared against exact numerical results for various values of the charge q. Naturally, the agreement is best at small values of q.

Table 2.1 The exponent δ characterizing the low-temperature dependence of the superfluid density n_s

$q/\sqrt{G}$	1	3	5
δ	1.025 ± 0.007	1.52 ± 0.03	1.78 ± 0.03

Table 2.2 Numerical *vs* analytical results for the normal and superfluid densities

$q/\sqrt{G}$	$\gamma_{\text{numerical}}$	$\gamma_{\text{analytical}}$	$\alpha_{\text{numerical}}$	$\alpha_{\text{analytical}}$
0.1	0.0020	0.0024	0.0225	0.024
0.5	0.0538	0.0597	0.552	0.589
1.0	0.187	0.239	2.196	2.356
2.0	0.684	0.955	8.678	9.425
3.0	1.325	2.15	20.35	21.21
5.0	2.522	5.97	52.90	58.90

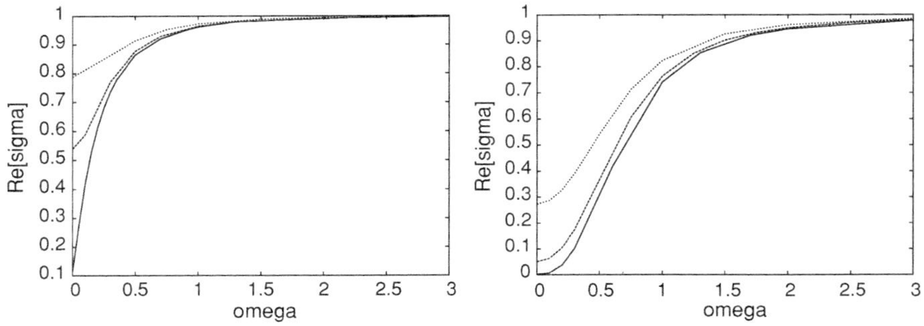

Fig. 2.1 The real part of the conductivity *vs* ω for $q/\sqrt{G} = 2$ (*left*) and $q/\sqrt{G} = 5$ (*right*) and $T = 0.0032, 0.032, 0.064$. The *lowest curve* corresponds to the lowest temperature

The normal, non-superconducting, component of the DC conductivity is

$$n_n = \lim_{\omega \to 0} \Re[\sigma(\omega)] . \tag{2.156}$$

Therefore,

$$\ln n_n = 2q^2 \int_{r_+}^{\infty} dr \Psi^2(r) r_* . \tag{2.157}$$

At low T,

$$n_n \sim T^{\gamma} , \quad \gamma = \frac{3q^2}{4\pi G} . \tag{2.158}$$

This analytic result and the prediction for the parameter α determining the critical behavior of the superfluid density are compared against exact numerical results in Table 2.2. Again, the agreement is best at small q.

Figures 2.1 and 2.2 show the frequency dependence of the real and imaginary, respectively, parts of the conductivity. The real part of the conductivity becomes smaller as we increase the charge q. Unfortunately, numerical instabilities also increase and we have not been able to produce reliable numerical results above $q/\sqrt{G} = 5$. The superconductor appears to be *gapless*. However, a gap is likely to

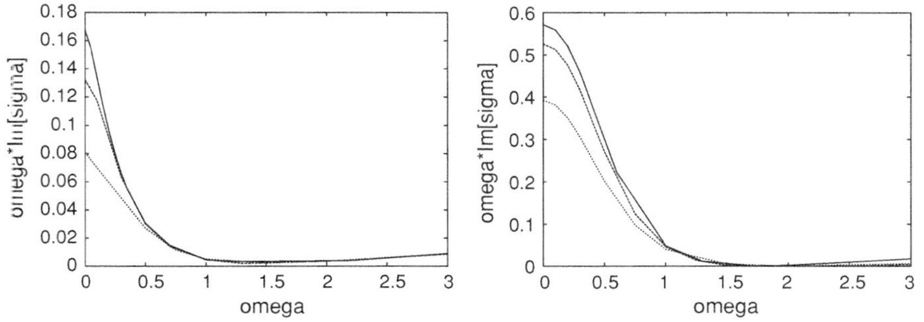

Fig. 2.2 2 The imaginary part of the conductivity multiplied by ω *vs* ω for $q/\sqrt{G} = 2$ (*left*) and $q/\sqrt{G} = 5$ (*right*) and $T = 0.0032, 0.032, 0.064$. The *uppermost curve* corresponds to the lowest temperature

develop above a certain value of the charge q, as indicated by the trend in the graphs as q increases.

2.5 Conclusion

The quasi-normal modes that govern perturbations of black holes in asymptotically AdS space are a powerful tool in understanding the hydrodynamic behavior of a gauge theory fluid at strong coupling. Here I focused on the analytic calculation of QNMs. I discussed both high overtones and low frequencies. I applied the results on gravitational perturbations to the understanding of the quark-gluon plasma produced in heavy ion collisions at RHIC and the LHC. I also considered hairy black holes whose electromagnetic perturbations allow one to analyze the conductivity of the dual conformal field theory and the phase transition to a superconducting state. I reviewed the case of a flat horizon and compared the results with those from black holes with hyperbolic horizon for which exact hairy solutions have been constructed (MTZ black holes [22]). In all these cases, only the low-lying QNMs were needed. It is unclear what physical role high overtones play.

Acknowledgement Research supported in part by the US Department of Energy under grant DE-FG05-91ER40627.

References

1. Cardoso, V., Lemos, J. P.S.: Scalar, electromagnetic and Weyl perturbations of BTZ black holes: quasi normal modes. Phys. Rev. D **63**, 124015 (2001)
2. Birmingham, D., Sachs, I., Solodukhin, S.N.: Conformal field theory interpretation of black hole quasi-normal modes. Phys. Rev. Lett. **88**, 151301 (2002)

3. Horowitz, G.T., Hubeny, V.E.: Quasinormal modes of AdS black holes and the approach to thermal equilibrium. Phys. Rev. D **62**, 024027 (2000)
4. Starinets, A.O.: Quasinormal modes of near extremal black branes. Phys. Rev. D **66**, 124013 (2002)
5. Konoplya, R.A.: On quasinormal modes of small Schwarzschild-Anti-de-Sitter black hole. Phys. Rev. D **66**, 044009 (2002)
6. Musiri, S., Siopsis, G.: Asymptotic form of quasi-normal modes of large AdS black holes. Phys. Lett. B **576**, 309 (2003)
7. Natário, J., Schiappa, R.: On the classification of asymptotic quasi-normal frequencies for d-dimensional black holes and quantum gravity. Adv. Theor. Math. Phys. **8**, 1001 (2004)
8. Musiri, S., Ness, S., Siopsis, G.: Perturbative calculation of quasi-normal modes of AdS Schwarzschild black holes. Phys. Rev. D **73**, 064001 (2006)
9. Cardoso, V., Konoplya, R.A., Lemos, J.P.S.: Quasi-normal frequencies of Schwarzschild black holes in AdS space-times: a complete study on the asymptotic behavior. Phys. Rev. D **68**, 044024 (2003)
10. Alsup, J., Siopsis, G.: Low-lying quasinormal modes of topological AdS black holes and hydrodynamics. Phys. Rev. D **78**, 086001 (2008)
11. Ishibashi, A., Kodama, H.: A master equation for gravitational perturbations of maximally symmetric black holes in higher dimensions. Prog. Theor. Phys. **110**, 701 (2003)
12. Siopsis, G.: Low frequency quasi-normal modes of AdS black holes. JHEP **0705**, 042 (2007)
13. Friess, J.J., Gubser, S.S., Michalogiorgakis, G., Pufu, S.S.: Expanding plasmas and quasinormal modes of anti-de Sitter black holes. JHEP **0704**, 080 (2007)
14. Michalogiorgakis, G., Pufu, S.S.: Low-lying gravitational modes in the scalar sector of the global AdS4 black hole. JHEP **0702**, 023 (2007)
15. Policastro, G., Son, D.T., Starinets, A.O.: From AdS/CFT correspondence to hydrodynamics. JHEP **0209**, 043 (2002)
16. PHENIX Collaboration, Adare, A., et al.: Scaling properties of azimuthal anisotropy in Au+Au and Cu+Cu collisions at $\sqrt{s_{NN}} = 200$ GeV. arXiv:nucl-ex/0608033
17. Arnold, P., Lenaghan, J., Moore, G.D., Yaffe, L.G.: Apparent thermalization due to plasma instabilities in quark-gluon plasma. Phys. Rev. Lett. **94**, 072302 (2005)
18. Baier, R., Mueller, A.H., Schiff, D., Son, D.T.: "Bottom-up" thermalization in heavy ion collisions. Phys. Lett. B **502**, 51 (2001)
19. Molnar, D., Gyulassy, M.: Saturation of elliptic flow at RHIC: results from the covariant elastic parton cascade model MPC. Nucl. Phys. A **697**, 495 (2002)
20. Hartnoll, S.A., Herzog, C.P., Horowitz, G.T.: Building an AdS/CFT superconductor. Phys. Rev. Lett. **101**, 031601 (2008)
21. Koutsoumbas, G., Papantonopoulos, E., Siopsis, G.: Exact gravity dual of a gapless superconductor. JHEP **0907**, 026 (2009)
22. Martinez, C., Troncoso, R., Zanelli, J.: Exact black hole solution with a minimally coupled scalar field. Phys. Rev. D **70**, 084035 (2004)
23. Hertog, T., Maeda, K.: Black holes with scalar hair and asymptotics in N=8 supergravity. JHEP **0407**, 051 (2004)

Chapter 3
Introduction to the AdS/CFT Correspondence

Philip C. Argyres

Abstract We review the basic properties of d-dimensional conformal field theories (CFTs) and describe their relation to quantum gravitational theories on $d+1$-dimensional anti-de Sitter (AdS_{d+1}) space–times. We briefly review the 't Hooft limit of $U(N)$ Yang–Mills theory, and then describe how an appropriate limit of type IIB superstring theory with D3-branes can be used to motivate a precise and computable correspondence between a four-dimensional conformal field theory and a quantum gravitational theory on $\mathrm{AdS}_5 \times S^5$. We then discuss an extension of this construction in which probe branes on the AdS space–time are included.

3.1 Introduction

These lectures review the foundations of the AdS/CFT correspondence, emphasizing the logical structure of the arguments that lead to this conjectured equivalence of a (non-gravitational) quantum field theory with a higher-dimensional theory of quantum gravity. Included are summaries of the basic ingredients in the correspondence: d-dimensional conformal field theories, anti-de Sitter (AdS) space–times, and the 't Hooft limit of Yang–Mills (YM) theories. A basic familiarity with string theory [1–3] is assumed. Although the framework for how computations in the AdS/CFT correspondence can be carried out is set up, the description of detailed and efficient computational techniques are left for other lectures in this volume.

P. C. Argyres (✉)
University of Cincinnati, P.O. Box 210011, Cincinnati, OH 45221-0011, USA
e-mail: argyres@physics.uc.edu

E. Papantonopoulos (ed.), *From Gravity to Thermal Gauge Theories: The Ads/CFT Correspondence,* Lecture Notes in Physics, 828,
DOI: 10.1007/978-3-642-04864-7_3,

A basic introduction to the AdS/CFT correspondence of a descriptive nature, together with an elementary introduction to string theory can be found in [4], while [5] gives a descriptive survey of extensions and some applications of the correspondence. Some more detailed reviews of the AdS/CFT correspondence and modern developments of the formalism for carrying out computations are [6–8].

3.2 CFTs

Scale-invariant quantum field theories are important as possible end-points of renormalization group flows on the space of cut-off effective field theories. Knowledge of them and their relevant deformations partially organizes the space of quantum field theories; see Fig. 3.1. Scaling a local or quasi-local quantum field theory to its infrared (IR), or long-wavelength, limit results in a local scale-invariant field theory, by definition. Scaling to the ultraviolet (UV), or short distances, need not always lead to a scale-invariant local field theory (*e.g.*, the non-local "little string theories" are a known exception; see [9] for a review), but many such examples are known, including all the asymptotically free and scale invariant Yang–Mills theories in four space–time dimensions.

It is believed [10] that unitary scale-invariant theories are also conformally invariant: the space–time symmetry group Poincaré×dilatations enlarges to the conformal group of symmetries. In what follows we will concentrate on such conformal field theories (CFTs) for which there exists a local, conserved, traceless energy–momentum tensor $T^{\mu\nu}(x)$

3.2.1 Conformal Algebra

The conformal group in d space–time dimensions is the group of reparameterizations which preserve the (flat) space–time metric, $g^{\mu\nu}$, up to a local scale factor, where $\mu, \nu \in \{1, \ldots, d\}$. (We will work in Euclidean signature where convenient.)

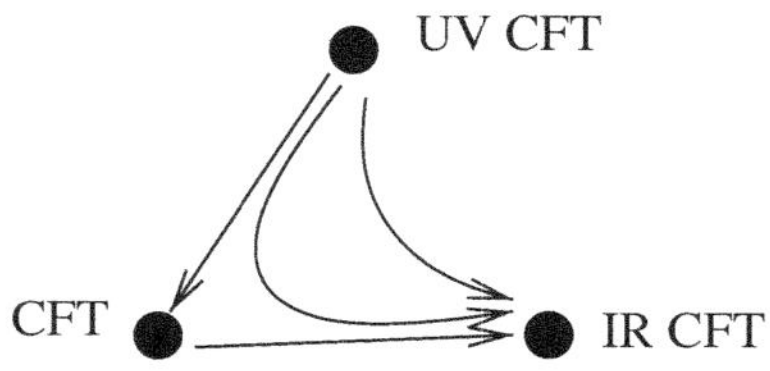

Fig. 3.1 Effective field theory RG flows from UV to IR CFTs, which appear as fixed points of RG flows. CFTs can also govern the behavior of flows at intermediate scales, as shown

For $d > 2$ its generators are (Lorentz) rotations, $M^{\mu\nu}$, plus translations, dilatations, and special conformal transformations which, respectively, act infinitesimally as

$$\begin{aligned} P^{\mu} : x^{\mu} &\rightarrow x^{\mu} + \alpha^{\mu}, \\ D : x^{\mu} &\rightarrow (1+\epsilon)x^{\mu}, \\ K_{\nu} : x^{\mu} &\rightarrow x^{\mu} + \epsilon_{\nu}(g^{\mu\nu}x^{2} - 2x^{\mu}x^{\nu}). \end{aligned} \tag{3.1}$$

They obey the algebra

$$[D, K_{\mu}] = iK_{\mu}, \quad [D, P_{\mu}] = -iP_{\mu}, \quad [P_{\mu}, K_{\nu}] = 2iM_{\mu\nu} - 2ig_{\mu\nu}D, \tag{3.2}$$

where the other commutators either vanish or follow from rotational invariance. By defining $2d+1$ additional rotation generators from P_{μ}, D, and K_{μ},

$$M_{\mu\, d+1} \equiv (K_{\mu} - P_{\mu})/2, \quad M_{\mu\, d+2} \equiv (K_{\mu} + P_{\mu})/2, \quad M_{d+1\, d+2} \equiv D, \tag{3.3}$$

we see that the d-dimensional conformal algebra is equivalent to $SO(d,2)$ in Minkowski signature or $SO(d+1,1)$ in Euclidean signature.

3.2.2 Local Field Operators

As in usual quantum field theory, we classify local field operators $\mathcal{O}(x^{\mu})$, by their transformation properties of the little group $SO(d) \times SO(2) \subset SO(d,2)$ of the conformal group. In radial quantization in Euclidean signature, the $SO(d)$ irreducible representation is the "spin" of the field, while the charge under the (infinite cover of the) $SO(2)$ subgroup is the scaling dimension, Δ, of the field,

$$\mathcal{O}_{\Delta}(\lambda x^{\mu}) = \lambda^{-\Delta}\mathcal{O}_{\Delta}(x^{\mu}) \quad \Leftrightarrow \quad [D, \mathcal{O}_{\Delta}(0)] = -i\Delta\mathcal{O}_{\Delta}(0) \tag{3.4}$$

In radial quantization we foliate Euclidean $\mathbf{R}^{d}$ by $(d-1)$-spheres, S^{d-1}, concentric at the origin, and define the Hilbert space of states of the CFT at a given radial slice. The dilatation operator, D, then generates the evolution of states to different radial slices. In this case there is a correspondence between local operators and states in the Hilbert space. First, we denote by $|0\rangle$ the conformal vacuum state, defined to be the state annihilated by all the generators of the conformal algebra. Then each local operator $\mathcal{O}(x^{\mu})$ defines a state at the origin (i.e., in the zero radius limit of the slicing) by $\mathcal{O}(0)|0\rangle$. In a CFT, by scale invariance this is a state on any size S^{d-1} around the origin. Conversely any state on an S^{d-1} created by the action of operators at smaller radii can be written as a state created by a local operator insertion at the origin by shrinking the sphere.

Primary operators are those annihilated by the special conformal generators, K^μ, at the origin. By translating such operators to arbitrary position, it follows that they obey the commutation relations

$$\begin{aligned} [P_\mu, \mathcal{O}_\Delta(x)] &= i\partial_\mu \mathcal{O}_\Delta(x), \\ [M_{\mu\nu}, \mathcal{O}_\Delta(x)] &= \left[i(x_\mu\partial_\nu - x_\nu\partial_\mu) + \Sigma^R_{\mu\nu} \right] \mathcal{O}_\Delta(x), \\ [D, \mathcal{O}_\Delta(x)] &= i(x^\mu\partial_\mu - \Delta)\mathcal{O}_\Delta(x), \\ [K_\mu, \mathcal{O}_\Delta(x)] &= \left[i(x^2\partial_\mu - 2x_\mu x^\nu\partial_\nu + 2x_\mu\Delta) - 2x^\nu\Sigma^R_{\mu\nu} \right] \mathcal{O}_\Delta(x). \end{aligned} \tag{3.5}$$

Here $\Delta \in \mathbf{R}$ is the conformal dimension of the primary operator, and $\Sigma^R_{\mu\nu}$ are the representation matrices of the irreducible spin R of the primary which acts on its spin indices (which we have suppressed).

Unitarity puts lower bounds on the conformal dimension, Δ, of primaries depending on their spin, R; see [11] for a review. In particular, for a scalar primary, $\Delta \geq (d-2)/2$, with equality only for a free field.

Descendant operators are those made by acting on a primary operator with P^μs. All local operators are found in this way. Note that acting with P^μ (K^μ) increases (decreases) the scaling dimension Δ of the operator by 1.

3.2.3 Conformal Correlators

Correlators of descendant fields can be derived from those of the primary fields by taking derivatives. Conformal invariance determines the form of the correlators of primaries in terms of their scaling dimensions and spins up to undetermined functions of conformal invariants made up of the positions of the insertions of the operators. For example, for scalar primaries, the 2-, 3-, and 4-point functions are determined to have the forms

$$\begin{aligned} \langle \mathcal{O}_{\Delta_1}(x_1)\mathcal{O}_{\Delta_2}(x_2) \rangle &= \delta_{1,2} \prod_{i<j}^{2} |x_{ij}|^{-\Delta}, \\ \langle \mathcal{O}_{\Delta_1}(x_1)\mathcal{O}_{\Delta_2}(x_2)\mathcal{O}_{\Delta_3}(x_3) \rangle &= c_{123} \prod_{i<j}^{3} |x_{ij}|^{\Delta - 2\Delta_i - 2\Delta_j}, \\ \langle \mathcal{O}_{\Delta_1}(x_1)\mathcal{O}_{\Delta_2}(x_2)\mathcal{O}_{\Delta_3}(x_3)\mathcal{O}_{\Delta_4}(x_4) \rangle &= c_{1234}(u,v) \prod_{i<j}^{4} |x_{ij}|^{\frac{1}{3}\Delta - \Delta_i - \Delta_j}. \end{aligned} \tag{3.6}$$

Here $x_{ij} \equiv x_i - x_j$ and $\Delta \equiv \sum_i \Delta_i$ are the sum of the dimensions of the operators in the correlator. For the 2-point function the Kronecker δ represents a choice of normalization of the primary fields. For the 3-point function, the c_{ijk} are undetermined constant coefficients, while for the 4-point function $c_{1234}(u, v)$ is an undetermined function of the two independent conformally invariant cross-ratios

$$u \equiv \frac{|x_{12}||x_{34}|}{|x_{13}||x_{24}|}, \qquad v \equiv \frac{|x_{14}||x_{23}|}{|x_{13}||x_{24}|}. \tag{3.7}$$

The number of independent conformal invariants grows with the number of insertion points, so higher-point correlators would appear to be even less constrained by conformal invariance.

However, the state-operator correspondence implies that the product of any two conformal primaries can be rewritten as a linear combination of conformal operators inserted at a nearby point (as long as there are no other insertions in between the two primaries). This expansion is called the operator product expansion (OPE) of the two primaries. For example, for two scalar primaries it has the form

$$\mathcal{O}_{\Delta_i}(x)\mathcal{O}_{\Delta_j}(0) = \sum_k c_{ijk}|x|^{-\Delta_i-\Delta_j+\Delta_k}(\mathcal{O}_{\Delta_k}(0) + \text{descendants}) \tag{3.8}$$

where the sum on the right side is over all primaries and their descendants. The coefficients appearing in front of the descendants of each primary are completely determined by conformal invariance. It is also easy to see that the coefficients c_{ijk} appearing in front of each primary on the right side are the same as the coefficients appearing in the 3-point functions.

In principle all n-point correlators are determined by the OPEs, since any n-point function can be replaced by an infinite sum of $(n-1)$-point functions by using the OPE for any two adjacent insertions. Thus, the CFT is completely determined by the data $\{\Delta_i, \text{spins}, c_{ijk}\}$ for all the primaries (labelled by i, j, k). However, this data is highly constrained: unitarity and associativity ("crossing symmetry") of the OPEs imply many non-trivial relations among the CFT data, so arbitrary lists of $\{\Delta_i, \text{spins}, c_{ijk}\}$ will not in general define a consistent CFT.

Since the associativity constraints from the OPEs are notoriously difficult to even state, let alone solve, we look for other approaches to constructing CFTs. The AdS/CFT correspondence is such an alternative approach.

It is based on a different, useful way of encoding the correlation functions of a CFT in terms of its partition function,

$$Z[\phi_{\Delta_i}] \equiv \left\langle \exp\left(\int d^dx\, \phi_{\Delta_i}(x)\mathcal{O}_{\Delta_i}(x)\right)\right\rangle_{\text{CFT}}. \tag{3.9}$$

This functional of the classical *sources* $\phi_{\Delta_i}(x)$ associated with each field operator in the CFT generates the correlation functions by taking derivatives of Z with respect to the sources:

$$\langle \mathcal{O}_{\Delta_1}(x_1)\mathcal{O}_{\Delta_2}(x_2)\dots\rangle = \left.\frac{\partial^n Z[\phi_{\Delta_i}]}{\partial\phi_{\Delta_1}(x_1)\partial\phi_{\Delta_2}(x_2)\cdots}\right|_{\phi_{\Delta_i}=0}. \tag{3.10}$$

The conformal invariance of the correlators is reflected in the conformal invariance of $Z[\phi_\Delta]$. In particular, under scaling,

$$\int d^d x\, \phi_\Delta(x)\mathcal{O}_\Delta(x) = \int d^d(\lambda x)\, \phi_\Delta(\lambda x)\mathcal{O}_\Delta(\lambda x) = \lambda^{d-\Delta} \int d^d x\, \phi_\Delta(\lambda x)\mathcal{O}_\Delta(x), \tag{3.11}$$

so Z is invariant under the following scaling transformation of the sources:

$$\phi_\Delta(x) \to \lambda^{d-\Delta}\phi_\Delta(\lambda x). \tag{3.12}$$

In general, the sources transform in field representations of the conformal group and $Z[\phi_\Delta]$ is an invariant combination of these fields.

3.3 AdS/CFT Correspondence

We now introduce a trick to generate conformally invariant $Z[\phi_\Delta]$s. It is not guaranteed to generate all CFTs, perhaps just a special subset of them. The idea is to copy the way we write actions as invariants of field representations of the Poincaré group, and apply it instead to the conformal group. Local field representations of the Poincaré group are functions on $\mathbf{R}^d$ valued in finite dimensional representations of the Lorentz group. Poincaré invariant actions are formed by taking translationally invariant integrals over $\mathbf{R}^d$ of scalar combinations of fields and their derivatives formed using the Minkowski (Euclidean) metric on $\mathbf{R}^d$. The key point in this familiar construction is that d-dimensional Minkowski (Euclidean) space is the space whose isometry group is the d-dimensional Poincaré group.

3.3.1 AdS Geometry

To apply the same strategy to construct conformal invariant functions we want to find the space whose isometry group is the d-dimensional conformal group $SO(d+1,1)$. (We will work in Euclidean signature for now.) Since $SO(d+1,1)$ is the $(d+2)$-dimensional Lorentz rotation group, flat $\mathbf{R}^{d+2}$ might work. But this space has additional translational symmetries beyond the Lorentz rotations. So we remove them by restricting to a $(d+1)$-dimensional "sphere" of radius R: this preserves the rotational symmetry, but will leave no translational symmetries.

To realize $SO(d+1,1)$, we really want a Lorentzian "sphere", that is, a hyperboloid

$$\boldsymbol{X}^2 + V_+^2 - V_-^2 = R^2, \tag{3.13}$$

where $\boldsymbol{X} := \{X_1, \ldots, X_d\}$ and $V_\pm$ are Cartesian coordinates on $\mathbf{R}^{d+1,1}$ with metric

$$ds^2 = d\boldsymbol{X}^2 + dV_+^2 - dV_-^2. \tag{3.14}$$

The surface in $\mathbf{R}^{d+1,1}$ described by (3.13) is called the $(d+1)$-dimensional (Euclidean) anti-de Sitter space, or AdS_{d+1}, of radius R.

Useful coordinates on this space are $\{\boldsymbol{x}, z\}$ defined by

$$\boldsymbol{X} = \frac{R}{z}\boldsymbol{x}, \quad V_\pm = \frac{1}{2}\left(z + \frac{\boldsymbol{x}^2 \pm R^2}{z}\right), \tag{3.15}$$

which parametrizes solutions of the hyperboloid constraint (3.13) for $\boldsymbol{x} \in \mathbf{R}^d$ and $z > 0$. In these coordinates (called Poincaré patch coordinates), the AdS metric reads

$$ds^2 = \frac{R^2}{z^2}(dz^2 + d\boldsymbol{x}^2). \tag{3.16}$$

Thus AdS_{d+1} is conformal to the upper-half space $z > 0$ of $\mathbf{R}^{d+1}$ see Fig. 3.2.

A closely related set of coordinates are $\boldsymbol{x}$ and $r = R^2/z$, in which

$$ds^2 = \frac{R^2}{r^2}dr^2 + \frac{r^2}{R^2}d\boldsymbol{x}^2. \tag{3.17}$$

r is often called the radial coordinate of the AdS. Here $r = \infty$ is the boundary of AdS, and $r = 0$ can be thought of as a horizon. Note that both are infinitely distant from any finite r.

To go to Minkowski signature, one simply changes the d-dimensional Euclidean factor $d\boldsymbol{x}^2$ in the AdS metric to

$$d\boldsymbol{x}^2 \to -dt^2 + d\boldsymbol{x}^2, \tag{3.18}$$

the d-dimensional Minkowski metric, where now on the right side $\boldsymbol{x} = (x_1, \ldots, x_{d-1})$. Minkowski-AdS has the interesting feature that though the boundary is radially infinitely far from any interior point, it can be reached in finite time by radial light-like signals.

We will work mostly in Euclidean-signature AdS from now on.

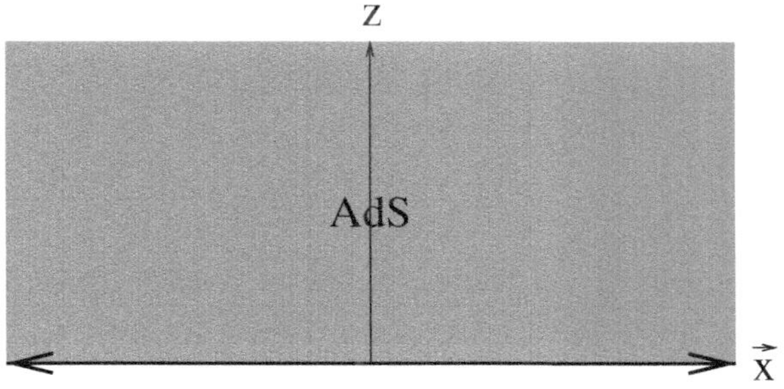

Fig. 3.2 AdS_{d+1} is conformal to the upper half-space of $\mathbf{R}^{d+1}$. In these coordinates, the boundary of AdS_{d+1} is conformal to $\mathbf{R}^d$ at $z = 0$

3.3.2 Partition Function

Any generally covariant function of (tensor) fields $\phi(z,\boldsymbol{x})$ on AdS will be conformally invariant. But these fields live in one more dimension (z) than we want. So we want to restrict to fields on a d-dimensional subspace of AdS_{d+1} while keeping general covariance in the full AdS_{d+1}.

The boundary, $\partial\mathrm{AdS} \simeq \{z{=}0,\boldsymbol{x}\}$, is special because the group of conformal isometries acts on it in the same way as the conformal group acts in d-dimensional space–time: translations acting as $\{z,\boldsymbol{x}\} \to \{z,\boldsymbol{x}+\alpha\}$, (Lorentz) rotations acting as $\{z,\boldsymbol{x}\} \to \{z,\Lambda\boldsymbol{x}\}$, and the scaling transformation $\{z,\boldsymbol{x}\} \to \{\lambda z,\lambda\boldsymbol{x}\}$ are all clearly isometries. (The special conformal transformations are more complicated, but not hard to work out.) So boundary values of (tensor) functions on AdS, $\lim_{z\to 0}\phi(z,\boldsymbol{x})$, transform as representations of the conformal group on space–time.

We can therefore try to identify any generally covariant (on AdS_{d+1}) function of the boundary values of fields as the partition function of a CFT (as a functional of the boundary values of those fields).

A way of writing a general class of such functions is as a functional integral over fields $\phi(z,\boldsymbol{x})$ on AdS_{d+1} with a generally covariant measure keeping the boundary values $\phi(0,\boldsymbol{x}) \sim \overline{\phi}(\boldsymbol{x})$ fixed:

$$Z[\overline{\phi}(\boldsymbol{x})] = \int\limits_{\phi|_\partial=\overline{\phi}} \mathcal{D}\phi(z,\boldsymbol{x})\, e^{-S[\phi(z,\boldsymbol{x})]} \tag{3.19}$$

Then, e.g., Z is invariant under scale transformations taking scalar fields $\phi(z,\boldsymbol{x}) \to \phi(\lambda z,\lambda\boldsymbol{x})$. In particular, if ϕ behaves near the boundary like

$$\phi(z,\boldsymbol{x}) \sim z^{d-\Delta}\overline{\phi}(\boldsymbol{x}) + \mathrm{O}(z^{d-\Delta+1}) \tag{3.20}$$

it follows that the boundary value $\overline{\phi}(\boldsymbol{x})$ transforms under scaling as

$$\overline{\phi}(\boldsymbol{x}) \to \lambda^{d-\Delta}\overline{\phi}(\lambda\boldsymbol{x}). \tag{3.21}$$

Comparing to the scaling of sources in the CFT, (3.12), it follows that $\overline{\phi} \equiv \overline{\phi}_\Delta$ is the source of a (scalar) primary operator $\mathcal{O}_\Delta$ of dimension Δ in the CFT. Similar remarks apply to non-scalar fields as well.

Since Z is given as a path integral, it defines a quantum theory if the ϕ fluctuate, i.e., if they have kinetic terms. For example, if ϕ is a free scalar on AdS so that it has action

$$S[\phi] = \frac{1}{2}\int\limits_{\mathrm{AdS}_{d+1}} \sqrt{g}\,\phi(-D_\mu D^\mu + m^2)\phi, \tag{3.22}$$

then it is not hard to check that its equation of motion has two independent solutions scaling near the boundary as $z^{d-\Delta}$ and z^Δ, so that asymptotically

$$\phi(z,\boldsymbol{x}) \sim z^{d-\Delta}\overline{\phi}(\boldsymbol{x}) + \cdots + z^\Delta\varphi(\boldsymbol{x}) + \cdots \tag{3.23}$$

with

$$\Delta = \frac{d}{2} + \sqrt{\left(\frac{d}{2}\right)^2 + m^2R^2}. \tag{3.24}$$

and $\overline{\phi}$ is the source for $\mathcal{O}_\Delta$ in the CFT. Note that in the limit where the mass of the scalar field goes to infinity—turning off the ϕ-fluctuations—the dimension Δenergy–momentum of the associated CFT operator also goes to infinity.

Every CFT has a local tensor, $T^{\mu\nu}(\boldsymbol{x})$, as a primary operator of dimension $\Delta = d$. It is sourced by $\int h_{\mu\nu}T^{\mu\nu}$, so $h_{\mu\nu}$ is the boundary value of a spin-2 field on AdS. An argument similar to the one for the scalar field shows that $\Delta = d$ for $T^{\mu\nu}$ implies that $m^2 = 0$ for $h_{\mu\nu}$. So the AdS theory must have a dynamical, massless spin-2 field: a graviton. This gives the basic AdS/CFT correspondence [12, 13].

The partition function of a quantum gravity theory on an asymptotically AdS_{d+1} space–time as a function of the boundary values of its fields is the partition function of a CFT_d with the boundary values acting as sources for the primary operators:

$$Z_{\text{qu-grav}}[\overline{\phi}] = Z_{\text{CFT}}[\overline{\phi}] \tag{3.25}$$

It is worth pointing out a few simple but important properties or modifications of the AdS/CFT correspondence as described so far.

- If we replace AdS_{d+1} by $\mathrm{AdS}_{d+1} \times X$ with X any space without boundary, the same procedure still works. The isometry group of X becomes a global internal symmetry of the CFT.
- If we define a partition function by the same procedure but cut off the AdS at $z = \epsilon > 0$ (i.e., make $z = \epsilon$ the boundary), we then keep Poincaré invariance, but break conformal invariance and locality on length scales smaller than ϵ.
- For negative mass-squared in the range $0 > m^2 > -d^2/4R^2$, the associated scaling dimension of the CFT primary is $d > \Delta > d/2$, which is real and above the unitarity bound. This corresponds to the fact [14, 15] that scalars with negative mass-squared in this range are stable in AdS.
- For $(d+2)/2 > \Delta > d/2$, both the $z^{d-\Delta}$ as well as the z^{Δ} asymptotic solution are normalizable, so we can use either as a source of an operator $\mathcal{O}_\Delta$ or $\mathcal{O}_{d-\Delta}$, respectively [16]. This allows us to describe operators with dimensions down to the unitarity bound $\Delta = (d-2)/2$.

3.3.3 Semi-classical Gravity Limit

The result for the AdS/CFT correspondence summarized by (3.25) immediately raises two questions: (1) Are there consistent quantum gravity theories in which we can compute the left side of (3.25) and (2) What class of CFTs do they correspond to?

The only examples of quantum gravity theories whose consistency we have confidence in are string theories. It is difficult to compute string theory partition functions on AdS backgrounds, except in the weak-coupling, low-energy limit, in which case it reduces to classical Einstein gravity coupled to other massless string fields. So we now describe the correspondence in more detail in this limit.

A gravitational theory on AdS_{d+1} has the low-energy effective action

$$\frac{1}{\kappa^2}\int d^{d+1}x\,\sqrt{g}\left(\mathcal{R}+\alpha'\mathcal{R}^2+\cdots\right) \tag{3.26}$$

where $\mathcal{R}$ is the Ricci scalar curvature and the dots denote an infinite series of higher-derivative terms built from curvature tensors and covariant derivatives (as well as other possible fields). The relative sizes of these terms define the basic length scales governing strength of gravity and size of higher-derivative terms, respectively:

$$\text{Planck length:}\quad \ell_p \sim \kappa^{\frac{2}{d-1}},\qquad \text{String length:}\quad \ell_s \sim \sqrt{\alpha'}. \tag{3.27}$$

In an AdS background $\mathcal{R} \sim R^{-2}$, so gravity is weak and the higher-derivative terms are small when

$$\ell_p \ll R, \quad \text{and} \quad \ell_s \ll R. \tag{3.28}$$

In this limit the semi-classical (saddle-point) approximation to the gravitational partition function is then

$$Z_{\text{qu-grav}}[\overline{\phi}] \sim \sum_{\{\phi_{cl}\}} e^{-S_{\text{Einstein}}[\phi_{cl}]} \tag{3.29}$$

where $\{\phi_{cl}\}$ are the classical field values (including the space–time metric) found by extremizing the action subject to the $\phi|_\partial = \overline{\phi}$ boundary conditions (including the boundary condition that the space–time is asymptotically AdS), and the sum is over all such extrema. S_{Einstein} is the Einstein action, i.e., (3.26) with all the α' and higher-derivative terms dropped.

In this semi-classical gravity limit, there is now a concrete calculational procedure for extracting correlation functions of the associated CFT via the AdS/CFT correspondence.

1. Find the dominant saddle point.
2. Each $(d+1)$-dimensional (or "bulk") field, ϕ, obeys a second order partial differential equation of motion on the (asymptotically) AdS space, with near-boundary asymptotic expansion

$$\phi(z,\boldsymbol{x}) \sim z^{d-\Delta}\overline{\phi}(\boldsymbol{x}) + \cdots + z^{\Delta}\varphi(\boldsymbol{x}) + \cdots \tag{3.30}$$

3. $\phi \equiv \phi_\Delta$ is associated with the source for an $\mathcal{O}_\Delta$ CFT primary.
4. With boundary conditions fixing $\overline{\phi}(\boldsymbol{x})$, then $\varphi(\boldsymbol{x})$ is determined by the equations of motion, so $\phi = \phi[\overline{\phi}]$.

5. Evaluate S_{Einstein} on these solutions to get, by the AdS/CFT correspondence (3.25),

$$S_{\text{Einstein}}[\overline{\phi}] = -\ln\big(Z_{\text{CFT}}[\overline{\phi}]\big) \equiv -W_{\text{CFT}}[\overline{\phi}]. \tag{3.31}$$

6. Then the "connected" correlators in the CFT are

$$\langle \mathcal{O}_{\Delta_1}(\boldsymbol{x}_1)\cdots\mathcal{O}_{\Delta_n}(\boldsymbol{x}_n)\rangle_{\text{CFT-conn}} = -\frac{\partial^n S_{\text{Einstein}}[\{\overline{\phi}\}]}{\partial\overline{\phi}_{\Delta_1}(\boldsymbol{x}_1)\cdots\partial\overline{\phi}_{\Delta_n}(\boldsymbol{x}_n)}\Bigg|_{\overline{\phi}=0}. \tag{3.32}$$

To actually compute $S_{\text{Einstein}}[\overline{\phi}]$ and its derivatives, one must regulate (by cutting off consistently at $z = \epsilon > 0$), renormalize (by adding local counterterms on the $z = \epsilon$ boundary) to preserve conformal invariance, then take the $\epsilon \longrightarrow 0$ limit to extract finite answers. See for example [8]. As an example of the kind of results that one finds, for the simplest case of a free massive scalar plus $-(g/4)\phi^4$ interaction term, with some work one finds [8]

$$\begin{aligned}
\langle \mathcal{O}_\Delta(\boldsymbol{x})e^{\int\overline{\phi}_\Delta\mathcal{O}_\Delta}\rangle_{\text{CFT}} &= (2\Delta - d)\varphi(\boldsymbol{x})\\
\langle \mathcal{O}_\Delta(\boldsymbol{x}_1)\mathcal{O}_\Delta(\boldsymbol{x}_2)\rangle_{\text{CFT}} &= (2\Delta - d)\frac{\Gamma(\Delta)}{\pi^{d/2}\Gamma(\Delta - \frac{d}{2})}\frac{1}{|x_{12}|^{2\Delta}}\\
\langle \mathcal{O}_\Delta(\boldsymbol{x}_1)\mathcal{O}_\Delta(\boldsymbol{x}_2)\mathcal{O}_\Delta(\boldsymbol{x}_3)\rangle &= 0\\
\langle \mathcal{O}_\Delta(\boldsymbol{x}_1)\mathcal{O}_\Delta(\boldsymbol{x}_2)\mathcal{O}_\Delta(\boldsymbol{x}_3)\mathcal{O}_\Delta(\boldsymbol{x}_4)\rangle &= \ldots\text{explicit but complicated}\ldots
\end{aligned}$$

Many other more difficult examples have been worked out in the literature, notably correlators of the energy momentum tensor and of global conserved currents in the CFT.

3.4 Large N

We now briefly review a few salient points about the large-N expansion of gauge theories. These will be important in the precise form of the AdS/CFT correspondence that will be discussed in the next section.

Consider a $U(N)$ Yang–Mills theory in the 't Hooft limit, i.e., take $N \rightarrow \infty$ keeping $\lambda \equiv g_{ym}^2 N$ fixed. The $U(N)$ YM action has the general form

$$\mathcal{L} \sim \frac{N}{\lambda}\ \text{tr}\big(d\phi d\phi + \phi^2 d\phi + \phi^4\big), \tag{3.33}$$

for fields ϕ in the adjoint representation of the gauge group. ϕ here represents either the vector boson or an adjoint scalar superpartner of the vector boson in a super–Yang–Mills (SYM) theory. In the 't Hooft limit the growth of the action

Fig. 3.3 Two sample Feynman diagrams and their associated N and λ weights. Note that the left diagram is planar, and can be interpreted as a triangulation of a disk; while the right one is not planar: it is the triangulation of a torus minus a disk

with N makes the theory look semi-classical, i.e., dominated by the saddle point contributions.

We can write the adjoint indices as ϕ^i_j with $i,j = 1,\ldots,N$, corresponding to the $N \otimes \overline{N}$ decomposition of the adjoint representation. Then $\langle \phi^i_j \phi^k_\ell \rangle \propto \delta^i_\ell \delta^j_k$, so we can notate propagators in Feynman diagrams as oppositely oriented double lines with each line tracking the fundamental or antifundamental indices, see Fig. 3.3. Associate to each interaction a vertex, to each propagator an edge, and to each loop a face, to form a triangulation of some oriented surface. Denote the number of vertices, edges, and faces, by E, V, and F, respectively.

Since each vertex has a factor of N/λ, each propagator λ/N, and each face N (from the associated trace over the fundamental indices), a given diagram has weight $N^{V-E+F}\lambda^{E-V}$. Since $V - E + F = 2 - 2g$, where g is the genus of the surface, and extending to diagrams with n external propagators (with appropriate normalization), we find that connected correlators have the general form

$$\langle \mathcal{O}_1 \cdots \mathcal{O}_n \rangle \sim \sum_{g=0}^{\infty} N^{2-2g-n} F_{g,n}(\lambda). \tag{3.34}$$

So correlators are dominated by the $g = 0$, or "planar", diagrams in the 't Hooft limit.

In this limit, 2-point functions $\sim N^0$, 3-point functions $\sim N^{-1}$, etc., so $1/N$ acts like an effective coupling constant in large-N YM (in addition to the 't Hooft coupling λ).

3.5 D3-Branes and $\mathrm{AdS}_5 \times S^5$

Now we will use a specific quantum gravity theory—string theory—to derive a specific example of the AdS/CFT correspondence. In particular we will derive the *Maldacena conjecture* [17]: type IIB string on $\mathrm{AdS}_5 \times S^5$ is equivalent to four-dimensional $\mathcal{N} = 4$ supersymmetric SU(N) YM. The argument can be organized into three basic steps.

1. Identify $SU(N)$ SYM as a sector of the low energy limit of IIB strings in the presence of N parallel D3-branes at weak coupling.
2. Identify the space–time geometry sourced by the D3-branes as approximately an $AdS_5 \times S^5$ "throat" glued into an asymptotically flat region whose gravitational modes decouple from the string modes on the throat.
3. Compare these two low energy descriptions to identify the SYM sector with the $AdS_5 \times S^5$ sector with a specific mapping of parameters.

The same type of argument can be used to find many other examples.

3.5.1 SYM from D-Branes

In IIB string theory place a stack of N parallel D3-branes extended along the $x = x_{0,1,2,3}$ directions, and take the limit in which they become coincident. At weak enough string coupling, $g_s \ll 1$, this stack is well-described as a 4-manifolds where oriented open strings may end, see Fig. 3.4. Label the D3-branes by indices $i, j, \ldots = 1, \ldots, N$. The lightest modes of open strings connecting the branes then carry adjoint $U(N)$ labels. They fill out a massless four-dimensional $\mathcal{N} = 4$ vector supermultiplet of the $U(N)$ SYM theory. The SYM coupling is the open string coupling, $g_{ym} \sim \sqrt{g_s}$.

For small enough g_s at fixed N, the gravitational back-reaction of the D3-branes is small, since even though the D3-brane tension τ_3 is proportional to Ng_s^{-1}, the gravitational coupling κ^2 is proportional to g_s^2, so the net strength of the D3-brane source goes as $\tau_3 \kappa^2 \propto g_s N$.

In the low energy limit, $E \ll \ell_s^{-1}$ or $E^2\alpha' \to 0$, only the massless string modes can be excited. These modes are the bulk (10-dimensional) supergravity modes of the closed strings and the (four-dimensional) $U(N)$ SYM modes of the open strings. The couplings between the four-d SYM and ten-d supergravity sectors are higher-derivative, and so vanish in the low energy limit.

Since the gravitational coupling $\kappa \sim g_s \alpha'^2$, the dimensionless coupling strength is $E^4\kappa$ which vanishes as the energy scale $E \to 0$. Thus the supergravity sector is IR-free. On the other hand, the SYM sector stays at fixed coupling, g_{ym}^2, since

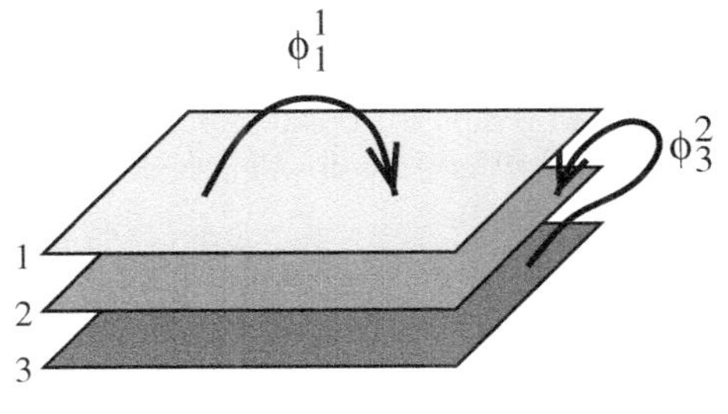

Fig. 3.4 A stack of nearly coincident D3-branes labelled by indices 1, 2, 3, and two sample open string modes labelled by the ordered pair of indices associated with the brane the string starts on and ends on

$\mathcal{N} = 4$ SYM is a CFT for all g_{ym}. In other words, the gauge coupling is an exactly marginal coupling so does not run with scale. (More precisely, the diagonal $U(1) \subset U(N)$ decouples and is free, corresponding to the overall translational degrees of freedom of the N D3-branes. So only the $SU(N)$ SYM factor stays at fixed non-zero coupling.) The net result of this analysis is that:

In the small g_s limit at fixed N, the low energy effective action of IIB strings in the presence of N coincident D3-branes consists of three decoupled sectors:

$$\begin{aligned}&(\text{free 10-d supergravity})\times\\&\quad(\text{free 4-d U(1) SYM})\times\\&\qquad(\text{4-d SU(N) SYM CFT with } g_{ym}^2 \sim g_s).\end{aligned} \tag{3.35}$$

3.5.2 D3-Brane Near Horizon Geometry

Now look at the same system in the $g_s \to 0$, $N \to \infty$ limit, keeping $g_s N \equiv \lambda$ fixed. In this limit the gravitational back-reaction of the D3-branes cannot be neglected, since the D3-brane tension times the gravitational coupling is proportional to $Ng_s^{-1} \times g_s^2 = \lambda$. In the low energy and $g_s \to 0$ limit, the N D3-branes source a classical supergravity solution with metric and RR 5-form field strength given by

$$\begin{aligned}ds^2 &= f^{-1/2}(-dt^2 + d\mathbf{x}^2) + f^{1/2}(dr^2 + r^2 d\Omega_5^2)\\ F_5 &= (1 + *)dt \wedge dx_1 \wedge dx_2 \wedge dx_3 \wedge df^{-1}\end{aligned} \tag{3.36}$$

where

$$f \equiv 1 + \frac{R^4}{r^4}, \quad \text{and} \quad R^4 \equiv 4\pi g_s N \alpha'^2. \tag{3.37}$$

For $r \gg R, f \to 1$, so the solution asymptotes to flat $\mathbf{R}^{9,1}$, while for $r \ll R$, $f^{1/2} \to R^2/r^2$, so the metric becomes

$$ds^2 \sim \frac{R^2}{r^2}dr^2 + \frac{r^2}{R^2}(-dt^2 + d\mathbf{x}^2) + d\Omega_5^2 \tag{3.38}$$

which we recognize as $\text{AdS}_5 \times S^5$ in Poincaré patch coordinates; compare to (3.17). The AdS boundary at $r = \infty$ is replaced by a transition to flat $\mathbf{R}^{9,1}$; see Fig. 3.5.

We now want to consider the low energy effective description of physics in this geometry. There are low energy scattering states of the massless supergravity modes in the $\mathbf{R}^{9,1}$ asymptotic region. There are also arbitrarily low-energy modes far down the AdS "throat". To see this, consider an object of fixed energy E_* as measured in the frame of a co-moving observer. From the point of view of an asymptotic observer at $r = \infty$ (where we are measuring all our observables) it has

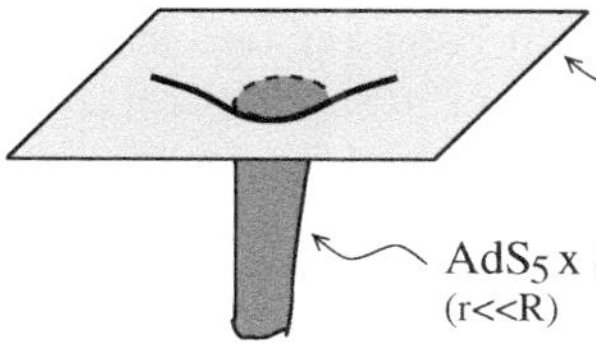

Fig. 3.5 A cartoon illustrating how the $\mathrm{AdS}_5 \times S^5$ "throat" geometry for $r \ll R$ is attached to the flat $\mathbf{R}^{9,1}$ geometry for $r \gg R$

a red-shifted energy $E = \sqrt{g_{tt}}E_* = f^{-1/4}E_*$. So $E \sim rE_*/R \to 0$ as $r \to 0$. Therefore states of *arbitrary* finite energy E_* in the $\mathrm{AdS}_5 \times \mathrm{S}^5$ throat are low-energy excitations.

Finally there are also a few massless modes corresponding (from the asymptotic $\mathbf{R}^{9,1}$ perspective) to the translational zero-modes of the whole throat. (They are the "singleton" boundary modes on the AdS.) Since they are the translational modes of the whole stack of D3-branes, they are equivalent to a free four-dimensional $U(1)$ SYM theory. In summary,

In the small g_s, fixed $\lambda = g_s N$ limit, the low energy effective action of IIB strings in the presence of N coincident D3-branes has three sectors:

$$\begin{aligned}&(\text{free 10-d supergravity})\times\\ &\quad(\text{free 4-d U(1) SYM})\times\\ &\quad(\text{IIB string theory on } \mathrm{AdS}_5 \times \mathrm{S}^5 \text{ with} R^4 = 4\pi\lambda\alpha'^2 \text{ and } \kappa = g_s\alpha'^2).\end{aligned} \tag{3.39}$$

Furthermore, these three sectors decouple in the low-energy limit, since the cross-section for an $\mathbf{R}^{9,1}$ supergravity wave of frequency ω to scatter off the throat $(r < R)$ region is $\sigma \sim \omega^3 R^8$, so vanishes in the low-energy $\omega \to 0$ limit. Likewise, as the low-energy throat modes are localized closer to $r = 0$, escape to the asymptotically flat region is energetically suppressed. Finally, low energy $\mathbf{R}^{9,1}$ gravitational waves can not excite the massive throat translational modes, while the associated singleton modes on the boundary of AdS decouple from bulk AdS modes. Thus all three sectors decouple.

3.5.3 Strong/Weak Duality

Comparing these two low energy descriptions, (3.35) and (3.39), leads to the Maldacena conjecture:

$$\begin{gathered}\text{Type IIB string theory on} \mathrm{AdS}_5 \times S^5\\ \text{is equivalent to}\\ \text{4-dimensional} \mathcal{N} = 4 \text{ supersymmetric } SU(N) \text{ YM}\\ \text{with parameters identified as}\\ 4\pi g_s = g_{ym}^2, \text{ and } R^4/\alpha'^2 = \lambda \equiv g_{ym}^2 N.\end{gathered}$$

It is just a conjecture because the low energy descriptions of (3.35) and (3.39) were in different limits (the first at fixed N, the second as $N \to \infty$). Note that $o(\alpha')$ string worldsheet corrections are $o(1/\sqrt{\lambda})$ corrections in the Yang–Mills theory, and at fixed λ, the $o(g_s)$ string loop corrections are $o(1/N)$ corrections on the Yang–Mills side. The N D3-brane near horizon AdS geometry came from a classical supergravity solution, i.e., it did not contain g_s or α' corrections, so it is not *a priori* clear that the AdS solution should still be valid in the regime where the SYM description was derived.

Thus there are different possible versions of the conjecture that one can imagine:

Weak: valid only for $g_sN \to \infty$: neither α' nor g_s corrections agree.
Medium: valid for all g_sN but only for $N \to \infty$: only α' corrections agree.
Strong: valid for all g_sN and all N: an exact equivalence.

The strong version of the conjecture allows *all* kinds of finite energy interior processes and objects in the AdS space–time, including space–times with different topologies (e.g., black holes). So in this version,

$$Z_{\text{CFT}} = \sum \forall \text{ asymptotic AdS geometries} \tag{3.40}$$

Most tests of the conjecture are by computing at large N quantities whose λ-dependence is determined by supersymmetry, though a few are also checks that $1/N$ corrections match as well.

In any case, the string theory on $\text{AdS}_5 \times S^5$ is currently only really calculable in the classical supergravity limit where $g_s \ll 1$ (so no string loops) and $\ell_s \ll R$ (so no α' corrections). In terms of YM parameters this means that $N \gg \lambda \gg 1$, which is the planar 't Hooft limit, but at *strong* 't Hooft coupling. On the other hand, the YM theory is only under perturbative control at small λ and finite N. A great deal of the power of Maldacena's conjecture comes not just from the fact that it is an explicit realization of the AdS/CFT conjecture, but also that weak coupling on one side of the equivalence is strong coupling on the other.

3.6 Extensions

There are many examples, refinements, extensions and deformations of the AdS/CFT correspondence; see the other lectures in this volume for a flavor. Two extensions which are basic and play an important role in many other applications involve

- turning on a finite temperature, and
- adding brane probes.

The finite temperature extension can be derived by the same argument as in the last section, but keeping a finite energy density on the D3-branes. In this case the supergravity solution become a black 3-brane, and the near-horizon limit is the AdS_5-black hole $\times S^5$ geometry. The AdS black hole Hawking temperature is the same as the temperature of the SYM theory. This extension of the AdS/CFT correspondence is straightforward in the sense that it does not change the underlying equivalence, but involves identifying certain excited states on the two sides of the correspondence. Nevertheless, this extension leads to rich and beautiful physics, such as the correspondence between the deconfinement transition in finite volume SYM and the instability of small black holes in AdS space [18].

The brane probe approximation in the AdS/CFT correspondence can be more subtle, so I will describe an example of it in more detail. This particular example can be motivated by the following question: How can one add fundamental (as opposed to adjoint) matter to the SYM theory in the AdS/CFT correspondence? In the weak coupling limit, we need string states with one end on the D3-brane stack (carrying a fundamental color index) and the other end elsewhere. Since fundamental strings end on D-branes, we should add other kinds of D-branes to the initial setup. We call these other branes "flavor branes" since they will label different flavors of fundamental matter.

In general it is hard to find supergravity solutions for the gravitational back-reaction of adding flavor branes. However, if the number, N_f, of flavor branes is much smaller than the number, N, of (color) D3-branes, this back-reaction can be ignored in the large N limit. To see this, recall that Newton's constant $\kappa^2 \sim g_s^2 \sim N^{-2}$ in the 't Hooft limit (where $\lambda = g_s N$ is kept fixed), and the tension of N_f Dp-branes is $N_f T^{\rm Dp}_{\mu\nu} \sim N_f g_s^{-1} \sim N_f N$, so the gravitational back-reaction of the flavor branes is $\kappa^2 N_f T^{\rm Dp}_{\mu\nu} \sim N_f/N$ which vanishes as $N \to \infty$ with N_f fixed. Thus such probe branes need only satisfy their classical equations of motion in the unperturbed background space–time generated by the N color branes. These equations come from extremizing the probe branes' world volume [or the DBI action if the brane's $U(1)$ field is turned on].

It was shown in [19, 20] that probe flavor D7-branes can be added in such a way as to keep half of the $\mathcal{N} = 4$ supersymmetry of the SYM theory. The required embedding of the flavor brane in $AdS_5 \times S^5$ is shown in Fig. 3.6. The dual theory is then interpreted to be $\mathcal{N} = 2\, SU(N)$ SYM with N_f massive quark hypermultiplets. Since the probe D7-branes only added new open string states—identified with the fundamental hypermultiplets on the YM side—the dual theory can no longer be a CFT, but must in fact have non-zero beta function. The AdS/CFT correspondence then implies that the dual string background should not be asymptotically AdS. This is reflected in the fact that a D7 brane is co-dimension 2 in 10 dimensions, and so no matter how light, its tension will deform the asymptotic structure of the space (by causing a deficit angle). Thus, strictly speaking, D7-branes can never satisfy the probe approximation, and the AdS/$\mathcal{N} = 2$ SYM correspondence is not exact, but is broken by N_f/N effects.

Higher co-dimension probes are less problematic. For example, fundamental string and D1-brane probes of the AdS_5 geometry have proven very useful, either as strings extending to the boundary of AdS whose worldsheets are dual to

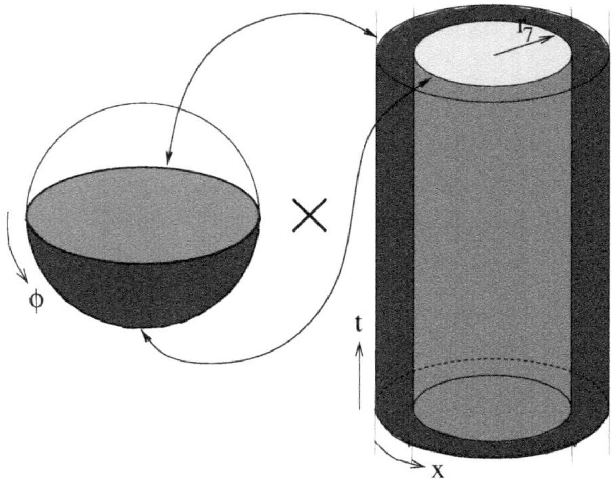

Fig. 3.6 The topology of the embedding of a probe D7-brane in $AdS_5 \times S^5$. The *solid cylinder* is the AdS factor, with t its axial direction, r its radial direction, and $\mathbf{x}$ its angular direction (s). The D7-brane fills the AdS cylinder from $r = r_7$ out to the boundary ($r = \infty$). The *sphere* is the S^5 factor, and the D7-brane fills an S^4 hemisphere. The D7-brane is embedded in such a way that constant r slices of the *cylinder* are associated with constant latitude ϕ S^3s, with the equatorial S^3 corresponding to $r = \infty$, and the degenerate point S^3 at the south pole of the S^5 to $r = r_7$

insertions of the non-local Wilson and 't Hooft loop operators in the CFT, or as strings ending on probe D7-branes dual to massive quark states in the $\mathcal{N} = 2$ SYM. Both of these scenarios have led to physically rich and active areas of development of the AdS/CFT correspondence following on the work of [21, 22] and [23, 24], respectively.

Acknowledgments It is a pleasure to thank the organizers for inviting me to lecture at the summer school, and to thank K. Skenderis and M. Rangamani for helpful comments.

References

1. Green, M.B., Schwarz, J.H., Witten, E.: Superstring Theory, vol. 2. Cambridge University Press, Cambridge (1987)
2. Polchinski, J.: String Theory, vol. 2, Cambridge University Press, Cambridge (1998)
3. Becker, K., Becker, M., Schwarz, J.H.: String Theory and M-Theory: A Modern Introduction. Cambridge University Press, Cambridge (2007)
4. Zwiebach, B.: A First Course in String Theory, 2nd edn. Cambridge University Press, Cambridge (2004)
5. Gubser, S.S., Karch, A.: From gauge-string duality to strong interactions: a Pedestrian's Guide. Ann. Rev. Nucl. Part. Sci. **59**, 145 (2009) [arXiv:0901.0935]
6. Aharony, O., Gubser, S.S., Maldacena, J.M., Ooguri, H., Oz, Y.: Large N field theories, string theory and gravity. Phys. Rept. **323**, 183 (2000) [arXiv:hep-th/9905111]

7. D'Hoker, E., Freedman, D.Z.: Supersymmetric gauge theories and the AdS/CFT correspondence. (TASI 2001 lecture notes) [arXiv:hep-th/0201253]
8. Skenderis, K.: Lecture notes on holographic renormalization. Class. Quant. Grav. **19**, 5849 (2002) [arXiv:hep-th/0209067]
9. Aharony, O.: A brief review of little string theories. Class. Quant. Grav. **17**, 929 (2000) [arXiv:hep-th/9911147]
10. Polchinski, J.: Scale and conformal invariance in quantum field theory. Nucl. Phys. B. **303**, 226 (1988)
11. Minwalla, S.: Restrictions imposed by superconformal invariance on quantum field theories. Adv. Theor. Math. Phys. **2**, 781 (1998) [arXiv:hep-th/9712074]
12. Witten, E.: Anti-de Sitter space and holography. Adv. Theor. Math. Phys. **2**, 253 (1998) [arXiv:hep-th/9802150]
13. Gubser, S.S., Klebanov, I.R., Polyakov, A.M.: Gauge theory correlators from noncritical string theory. Phys. Lett. **B428**, 105 (1998) [arXiv:hep-th/9802109]
14. Breitenlohner, P., Freedman, D.Z.: Positive energy in anti-de Sitter backgrounds and gauged extended supergravity. Phys. Lett. **115B**, 197 (1982)
15. Breitenlohner, P., Freedman, D.Z.: Stability in gauged extended supergravity. Ann. Phys. **144**, 249 (1982)
16. Klebanov, I.R., Witten, E.: AdS/CFT correspondence and symmetry breaking. Nucl. Phys. B. **556**, 89 (1999) [arXiv:hep-th/9905104]
17. Maldacena, J.M.: The large N limit of superconformal field theories and supergravity. Adv. Theor. Math. Phys. **2**, 231 (1998) [Int. J. Theor. Phys. **38**, 1113 (1999)] [arXiv:hep-th/9711200]
18. Witten, E.: Anti-de Sitter space, thermal phase transition, and confinement in gauge theories. Adv. Theor. Math. Phys. **2**, 505 (1998) [arXiv:hep-th/9803131]
19. Karch, A., Katz, E.: Adding flavor to AdS/CFT. JHEP 0206, 043 (2002) [arXiv:hep-th/0205236]
20. Babington, J., Erdmenger, J., Evans, N.J., Guralnik, Z., Kirsch, I.: Chiral symmetry breaking and pions in non-supersymmetric gauge/gravity duals. Phys. Rev. D. **69**, 066007 (2004) [arXiv:hep-th/0306018]
21. Maldacena, J.: Wilson loops in large N eld theories. Phys. Rev. Lett. **80**, 4859 (1998) [arXiv:hep-th/9803002]
22. Rey, S.J., Yee, J.T.: Macroscopic strings as heavy quarks in large N gauge theory and anti-de Sitter supergravity. Eur. Phys. J. C. **22**, 379 (2001) [arXiv:hep-th/9803001]
23. Herzog, C.P., Karch, A., Kovtun, P., Kozcaz, C., Yaffe, L.G.: Energy loss of a heavy quark moving through N = 4 supersymmetric Yang–Mills plasma. JHEP **0607**, 013 (2006) [arXiv:hep-th/0605158]
24. Gubser, S.S.: Drag force in AdS/CFT. Phys. Rev. D. **74**, 126005 (2006) [arXiv:hep-th/0605182]

Part II
Holography and the AdS/CFT Correspondence

Chapter 4
Improved Holographic QCD

U. Gursoy, E. Kiritsis, Liuba Mazzanti, G. Michalogiorgakis and Francesco Nitti

Abstract We provide a review to holographic models based on Einstein-dilaton gravity with a potential in five dimensions. Such theories, for a judicious choice of potential are very close to the physics of large-N YM theory both at zero and finite temperature. The zero temperature glueball spectra as well as their finite temperature thermodynamic functions compare well with lattice data. The model can be used to calculate transport coefficients, like bulk viscosity, the drag force and jet quenching parameters, relevant for the physics of the Quark–Gluon Plasma.

U. Gursoy (✉)
Institute for Theoretical Physics, Utrecht University, Leuvenlaan 4, 3584 CE Utrecht, The Netherlands

E. Kiritsis
Department of Physics, Crete Center for Theoretical Physics, University of Crete, 71003, Heraklion, Greece

L. Mazzanti
Department of Particle Physics and IGFAE, University of Santiago de Compostela, 15782, Santiago de Compostela, Spain

G. Michalogiorgakis
Department of Physics, Purdue University, 525 Northwestern Avenue, West Lafayette, IN 47907-2036, USA

F. Nitti
APC, Université Paris 7, (UMR du CNRS 7164), Bâtiment Condorcet, 75205, Paris Cedex 13, France

E. Papantonopoulos (ed.), *From Gravity to Thermal Gauge Theories: The Ads/CFT Correspondence,* Lecture Notes in Physics, 828,
DOI: 10.1007/978-3-642-04864-7_4,

4.1 Introduction

The experimental efforts at RHIC [1–4] have provided a novel window in the physics of the strong interactions. The consensus on the existing data is that shortly after the collision, a ball of quark–gluon plasma (QGP) forms that is at thermal equilibrium, and subsequently expands until its temperature falls below the QCD transition (or crossover) where it finally hadronizes. Relativistic hydrodynamics describes very well the QGP [5], with a shear-viscosity to entropy density ratio close to that of $\mathcal{N} = 4$ SYM [6, 7]. The QGP is at strong coupling, and it necessitates a treatment beyond perturbative QCD approaches, [8–10]. Moreover, although the shear viscosity from $\mathcal{N} = 4$ seems to be close to that "measured" by experiment, lattice data indicate that in the relevant RHIC range $1 \leq \frac{T}{T_c} \leq 3$ the QGP seems not to be a fully conformal fluid. Therefore the bulk viscosity may play a role near the phase transition [11–13]. The lattice techniques have been successfully used to study the thermal behavior of QCD, however they are not easily extended to the computation of hydrodynamic quantities. They can be used however, together with parametrizations of correlators in order to pin down parameters [12, 13]. On the other hand, approaches based on holography have the potential to address directly the real-time strong coupling physics relevant for experiment.

In the bottom-up holographic model of AdS/QCD [14, 15], the bulk viscosity is zero as conformal invariance is essentially not broken (the stress tensor is traceless). In the soft-wall model [16], no reliable calculation can be done for glue correlators and therefore transport coefficients are ill-defined. Similar remarks hold for other phenomenologically interesting observables as the drag force and the jet quenching parameter [17–21].

Top-down holographic models of QCD displaying all relevant features of the theory have been difficult to obtain. Bottom-up models based on AdS slices [22] have given some insights mostly in the meson sector [14, 15] but necessarily lack many important holographic features of QCD. A hybrid approach has been advocated [23–25] combining features of bottom-up and top-down models. An similar approach was proposed independently in [26]. Such an approach, called Improved Holographic QCD (or IHQCD for short) is essentially a five-dimensional dilaton-gravity system with a non-trivial dilaton potential. Flavor can be eventually added in the form of N_f space-time filling $D4 - \overline{D4}$ brane pairs, supporting $U(N_f)_L \times U(N_f)_R$ gauge fields and a bi-fundamental scalar [27]. The UV asymptotics of the potential are fixed by QCD perturbation theory, while the IR asymptotics of the potential can be fixed by confinement and linear glueball asymptotics. An analysis of the finite temperature behavior [28, 29] has shown that the phase structure is exactly what one would expect from YM. A potential with a single free parameter tuned to match the zero temperature glueball spectrum was able to agree with the thermodynamic behavior of glue to a good degree [28]. Similar results, but with somewhat different potentials were also obtained in [26, 30].

In [26, 28, 29] it was shown that Einstein-dilaton gravity with a strictly monotonic dilaton potential that grows sufficiently fast, generically shares the same phase structure and thermodynamics of finite-temperature pure Yang–Mills theory at large N_c. There is a deconfinement phase transition (dual to a Hawking-Page phase transition between a black hole and thermal gas background on the gravity side), which is generically first order. The latent heat scales as N_c^2. In the deconfined gluon-plasma phase, the free energy slowly approaches that of a free gluon gas at high temperature, and the speed of sound starts from a small value at T_c and approaches the conformal value $c_s^2 = 1/3$ as the temperature increases. The deviation from conformal invariance is strongest at T_c, and is signaled by the presence of a non-trivial gluon condensate, which on the gravity side emerges as a deviation of the scalar solution that behaves asymptotically as r^4 close to the UV boundary. In the CP-violating sector, the topological vacuum density $\mathrm{Tr}F\tilde{F}$ has zero expectation value in the deconfined phase, in agreement with lattice results [31] and large-N_c expectations.

The analysis performed in [29] was completely general and did not rely on any specific form of the dilaton potential $V(\lambda)$. A detailed analysis of an explicit model in [32] shows that the thermodynamics matches *quantitatively* the thermodynamics of pure Yang–Mills theory. The (dimensionless) free energy, entropy density, latent heat and speed of sound, obtained on the gravity side by numerical integration of the 5D field equations, can be compared with the corresponding quantities, calculated on the lattice for pure Yang–Mills at finite-T, resulting in excellent agreement, for the temperature range that is accessible by lattice techniques. The same model also shows a good agreement with the lattice calculation of glueball mass ratios at zero temperature, and the value of the deconfining critical temperature (in units of the lowest glueball mass) is also in good agreement with the lattice results.

In short, the model we present gives a good phenomenological holographic description of most static properties[1] (spectrum and equilibrium thermodynamics) of large-N_c pure Yang–Mills, as computed on the lattice, for energies up to several times T_c. Thus it constitutes a good starting point for the computation of dynamical observables in a *realistic* holographic dual to QCD (as opposed to e.g. $\mathcal{N} = 4$ SYM), such as transport coefficients and other hydrodynamic properties that are not easily accessible by lattice techniques, at energies and temperatures relevant for relativistic heavy-ion collision experiments. We will report on such a calculation in the near future.

The vacuum solution in this model is described in terms of two basic bulk fields, the metric and the dilaton. These are not the only bulk fields however, as the bulk theory is expected to have an a priori infinite number of fields, dual to all possible YM operators. In particular we know from the string theory side that there are a few other low mass fields, namely the RR axion (dual to the QCD θ-angle) the NSNS and RR two forms B_2 and C_2 as well as other higher-level fields. With the exception of the RR axion, such fields are dual to higher-dimension and/or higher-spin operators of YM. Again, with the exception of the RR axion, they are not expected to play an important role into the structure of the vacuum and this is

[1] There are very few observables also that are not in agreement with YM. They are discussed in detail in [25].

why we neglect them when we solve the equations of motion. However, they are going to generate several new towers of glueball states beyond those that we discuss in this paper (namely the 0^{++} glueballs associated to dilaton fluctuations, 2^{++} glueballs associated to graviton fluctuations and 0^{-+} glueballs associated to RR axion fluctuations). Such fields can be included in the effective action and the associated glueball spectra calculated. Since we do not know the detailed structure of the associated string theory, their effective action will depend on more semi-phenomenological functions like $Z(\lambda)$ in (4.45). These functions can again be determined in a way similar to $Z(\lambda)$. In particular including the B_2 and C_2 field will provide 1^{+-} glueballs among others. Fields with spin greater than 2 are necessarily stringy in origin. We will not deal further with extra fields, like B_2 and C_2 and other as they are not particularly relevant for the purposes of this model, namely the study of finite temperature physics in the deconfined case. We will only consider the axion, as its physics is related to the CP-odd sector of YM with an obvious phenomenological importance.

It is well documented that string theory duals of YM must have strong curvatures in the UV regime. This has been explained in detail in [25] where it was also argued, that although the asymptotic AdS boundary geometry is due to the curvature non-linearities of the associated string theory, the inwards geometry is perturbative around AdS, with logarithmic corrections, generating the YM perturbation theory. The present model is constructed so that it takes the asymptotic AdS geometry for granted, by introducing the associated vacuum energy by hand, and simulates the perturbative YM expansion by an appropriate dilaton potential. In the IR, we do not expect strong curvatures in the string frame, and indeed the preferred backgrounds have this property. In this sense the model contains in itself the relevant expected effects that should arise from strong curvatures in all regimes. These issues have been explained in [23, 24] and in more detail in [25].

A different and interesting direction is the use of such models to study the expansion of the plasma and the associated dynamics. Such a context is similar to what happens on cosmology, especially the one related to the Randall–Sundrum setup. Indeed in this case the expansion can be found by following the geodesic motion of probe branes in the relevant background [33–36]. This generalizes to more complicated backgrounds [36] like the ones studied here.

Once we have a holographic model we trust, we should calculate observables , like transport coefficients that are hard to calculate on the lattice. A first class of transport coefficients are viscosity coefficients.[2] A general fluid is characterized by two viscosity coefficients, the shear η and the bulk viscosity ζ. The shear viscosity in strongly coupled theories described by gravity duals was shown to be universal [6, 7]. In particular, the ratio η/s, with s the entropy density, is equal to $\frac{1}{4\pi}$. This is correlated to the universality of low-energy scattering of gravitons from black

[2] These are the leading transport coefficients in the derivative expansion. There are subleading coefficients that have been calculated recently for $\mathcal{N} = 4$ SYM [37, 38]. However, at the present level of accuracy, they cannot affect substantially the comparison to experimental data [5].

holes. It is also known that deviations from this value can only be generated by higher curvature terms that contain the Riemann tensor (as opposed to the Ricci tensor of the scalar curvature). In QCD, as the theory is strongly coupled in the temperature range $T_c \leq T \leq 3T_c$, we would expect that $\eta/s \simeq \frac{1}{4\pi}$. Recent lattice calculations [39] agree with this expectations although potential systematic errors in lattice calculations of transport coefficients can be large.

Conformal invariance forces the bulk viscosity to vanish. Therefore the $\mathcal{N} = 4$ SYM plasma, being a conformal fluid, has vanishing bulk viscosity. QCD on the other hand is not a conformal theory. The classical theory is however conformally invariant and asymptotic freedom implies that conformal invariance is a good approximation in the UV. This would suggest that the bulk viscosity to entropy ratio is negligible at large temperatures. However it is not expected to be so in the IR: as mentioned earlier lattice data indicate that in the relevant RHIC range $1 \leq \frac{T}{T_c} \leq 3$ the QGP seems not to be a fully conformal fluid. Therefore the bulk viscosity may play a role near the phase transition.

So far there have been two approaches that have calculated the bulk viscosity in YM/QCD [12, 40–43] and have both indicated that the bulk viscosity rises near the phase transition as naive expectation would suggest. The first used the method of sum rules in conjunction with input from Lattice thermodynamics [40–43]. It suggested a dramatic rise of the bulk viscosity near T_c although the absolute normalization of the result is uncertain. The reason is that this method relies on an ansatz for the density associated with stress-tensor two point functions that are otherwise unknown.

The second method [12] relies on a direct computation of the density at low frequency of the appropriate stress-tensor two-point function. As this computation is necessarily Euclidean, an analytic continuation is necessary. The values at a finite number of discrete Matsubara frequencies are not enough to analytically continue. An ansatz for the continuous density is also used here, which presents again a potentially large systematic uncertainty.

Calculations in IHQCD support a rise of the bulk viscosity near T_c, but the values are much smaller than previously expected. Studies of how this affects hydrodynamics at RHIC [44] suggest that this implies a fall in radial and elliptic flow.

Another class of interesting experimental observables is associated with quarks, and comes under the label of "jet quenching". Central to this is the expectation that an energetic quark will loose energy very fast in the quark–gluon plasma because of strong coupling. This has as a side effect that back-to back jets are suppressed. Moreover if a pair of energetic quarks is generated near the plasma boundary then one will exit fast the plasma and register as an energetic jet, while the other will thermalize and its identity will disappear. This has been clearly observed at RHIC and used to study the energy loss of quarks in the quark–gluon plasma.

Heavy quarks are of extra importance, as their mass masks some low-energy strong interaction effects, and can be therefore cleaner probes of plasma energy loss. There are important electron observables at RHIC [45] that can probe heavy-quark energy loss in the strongly coupled quark–gluon plasma. Such observables are also expected to play an important role in LHC [46].

A perturbative QCD approach to calculate the energy loss of a heavy quark in the plasma has been pursued by calculating radiative energy loss [47–49]. However its application to the RHIC plasma has recently raised problems, based on comparison with data. A phenomenological coefficient used in such cases is known as the jet quenching coefficient $\hat{q}$, and is defined as the rate of change of the average value of transverse momentum square of a probe. Current fits [45, 50] indicate that a value of order 10 $\mathrm{GeV}^2/\mathrm{fm}$ or more is needed to describe the data while perturbative approaches are trustworthy at much lower values.

Several attempts were made to compute quark energy loss in the holographic context, relevant for $\mathcal{N} = 4$ SYM.[3] In some of them [18, 52] the jet-quenching coefficient $\hat{q}$ was calculated via its relationship to a light-like Wilson loop. Holography was then used to calculate the appropriate Wilson loop. The $\hat{q}$ obtained scales as $\sqrt{\lambda}$ and as the third power of the temperature,

$$\hat{q}_{\mathrm{conformal}} = \frac{\Gamma\left[\frac{3}{4}\right]}{\Gamma\left[\frac{5}{4}\right]} \sqrt{2\lambda}\pi^{\frac{3}{2}} T^3. \tag{4.1}$$

A different approach chooses to compute the drag force acting a string whose UV end-point (representing an infinitely heavy quark) is forced to move with constant velocity v [17, 19–21] in the context of $\mathcal{N} = 4$ SYM plasma. The result for the drag force is

$$F_{\mathrm{conformal}} = \frac{\pi}{2} \sqrt{\lambda} T^2 \frac{v}{\sqrt{1 - v^2}} \tag{4.2}$$

and is calculated by first studying the equilibrium configuration of the appropriate string world-sheet string and then calculating the momentum flowing down the string. This can be the starting point of a Langevin evolution system, as the process of energy loss has a stochastic character, as was first pointed out in [53] and more recently pursued in [54–60].

Such a system involves a classical force, that in this case is the drag force, and a stochastic noise that is taken to be Gaussian and which is characterized by a diffusion coefficient. There are two ingredients here that are novel. The first is that the Langevin evolution must be relativistic, as the quarks can be very energetic. Such relativistic systems have been described in the mathematical physics literature [61, 62] and have been used in phenomenological analyses of heavy-ion data [50]. They are known however to have peculiar behavior, since demanding an equilibrium relativistic Boltzmann distribution, provides an Einstein relation that is pathological at large temperatures. Second, the transverse and longitudinal diffusion coefficients are not the same [57]. A first derivation of such Langevin dynamics from holography was given in [57]. This has been extended in [60] where the thermal-like noise was associated and interpreted in terms of the world-sheet horizon that develops on the probe string.

[3] Most are reviewed in [51].

Most of the transport properties mentioned above have been successfully computed in $\mathcal{N} = 4$ SYM and a lot of debate is still waged as to how they can be applied to QCD in the appropriate temperature range [19, 20, 63–65]. A holographic description of QCD has been elusive, and the best we have so far have been simple bottom up models.

In the simplest bottom-up holographic model known as AdS/QCD [14, 15], the bulk viscosity is zero as conformal invariance is essentially not broken (the stress tensor is traceless), and the drag force and jet quenching essentially retain their conformal values.

In the soft-wall model [16], no reliable calculation can be done for glue correlators and therefore transport coefficients are ill-defined, as bulk equations of motion are not respected. Similar remarks hold for other phenomenologically interesting observables as the drag force and the jet quenching parameter.

The shear viscosity of IHQCD is the same as that of $\mathcal{N} = 4$ SYM, as the model is a two derivative model. Although this is not a good approximation in the UV of QCD, it is expected to be a good approximation in the energy range $T_c \leq T \leq 5T_c$. The bulk viscosity in IHQCD rises near the phase transition but ultimately stays slightly below the shear viscosity. There is a general holographic argument that any (large-N) gauge theory that confines color at zero temperature should have an increase in the bulk viscosity-to-entropy density ratio close to T_c.

The drag force on heavy quarks, and the associated diffusion times, can be calculated and found to be momentum depended as anticipated from asymptotic freedom. Numerical values of diffusion times are in the region dictated by phenomenological analysis of heavy-ion data. The medium-induced corrections to the quark mass (needed for the diffusion time calculation) can be calculated, and they result in a mildly decreasing effective quark mass as a function of temperature. This is consistent with lattice results. Finally, the jet-quenching parameter can be calculated and found to be comparable at T_c to the one obtained by extrapolation from $\mathcal{N} = 4$ SYM. Its temperature dependence is however different and again reflects the effects of asymptotic freedom.

4.2 The 5D Model

The holographic dual of large N_c Yang Mills theory, proposed in [23, 24], is based on a five-dimensional Einstein-dilaton model, with the action[4]:

$$S_5 = -M_p^3 N_c^2 \int d^5x \sqrt{g} \left[R - \frac{4}{3}(\partial\Phi)^2 + V(\Phi) \right] + 2M_p^3 N_c^2 \int_{\partial M} d^4x \sqrt{h} K. \quad (4.3)$$

[4] Similar models of Einstein-dilaton gravity were proposed independently in [26] to describe the finite temperature physics of large N_c YM. They differ in the UV as the dilaton corresponds to a relevant operator instead of the marginal case we study here. The gauge coupling e^{Φ} also asymptotes to a constant instead of zero in such models.

Here, M_p is the five-dimensional Planck scale and N_c is the number of colors. The last term is the Gibbons–Hawking term, with K being the extrinsic curvature of the boundary. The effective five-dimensional Newton constant is $G_5 = 1/(16\pi M_p^3 N_c^2)$, and it is small in the large-N_c limit.

Of the 5D coordinates $\{x_i, r\}_{i=0\ldots3}$, x_i are identified with the 4D space-time coordinates, whereas the radial coordinate r roughly corresponds to the 4D RG scale. We identify $\lambda \equiv e^{\Phi}$ with the running 't Hooft coupling $\lambda_t \equiv N_c g_{YM}^2$, up to an *a priori* unknown multiplicative factor,[5] $\lambda = \kappa\lambda_t$.

The dynamics is encoded in the dilaton potential,[6] $V(\lambda)$. The small-λ and large-λ asymptotics of $V(\lambda)$ determine the solution in the UV and the IR of the geometry respectively. For a detailed but concise description of the UV and IR properties of the solutions the reader is referred to Sect. 2 of [29]. Here we will only mention the most relevant information:

1. For small λ, $V(\lambda)$ is required to have a power-law expansion of the form:

$$V(\lambda) \sim \frac{12}{\ell^2}(1 + v_0\lambda + v_1\lambda^2 + \cdots), \qquad \lambda \to 0. \tag{4.4}$$

   The value at $\lambda = 0$ is constrained to be finite and positive, and sets the UV AdS scale ℓ. The coefficients of the other terms in the expansion fix the β-function coefficients for the running coupling $\lambda(E)$. If we identify the energy scale with the metric scale factor in the Einstein frame, as in [23, 24], we obtain:

$$\begin{aligned} \beta(\lambda) &\equiv \frac{d\lambda}{d\log E} = -b_0\lambda^2 - b_1\lambda^3 + \cdots \\ b_0 &= \frac{9}{8}v_0, \quad b_1 = \frac{9}{4}v_1 - \frac{207}{256}v_0^2. \end{aligned} \tag{4.5}$$

2. For large λ, confinement and the absence of bad singularities[7] require:

$$V(\lambda) \sim \lambda^{2Q}(\log\lambda)^P \quad \lambda \to \infty, \quad \begin{cases} 2/3 < Q < 2\sqrt{2}/3, & P \text{ arbitrary} \\ Q = 2/3, & P \geq 0 \end{cases}. \tag{4.6}$$

   In particular, the values $Q = 2/3, P = 1/2$ reproduce an asymptotically-linear glueball spectrum, $m_n^2 \sim n$, besides confinement. We will restrict ourselves to this case in what follows.

---

[5] This relation is well motivated in the UV, although it may be modified at strong coupling (see Sect. 4.3. The quantities we will calculate do not depend on the explicit relation between λ and λ_t.

[6] With a slight abuse of notation we will denote $V(\lambda)$ the function $V(\Phi)$ expressed as a function of $\lambda \equiv e^{\Phi}$.

[7] We call "bad singularities" those that do not have a well defined spectral problem for the fluctuations without imposing extra boundary conditions.

In the large N_c limit, the canonical ensemble partition function of the model just described, can be approximated by a sum over saddle points, each given by a classical solution of the Einstein-dilaton field equations:

$$\mathcal{Z}(\beta) \simeq e^{-\mathcal{S}_1(\beta)} + e^{-\mathcal{S}_2(\beta)} + \cdots \tag{4.7}$$

where $\mathcal{S}_i$ are the euclidean actions evaluated on each classical solution with a fixed temperature $T = 1/\beta$, i.e. with euclidean time compactified on a circle of length β. There are two possible types of Euclidean solutions which preserve three-dimensional rotational invariance. In conformal coordinates these are:

1. *Thermal gas solution*,

$$ds^2 = b_o^2(r)\left(dr^2 + dt^2 + dx_m dx^m\right), \quad \Phi = \Phi_o(r), \tag{4.8}$$

   with $r \in (0, \infty)$ for the values of P and Q we are using;
2. *Black-hole solutions*,

$$ds^2 = b(r)^2\left[\frac{dr^2}{f(r)} + f(r)dt^2 + dx_m dx^m\right], \quad \Phi = \Phi(r), \tag{4.9}$$

   with $r \in (0, r_h)$, such that $f(0) = 1$, and $f(r_h) = 0$.

In both cases Euclidean time is periodic with period β_o and β respectively for the thermal gas and black-hole solution, and three-space is taken to be a torus with volume V_{3o} and V_3 respectively, so that the black-hole mass and entropy are finite.[8]

The black holes are dual to a deconfined phase, since the string tension vanishes at the horizon, and the Polyakov loop has non-vanishing expectation value [66, 67]. On the other hand, the thermal gas background is confining.

The thermodynamics of the deconfined phase is dual to the 5D black-hole thermodynamics. The free energy, defined as

$$\mathcal{F} = E - TS, \tag{4.10}$$

is identified with the black-hole on-shell action; as usual, the energy E and entropy S are identified with the black-hole mass, and one fourth of the horizon area in Planck units, respectively.

The thermal gas and black-hole solutions with the same temperature differ at $O(r^4)$:

$$b(r) = b_o(r)\left[1 + \mathcal{G}\frac{r^4}{\ell^3} + \cdots\right], \quad f(r) = 1 - \frac{C}{4}\frac{r^4}{\ell^3} + \cdots \quad r \to 0, \tag{4.11}$$

[8] The periods and three-space volumes of the thermal gas solution are related to the black-hole solution values by requiring that the geometry of the two solutions are the same on the (regulated) boundary. See [29] for details.

where $\mathcal{G}$ and C are constants with units of energy. As shown in [29] they are related to the enthalpy TS and the gluon condensate $\langle \mathrm{Tr}F^2 \rangle$:

$$C = \frac{TS}{M_p^3 N_c^2 V_3}, \quad \mathcal{G} = \frac{22}{3(4\pi)^2} \frac{\langle \mathrm{Tr}F^2 \rangle_T - \langle \mathrm{Tr}F^2 \rangle_o}{240 M_p^3 N_c^2}. \tag{4.12}$$

Although they appear as coefficients in the UV expansion, C and $\mathcal{G}$ are determined by regularity at the black-hole horizon. For T and S the relation is the usual one,

$$T = -\frac{\dot{f}(r_h)}{4\pi}, \quad S = \frac{\mathrm{Area}}{4G_5} = 4\pi \left(M_p^3 N_c^2 V_3\right) b^3(r_h). \tag{4.13}$$

For $\mathcal{G}$ the relation with the horizon quantities is more complicated and cannot be put in a simple analytic form. However, as discussed in [29], for each temperature there exist only specific values of $\mathcal{G}$ (each corresponding to a different black hole) such that the horizon is regular.

At any given temperature there can be one or more solutions: the thermal gas is always present, and there can be different black holes with the same temperature. The solution that dominates the partition function at a certain T is the one with smallest free energy. The free energy difference between the black hole and thermal gas was calculated in [29] to be:

$$\frac{\mathcal{F}}{M_p^3 N_c^2 V_3} = \frac{\mathcal{F}_{BH} - \mathcal{F}_{th}}{M_p^3 N_c^2 V_3} = 15\mathcal{G} - \frac{C}{4}. \tag{4.14}$$

For a dilaton potential corresponding to a confining theory, like the one we will assume, the phase structure is the following [29]:

1. There exists a minimum temperature $T_{\min}$ below which the only solution is the thermal gas.
2. Two branches of black holes ("large" and "small") appear for $T \geq T_{\min}$, but the ensemble is still dominated by the confined phase up to a temperature $T_c > T_{\min}$
3. At $T = T_c$ there is a first order phase transition to the *large* black-hole phase. The system remains in the black-hole (deconfined) phase for all $T > T_c$.

In principle there could be more than two black-hole branches, but this will not happen with the specific potential we will use.

4.3 Scheme Dependence

There are several sources of scheme dependence in any attempt to solve a QFT. Different parametrizations of the coupling constant (here λ) give different descriptions. However, physical statements must be invariant under such a change.

In our case, reparametrizations of the coupling constant are equivalent to radial diffeomorphisms as we could use λ as the radial coordinate.

In the holographic context, scheme dependence related to coupling redefinitions translates into field redefinitions for the bulk fields. As the bulk theory is on-shell, all on-shell observables (that are evaluated at the single boundary of space-time) are independent of the field redefinitions showing that scheme-independence is expected. Invariance under radial reparametrizations of scalar bulk invariants is equivalent to RG invariance. Because of renormalization effects, the boundary is typically shifted and in this case field redefinitions must be combined with appropriate radial diffeomorphisms that amount to RG-transformations.

Another source of scheme dependence in our setup comes from the choice of the energy function. Again we may also consider this as a radial coordinate and therefore it is subject to coordinate transformations. A relation between λ and E is the β-function,

$$\frac{d\lambda}{d\log E} = \beta(\lambda). \tag{4.15}$$

β by definition transforms as a vector under λ reparametrizations and as a form under E reparametrizations. $\beta(\lambda)$ can therefore be thought of as a vector field implementing the change of coordinates from λ to E and vice-versa.

Physical quantities should be independent of scheme. They are quantities that are fully diffeomorphism invariant. If the gravitational theory had no boundary there would be no diffeomorphism invariant quantities, except for possible topological invariants. Since we have a boundary, diffeomorphism invariant quantities are defined at the boundary.

Note that scalar quantities are not invariant. To be invariant they must be scalar and constant. We therefore need to construct scalar functions that are invariant under changes of radial coordinates.

We can fix this reparametrization invariance by picking a very special frame. For example choosing the (string) metric in the conformal frame

$$ds^2 = e^{2A}\left[dr^2 + dx^\mu dx_\mu\right], \quad \lambda(r) \tag{4.16}$$

or in the domain-wall frame

$$ds^2 = du^2 + e^{2A}dx^\mu dx_\mu, \quad \lambda(u) \tag{4.17}$$

fixes the radial reparametrizations almost completely. In conformal frame, common scalings of r, x^μ are allowed, corresponding to constant shifts of $A(r)$.

Eventually we are led to calculate and compare our results to other ways of calculating (like the lattice). Some outputs are easier to compare (for example correlators). Others are much harder as they are not invariant (like the value of the coupling at a given energy scale).

In the UV such questions are well understood. The asymptotic energy scale is fixed by comparison to conformal field theory examples. This is possible because the space is asymptotically AdS_5.[9]

The coupling constant is also fixed to leading order from the coupling of the dilaton to D_3 branes (up to an overall multiplicative factor). Subleading (in perturbation theory) redefinitions of the coupling constant and the energy lead to changes in the β-function beyond two loops.

More in detail, as it has been described in [23–25], the general form of the kinetic term for the gauge fields on a D_3 brane is expected to be:

$$S_{F^2} = e^{-\Phi} Z(R,\xi) Tr[F^2], \quad \xi \equiv -e^{2\Phi}\frac{F_5^2}{5!} \tag{4.18}$$

where $Z(R,\xi)$ is an (unknown) function of curvature R and the five-form field strength, ξ. At weak background fields, $Z \simeq -\frac{1}{4} + \cdots$. In the UV regime, expanding near the boundary in powers of the coupling $\lambda \equiv N_c e^{\Phi}$ we obtain, [25]

$$S_{F^2} = N_c\, Tr[F^2]\frac{1}{\lambda}\left[Z(R_*,\xi_*) - \frac{Z_\xi(R_*,\xi_*)}{F_{\xi\xi}(R_*,\xi_*)\sqrt{\xi_*}}\frac{\lambda}{\ell} + \mathcal{O}(\lambda^2)\right] \tag{4.19}$$

where $F(R,\xi)$ is the bulk effective action and R_*, ξ_* are the boundary values for these parameters. Therefore the true 't Hooft coupling of QCD is

$$\lambda_{\text{t Hooft}} = -\frac{\lambda}{Z(R_*,\xi_*)}\left[1 + \frac{Z_\xi(R_*,\xi_*)}{Z(R_*,\xi_*)F_{\xi\xi}(R_*,\xi_*)\sqrt{\xi_*}}\frac{\lambda}{\ell} + \mathcal{O}(\lambda^2)\right]. \tag{4.20}$$

In the IR, more important changes can appear between our λ and other definitions as for example in lattice calculations.

In the region of strong coupling we know much less in order to be guided concerning the correct definition of the energy. We can obtain some hints however by comparing with lattice results.[10] In particular, based on lattice calculations using the Schröndiger functional approach [68], it is argued that at long distance L the 't Hooft coupling constant scales as

$$\lambda_{\text{lat}} \sim e^{mL}, \quad m \simeq \frac{3}{4} m_{0^{++}}. \tag{4.21}$$

---

[9] As the dilaton is now not constant there is a non-trivial question: in which frame is the metric AdS. In [25] it was argued that this should be the case in the string frame. The difference of course between the string and Einstein frame is subleading in the UV as the coupling constant vanishes logarithmically. But this may not be the case in the IR where we have very few criteria to check. In the model we are using we impose that the space is asymptotically AdS in the Einstein frame as this is the only choice consistent with the whole framework.

[10] We would like to thank K. Kajantie for asking the question, suggesting to compare with lattice data, and providing the appropriate references.

This was based on a specific definition of the coupling constant, and length scale on the lattice as well as on numerical data, and some general expectations on the fall-off of correlations in a massive theory. This suggests an IR β function of the form

$$L\frac{d\lambda}{dL} = \lambda \log\frac{\lambda}{\lambda_0}, \quad \lambda = \lambda_0\, e^{mL}. \tag{4.22}$$

On the other hand our β-function at strong coupling uses the UV definition of energy, $\log E = A_E$ (the scale factor in the Einstein frame), $E \sim 1/L$ and is

$$L\frac{d\lambda}{dL} = \frac{3}{2}\lambda\left[1 + \frac{3}{4}\frac{a-1}{a}\frac{1}{\log\lambda} + \cdots\right], \quad \lambda \simeq \left(\frac{L}{L_0}\right)^{\frac{3}{2}}. \tag{4.23}$$

where a is a parameter in the IR asymptotics of the potential. The case we consider as best fitting YM is $a = 2$ as then the asymptotic glueball trajectories are linear.

Consider now taking as length scale the string scale factor e^{A_s} in the IR.[11] Since it increases, it is consistent to consider it as a monotonic function of length. From its relation to the Einstein scale factor $A_s = A_E + \frac{2}{3}\log\lambda$ and (4.23) we obtain

$$\frac{d\lambda}{dA_s} = \frac{2a}{a-1}\lambda\log\lambda + \cdots \tag{4.24}$$

Therefore if we define as length scale in the IR

$$\log L = \frac{2a}{a-1}A_s \to L = \left(e^{A_s}\right)^{\frac{2a}{a-1}} \tag{4.25}$$

we obtain a running of the coupling compatible with the given lattice scheme. Note however that $L = (e^{A_s})^{\frac{2a}{a-1}}$ cannot be a global choice but should be only valid in the IR. The reason is that this function is not globally monotonic.

We conclude this section by restating that physical observables are independent of scheme. But observables like the 't Hooft coupling constant do depend on schemes, and it is obvious that our scheme is very different from lattice schemes in the IR.

4.4 The Potential and the Parameters of the Model

We will make the following ansatz for the potential,[12]

$$V(\lambda) = \frac{12}{\ell^2}\left\{1 + V_0\lambda + V_1\lambda^{4/3}\left[\log\left(1 + V_2\lambda^{4/3} + V_3\lambda^2\right)\right]^{1/2}\right\}, \tag{4.26}$$

[11] The string scale factor is not a monotonic function on the whole manifold [23, 24], and this is the reason that it was not taken as a global energy scale. In particular in the UV, e^{A_s} decreases until it reaches a minimum. The existence of the minimum is crucial for confinement. After this minimum e^{A_s} increases and diverges at the IR singularity.

[12] Further studies of IHQCD with different potentials can be found in [69].

which interpolates between the two asymptotic behaviors (4.4) for small λ and (4.6) for large λ, with $Q = 2/3$ and $P = 1/2$. Not all the parameters entering this potential have physical relevance. Below we will discuss the independent parameters of the model, and their physical meaning.

4.4.1 The Normalization of the Coupling Constant λ

As discussed in the previous section, the relation between the bulk field $\lambda(r)$ and the physical QCD 't Hooft coupling $\lambda_t = g_{YM}^2 N_c$ is a priori unknown. In the UV, the identification of the $D3$-brane coupling to the dilaton implies that the relation is linear, and depends on an a priori unknown coefficient κ, defined as:

$$\lambda = \kappa\lambda_t. \tag{4.27}$$

The coefficient κ can in principle be identified by relating the perturbative UV expansion of the Yang–Mills β-function, to the holographic β-function for the bulk field λ:

$$\beta(\lambda_t) = -\beta_0\lambda_t^2 - \beta_1\lambda_t^3 + \cdots, \quad \beta_0 = \frac{22}{3(4\pi)^2}, \quad \beta_1 = \frac{51}{121}\beta_0^2, \ldots \tag{4.28}$$

$$\beta(\lambda) = -b_0\lambda^2 - b_1\lambda^3 + \cdots, \quad b_0 = \frac{9}{8}v_0, \quad b_1 = \frac{9}{4}v_1 - \frac{207}{256}v_0^2 \cdots \tag{4.29}$$

The two expressions (4.28) and (4.29) are consistent with a linear relation as in (4.27), and expanding the identity $\kappa\beta_t(\lambda_t) = \beta(\kappa\lambda_t)$ to lowest order leads to:

$$\kappa = \beta_0/b_0. \tag{4.30}$$

Therefore, to relate the bulk field λ to the true coupling λ_t one looks at the linear term in the expansion of the potential. More generally, the other β function coefficients are related by $\beta_n = \kappa^{n+1}b_n$, and the combinations $b_n/b_0^{n+1} = \beta_n/\beta_0^{n+1}$ are κ-independent (however they are scheme-dependent for $n \geq 2$).

As discussed in Sect. 4.3, the introduction of the coefficient κ amounts to a field redefinition and therefore its precise value does not affect physical (scheme-independent) quantities. In this sense, κ is not a parameter that can be fixed by matching some observable computed in the theory. Assuming the validity of the relation (4.27), we could eventually fix κ by matching a RG-invariant (but scheme-dependent) quantity, e.g. λ at a given energy scale.

However, as we discuss later in this section, rescaling λ in the potential (thus changing κ) affects other parameters in the models, that are defined in the string frame, e.g. the fundamental string length ℓ_s: if we hold the physical QCD string tension fixed, the ratio (ℓ_s/ℓ) scales with degree $-2/3$ under a rescaling of κ.

An important point to keep in mind, is that the simple linear relation (4.27) may be modified at strong coupling, but again this does not have any effect on physical observables. *As long as we compute RG-invariant and scheme-independent quantities, knowledge of the exact relationship* $\lambda = F(\lambda_t)$ *is unnecessary.*

4.4.2 The AdS Scale ℓ

This is set by the overall normalization of the potential, and its choice is equivalent to fixing the unit of energy. It does not enter dimensionless physical quantities. As usual the AdS length at large N_c is much larger than the Planck length ($\ell_p \sim 1/(M_p N_c^{2/3})$, independently of the 't Hooft coupling.

4.4.3 The UV Expansion Coefficients of $V(\lambda)$

They can be fixed order by order by matching the Yang–Mills β-function. We impose this matching up to two-loops in the perturbative expansion, i.e. $O(\lambda^3)$ in (λ). One could go to higher orders by adding additional powers of λ inside the logarithm, but since our purpose is not to give an accurate description of the theory in the UV, we choose not to introduce extra parameters.[13]

Identifying the energy scale with the Einstein frame scale factor, $\log E \equiv \log b(r)$, we have the relation (4.29) between the β-function coefficients and the expansion parameters of $V(\lambda)$, with

$$v_0 = V_0, \quad v_1 = V_1\sqrt{V_2}. \tag{4.31}$$

The term proportional to V_2 in (4.26) is needed to reproduce the correct value of the quantity $b_1/b_0^2 = \beta_1/\beta_0^2 = 51/121$, which is invariant under rescaling of λ. Thus, V_2 is not a free parameter, but is fixed in terms of V_0 and V_1 by:

$$V_2 = b_0^4\left(\frac{23 + 36\, b_1/b_0^2}{81 V_1}\right)^2, \quad b_0 = \frac{9}{8}V_0, \quad \frac{b_1}{b_0^2} = \frac{51}{121}. \tag{4.32}$$

As explained earlier in this section, when discussing the normalization of the coupling, fixing the coefficient V_0 is the same as fixing the normalization κ through (4.30). As we argued, the actual value of κ should not have any physical consequences, so it is tempting to set $V_0 = 1$ by a field redefinition, $\lambda \to \lambda/V_0$ and eliminate this parameter altogether.

In fact, most of the quantities we will compute are not sensitive to the value of V_0, but for certain quantities, such as the string tension, some extra care is needed. In general, we can ask whether two models of the same form (4.3), but with

[13] Moreover, higher order β-function coefficients are known to be scheme-dependent.

different potentials $V(\lambda)$ and $\tilde{V}(\lambda)$, such that $\tilde{V}(\lambda) = V(\alpha\lambda)$ for some constant α, lead to different physical predictions. As we can change from one model to the other simply by a field redefinition $\lambda \to \alpha\lambda$ (this has no effect on the other terms in the action in the Einstein frame, (4.3), clearly the two potentials lead to the same result for any physical quantity that can be computed unambiguously from the Einstein frame action, e.g. dimensionless ratios between glueball masses, critical temperature, latent heat etc.

However a rescaling of λ does affect the string frame metric, since the latter explicitly contains factors of λ: $b_s(r) = b(r)\lambda^{2/3}$ [23, 24] thus, under the rescaling $\lambda \to \alpha\lambda$, $b_s(r) \to \alpha^{2/3} b_s(r)$. This means that any dimensionless ratio of two quantities, such that one of them remains fixed in the string frame and the other in the Einstein frame, will depend on α. An example of this is the ratio ℓ_s/ℓ, where ℓ_s is the string length, that we will discuss shortly.

Therefore, we can safely perform a field redefinition and set V_0 to a given value, as long as we are careful when computing quantities that depend explicitly on the fundamental string length.

Bearing this caveat in mind, we will choose a normalization such that $b_0 = \beta_0$, i.e.

$$V_0 = \frac{8}{9}\beta_0, \tag{4.33}$$

so that the normalization of λ in the UV matches the physical Yang–Mills coupling. With this choice, out of the four free parameters V_i appearing in (4.26) only V_1 and V_3 play a non-trivial role (V_2 being fixed by (4.32)).

4.4.4 The 5D Planck Scale M_p

M_p appears in the overall normalization of the 5D action (4.3). Therefore it enters the overall scale of quantities derived by evaluating the on-shell action, e.g. the free energy and the black-hole mass. It also sets the conversion factor between the entropy and horizon area. M_p cannot be fixed directly as we lack a detailed underlining string theory for YM. To obtain quantitative predictions, M_p must be fixed in terms of the other dimension-full quantity of the model, namely the AdS scale ℓ. As shown in [29] this can be done by imposing that the high-temperature limit of the black-hole free energy be that of a *free* gluon gas with the correct number of degrees of freedom.[14] This requires:

$$(M_p\ell)^3 = \frac{1}{45\pi^2}. \tag{4.34}$$

[14] Note that this is conceptually different from the $\mathcal{N} = 4$ case. There, near the boundary, the theory is strongly coupled and this number must be calculated in string theory. It is different by a factor of 3/4 from the free sYM answer. Here near the boundary the theory is free. Therefore the number of degrees of freedom can be directly inferred.

4.4.5 The String Length

In the non-critical approach the relation between the string length ℓ_s and the 5D Planck length (or the AdS length ℓ) is not known from first principles. The string length does not appear explicitly in the two-derivative action (4.3), but it enters quantities like the static quark–antiquark potential. The ratio ℓ_s/ℓ can be fixed phenomenologically to match the lattice results for the confining string tension.

More in detail, the relation between the fundamental and the confining string tensions T_f and σ is given by:

$$\sigma = T_f\, b^2(r_*)\lambda^{4/3}(r_*), \tag{4.35}$$

where r_* is the point where the string frame scale factor, $b_s(r) \equiv b(r)\lambda^{2/3}(r)$, has its minimum. Fixing the confining string tension by comparison with the lattice result we can find T_f (more precisely, the dimensionless quantity $T_f\ell^2$, since the overall scale of the metric depends on ℓ). The string length is in turn given by $\ell_s/\ell = 1/\sqrt{2\pi T_f \ell^2}$.

As is clear from (4.35), rescaling $\lambda \to \alpha\lambda$, keeping the value of the QCD string tension σ and of the AdS scale ℓ fixed, affects the fundamental string length in AdS units as $\ell_s/\ell \to \alpha^{-2/3}(\ell_s/\ell)$. Therefore two models a and b, *defined in the Einstein frame* by (4.3), but with potentials related by $V_b(\lambda) = V_a(\alpha\lambda)$, must have different fundamental string tensions in order to reproduce the same result for the QCD string tension. The quantity ℓ_s/ℓ therefore depends on the value of V_0.

4.4.6 Integration Constants

Besides the parameters appearing directly in the gravitational action, there are also other physically relevant quantities that label different solutions to the 5th order system of field equations. Any solution is characterized by a scale Λ, the temperature T and a value for the gluon condensate $\mathcal{G}$, that correspond to three of the five independent integration constants.[15]

Regularity at the horizon fixes $\mathcal{G}$ as a function of T, so that effectively the gluon condensate is a temperature-dependent quantity.

The quantity Λ controls the asymptotic form of the solution, as it enters the dilaton running in the UV: $\lambda \simeq -(b_0 \log r\Lambda)^{-1}$. It can be defined in a reparametrization invariant way as:

[15] The remaining two are the value $f(0)$ which should be set to one for the solution (4.9) to obey the right UV asymptotics, and an unphysical degree of freedom in the reparametrization of the radial coordinate.

$$\Lambda = \ell^{-1} \lim_{\lambda \to 0} \left\{ b(\lambda) \frac{\exp\left[-\frac{1}{b_0 \lambda}\right]}{\lambda^{b_1/b_0^2}} \right\}, \tag{4.36}$$

and it is fixed once we specify the value of the scale factor $b(\lambda)$ at a given λ_0.

Every choice of Λ corresponds to an inequivalent class of solutions, that differ by UV boundary conditions. Each class is thermodynamically isolated, since solutions with different Λ's have infinite action difference. Thus, in the canonical partition sum we need to consider only solutions with a fixed value of Λ. However, this choice is merely a choice of scale, as solutions with different Λ's will give the same predictions for any dimensionless quantity. In short, Λ is the holographic dual to the QCD strong coupling scale: it is defined by the initial condition to the holographic RG equations, and does not affect dimensionless quantities such as mass ratios, etc. Therefore, as long as all solutions we consider obey the same UV asymptotics, the actual value of Λ is immaterial, since the physical units of the system can always be set by fixing ℓ.

To summarize, the only *nontrivial phenomenological* parameters we have at our disposal are V_1 and V_3 appearing in (4.26). The other quantities that enter our model are either fixed by the arguments presented in this section, or they only affect trivially (e.g. by overall rescaling that can be absorbed in the definition of the fundamental string scale) the physical quantities.

In the next section we present a numerical analysis of the solutions and thermodynamics of the model defined by (4.26), and show that for an appropriate choice of the parameters it reproduces the lattice results for the Yang–Mills deconfinement transition and high-temperature phase as well as the zero temperature glueball data.

4.5 Matching the Thermodynamics of Large-N_c YM

Assuming a potential of the form (4.26), we look for values of the parameters such that the thermodynamics of the 5D model match the lattice results for the thermodynamics of 4D YM. As explained in Sect. 4.3, we set V_0 and V_2 as in (4.33 and 4.32), respectively, with $b_0 = \beta_0 = 22/3(4\pi)^{-2}$.

We then vary V_1 and V_3 only. We fix these parameters by looking at thermodynamic quantities corresponding to the latent heat per unit volume, and the pressure at one value of the temperature above the transition, which we take as $2T_c$.

It is worth remarking that V_1 and V_3 are phenomenological parameters that we use to fit *dimensionless* QCD quantities. The single (dimension-full) parameter of pure Yang–Mills, the strong coupling scale, is an extra input that fixes the overall energy scale of our solution.

Using the numerical method explained in [32], for each set of parameters (V_1, V_3) we numerically generate black-hole solutions for a range of values of λ_h, then from the metric at the horizon and its derivative we extract the temperature and entropy functions $T(\lambda_h)$ and $S(\lambda_h)$, and the function $\mathcal{F}(\lambda_h)$ from the integrated form of the first law,

$$\mathcal{F}(\lambda_h) = \int_{\lambda_h}^{+\infty} \mathrm{d}\bar{\lambda}_h\, S(\bar{\lambda}_h) \frac{\mathrm{d}T(\bar{\lambda}_h)}{\mathrm{d}\bar{\lambda}_h}. \tag{4.37}$$

Here $S(\lambda_h)$ is given by (4.13) and both the large black hole and small black-hole branches are needed in order the get the full result for the free energy. This is because the integral in (4.37) extends to $+\infty$, entering deeply in the small black-hole branch.

The behavior of the thermodynamic functions is shown in Figs. 4.1, 4.2 and 4.3, for the best fit parameter values that we discuss below. One can see the existence of a minimal temperature $T_{\min} = T(\lambda_{\min})$, and a critical value λ_c where $\mathcal{F}$ changes sign. The resulting function $\mathcal{F}(T)$ is shown in Fig. 4.1.

The phase transition is first order, and the latent heat per unit volume L_h, normalized by $N_c^2 T_c^4$, is given by the derivative of the curve in Fig. 4.4 at $T/T_c = 1$.

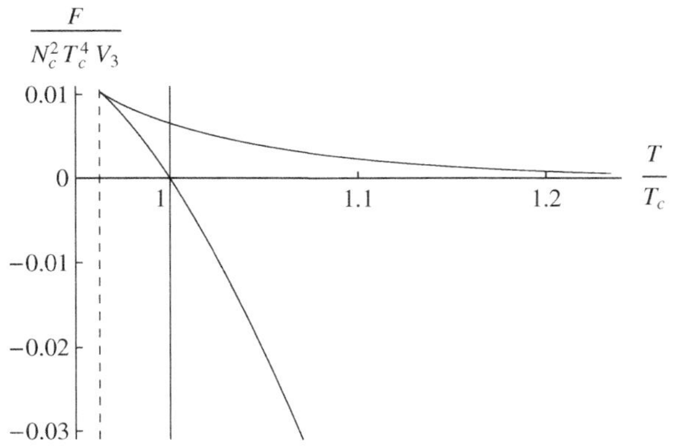

Fig. 4.1 The free energy density (in units of T_c) as a function of T/T_c, for $V_1 = 14$ and $V_3 = 170$. The *vertical lines* correspond to the critical temperature (*solid*) and the minimum black-hole temperature (*dashed*)

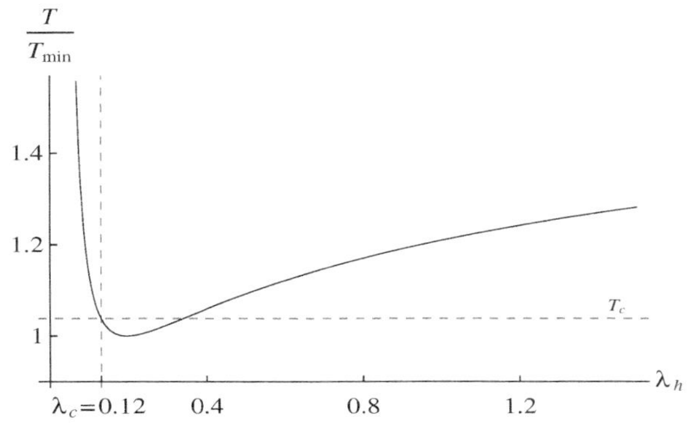

Fig. 4.2 Temperature in units of $T_{\min}$, as a function of λ_h, for $V_1 = 14$ and $V_3 = 170$. The *dashed horizontal* and *vertical lines* indicate the critical temperature and the critical value of the dilaton field at the horizon

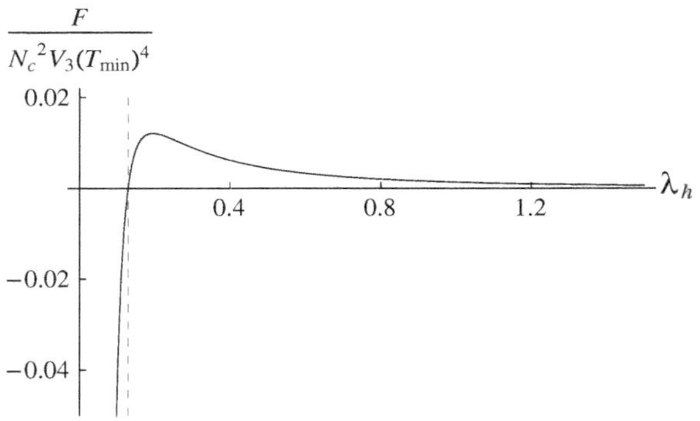

Fig. 4.3 The free energy density in units of T_{min}, as a function of λ_h

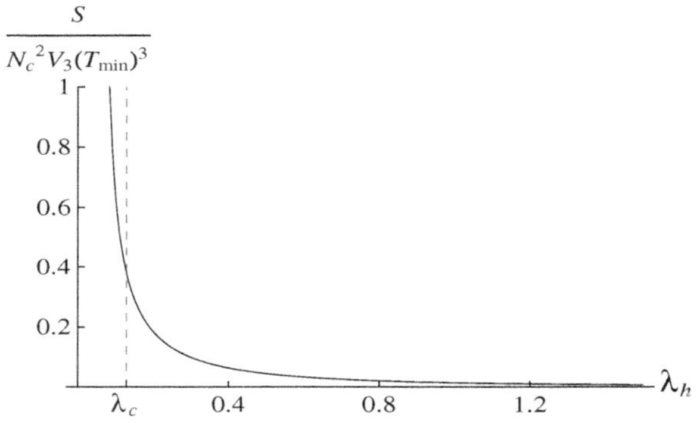

Fig. 4.4 Entropy density in units of T_{min}, as a function of λ_h

Equivalently, L_h is proportional to the jump in the entropy density $s = S/V_3$ at the phase transition from the thermal gas (whose entropy is of $O(1)$, in the limit $N_c \to \infty$) to the black hole (whose entropy scales as N_c^2 in the same limit): thus, in the large N_c limit,

$$L_h \equiv T\Delta s \simeq T_c s(\lambda_c) \tag{4.38}$$

up to terms of $O(1/N_c^2)$.

To fix V_1 and V_3 we compare our results to the data of G. Boyd et al. [70]. The relevant quantities to compare are the dimensionless ratios $p(T)/T^4$, $e(T)/T^4$ and $s(T)/T^3$, where $p = \mathcal{F}/V_3$ is the pressure, and $e = p + Ts$ is the energy density. Lattice results for these functions are available in the range $T = T_c \sim 5T_c$, and can be seen in Fig. 7 of [70]. The analysis of [70] correspond to $N_c = 3$, but one expects that the thermodynamic functions do not change to much for large N_c.[16]

---

[16] See e.g. [71–73], in which results for $N_c = 8$ do not different significantly from those for $N_c = 3$ as well as the recent high-precision data by Panero [74, 75].

An additional quantity of relevance is the value for the "dimensionless" latent heat per unit volume, L_h/T_c^4 which for large N_c was found in [76] to be $(L_h/T_c^4)_{lat} = 0.31N_c^2$. The result for $N_c^2 = 3$ is slightly lower ($\simeq 0.28N_c^2$).

As already noted in [28, 29], the qualitative features of the thermodynamic functions are generically reproduced in our setup: the curves $3p(T)/T^4$, $e(T)/T^4$ and $3s(T)/4T^3$ increase starting at T_c, then (very slowly) approach the constant free field value $\pi^2 N_c^2/15$ (given by the Stefan–Boltzmann law) as T increases. By computing the thermodynamic functions for various sets of values of V_1 and V_3 we obtain that:

(1) V_1 roughly controls the height reached by the curves $p(T)/T^4$, $e(T)/T^4$ and $s(T)/T^3$ at large T/T_c ($\sim$ a few): for larger V_1 the curves approach the free field limit faster;
(2) V_3 does not affect much the height of the curves at large T/T_c, but on the other hand it changes the latent heat, which is increasing as V_3 decreases.

The best fit corresponds to the values

$$V_1 = 14 \quad V_3 = 170. \tag{4.39}$$

Below we discuss the values of various physical quantities (both related to thermodynamics, and to zero-temperature properties) obtained with this choice of parameters.

4.5.1 Latent Heat and Equation of State

The comparison between the curves $p(T)/T^4$, $e(T)/T^4$ and $s(T)/T^3$ obtained in our models with (4.39), and the lattice results [70] is shown in Fig. 4.5. The match

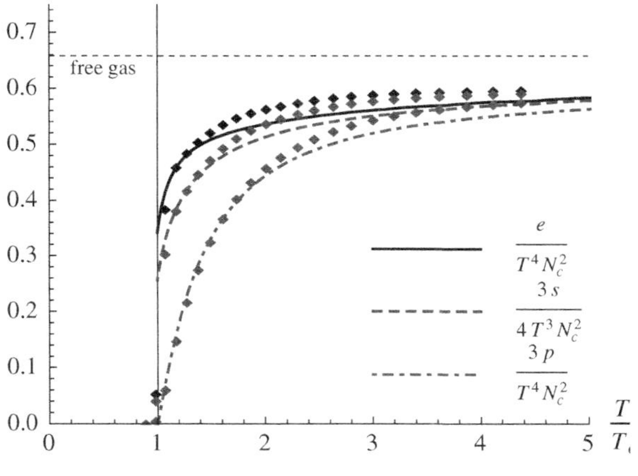

Fig. 4.5 Temperature dependence of the dimensionless thermodynamic densities, normalized such that they reach the common limiting value $\pi^2/15$ (*dashed horizontal line*) as $T \to \infty$. The *dots* correspond to the lattice data for $N_c = 3$ [70]

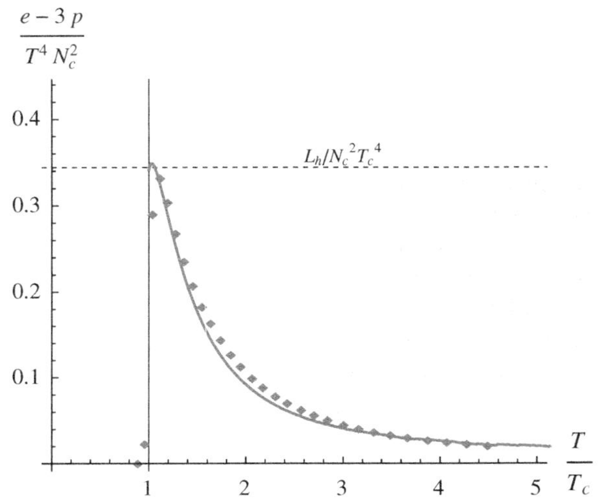

Fig. 4.6 The trace anomaly as a function of temperature in the deconfined phase of the holographic model (*solid line*) and the corresponding lattice data [70] for $N_c = 3$ (*dots*). The peak in the lattice data slightly above T_c is expected to be an artifact of the finite lattice volume. In the infinite volume limit the maximum value of the curve is at T_c, and it equals $L_h/N_c^2T_c^4$

is remarkably good for $T_c < T < 2T_c$, and deviates slightly from the lattice data in the range up to $5T_c$.

The latent heat we obtain is:

$$L_h/T_c^4 = 0.31N_c^2, \tag{4.40}$$

which matches the lattice result for $N_c \to \infty$ [76].

An interesting quantity is the *trace anomaly* $(e - 3p)/T^4$, (also known as *interaction measure*), that indicates the deviation from conformality, and it is proportional to the thermal gluon condensate. The trace anomaly in our setup is shown, together with the corresponding lattice data, in Fig. 4.6, and the agreement is again very good. Our results agree even better with recent high-precision lattice calculations of the thermodynamics functions done by Panero at different values of N_c up to $N_c = 8$, [74, 75]. In Fig. 4.7 a comparison (taken from [74, 75]) of the normalized interaction measure with lattice results for different N_c is shown.

We also compute the specific heat per unit volume c_v, and the speed of sound c_s in the deconfined phase, by the relations

$$c_v = -T\frac{\partial^2 F}{\partial T^2}, \quad c_s^2 = \frac{s}{c_v}. \tag{4.41}$$

These are shown in Figs. 4.8 and 4.9 respectively. The speed of sound is shown together with the lattice data, and the agreement is remarkable.

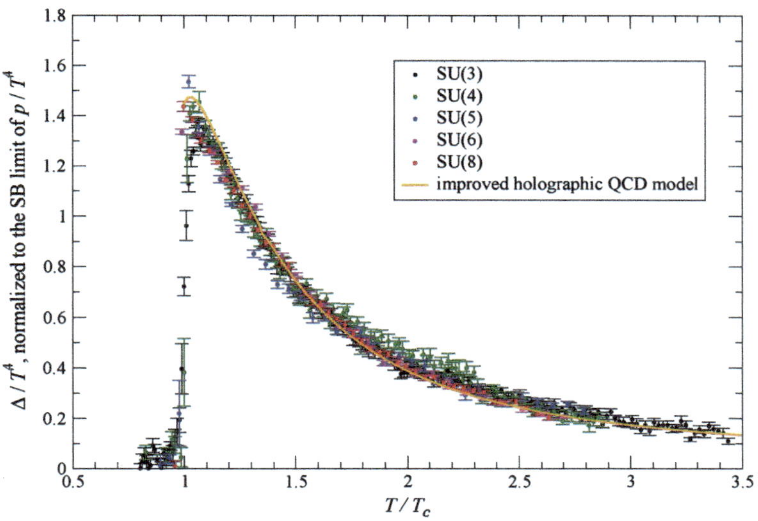

Fig. 4.7 The rescaled trace anomaly (so that it is N_c-independent) as a function of temperature in the deconfined phase of the holographic model (*solid line*) and the corresponding recent high precision lattice data taken from [74, 75] for different N_c. The errors shown are statistical only

Fig. 4.8 The specific heat (divided by T^3), as a function of temperature, in the deconfined phase of the holographic model

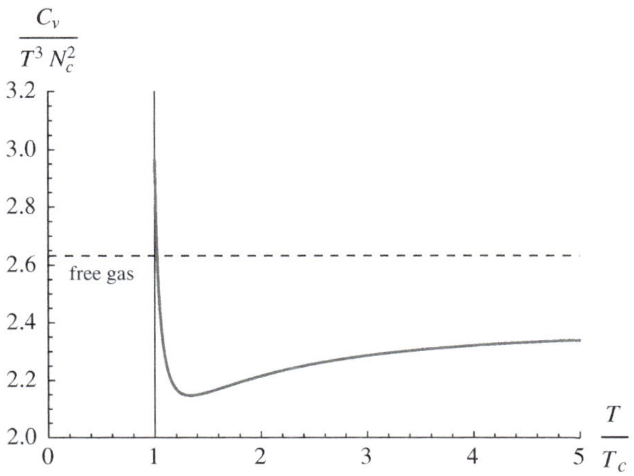

4.5.2 *Glueball Spectrum*

In [23, 24], the single phenomenological parameters of the potential was fixed by looking at the zero-temperature spectrum, i.e. by computing various glueball mass ratios and comparing them to the corresponding lattice results. The masses are computed by deriving the effective action for the quadratic fluctuations around the background [77] and subsequently reducing the dynamics to four dimensions.

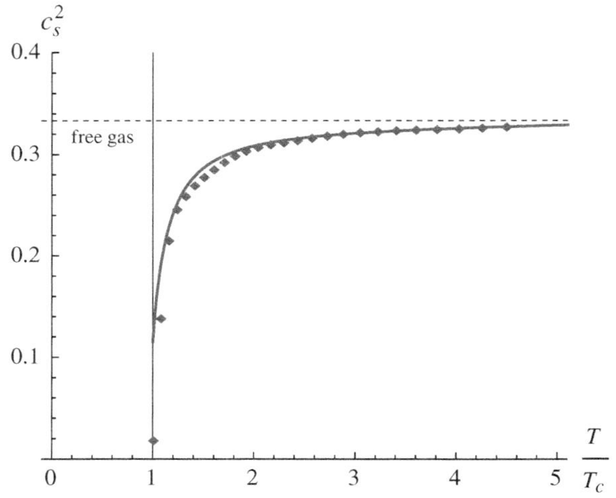

Fig. 4.9 The speed of sound in the deconfined phase, as a function of temperature, for the holographic model (*solid line*) and the corresponding lattice data [70] for $N_c = 3$ (*dots*). The *dashed horizontal line* indicates the conformal limit $c_s^2 = 1/3$

The associated thermodynamics for this potential was studied in [28] which was in qualitative agreement with lattice QCD results, but not in full quantitative agreement. This is due to the fact that the thermodynamics depends more on the details of the potential than the glueball spectrum for the main Regge trajectories. Here we use the potential (4.26), but with the two phenomenological parameters V_1 and V_3 already determined by the thermodynamics (4.39).

The glueball spectrum is obtained holographically as the spectrum of normalizable fluctuations around the zero-temperature background. As explained in the introduction, and motivated in [23–25], here we consider explicitly the 5D metric, one scalar field (the dilaton), and one pseudoscalar field (the axion). As a consequence, the only normalizable fluctuations above the vacuum correspond to spin 0 and spin 2 glueballs[17] (more precisely, states with $J^{PC} = 0^{++}, 0^{-+}, 2^{++}$), each species containing an infinite discrete tower of excited states.

In 4D YM there are many more operators generating glueballs, corresponding to different values of J^{PC}, that are not considered here. These are expected to correspond holographically to other fields in the noncritical string spectrum (e.g. form fields, which may yield spin 1 and CP-odd spin 2 states) and to higher string states that provide higher-spin glueballs. As the main focus is in reproducing the YM thermodynamics in detail rather than the entire glueball spectrum, we choose not to include these states.[18] Therefore we only compare the mass spectrum obtained in our model to the lattice results for the lowest $0^{++}, 0^{-+}, 2^{++}$ glueballs

---

[17] Spin 1 excitations of the metric can be shown to be non-normalizable.

[18] A further reason is that, unlike the scalar and (to some extent) the pseudoscalar sector that we are considering, the action governing the higher Regge slopes is less and less universal as one goes to higher masses. Only a precise knowledge of the underline string theory is expected to provide detailed information for such states.

Table 4.1 Glueball masses

	HQCD	$N_c = 3$ [78, 79]	$N_c = \infty$ [80]
$m_{0^{*++}}/m_{0^{++}}$	1.61	1.56(11)	1.90(17)
$m_{2^{++}}/m_{0^{++}}$	1.36	1.40(4)	1.46(11)

and their available excited states. These are limited to one for each spin 0 species, and none for the spin 2, in the study of [78, 79], which is the one we use for our comparison. This provides two mass ratios in the CP-even sector and two in the CP-odd sector.

The glueball masses are computed by first solving numerically the zero-temperature Einstein's equations, by setting $f(r) = 1$, and using the resulting metric and dilaton to setup an analogous Schrödinger problem for the fluctuations [23, 24]. The results for the parity-conserving sector are shown in Table 4.1, and are in good agreement with those reported by [78, 79] for $N_c = 3$, whereas the results reported by [80] for large N_c are somewhat larger. The CP-violating sector (axial glueballs) will be discussed separately.

We should add that there are other lattice studies (see e.g. [81]) that report additional excited states. Our mass ratios offer a somewhat worse fit of the mass ratios found in [81] (whose results are not entirely compatible with those of [78, 79] for the states the two studies have in common). We should stress however that reproducing the detailed glueball spectrum is secondary here since the main focus is thermodynamics. However, the comparison of our spectrum to the existing lattice results shows that our model provides a good global fit to 4D YM also with respect to quantities beyond thermodynamics.

Unlike the various mass ratios, the value of any given mass in AdS-length units (e.g. $m_{0++}\ell$) *does depend* on the choice of integration constants in the UV, i.e. on the value of $b_{\rm UV}$ and $\lambda_{\rm UV}$. Therefore its numerical value does not have an intrinsic meaning. However it can be used as a benchmark against which all other dimension-full quantities can be measured (provided one always uses the same UV b.c.). On the other hand, given a fixed set of initial conditions, asking that m_{0++} matches the physical value (in MeV) obtained on the lattice, fixes the value of ℓ hence the energy unit.

4.5.3 Critical Temperature

The thermodynamic quantities we have discussed so far, are dimensionless ratios, in units of the critical temperature. To compute T_c, we need an extra dimension-full quantity which can be used independently to set the unit of energy. In lattice studies this is typically the confining string tension σ in the $T = 0$ vacuum, with a value of around $(440\,\text{MeV})^2$, and results are given in terms of the dimensionless ratio $T_c/\sqrt{\sigma}$. In our case we cannot compute σ directly, since it depends on the *fundamental* string tension, which is a priori unknown. Instead, we take the mass m_{0++} of the lowest-lying glueball state as a reference.

We compute m_{0++} with the potential (4.26), with V_1 and V_3 fixed as in (4.39), then compare T_c/m_{0++} to the same quantity obtained on the lattice. For the lattice result, we take the large N_c result of [76], $T_c/\sqrt{\sigma} = 0.5970(38)$, and combine it with the large N_c result for the lowest-lying glueball mass [80], $m_{0++}/\sqrt{\sigma} = 3.37(15)$. The two results are in fair agreement, without need to adjust any extra parameter:

$$\left(\frac{T_c}{m_0}\right)_{\text{hQCD}} = 0.167, \quad \left(\frac{T_c}{m_0}\right)_{\text{lattice}} = 0.177(7). \tag{4.42}$$

In physical units, the critical temperature we obtain is given by

$$T_c = 0.56\,\sqrt{\sigma} = 247\,\text{MeV}. \tag{4.43}$$

4.5.4 String Tension

The fundamental string tension $T_f = \frac{1}{2\pi\ell_s^2}$ cannot be computed from first principles in our model, but can be obtained using as extra input the lattice value of the confining string tension σ, at $T = 0$. The fundamental and confining string tensions are related by (4.35).

As for the critical temperature, we can relate T_f to the value of the lowest-lying glueball mass, by using the lattice relation $\sqrt{\sigma} = \frac{m_{0++}}{3.37}$ [80]. Since what we actually compute numerically is $m_{0++}\ell$, this allows us to obtain the string tension T_f (and fundamental string length $\ell_s = 1/\sqrt{2\pi T_f}$ in AdS units:

$$T_f\ell^2 = 0.19, \quad \ell_s/\ell = 0.15. \tag{4.44}$$

This shows that the fundamental string length in our model is about an order of magnitude smaller than the AdS length. The meaning of this fact is a little more complicated conceptually, as the discussion in [25] indicates. Also, we should stress that, as discussed in Sect. 4.4, this result depends on our choice of the overall normalization of λ: changing the potential by $\lambda \to \kappa\lambda$ will yield different numerical values in (4.44) without affecting the other physical quantities.

Another related observable is the spatial string tension. It is calculated from the expectation value of the rectangular Wilson loop which stretches in spatial dimensions only. This has been calculated on the lattice [82], as well as using the high-temperature (resumed) perturbative expansion plus a zero-temperature calculation of the string tension in three-dimensional YM theory [83]. The two calculation agree reasonably well.

The spatial string tension at finite temperature can be calculated in IHQCD [84] by calculating the relevant Wilson loop. Very good agreement was found with the lattice calculations, especially at temperatures not far from the phase transition.

Finally, several calculations of quark–antiquark potentials exist. At zero temperature the long distance asymptotics of the quark potential was calculated in [23, 24] and used to classify the dilaton potentials as a function of the confinement property. The full quark potential including the short distance behavior was computed in [85]. There a comparison to the Cornell potential was done as well as with quarkonium spectra finding excellent agreement with data. The issue of quarkonium potentials from IHQCD-like theories was also recently discussed in [86].

Finally the Polyakov loop was recently computed [87, 88] in similar Einstein dilaton models that were studied first in [26].

4.5.5 CP-Odd Sector

The CP-odd sector of pure Yang–Mills is described holographically by the addition of a bulk pseudoscalar field $a(r)$ (the *axion*) with action[19]:

$$S_{\text{axion}} = \frac{M_p^3}{2} \int d^5 x Z(\lambda) \sqrt{-g} (\partial^\mu a)(\partial_\mu a). \tag{4.45}$$

The field $a(r)$ is dual to the topological density operator $\mathrm{Tr}F\tilde{F}$. The prefactor $Z(\lambda)$ is a dilaton-dependent normalization. The axion action is suppressed by a factor $1/N_c^2$ with respect to the action (4.3) for the dilaton and the metric, meaning that in the large-N_c limit one can neglect the back-reaction of the axion on the background.

As shown in [23, 24], requiring the correct scaling of $a(r)$ in the UV, and phenomenologically consistent axial glueball masses, constrain the asymptotics of $Z(\lambda)$ as follows:

$$Z(\lambda) \sim Z_0,\ \lambda \to 0; \quad Z(\lambda) \sim \lambda^4,\ \lambda \to \infty, \tag{4.46}$$

where Z_0 is a constant. As a simple interpolating function between these large- and small- λ asymptotics we can take the following:

$$Z(\lambda) = Z_0(1 + c_a \lambda^4). \tag{4.47}$$

The parameter Z_0 can be fixed by matching the topological susceptibility of pure Yang–Mills theory, whereas c_a can be fixed by looking at the axial glueball mass spectrum.

---

[19] This action was justified in [23–25]. The dilaton dependent coefficient $Z(\lambda)$ is encoding both the dilaton dependence as well as the UV curvature dependence of the axion kinetic terms in the associated string theory. We cannot determine it directly from the string theory, but we pin it down by a combination of first principles and lattice input, as we explain further below.

4.5.5.1 Axial Glueballs

As in [23, 24], we can fix c_a by matching to the lattice results the mass ratio m_{0-+}/m_{0++} between the lowest-lying axial and scalar glueball states. This is independent of the overall coefficient Z_0 in (4.47). The lattice value $m_{0-+}/m_{0++} = 1.49$ [78, 79] is obtained for:

$$c_a = 0.26. \tag{4.48}$$

With this choice, the mass of the first excited axial glueball state is in good agreement with the corresponding lattice result [78, 79]:

$$\left(\frac{m_{0-+*}}{m_{0++}}\right)_{\text{hQCD}} = 2.10 \quad \left(\frac{m_{0-+*}}{m_{0++}}\right)_{\text{lattice}} = 2.12(10). \tag{4.49}$$

4.5.5.2 Topological Susceptibility

In pure Yang–Mills, the topological χ susceptibility is defined by:

$$E(\theta) = \frac{1}{2}\chi\theta^2, \tag{4.50}$$

where $E(\theta)$ is the vacuum energy density in presence of a θ-parameter. $E(\theta)$ can be computed holographically by solving for the axion profile $a(r)$ on a given background, and evaluating the action (4.45) on-shell.

In the deconfined phase, the axion profile is trivial, implying a vanishing topological susceptibility [29]. This is in agreement with large-N_c arguments and lattice results [31].

In the low-temperature phase, the axion acquires a non-trivial profile,

$$a(r) = a_{\text{UV}}\,\frac{F(r)}{F(0)}, \quad F(r) \equiv \int_r^\infty \frac{dr}{Z(\lambda(r))e^{3A(r)}}. \tag{4.51}$$

This profile is shown, for the case at hand, in Fig. 4.10, where the axion is normalized to its UV value.

The topological susceptibility is given by [23, 24]:

$$\chi = M_p^3 F(0)^{-1} = M_p^3 \left[\int_0^\infty \frac{dr}{e^{3A(r)}Z(r)}\right]^{-1}, \tag{4.52}$$

where $Z(r) \equiv Z(\lambda(r))$. Evaluating this expression numerically with $Z(\lambda)$ as in (4.47), and $c_a = 0.26$ (to match the axial glueball spectrum), we can determine the coefficient Z_0 by looking at the lattice result for χ. For $N_c = 3$ [89] obtained $\chi = (191\,\text{MeV})^4$, which requires $Z_0 = 133$.

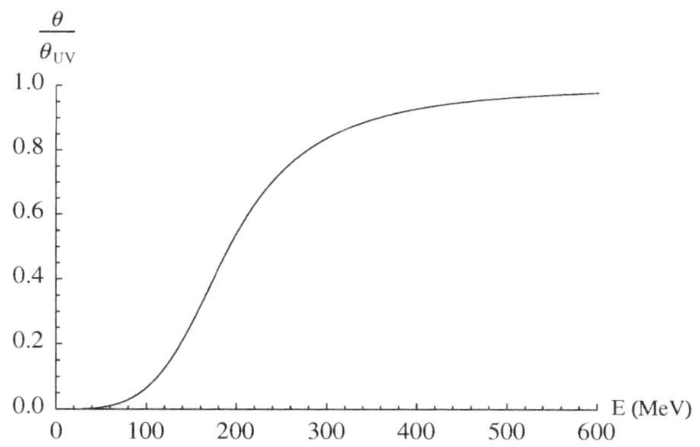

Fig. 4.10 Axion profile in the radial direction. The x-axis is taken to be the energy scale, $E(r) = E_0 b(r)$, where the unit E_0 is fixed to match the lowest glueball mass

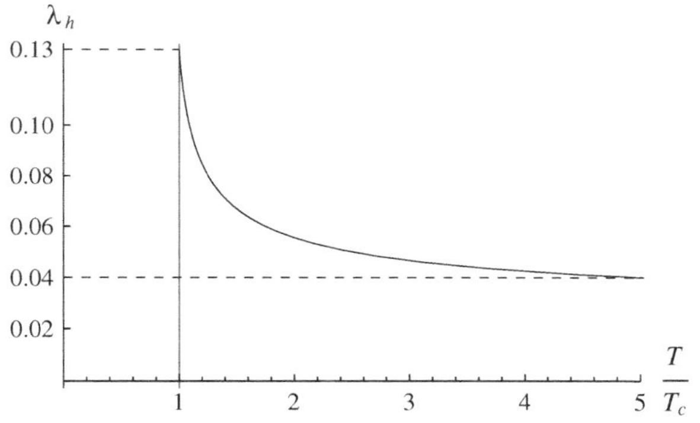

Fig. 4.11 The coupling at the horizon as a function of temperature in the range T_c–$5T_c$

In Table 4.4 we present a summary of the various physical quantities discussed in this section, as obtained in our holographic model, and their comparison with the lattice results for large N_c (when available) and for $N_c = 3$. The quantities shown in the upper half of the table are the ones that were used to fix the free parameters (reported in the last column) of the holographic model.

4.5.6 Coupling Normalization

Finally, we can relate the field $\lambda(r)$ to the running 't Hooft coupling. All other quantities we have discussed so far are scheme-independent and RG-invariant. This is not the case for the identification of the physical YM 't Hooft coupling, which is scheme dependent.

In the black-hole phase we can take $\lambda_h \equiv \lambda(r_h)$ as a measure of the temperature-dependent coupling. In Fig. 4.11 we show λ_h as a function of the temperature in the range T_c to $5T_c$.

As a reference, we may take the result of [70], that found $g^2(5T_c) \simeq 1.5$ for $N_c = 3$, which translates to $\lambda_t(5T_c) \simeq 5$. On the other hand, if we make the assumption that the identification $\lambda = \lambda_t$ is valid at all scales (not only in the UV), we find in our model $\lambda_t(5T_c) \simeq 0.04$ (see Fig. 4.11), i.e. a factor of 100 smaller than the lattice result.

This discrepancy is almost certainly due to the identification (4.27) being very different from lattice at strong coupling.

4.6 Bulk Viscosity

The bulk viscosity ζ is an important probe of the quark–gluon plasma. Its profile as a function of T reveals information regarding the dynamics of the phase transition. In particular, both from the low-energy theorems and lattice studies [12, 40, 41], there is evidence that ζ increases near T_c.

For a viscous fluid the shear η and bulk ζ viscosities are defined via the rate of entropy production as

$$\frac{\partial s}{\partial t} = \frac{\eta}{T}\left[\partial_i v_j + \partial_j v_i - \frac{2}{3}(\partial \cdot v)\delta_{ij}\right]^2 + \frac{\zeta}{T}(\partial \cdot v)^2. \tag{4.53}$$

Therefore, in a holographic setup, the bulk viscosity can be defined as the response of the diagonal spatial components of the stress–energy tensor to a small fluctuation of the metric. It can be directly related to the retarded Green's function of the stress–energy tensor by Kubo's linear response theory (Table 4.2):

Table 4.2 Collected in this table is the complete set of physical quantities that we computed in our model and compared with data

	HQCD	Lattice $N_c = 3$	Lattice $N_c \to \infty$	Parameter
$[p/(N_c^2 T^4)]_{T=2T_c}$	**1.2**	**1.2**	–	$V1 = 14$
$L_h/(N_c^2 T_c^4)$	**0.31**	0.28 [70]	**0.31** [76]	$V3 = 170$
$[p/(N_c^2 T^4)]_{T\to+\infty}$	$\pi^2/45$	$\pi^2/45$	$\pi^2/45$	$M_p\ell = [45\pi^2]^{-1/3}$
$m_{0^{++}}/\sqrt{\sigma}$	**3.37**	3.56 [78, 79]	**3.37** [80]	$\ell_s/\ell = 0.15$
$m_{0^{-+}}/m_{0^{++}}$	**1.49**	**1.49** [78, 79]	–	$c_a = 0.26$
χ	$(191\,\mathrm{MeV})^4$	$(191\,\mathrm{MeV})^4$ [89]	–	$Z_0 = 133$
$T_c/m_{0^{++}}$	0.167	–	0.177(7)	
$m_{0^{*++}}/m_{0^{++}}$	1.61	1.56(11)	1.90(17)	
$m_{2^{++}}/m_{0^{++}}$	1.36	1.40(4)	1.46(11)	
$m_{0^{*-+}}/m_{0^{++}}$	2.10	2.12(10)	–	

The upper half of the table contains the quantities that we used as input (shown in boldface) for the holographic QCD model (HQCD). Each quantity can be roughly associated to one parameter of the model (last column). The lower half of the table contains our "postdictions" (i.e. quantities that we computed after all the parameters were fixed) and the comparison with the corresponding lattice results. The value we find for the critical temperature corresponds to $T_c = 247\,\mathrm{MeV}$

$$\zeta = -\frac{1}{9}\lim_{\omega\to 0}\frac{1}{\omega} ImG_R(\omega,0), \tag{4.54}$$

where $G_R(w,\boldsymbol{p})$ is the Fourier transform of retarded Green's function of the stress–energy tensor:

$$G_R(w,\boldsymbol{p}) = -i\int d^3xdte^{i\omega t - i\boldsymbol{p}\cdot\boldsymbol{x}}\theta(t)\sum_{i,j=1}^{3}\langle[T_{ii}(t,\boldsymbol{x}),T_{jj}(0,0)]\rangle. \tag{4.55}$$

A direct computation of the RHS on the lattice is non-trivial as it requires analytic continuation to Lorentzian space-time. In Refs. [40, 41] the low energy theorems of QCD, as well as (equilibrium) lattice data at finite temperature were used in order to evaluate a particular moment of the spectral density of the relevant correlator. using a parametrization of the spectral density via two time-dependent constants, one of which is the bulk viscosity a relation for their product was obtained as a function of temperature. This can be converted to a relation for ζ, assuming the other constant varies slowly with temperature.

The conclusion was that ζ/s increases near T_c. Another conclusion is that the fermionic contributions to ζ are small compared to the glue contributions.

The weak point of the approach of [41], is that it requires an ansatz on the spectrum of energy fluctuations, and further assumptions on the other parameters. which are not derived from first principles.

A direct lattice study of the bulk viscosity was also made in [12]. Here, the result is also qualitatively similar 4.12. However, the systematic errors in this computation are large especially near T_c, mostly due to the analytic continuation that one has to perform after computing the Euclidean correlator on the lattice.

The results of references [40, 41] and the assumptions of the lattice calculation have been recently challenged in [90].

4.6.1 The Holographic Computation

The holographic approach offers a new way of computing the bulk viscosity. In the holographic set-up, ζ is obtained from (4.54). Using the standard AdS/CFT prescription, the two point-function of the energy-momentum tensor can be read off from the asymptotic behavior of the metric perturbations $\delta g_{\mu\nu}$. This is similar in spirit to the holographic computation of the shear viscosity [6], but it is technically more involved. A recent treatment of the fluctuation equation governing the scalar mode of a general Einstein-Dilaton system can be found in [91]. Here, we follow the method proposed by [92].

As explained in [92], one only needs to examine the equations of motion in the gauge $r = \Phi$, where the radial coordinate is equal to the dilaton. In our type of metrics, the applicability of this method requires some clarifications, that we provide in [93]. Using $SO(3)$ invariance and the five remaining gauge degrees of freedom the metric perturbations can be diagonalized as

$$\delta g = \mathrm{diag}(g_{00}, g_{11}, g_{11}, g_{11}, g_{55}), \tag{4.56}$$

where

$$g_{00} = -e^{2A} f[1 + h_{00}(\Phi)e^{-i\omega t}], \quad g_{11} = e^{2A}[1 + h_{11}(\Phi)e^{-i\omega t}], \tag{4.57}$$
$$g_{55} = \frac{e^{2B}}{f}[1 + h_{55}(\Phi)e^{-i\omega t}],$$

where the functions A and B emerge from the metric

$$ds^2 = e^{2A(\Phi)}(-f dt^2 + dx_m dx^m) + e^{2B(\Phi)} \frac{d\Phi^2}{f}. \tag{4.58}$$

Here, the fluctuations are taken to be harmonic functions of t while having an arbitrary dependence on Φ.

The bulk viscosity depends only on the correlator of the diagonal components of the metric and so it suffices to look for the asymptotics of h_{11}. Interestingly, in the $r = \Phi$ gauge this decouples from the other components of the metric and satisfies the following equation[20]

$$h_{11}'' - \left(-\frac{8}{9A'} - 4A' + 3B' - \frac{f'}{f}\right) h_{11}' - \left(-\frac{e^{2B-2A}}{f^2}\omega^2 + \frac{4f'}{9fA'} - \frac{f'B'}{f}\right) h_{11} = 0\,. \tag{4.59}$$

One needs to impose two boundary conditions. First, we require that only the infalling condition survives at the horizon:

$$h_{11} \to c_b(\Phi_h - \Phi)^{-\frac{i\omega}{4\pi T}}, \quad \Phi \to \Phi_h, \tag{4.60}$$

where c_b is a normalization factor. The second boundary condition is that h_{11} has unit normalization on the boundary:

$$h_{11} \to 1, \quad \Phi \to -\infty. \tag{4.61}$$

Having solved for $h_{11}(\Phi)$, Kubo's formula (4.54) and a wise use of the AdS/CFT prescription to compute the stress–energy correlation function [92] determines the ratio of bulk viscosity as follows.

The AdS/CFT prescription relates the imaginary part of the retarded T_{ii} Green's function to the number flux of the h_{11} gravitons $\mathcal{F}$ [92]:

$$Im\, G_R(\omega, 0) = -\frac{\mathcal{F}}{16\pi G_5} \tag{4.62}$$

---

[20] Difference in the various numerical factors in this equation w.r.t [92] is due to our different normalization of the dilaton kinetic term.

where the flux can be calculated as the Noether current associated to the $U(1)$ symmetry $h_{11} \to e^{i\theta} h_{11}$ in the gravitational action for fluctuations. One finds,

$$\mathcal{F} = i \frac{e^{4A-B} f}{3A'^2} [h_{11}^* h_{11}' - h_{11} h_{11}^{*'}]. \tag{4.63}$$

As $\mathcal{F}$ is independent of the radial variable, one can compute it at any Φ, most easily near the horizon, where h_{11} takes the form (4.60). Using also the fact that $(dA/d\Phi)(\Phi_h) = -8V(\Phi_h)/9V'(\Phi_h)$, one finds

$$\mathcal{F}(\phi) = \frac{27}{32} \emptyset |c_b(\emptyset)|^2 e^{3A(\Phi_h)} \frac{V'(\Phi_h)^2}{V(\Phi_h)}. \tag{4.64}$$

Then, (4.54 and 4.62) determine the ratio of bulk viscosity and the entropy density as,

$$\frac{\zeta}{s} = \frac{3}{32\pi} \left(\frac{V'(\Phi_h)}{V(\Phi_h)} \right)^2 |c_b|^2. \tag{4.65}$$

In the derivation we use the Bekenstein–Hawking formula for the entropy density, $s = \exp 3A(\Phi_h)/4G_5$.

To find ζ we need to find c_b only in the limit $\omega \to 0$. The computation is performed by numerically solving equation (4.59) with the appropriate boundary conditions. There are two separate methods that one can employ to determine the quantity c_b:

1. One can solve (4.59) numerically with a fixed ω/T, but small enough so that c_b reaches a fixed value. The method is valid also for finite values of ω. From a practical point of view, it is easier to solve (4.59) with the boundary condition (4.60) with a unit normalization factor, read off the value on the boundary $h_{11}(-\infty)$ from the solution and finally use the symmetry of (4.59) under constant scalings of h_{11} to determine $|c_b| = 1/|h_{11}(-\infty)|$.
2. An alternative method of computation that directly extracts the information at $\omega = 0$ follows from the following trick [92]. Instead of solving (4.59) for small but finite ω, one can instead solve it for $\omega = 0$. This is a simpler equation, yet complicated enough to still evade analytic solution. Let us call this solution h_{11}^0. One numerically solves it by fixing the boundary conditions on the boundary: $h_{11}^0(-\infty) = 1$ and the derivative $dh_{11}^0/d\Phi(-\infty)$ is chosen such that h_{11} is regular at the horizon. Matching this solution to the expansion of (4.60) for small ω than yields $|c_b| = h_{11}^0(\Phi_h)$.

Both methods were used to obtain ζ/s as a function of T and checked that they yield the same result. As explained in [29], most of the thermodynamic observables are easily computed using the method of scalar variables [29, 93].

The results are presented in Fig. 4.12. This figure gives a comparison of the curve obtained by the holographic calculation sketched above by solving (4.59) and the lattice data of [12]. We also show $\eta/s = 1/4\pi$ in this figure for comparison.

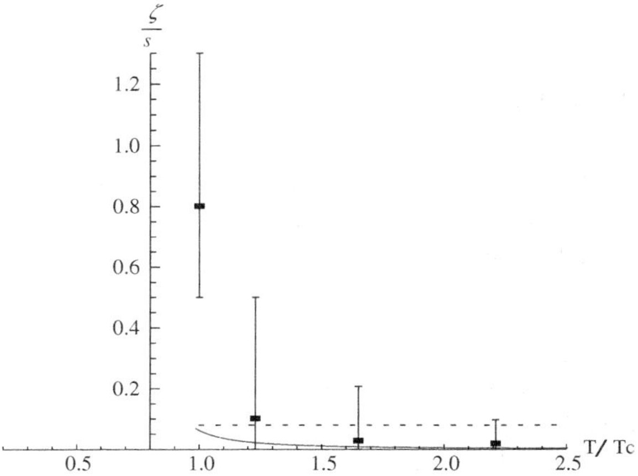

Fig. 4.12 Plot of ζ/s (*continuous line*) calculated in Improved Holographic QCD model. This is compared with the lattice data of [12] that are shown as *boxes*. The *horizontal dashed line* is indicating the (universal) value of $\frac{\eta}{s}$ for comparison

The result is qualitatively similar to the lattice result where ζ/s increases as T approaches T_c, however the rate of increase is slower than the lattice. As a result, we obtain a value $\zeta/s(T_c) \approx 0.06$ that is an order of magnitude smaller than the lattice result [12] which is 0.8. Note however that the error bars in the lattice evaluation are large near T_c and do not include all possible systematic errors from the analytic continuation.

We should note the fact that the holographic calculation gives a smaller value for the bulk viscosity near T_c than the lattice calculation is generic and has been found for other potentials with similar IR asymptotics [92]. The fact that the value of ζ/s near T_c is correlated with the IR asymptotics of the potential will be shown further below.

Another fact that one observes from Fig. 4.12 is that ζ/s vanishes in the high T limit. This reflects the conformal invariance in the UV and can be shown analytically as follows. ζ/s is determined by formula (4.65). In the high T limit, (corresponding to $\lambda_h \to 0$, near the boundary), the fluctuation coefficient $|c_b| \to 1$. This is because of the boundary condition $h_{11}(\lambda = 0) = 1$. We use the relation between T and λ_h in the high T limit [29],

$$\lambda_h \to (\beta_0 \log(\pi T/\Lambda))^{-1}. \tag{4.66}$$

Substitution in (4.65) leads to the result,

$$\left.\frac{\zeta}{s}\right|_{\text{large}} \to \frac{2}{27\pi} \frac{1}{\log^2(\pi T/\Lambda)}, \quad \text{as } T \to \infty. \tag{4.67}$$

As s itself diverges as T^3 in this limit—it corresponds to an ideal gas—we learn that ζ also diverges as $T^3/\log^2(T)$. Divergence at high T is expected from the bulk-viscosity of an *ideal gas*. We do not expect however the details of the asymptotic result to match with the pQCD result, for the same reasons that the shear-viscosity-to-entropy ratio does not, [25]. We note however, that the asymptotic T-dependence is very similar to the pQCD result [94]:

$$\zeta/s \propto \log^{-2}(\pi T/\Lambda)\log^{-1}\log(\pi T/\Lambda)\ . \tag{4.68}$$

4.6.2 Holographic Explanation for the Rise of ζ/s Near T_c

With the same numerical methods, one can also compute the ratio ζ/s on the small black-hole branch. As this solution has a smaller value of the action than the large black-hole solution, it is a subleading saddle point in the phase space of the theory, hence bears no direct significance for an holographic investigation of the quark–gluon plasma. However, as we show below, the existence of this branch provides a holographic explanation for the peak in ζ/s in the quark–gluon plasma, near T_c.

From the practical point of view, we find the second numerical method above (solving the fluctuation equation at $\omega = 0$) easier in the range of λ_h that corresponds to the small black hole. The result is shown in Fig. 4.13a. The presence of two branches for $T > T_{\rm min}$, is made clear in this figure. See also Fig. 4.13b for the respective ranges of λ_h that correspond to small and large BHs. In Fig. 4.13a, ζ/s on the large BH branch is depicted with a solid curve and the small BH branch is depicted with a dashed curve. We observe that ζ/s keeps increasing on the large-BH branch as T is lowered, up to the temperature $T_{\rm min}$ where the small and large BH branches merge.[21] On the other hand, on the small BH branch ζ/s keeps increasing as the T is increased, up to a certain $T_{\rm max}$ that lies between $T_{\rm min}$ and T_c, see Fig. 4.14. From this point onwards, ζ/s decreases with increasing T.

A simple fact that can be proved analytically is that the derivative of ζ/s diverges at $T_{\rm min}$. This is also clear from Fig. 4.14. This is shown by inspecting (4.65). The T derivative is determined as $d/dT = (dT/d\lambda_h)d/d\lambda_h$. Whereas the derivative w.r.t λ_h is everywhere smooth,[22] the factor $dT/d\lambda_h$ diverges at $T_{\rm min}$ by definition, see Fig. 4.13b.

Therefore, the presence of a $T_{\rm min}$ where the large and the small black holes meet, in other words, the presence of a small black-hole branch is responsible for the increase of ζ/s near $T_{\rm min}$. As in most of the holographic constructions that we analyzed, and specifically in the example we present here, T_c and $T_{\rm min}$ are close to one another, this fact implies a rise in the bulk viscosity near T_c. This proposal, combined with the fact that *the existence of a small BH branch and color confinement in the dual gauge theory at zero T are in one-to-one correspondence* [29], suggests that *in confining large-N gauge theories, there will be a peak in the ratio ζ/s close to T_c.*

Another fact that can be shown analytically is that ζ/s asymptotes to a finite value as $T \to \infty$ in the small black-hole branch.[23] We find that,

---

[21] As far as the thermodynamics of the gluon plasma is concerned, the temperatures below T_c (on the large BH branch) has little importance, because for $T < T_c$ the plasma is in the confined phase.

[22] Note that c_b is also a function of λ_h. As both the fluctuation (4.59) and the boundary conditions are smooth at $\lambda_h = \lambda_{\rm min}$, one concludes that c_b also is smooth at this point.

[23] See the discussion at appendix B of [93].

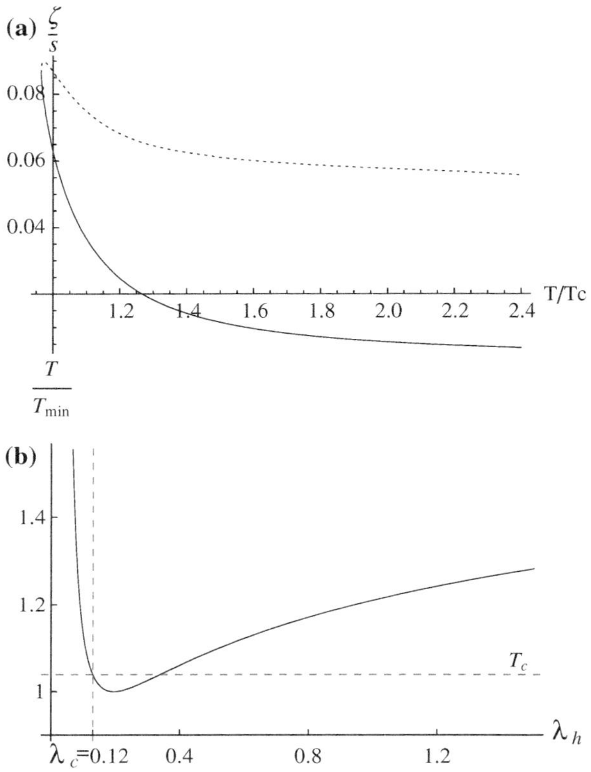

Fig. 4.13 **a** Numerical evaluation of ζ/η both on the large-BH branch (the *solid curve*) and on the small BH branch (the *dashed curve*). T_m denotes $T_{\rm min}$. **b** The two branches of black-hole solutions, that correspond to different ranges of λ_h. The large BH corresponds to $\lambda_h < \lambda_{\rm min}$ and the small BH corresponds to $\lambda_h > \lambda_{\rm min}$

$$\left.\frac{\zeta}{s}\right|_{\rm small} \to \frac{1}{6\pi}, \quad \text{as } T \to \infty. \tag{4.69}$$

As the entropy density vanishes in this limit [29], we conclude that ζ should vanish with the same rate.

For a general potential with strong coupling asymptotics

$$V(\lambda) \sim \lambda^Q \quad \text{as } \lambda \to \infty, \tag{4.70}$$

taking into account (4.65), (4.69) is modified to

$$\left.\frac{\zeta}{s}\right|_{\rm small} \to \frac{3Q^2}{32\pi}, \quad \text{as } r_h \to r_0. \tag{4.71}$$

where r_0 is the position of the singularity in the zero temperature solution.

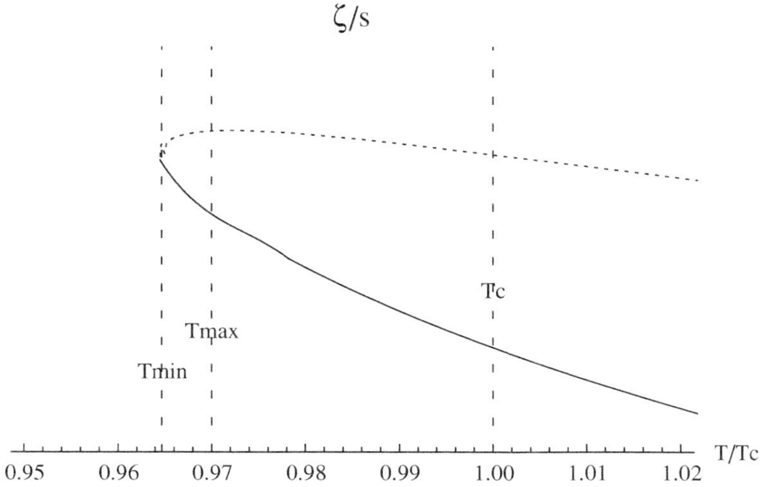

Fig. 4.14 An *inset* from the Fig. 4.13 around the maximum of ζ/s

For confining theories, the limit $r_h \to r_0$ corresponds to $T \to \infty$ on the small BH branch. However, one can show that the result (4.71) holds quite generally, regardless of whether the zero T theory confines or not. In particular, for the non-confining theories—that is either when $Q<4/3$ or when $Q = 4/3$ but the subleading term in the potential vanishes at the singularity—there is only the large black-hole branch and the limit $r_h \to r_0$ corresponds to the zero T limit of this BH. Thus, we also learn that there exist holographic models that correspond to non-confining gauge theories whose zero T limit yield a constant ζ/s. This constant approaches zero as $Q \to 0$, i.e. in the limiting AdS case.

We also see that the asymptotic value of ζ/s in the small BH branch is close to the value of ζ/s near T_c. We shall give an explanation of this fact in the next subsection. Using the asymptotic formula (4.71), the fact that $Q > \frac{4}{3}$ for confinement and $Q \leq \frac{4\sqrt{2}}{3}$ for the IR singularity to be good and repulsive we may obtain a range of values where we expect ζ/s to vary, namely

$$\frac{1}{6\pi} \leq \frac{\zeta}{s}\bigg|_{\text{small,asymptotic}} \leq \frac{1}{3\pi}. \tag{4.72}$$

A final observation concerns the coefficient $c_b(\lambda_h)$ in (4.65). This part is the only input from the solution of the fluctuation equation, the rest of (4.65) is fixed by the dilaton potential entirely. We plot the numerical result for c_b in Fig. 4.15 as a function of the coupling at the horizon λ_h.

First of all, Fig. 4.15 provides a check that, the approximate bound of [92] $|c_b| \geq 1$, is satisfied in the entire range. One also observes c_b approaches to 1 in the IR and UV asymptotics. These facts can be understood analytically: In the UV

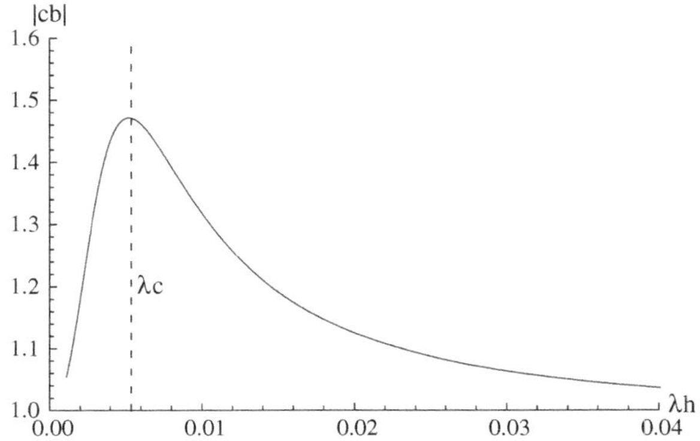

Fig. 4.15 The coefficient $|c_b|$ of (4.65) as a function of λ_h

(near the boundary) it is because of the boundary condition $c_b = 1$. In the IR, it is more subtle, and it is explained in appendix B of [93].

Finally, we observe that the deviation of c_b from the asymptotic value 1 is maximum around the phase transition point λ_c. In fact, we numerically observed that the top of the curve in Fig. 4.15 coincides with λ_c to a very high accuracy. Whether this is just a coincidence or not, remains to be clarified.

4.6.3 The Adiabatic Approximation

Motivated by the Chamblin-Reall solutions [95], Gubser et al. [26] proposed an approximate adiabatic formula for the speed of sound. In the case when V'/V is a slowly varying function of Φ [26] proposes the following formulae for the entropy density and the temperature:

$$\log s = -\frac{8}{3}\int^{\Phi_h} d\Phi \frac{V}{V'} + \cdots, \tag{4.73}$$

$$\log T = \int^{\Phi_h} d\Phi \left(\frac{1}{2}\frac{V'}{V} - \frac{8}{9}\frac{V}{V'}\right) \cdots, \tag{4.74}$$

where the ellipsis denote contributions slowly varying in Φ_h.[24]

It is very useful to reformulate this approximation using the method of scalar variables, which in turn allows us to extract the general T dependence of most of the thermodynamic observables in an approximate form. Here, we apply this formalism to the computation of ζ/s. The method of scalar variables and the details of the adiabatic approximation in the scalar variables are given in [93].

For the scalar variable $X = \frac{\Phi'}{3A'}$ the adiabatic approximation means

---

[24] Various coefficients in these equations differ from [26] due to our different normalization of the dilaton kinetic term.

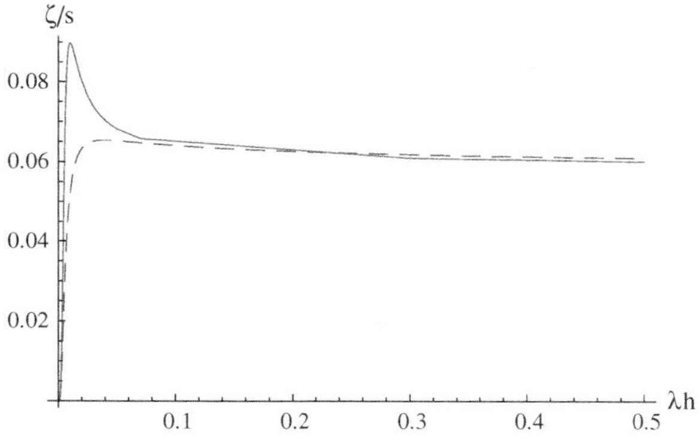

Fig. 4.16 Comparison of the exact ζ/s with the adiabatic approximation in the variable λ_h. *Solid* curve is the full numerical result and the *dashed* curve follows from (4.76)

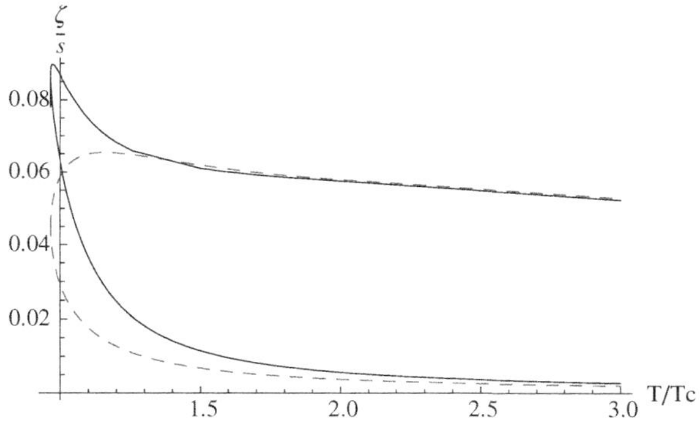

Fig. 4.17 Comparison of the exact ζ/s with the adiabatic approximation in variable T. *Solid* curve is the full numerical result and the *dashed* curve follows from (4.76)

$$X(\Phi) \approx -\frac{3}{8}\frac{V'(\Phi)}{V(\Phi)}. \tag{4.75}$$

The fluctuation (4.59) greatly simplifies with (4.75). In fact, as shown in [93], the solution becomes independent of Φ. With unit normalization on boundary, the adiabatic solution in the entire range of $\Phi \in (-\infty, \Phi_h)$ becomes $h_{adb}(\Phi) = 1$. Consequently, the coefficient c_b in (4.65) becomes unity, hence:

$$\left.\frac{\zeta}{s}\right|_{adb} = \frac{3}{32\pi}\left(\frac{V'(\Phi_h)}{V(\Phi_h)}\right)^2. \tag{4.76}$$

We plot this function in λ_h in Fig. 4.16, where we also provide the exact numerical result for comparison. Note that in Fig. 4.16 the whole large black-hole branch has been compressed at the left of the figure for $\lambda_h \lesssim 0.04$ The same functions in the variable T/T_c are plotted in Fig. 4.17.

The validity of the adiabatic approximation (4.75), is determined by the rate at which V'/V varies with Φ. In particular, the approximation becomes exact in the limits where V'/V becomes constant. This happens for a constant potential or a potential that is a single power of λ (exponential in Φ). This is the case in the UV ($\Phi \to -\infty$, where the potential becomes a constant) and the IR ($\Phi \to +\infty$ where the potential becomes a power law.). Therefore (4.76) allows us to extract the analytic behavior of ζ/s in the limits $\Phi_h \to \pm\infty$.

The numerical values one obtains from (4.76) in the intermediate region may differ from the exact result (4.65) considerably, especially near T_c. However, we expect that the general shape will be similar.

Finally, the adiabatic approximation hints at why, in the particular background that we study, ζ/s at T_c is close to the limit value (4.69): In order to see this we rewrite (4.76) as

$$\left.\frac{\zeta}{s}\right|_{adb} = \frac{2}{3\pi}X^2. \tag{4.77}$$

In the limit (4.69) we have $X \to -1/2$. The only other point where $X = -1/2$, is at the minimum of the string frame scale factor Φ_*. This is the point where the confining string saturates [23, 24]. On the other hand, we expect on general physical grounds that the de-confinement phase transition happens near this point, i.e. $\Phi_c \approx \Phi_*$. Thus, the adiabatic formula predicts that $\zeta/s(\Phi_c)$ be close to the limit value $1/6\pi$.[25]

4.6.4 Buchel's Bound

In [96], Buchel proposed a bound for the ratio of the bulk and shear viscosities, motivated by certain well-understood holographic examples. In four space-time dimensions the Buchel bound reads,

$$\frac{\zeta}{\eta} \geq 2\left(\frac{1}{3} - c_s^2\right). \tag{4.78}$$

We note that the bound is proposed to hold in the entire range of temperature from T_c to ∞. This bound is trivially satisfied for exact conformal theories such as $\mathcal{N} = 4$ YM, and saturated in theories on *Dp* branes [96, 97]. With the numerical evaluation at hand, we can check (4.78) in our case. In Fig. 4.18a we plot the LHS and RHS of the bound.[26] We clearly see that the bound is satisfied for all

[25] This argument may break down for two (dependent) reasons: First of all the adiabatic approximation becomes lees good near Φ_c. This is because, in this region V'/V varies relatively more rapidly as a function of Φ. Secondly, precisely because of this, even though Φ_c is not far away from Φ_* the difference can result in a considerable change in the value of ζ/s through (4.76).

[26] Since this theory contains two derivatives only, $\frac{\eta}{s}$ has the universal value $1/4\pi$.

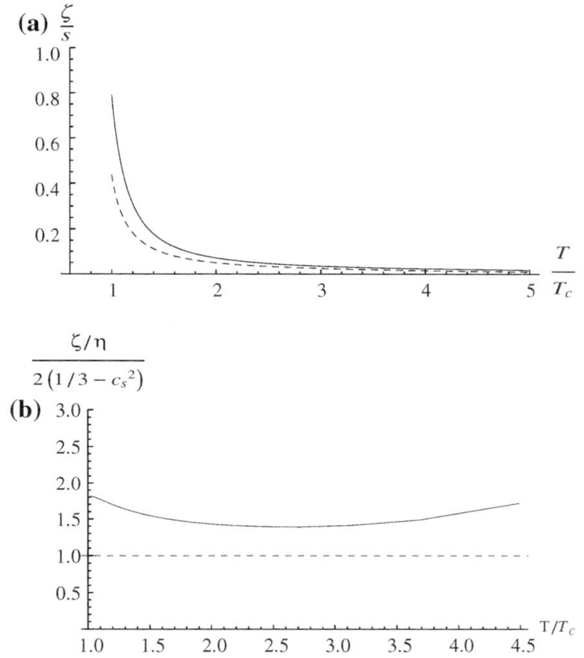

Fig. 4.18 **a** Comparison of ζ/η (*solid line*) and the RHS of (4.78) (*dashed line*), obtained using the speed of sound of the IHQCD model [22]. **b** Plot of the function $C(T)$ defined in (4.79) as a function of temperature. The *horizontal dashed line* indicates where Buchel's bound is saturated. We see that the bound is satisfied in the entire range of temperatures

temperatures. As expected, both the LHS and the RHS of (4.78) vanishes in the high T conformal limit.

A clear picture of Buchel's bound is obtained by defining the function:

$$C(T) = \frac{\zeta/\eta}{2\left(1/3 - c_s^2\right)}, \tag{4.79}$$

in terms of which the bound is simply $C > 1$. In Fig. 4.18b we show the function $C(T)$ obtained numerically in our IHQCD model, between T_c and $5T_c$. The values of this function are mildly dependent on temperature, and are between 1.5 and 2, the same range of values that were recently considered in the hydrodynamic codes by Heinz and Song [98].

We may also investigate the fate of the bound at large T. using the asymptotics of ζ/s in (4.67)

$$\left.\frac{\zeta}{s}\right|_{\text{large}} = \frac{2}{27\pi}\frac{1}{\log^2(\pi T/\Lambda)} + \cdots, \quad \text{as } T \to \infty. \tag{4.80}$$

and

$$\frac{1}{c_s^2} - 3 = \frac{4}{3} \frac{1}{\log^2\left(\frac{T}{T_c}\right)} + \frac{32b}{9} \frac{\log\left(\log\left(\frac{T}{T_c}\right)\right)}{\log^3\left(\frac{T}{T_c}\right)} + \cdots \tag{4.81}$$

from [29], that can be rewritten as

$$\frac{1}{3} - c_s^2 = \frac{4}{27} \frac{1}{\log^2\left(\frac{T}{T_c}\right)} + \frac{32b}{81} \frac{\log\left(\log\left(\frac{T}{T_c}\right)\right)}{\log^3\left(\frac{T}{T_c}\right)} + \cdots \tag{4.82}$$

where $b = \frac{b_1}{b_0^2} = \frac{3 \cdot 34}{2 \cdot 121}$ is the ratio of the two-loop to the one-loop squared β-function coefficients in large-N_c YM.

Since in this class of models $\eta/s = 1/4\pi$ exactly we obtain

$$\lim_{T \to \infty} \frac{\zeta/\eta}{2(1/3 - c_s^2)} = 1 \tag{4.83}$$

in agreement with a recently derived general formula, in Einstein dilaton gravity [99]

$$\lim_{T \to \infty} \frac{\zeta/\eta}{2(1/3 - c_s^2)} = 2\pi \frac{4 - \Delta}{4 - 2\Delta} \cot\left(\frac{\pi\Delta}{4}\right) \tag{4.84}$$

where Δ is the scaling dimensions of the scalar operator in the UV, that is marginal in our case.

It has also been suggested recently [100–102] that the speed of sound squared, in Einstein dilaton gravity asymptotes to 1/3 at high temperatures from below. This is evident in our asymptotic formula (4.82), although the formulae in [100–102] fail to capture correctly the marginal case that is relevant here.

4.7 The Drag Force on Strings and Heavy Quarks

We will now consider an (external) heavy quark moving through an infinite volume of gluon plasma with a fixed velocity v at a finite temperature T [17, 19, 20]. The quark feels a drag force coming from its interaction with the plasma and an external force has to be applied in order for it to keep a constant velocity. In a more realistic set up one would like to describe the deceleration caused by the drag.

The heavy external quark can be described by a string whose endpoint is at the boundary. One can accommodate flavor by introducing D-branes, but we will not do this here. A first step is to describe the classical string "trailing" the quark.

We consider the Nambu–Goto action on the world-sheet of the string.

$$S_{NG} = -\frac{1}{2\pi\ell_s^2}\int d\sigma d\tau \sqrt{det(-g_{MN}\partial_\alpha X^M \partial_\beta X^N)}, \tag{4.85}$$

where the metric is the string frame metric. The ansatz we are going to use to describe the trailing string is [19, 20]

$$X^1 = vt + \xi(r), \quad X^2 = X^3 = 0, \tag{4.86}$$

along with the gauge choice

$$\sigma = r, \quad \tau = t, \tag{4.87}$$

where r is the (radial) holographic coordinate. The string is moving in the X^1 direction.

This is a "steady-state" description of the moving quark as acceleration and deceleration are not taken into account. For a generic background the action of the string becomes

$$S = -\frac{1}{2\pi\ell_s^2}\int dtdr \sqrt{-g_{00}g_{rr} - g_{00}g_{11}\xi'^2 - g_{11}g_{rr}v^2}. \tag{4.88}$$

Note that g_{00} is negative, and we should check whether our solution produces a real action. For example a straight string stretching from the quark to the horizon is a solution to the equations of motion but has imaginary action.

We note that the action does not depend on ξ but only its derivative, therefore the corresponding "momentum" is conserved

$$\pi_\xi = -\frac{1}{2\pi\ell_s^2}\frac{g_{00}g_{11}\xi'}{\sqrt{-g_{00}g_{rr} - g_{00}g_{11}\xi'^2 - g_{11}g_{rr}v^2}}. \tag{4.89}$$

We solve for ξ' to obtain

$$\xi' = \frac{\sqrt{-g_{00}g_{rr} - g_{11}g_{rr}v^2}}{\sqrt{g_{00}g_{11}\left(1 + g_{00}g_{11}/(2\pi\ell_s^2\pi_\xi)^2\right)}}. \tag{4.90}$$

The numerator changes sign at some finite value of the fifth coordinate r_s. For the solution to be real, the denominator has to change sign at the same point. We therefore determine r_s via the equation

$$g_{00}(r_s) + g_{11}(r_s)v^2 = 0, \tag{4.91}$$

and the constant momentum

$$\pi_\xi^2 = -\frac{g_{00}(r_s)g_{11}(r_s)}{(2\pi\ell_s^2)^2} \tag{4.92}$$

Writing the string-frame metric as

$$ds^2 = e^{2A_s}\left[\frac{dr^2}{f} - f\,dt^2 + dx\cdot dx\right] \tag{4.93}$$

(4.91) becomes

$$v^2 = f(r_s) \tag{4.94}$$

The induced world-sheet metric is therefore

$$g_{\alpha\beta} = e^{2A_s(r)}\begin{pmatrix} -(f(r)-v^2) & \frac{e^{2A_s(r_s)}v^2}{f(r)}\sqrt{\frac{f(r)-v^2}{e^{4A_s(r)}f(r)-e^{4A_s(r_s)}v^2}} \\ \frac{e^{2A_s(r_s)}v^2}{f(r)}\sqrt{\frac{f(r)-v^2}{e^{4A_s(r)}f(r)-e^{4A_s(r_s)}v^2}} & \frac{e^{4A_s(r)}f^2(r)-v^4e^{4A_s(r_s)}}{f^2(r)[e^{4A_s(r)}f(r)-v^2e^{4A_s(r_s)}]} \end{pmatrix} \tag{4.95}$$

We can change the time coordinate to obtain a diagonal induced metric $t = \tau + \zeta(r)$ with

$$\zeta' = \frac{e^{2A_s(r_s)}v^2}{f(r)\sqrt{(f(r)-v^2)(e^{4A_s(r)}f(r)-e^{4A_s(r_s)}v^2)}}$$

The new metric is

$$ds^2 = e^{2A_s(r)}\left[-(f(r)-v^2)d\tau^2 + \frac{e^{4A_s(r)}}{(e^{4A_s(r)}f(r)-e^{4A_s(r_s)}v^2)}dr^2\right] \tag{4.96}$$

and near $r = r_s$ it has the expansion

$$\begin{aligned} ds^2 = &\left[-f'(r_s)e^{2A_s(r_s)}(r-r_s) + \mathcal{O}((r-r_s)^2)\right]d\tau^2 \\ &+ \left[\frac{e^{2A_s(r_s)}}{(4v^2A_s'(r_s)+f'(r_s))(r-r_s)} + \mathcal{O}(1)\right]dr^2 \end{aligned} \tag{4.97}$$

This is a world-sheet black-hole metric with horizon at the turning point $r = r_s$.

4.7.1 The Drag Force

The drag force on the quark can be determined by calculating the momentum that is lost by flowing along the string into the horizon:

$$F_{\text{drag}} = \frac{dp_1}{dt} = -\frac{1}{2\pi\ell_s^2}\frac{g_{00}g_{11}\zeta'}{\sqrt{-g}} = \pi_\xi. \tag{4.98}$$

This can be obtained by considering the world-sheet Noether currents Π_M^α and expressing the loss of momentum as $\Delta P_{x_1}^z = \int \Pi_1^r$. This may be evaluated at any value of r, but it is more convenient to evaluate it at $r = r_s$.

We finally find that

$$F_{\text{drag}} = -\frac{1}{2\pi\ell_s^2}\sqrt{-g_{00}(r_s)g_{11}(r_s)}. \tag{4.99}$$

Using the form (4.93) of our finite-temperature metric in the string frame we finally obtain

$$F_{\text{drag}} = -\frac{e^{2A_s(r_s)}\sqrt{f(r_s)}}{2\pi\ell_s^2} = -\frac{e^{2A(r_s)}\sqrt{f(r_s)}\lambda(r_s)^{4/3}}{2\pi\ell_s^2}, \tag{4.100}$$

where in the second equality we expressed the force in terms of the Einstein-frame scale factor and the "running" dilaton. Substituting from (4.94) we obtain

$$F_{\text{drag}} = -\frac{v\, e^{2A_s(r_s)}}{2\pi\ell_s^2} = -\frac{v\, e^{2A(r_s)}\lambda(r_s)^{4/3}}{2\pi\ell_s^2}, \tag{4.101}$$

Before proceeding further, we will evaluate the drag force for the conformal case of $\mathcal{N} = 4$ SYM where

$$e^{A_s} = \frac{\ell}{r}, \quad v^2 = f(r_s) = 1 - (\pi T r_s)^4, \quad \frac{\ell^2}{\ell_s^2} = \sqrt{\lambda}. \tag{4.102}$$

Substituting in (4.101) we obtain [17–20]

$$F_{\text{conf}} = \frac{\pi}{2}\sqrt{\lambda}\, T^2 \frac{v}{\sqrt{1-v^2}}. \tag{4.103}$$

Moving on to YM, to compute the drag force from (4.101) we must first determine ℓ_s in the IHQCD model. In this setup there is no analog of the $\mathcal{N} = 4$ SYM relation (4.102) between ℓ, ℓ_s and λ. Rather, the fundamental string length ℓ_s is determined in a bottom-up fashion, by matching the effective string tension to the QCD string tension σ_c derived from the lattice calculations. We obtain

$$\sigma_c = \frac{1}{2\pi\ell_s^2} e^{2A_{s,o}(r_*)} = \frac{1}{2\pi\ell_s^2} e^{2A_o(r_*)}\lambda_o(r_*)^{4/3}, \tag{4.104}$$

where r_* is the point where the zero-temperature string scale factor (at T = 0) $A_{s,o}(r)$ has a minimum. For a typical value of $\sigma_c \sim (440\,\text{MeV})^2$ [78, 79] we find

$$\ell_s = 6.4\,\ell, \tag{4.105}$$

where ℓ is the radius of the asymptotic AdS space.

On the other hand, unlike in $\mathcal{N} = 4$ SYM, in the IHQCD model the value of the coupling $\lambda(r_s)$ in (4.101) is not an extra parameter to be fixed by hand, but rather

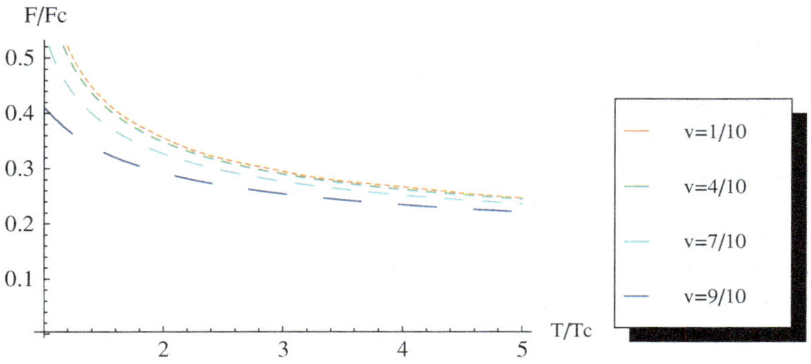

Fig. 4.19 The ratio of the drag force in improved holographic QCD to the drag force in $\mathcal{N}=4$ SYM is shown. The ratio is computed for different velocities as a function of temperature. The 't Hooft coupling for the $\mathcal{N}=4$ SYM theory is taken to be 5.5. We chose this value as it is considered in the central region of possible values for the 't Hooft coupling. It is seen that as the velocity increases the value of the ratio decreases

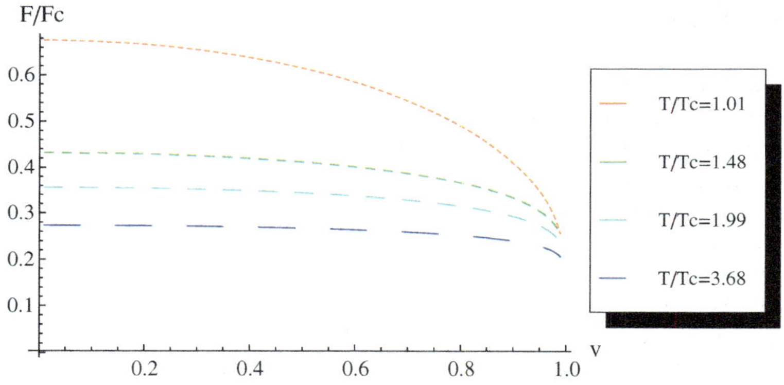

Fig. 4.20 The ratio of the drag force in improved holographic QCD to the drag force in $\mathcal{N}=4$ SYM is shown. The ratio is computed for different temperatures as a function of velocity. The 't Hooft coupling for the $\mathcal{N}=4$ SYM theory is taken to be 5.5. As temperature increases the value of the ratio decreases

it is determined dynamically together with the background metric (Figs. 4.19 and 4.20).

4.7.2 The Relativistic Asymptotics

When $v \to 1$ then $r_s \to 0$ and we approach the boundary. Near the boundary ($r \to 0$) we have the following asymptotics of the scale factor and the 't Hooft coupling [23, 24]

$$f(r) \simeq 1 - \frac{\pi T\, e^{3A(r_h)}}{\ell^3} r^4 \left[1 + \mathcal{O}\left(\frac{1}{\log(\Lambda r)}\right)\right] + \mathcal{O}(r^8),$$
$$e^{A(r)} = \frac{\ell}{r}\left[1 + \mathcal{O}\left(\frac{1}{\log(\Lambda r)}\right)\right] + \cdots \tag{4.106}$$

and

$$\lambda(r) = -\frac{1}{\beta_0 \log(r\Lambda)} + \mathcal{O}(\log(r\Lambda)^{-2}) \tag{4.107}$$

where r_h is the position of the horizon.

We therefore obtain for the turning point

$$r_s \simeq \left[\frac{\ell^3(1-v^2)}{\pi T e^{3A(r_h)}}\right]^{\frac{1}{4}} \left[1 + \mathcal{O}\left(\frac{1}{\log(1-v^2)}\right)\right], \quad \lambda(r_s) \simeq -\frac{4}{\beta_0 \log[1-v^2]} + \cdots \tag{4.108}$$

and the drag force

$$F_{\text{drag}} \simeq -\frac{\sqrt{\pi T \ell b^3(r_h)\lambda^{\frac{8}{3}}(r_s)}}{2\pi\ell_s^2}\frac{v}{\sqrt{1-v^2}} + \cdots \tag{4.109}$$

We also use

$$e^{3A(r_h)} = \frac{s(T)}{4\pi M_p^3 N_c^2} = \frac{45\pi\ell^3 s(T)}{N_c^2} \tag{4.110}$$

where $s(T)$ the entropy per unit three-volume, and we write the relativistic asymptotics of the drag force as,

$$F_{\text{drag}} \simeq -\frac{\sqrt{\pi T \ell b^3(r_h)}}{2\pi\ell_s^2}\frac{v}{\sqrt{1-v^2}\left(-\frac{\beta_0}{4}\log[1-v^2]\right)^{\frac{4}{3}}} + \cdots \tag{4.111}$$
$$= -\frac{\ell^2}{\ell_s^2}\sqrt{\frac{45\, Ts(T)}{4N_c^2}}\frac{v}{\sqrt{1-v^2}\left(-\frac{\beta_0}{4}\log[1-v^2]\right)^{\frac{4}{3}}} + \cdots$$

The force is proportional to the relativistic momentum combination $v/\sqrt{1-v^2}$ modulo a power of $\log[1-v^2]$. This factor is present because, as argued in [25] the asymptotic metric is AdS in the Einstein frame instead of the string frame. Its effects are not important phenomenologically.

4.7.3 The Non-relativistic Asymptotics

We now consider the opposite limit, $v \to 0$. In this case the turning point asymptotes to the horizon, $r_s \to r_h$ and we have the expansion

$$f(r) \simeq 4\pi T(r_h - r) + \mathcal{O}((r_h - r)^2), \quad r_s = r_h - \frac{v^2}{4\pi T} + \mathcal{O}(v^4) \tag{4.112}$$

and

$$\begin{aligned} F_{\text{drag}} &\simeq -\frac{e^{2A(r_h)}\lambda(r_h)^{\frac{4}{3}}}{2\pi\ell_s^2} v \left[1 - \frac{v^2}{2\pi T}A'(r_h) - \frac{v^2}{3\pi T}\frac{\lambda'(r_h)}{\lambda(r_h)} + \mathcal{O}(v^4)\right] \\ &\simeq -\frac{\ell^2}{\ell_s^2}\left(\frac{45\pi\, s(T)}{N_c^2}\right)^{\frac{2}{3}} \frac{\lambda(r_h)^{\frac{4}{3}}}{2\pi} v + \mathcal{O}(v^3) \end{aligned} \tag{4.113}$$

where primes are derivatives with respect to the conformal coordinate r.

4.7.4 The Diffusion Time

For a heavy quark with mass M_q we may rewrite (4.103) as

$$F_{\text{conf}} \equiv \frac{dp}{dt} = -\frac{1}{\tau}p, \quad p = \frac{M_q v}{\sqrt{1 - v^2}} \tag{4.114}$$

where the first equation defines the diffusion time τ. In the conformal case, the diffusion time is constant,

$$\tau_{\text{conf}} = \frac{2M_q}{\pi\sqrt{\lambda}\, T^2}. \tag{4.115}$$

This is not anymore the case in QCD, where τ defined as above is momentum dependent. We may still define it as in (4.114) in which case we obtain the following limits (Fig. 4.21):

$$\lim_{p\to\infty} \tau = M_q \frac{\ell_s^2}{\ell^2} \sqrt{\frac{4N_c^2}{45\, Ts(T)}} \left(\frac{\beta_0}{4}\log\frac{p^2}{M_q^2}\right)^{\frac{4}{3}} + \cdots \tag{4.116}$$

$$\lim_{p\to 0} \tau = M_q \frac{\ell_s^2}{\ell^2}\left(\frac{N_c^2}{45\pi\, s(T)}\right)^{\frac{2}{3}} \frac{2\pi}{\lambda(r_h)^{\frac{4}{3}}} + \cdots \tag{4.117}$$

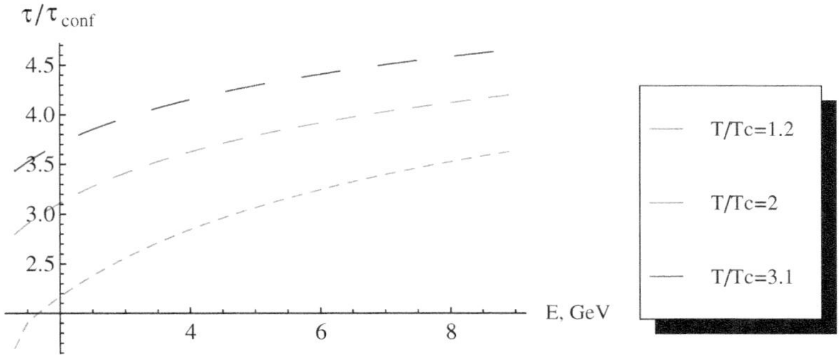

Fig. 4.21 The ratio of the diffusion time in the Improved Holographic QCD model to the diffusion time in $\mathcal{N} = 4$ SYM is shown. The 't Hooft coupling for $\mathcal{N} = 4$ SYM is taken to be $\lambda = 5.5$. The heavy quark has a mass of $M_q = 1.3\,\text{GeV}$. Note that with the definition of the diffusion time in (4.114) the ratio is the inverse of the ratio of the forces. A similar plot is valid for the bottom quark as well, as the mass drops out of the ratio. although the energy scales are different. In this plot the x-axis is taken to be in MeV units. As temperature increases the ratio also increases

4.7.5 Including the Correction to the Quark Mass

In order to estimate the diffusion time of a quark of finite rest mass, we must take into account the fact that the mass of the quark receives medium-induced corrections. In other words, the mass appearing in (4.114) is a temperature-dependent quantity, $M_q(T) \neq M_q(T = 0)$. The ratio $M_q(T)/M_q(0)$ can be estimated holographically by representing a static quark of finite mass by a static, straight string[27] stretched along the radial direction starting at a point $r = r_q \neq 0$. At zero temperature, the IR endpoint of the string can be taken as the "confinement" radius, r_*, where the string frame metric reaches its minimum value; At finite temperature, the string ends in the IR at the BH horizon.[28] The masses of the quark at zero and finite T are related to the world-sheet action evaluated on the static solution $(\tau = t, \sigma = r)$:

$$M_q(0) = \frac{\ell}{2\pi\ell_s^2} \int_{r_q}^{r_*} dr\, e^{2A_o(r)} \lambda_o^{4/3}(r)\,, \quad M_q(T) = \frac{\ell}{2\pi\ell_s^2} \int_{r_q}^{r_h} dr\, e^{2A(r)} \lambda^{4/3}(r). \quad (4.118)$$

---

[27] This representation ignores the fact that the *kinetic* mass of a moving quark may be different from the static mass [17].

[28] It would stop at the confinement radius if the latter were closer to the boundary than the horizon, i.e. if $r_*(T) < r_h(T)$. However, in the model we are considering, in the large BH branch we find that the relation $r_h < r_*$ is always satisfied.

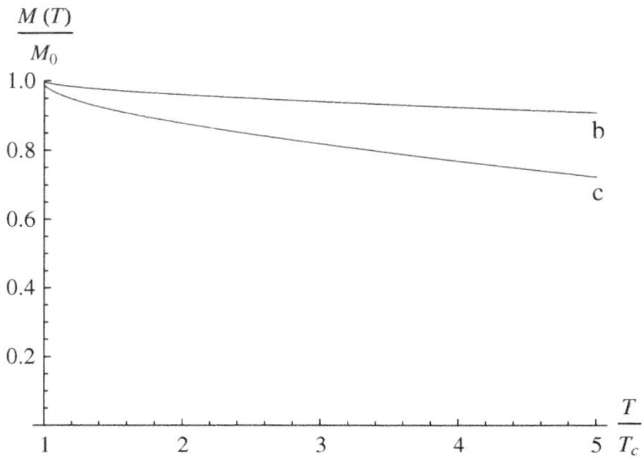

Fig. 4.22 Ratios between the thermal mass and the rest mass of the Charm (curve labelled "c") and Bottom (curve labelled "b") quarks, as a function of temperature

The value r_q can be fixed numerically by matching $M_q(0)$ to the physical quark mass, and translating the fundamental string tension in physical units by using the relation (4.104), with $\sigma_c = (440\,\mathrm{MeV})^2$. This makes $M_q(T)$ a function of $M_q(0)$. The ratios $M_q(T)/M_q(0)$ we found numerically in the model under consideration is shown in Fig. 4.22 for the Charm ($M(0) = 1.5\,\mathrm{GeV}$) and Bottom ($M(0) = 4.5\,\mathrm{GeV}$) quarks. The fact that, in the deconfined plasma, the quark mass decreases with increasing temperature is a direct consequence of the holographic framework,[29] since for higher temperature, the distance to the horizon is smaller. An indication that this result may be in the right direction comes from the lattice computation of the shift in the position of the quarkonium resonance peak at finite temperature [104]: in the deconfined phase the charmonium peak moves to lower mass at higher temperature. Our result for the medium-induced shift in the constituent quark mass is consistent with these observations.

We can now write the diffusion time from (4.101) and (4.114) as:

$$\tau(T,v) = \frac{M_q(T)}{\sigma_c\sqrt{1-v^2}}\left(\frac{\lambda_o(r_*)}{\lambda(r_s)}\right)^{4/3} e^{2A_o(r_*)-2A(r_s)}, \tag{4.119}$$

where once again we have eliminated the fundamental string length using (4.104). Given a set of zero- and finite-temperature solutions, (4.119) can be evaluated numerically for different values of the velocity and different quark masses. The results for the Charm ($M_q(0) = 1.5\,\mathrm{GeV}$) and Bottom ($M = 4.5\,\mathrm{GeV}$) quarks are displayed in Fig. 4.23.

---

[29] For a possible field theoretical explanation of this phenomenon, see [103].

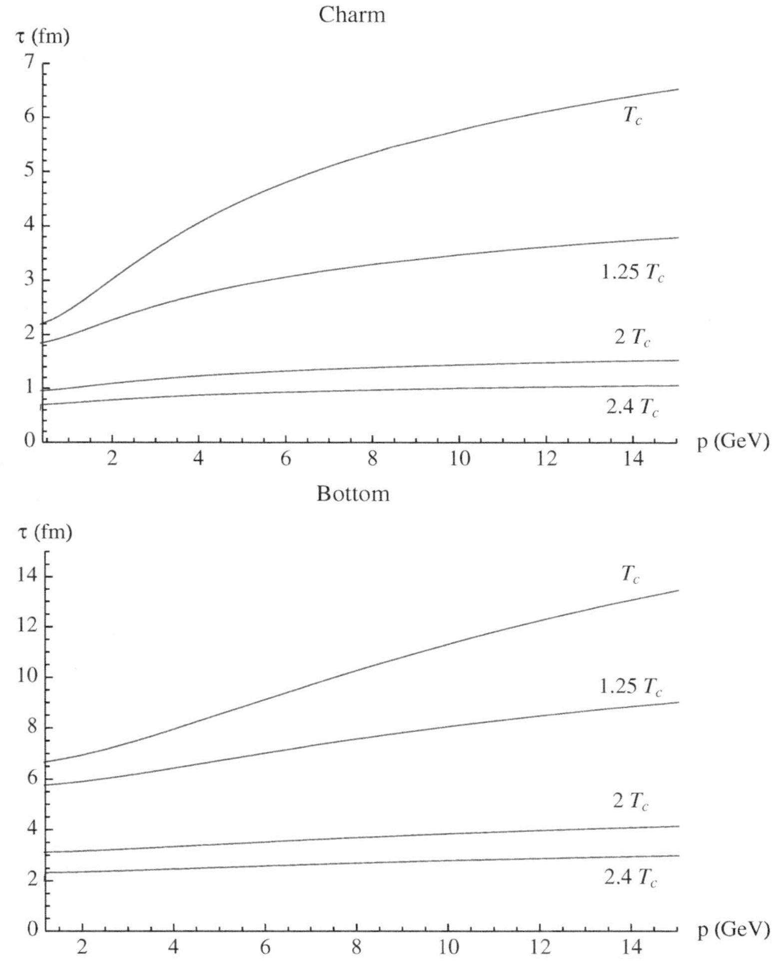

Fig. 4.23 Diffusion time for the Charm and Bottom quarks, as a function of energy, for different ratios of the temperature to the IHQCD transition temperature T_c

4.7.6 Temperature Matching and Diffusion Time Estimates

An important question is how we should choose the temperature in our holographic model in order to compare our results with heavy-ion collision experiments. This is nontrivial, since our setup is designed to describe pure $SU(N_c)$ Yang–Mills, whereas at RHIC temperatures there are three light quark flavors that become relevant. As a consequence, the critical temperatures and the number of degrees of freedom of the two theories are not the same: for pure $SU(N_c)$ Yang Mills we have $N_c^2 - 1$ degrees of freedom and a critical temperature around

260 MeV; For $SU(N_c)$ QCD with N_f flavors the number of degrees of freedom is $N_c^2 - 1 + N_cN_f$, and the transition temperature is lower, around 180 MeV.

In IHQCD, the transition temperature in physical units was calculated to be $T_c = 247$ MeV [32], i.e. close to the lattice result for the pure YM deconfining temperature. From now on, this is the value we will mean when we refer to T_c. This is also close to the temperature of QGP at RHIC, which we will denote $T_{\rm QGP}$, and is estimated to be around 250 MeV. Since this value is uncertain, below we give our results for a range of temperatures between 200 and 400 MeV. The higher temperatures will be relevant for the LHC heavy-ion collision experiments (see e.g. [105]).

Based on these considerations, there are different ways of fixing the temperature (see e.g. the recent review [51]): one *direct* and two *alternative* schemes (that we call the *energy* and *entropy* scheme).

- *Direct scheme*: The temperature of the holographic model is identified with the temperature of the QGP in the experimental situation (at RHIC or LHC), $T^{\rm (dir)}_{\rm ihqcd} = T_{\rm QGP}$.
- *Energy scheme*: One matches the energy densities, rather than the temperatures. The energy density at RHIC is approximately (treating the QCD plasma as a free gas.[30]) $\epsilon_{\rm QGP} \simeq (\pi^2/15)(N_c^2 - 1 + N_cN_f)(T_{\rm QGP})^4$. For $N_c = N_f = 3$, asking that our energy density matches this value requires us to consider the holographic model at temperature $T^{(\epsilon)}_{\rm ihqcd}$ given by

$$\epsilon_{\rm ihqcd}(T^{(\epsilon)}_{\rm ihqcd}) \simeq 11.2(T_{\rm QGP})^4 \tag{4.120}$$

- *Entropy scheme*: Instead of matching the energy densities, alternatively one can match the entropy density s, which for the QGP, in the free gas approximation,$\sigma_{\rm QGP} \simeq 4\pi^2/45(N_c^2 - 1 + N_cN_f)(T_{\rm QGP})^4$. This leads to the identification:

$$s_{\rm ihqcd}(T^{(s)}_{\rm ihqcd}) = 14.9(T_{\rm QGP})^3 \tag{4.121}$$

The temperature translation table between the various schemes is shown in Table 4.3. In that table, $T_c = 247$ MeV is the deconfining temperature of the holographic model.

In Fig. 4.24 we show the comparison between the diffusion times, as a function of initial quark momentum, in the different schemes for the Charm and Bottom quarks, at the temperature $T_{\rm QGP} = 250$ MeV.

The results for the diffusion times at different temperatures, computed at a reference heavy quark initial momentum $p \approx 10$ GeV, are displayed in Tables 4.4 and 4.5. We see that there is little practical difference between the *entropy* and

[30] This is itself an approximation, since as we know both from experiment and in our holographic model, the plasma is strongly coupled up to temperatures of a few T_c.

Table 4.3 Translation table between different temperature identification schemes

T_{QGP} (MeV)	T_{QGP}/T_c	$T^{(\epsilon)}_{\text{ihqcd}}$ (MeV)	$T^{(\epsilon)}_{\text{ihqcd}}/T_c$	$T^{(s)}_{\text{ihqcd}}$ (MeV)	$T^{(s)}_{\text{ihqcd}}/T_c$
190	0.77	259	1.05	274	1.11
220	0.89	290	1.18	302	1.23
250	1.01	325	1.31	335	1.35
280	1.13	361	1.46	368	1.49
310	1.26	398	1.61	402	1.63
340	1.38	434	1.76	437	1.77
370	1.50	471	1.90	472	1.91
400	1.62	508	2.06	507	2.05

The first two columns display temperatures in the direct scheme, (in which the temperature of the holographic model matches the physical QGP temperature) and the corresponding ratio to the IHQCD critical temperature, that was fixed by YM lattice results at $T_c = 247\,\text{MeV}$ [32]. The third and fourth columns display the corresponding temperatures (and respective ratios to T_c) in the energy scheme, and the last two in the entropy scheme

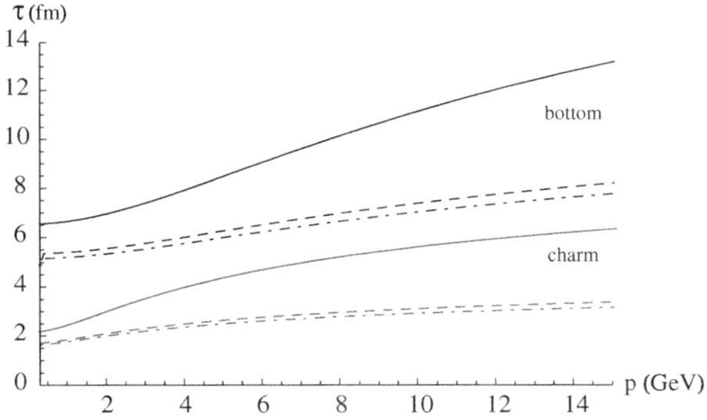

Fig. 4.24 Diffusion times for the Charm and Bottom quarks, as a function of initial momentum, at $T_{\text{QGP}} = 250\,\text{MeV}$. The *different lines* represent the in the *direct* scheme (*solid*), *energy* scheme (*dashed*) and *entropy* scheme (*dash-dotted*), all corresponding to the same temperature $T_{\text{QGP}} = 250\,\text{MeV}$

Table 4.4 The diffusion times for the charm quark are shown for different temperatures, in the three different schemes

T_{QGP}, MeV	τ_{charm} (fm/c) (direct)	τ_{charm} (fm/c) (energy)	τ_{charm} (fm/c) (entropy)
220	–	4.0	3.6
250	5.7	3.1	3.0
280	4.3	2.6	2.5
310	3.5	2.1	2.1
340	2.9	1.8	1.8
370	2.5	1.5	1.5
400	2.1	1.3	1.3

Diffusion times have been evaluated with a quark initial momentum fixed at $p \approx 10\,\text{GeV}$

Table 4.5 Diffusion times for the bottom quark are shown for different temperatures, in the three different schemes

T_{QGP} (MeV)	τ_{bottom} (fm/c) (direct)	τ_{bottom} (fm/c) (energy)	τ_{bottom} (fm/c) (entropy)
220	–	8.9	8.4
250	11.4	7.5	7.1
280	10.1	6.3	6.1
310	8.6	5.4	5.3
340	7.5	4.7	4.7
370	6.6	4.1	4.1
400	5.8	3.6	3.6

Diffusion times have been evaluated with a quark initial momentum fixed at $p \approx 10\,\mathrm{GeV}$

energy schemes; on the other hand the difference between the *direct* scheme and the two alternative schemes can be quite substantial.

4.8 Jet Quenching Parameter

In this section we discuss the jet quenching parameter in the class of holographic models under consideration, and we estimate its numerical value for the concrete model with potential (4.26) and parameters fixed as in [32]. For the holographic computation, we will follow [18, 52]. There is another method available [57], but we will not use it here.

The jet-quenching parameter $\hat{q}$ provides a measure of the dissipation of the plasma and it has been associated to the behavior of a Wilson loop joining two light-like lines. We consider two light-like lines which extend for a distance L^- and are situated distance L apart in a transverse coordinate. Then $\hat{q}$ is given by the large L^+ behavior of the Wilson loop

$$W \sim e^{-\frac{1}{4\sqrt{2}}\hat{q}L^- L^2}\,. \tag{4.122}$$

We consider the bulk string frame metric

$$ds^2 = e^{2A_s(r)}\left(-f(r)dt^2 + d\vec{x}^2 + \frac{dr^2}{f(r)}\right). \tag{4.123}$$

To address the problem of the Wilson loop we make a change of coordinates to light cone coordinates for the boundary theory

$$x^+ = x_1 + t \quad x^- = x_1 - t \tag{4.124}$$

for which the metric becomes

$$ds^2 = e^{2A_s}\left(dx_2^2 + dx_3^2 + \frac{1}{2}(1-f)(dx_+^2 + dx_-^2) + (1+f)dx_+dx_- + \frac{dr^2}{f}\right). \tag{4.125}$$

The Wilson loop in question stretches across x_2, and lies at a constant x_+, x_3. It is convenient to choose a world-sheet gauge in which

$$x_- = \tau, \quad x_2 = \sigma. \tag{4.126}$$

Then the action of the string stretching between the two lines is given by

$$S = \frac{1}{2\pi\ell_s^2}\int d\sigma d\tau \sqrt{-det(g_{MN}\partial_\alpha X^M \partial_\beta X^N)} \tag{4.127}$$

and assuming a profile of $r = r(\sigma)$ we obtain

$$S = \frac{L^-}{2\pi\ell_s^2}\int dx_2\, e^{2A_s}\sqrt{\frac{(1-f)}{2}\left(1+\frac{r'^2}{f}\right)}. \tag{4.128}$$

The integrand does not depend explicitly on x_2, so there is a conserved quantity, c:

$$r'\frac{\partial\mathcal{L}}{\partial r'} - \mathcal{L} = \frac{c}{\sqrt{2}}, \tag{4.129}$$

which leads to

$$r'^2 = f\left(\frac{e^{4A_s}(1-f)}{c^2} - 1\right). \tag{4.130}$$

A first assessment of this relation involves determining the zeros and the region of positivity of the right-hand side. f is always positive and vanishes at the horizon. For the second factor we need the asymptotics of $e^{4A_s}(1-f)$. This factor remains positive and bounded from below in the interior and up to the horizon. It vanishes however logarithmically near the boundary as

$$e^{4A_s}(1-f) = \pi T\ell e^{3A(r_h)}\left(-\frac{1}{\beta_0\log(\Lambda r)}\right)^{\frac{8}{3}}\left[1+\mathcal{O}\left(\frac{1}{\log(\Lambda r)}\right)\right]. \tag{4.131}$$

This is unlike the conformal case where we obtain a constant

$$e^{4A_s}(1-f)\big|_{\text{conformal}} = (\pi T\ell)^4. \tag{4.132}$$

Because of this, for fixed c, there is a region near the boundary where r'^2 becomes negative. At this stage we will avoid this region, by using a modified boundary at $r = \epsilon$. We will later show that this gymnastics will be irrelevant for

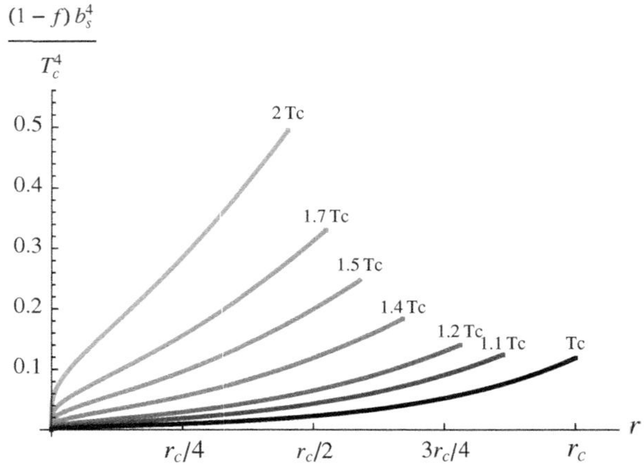

Fig. 4.25 The combination $(1-f)e^{4A_s}$ is plotted as a function of the radial distance, for several temperatures. The radial distance is given in units of the horizon position r_c for the black hole at the critical temperature T_c. All curves stop at the corresponding horizon position

the computation of the jet quenching parameter, as it involves effectively the limit $c \to 0$ (Fig. 4.25).

We will place the modified boundary $r = \epsilon$ a bit inward from the place $r = r_{\min}$ where the factor $\frac{e^{4A_s}(1-f)}{c^2} - 1$ vanishes:

$$e^{4A_s(r_{\min})}(1 - f(r_{\min})) = c^2. \tag{4.133}$$

Therefore we choose $r_{\min} < \epsilon$.

Then, in the range $\epsilon < r < r_h$ the factor $\frac{e^{4A_s}(1-f)}{c^2} - 1$ is positive for sufficiently small c. In this same range, r' vanishes only at $r = r_h$. This is the true turning point of the string world-sheet. This is also what happens in the conformal case. It is also intuitively obvious that the relevant Wilson loop must sample also the region near the horizon.

The constant c is determined by the fact that the two light-like Wilson loops are a $x_2 = L$ distance apart.

$$\frac{L}{2} = \int_{\epsilon}^{r_h} \frac{c\,dr}{\sqrt{f(e^{4A_s}(1-f) - c^2)}} \,. \tag{4.134}$$

The denominator vanishes at the turning point. The singularity is integrable.[31] Therefore, as we are interested in the small L region, it is obvious from the expression above that that c must also be small in the same limit.

This relation can then be expanded in powers of c as

$$\frac{L}{2c} = \int_{\epsilon}^{r_h} \frac{e^{-2A_s} dr}{\sqrt{f(1-f)}} + \frac{c^2}{2} \int_{\epsilon}^{r_h} \frac{e^{-6A_s} dr}{\sqrt{f(1-f)^3}} + \mathcal{O}(c^4)\,. \tag{4.135}$$

[31] Even if we choose $\epsilon = r_{\min}$, the new singularity at $r = r_{\min}$ is also integrable as suggested from (4.131).

Therefore to leading order in L

$$c = \frac{L}{2\int_{\epsilon}^{r_h} \frac{e^{-2A_s} dr}{\sqrt{f(1-f)}}} + \mathcal{O}(L^3).. \tag{4.136}$$

We are now ready to evaluate the Nambu–Goto action of the extremal configuration we have found. Starting from (4.128), we substitute r' from (4.130), and change integraLn variable from $x_2 \to r$ to obtain

$$S = \frac{2L^-}{2\pi\ell_s^2} \int_{\epsilon}^{r_h} dr \frac{e^{4A_s}(1-f)}{\sqrt{2f(e^{4A_s}(1-f) - c^2)}} . \tag{4.137}$$

As in [18, 52], we subtract from (4.137) the action of two free string straight world-sheets that hang down to the horizon. To compute this action a convenient choice of gauge is $x_- = \tau, r = \sigma$. The action of each sheet is

$$S_0 = \frac{L^-}{2\pi\ell_s^2} \int_{\epsilon}^{r_h} dr \sqrt{g_- g_{rr}} = \frac{L^-}{2\pi\ell_s^2} \int_{\epsilon}^{r_h} dr\, e^{2A_s} \sqrt{\frac{1-f}{2f}} . \tag{4.138}$$

The subtracted action is therefore:

$$S_r = S - 2S_0 = \frac{L^- c^2}{2\pi\ell_s^2} \int_{\epsilon}^{r_h} \frac{dr}{e^{2A_s}\sqrt{f(1-f)}} + \mathcal{O}(c^4) , \tag{4.139}$$

Using now (4.136) to substitute c we finally obtain

$$S_r = \frac{L^- L^2}{8\pi\ell_s^2} \frac{1}{\int_{\epsilon}^{r_h} \frac{dr}{e^{2A_s}\sqrt{f(1-f)}}} + \mathcal{O}(L^4) . \tag{4.140}$$

So far we have evaluated the relevant Wilson loop in the fundamental representation (by using probe quarks). On the other hand, the Wilson loop that defines the jet-quenching parameter is an adjoint one. We can obtain it in the large-N_c limit from the fundamental using $tr_{\text{Adjoint}} = tr^2_{\text{Fundamental}}$. We finally extract the jet-quenching parameter as

$$\hat{q} = \frac{\sqrt{2}}{\pi\ell_s^2} \frac{1}{\int_{\epsilon}^{r_h} \frac{dr}{e^{2A_s}\sqrt{f(1-f)}}} . \tag{4.141}$$

We are now ready to remove the cutoff. As the integral appearing is now well-defined up to the real boundary $r = 0$ we may rewrite it as

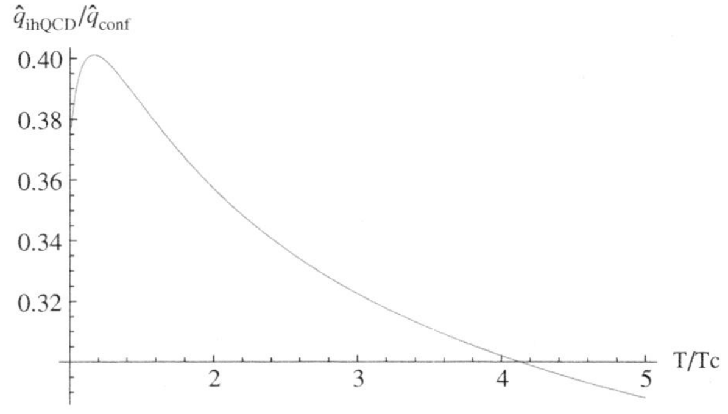

Fig. 4.26 The ratio of the jet quenching parameter in our model to the jet quenching parameter in $\mathcal{N}=4$ is shown. The integral present in (4.141) has been numerically calculated from an effective cutoff at $r = r_h/1000$. The jet quenching parameter in $\mathcal{N}=4$ SYM has been calculated with $\lambda_{\text{'t Hooft}} = 5.5$

$$\int_{\epsilon}^{r_h} \frac{e^{-2A_s} dr}{\sqrt{f(1-f)}} = \int_{0}^{r_h} \frac{e^{-2A_s} dr}{\sqrt{f(1-f)}} - I(\epsilon), \quad I(\epsilon) = \int_{0}^{\epsilon} \frac{e^{-2A_s} dr}{\sqrt{f(1-f)}}. \tag{4.142}$$

In [93] we obtain the small ϵ estimate of $I(\epsilon)$ that vanishes as $\sim \epsilon(\log\,\epsilon)^{\frac{4}{3}}$. $\sim \epsilon(\log\,\epsilon)^{\frac{4}{3}}$We may finally write[32]

$$\hat{q} = \frac{\sqrt{2}}{\pi \ell_s^2} \frac{1}{\int_0^{r_h} \frac{dr}{e^{2A_s}\sqrt{f(1-f)}}}. \tag{4.143}$$

From (4.143) we obtain, in the conformal case:

$$\hat{q}_{\text{conformal}} = \frac{\Gamma\left[\frac{3}{4}\right]}{\Gamma\left[\frac{5}{4}\right]} \sqrt{2\lambda} \pi^{\frac{3}{2}} T^3. \tag{4.144}$$

The conformal value, for the median value of $\lambda = 5.5$ and $T \simeq 250\,\text{MeV}$ gives $\hat{q}_{\text{conformal}} \simeq 1.95\,\text{GeV}^2/\text{fm}$ where we used the conversion $1\,\text{GeV} \simeq 5\,\text{fm}^{-1}$.

Numerical evaluation of (4.143) in the non-conformal IHQCD setup[33] gives us a value of $\hat{q}$ which is lower (at a given temperature) than the conformal value, as shown in Figs. 4.26, 4.27 and 4.28. Tables 4.6, 4.7, 4.8, and 4.9 display the numerical values of the jet quenching parameter at different temperatures in the experimentally relevant range, in different temperature matching schemes.

---

[32] In practise, the previous discussion including regularizing the UV is academic. The numerical calculation is done with a finite cutoff where the boundary conditions for the couplings are imposed.

[33] In this case, the value of ℓ_s appearing in (4.143) is fixed as explained in Sect. 4.4.

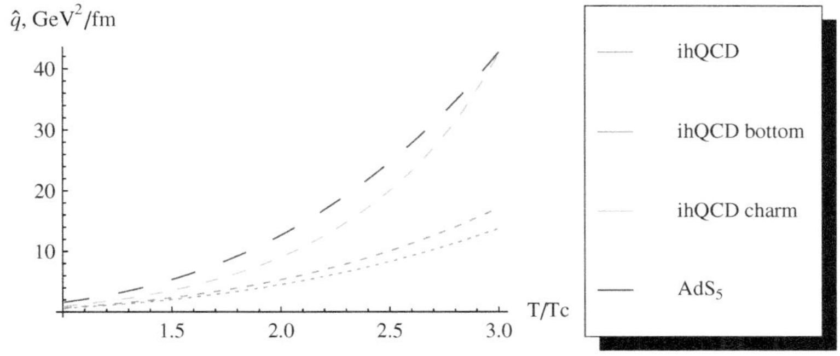

Fig. 4.27 The jet quenching parameter $\hat{q}$ for the Improved Holographic QCD model and $\mathcal{N} = 4$ SYM is shown in units of GeV^2/fm for a region close to $T = T_c$. The *smallest dashed curve* is the ihQCD result with an effective cutoff of $r_{\text{cutoff}} = r_h/1000$. The *small dashed curve* is the ihQCD result with the cutoff from the mass of the Bottom quark. The *medium dashed curve* has a cutoff coming from the Charm mass and *largest dashed curve* is the $\mathcal{N} = 4$ SYM result

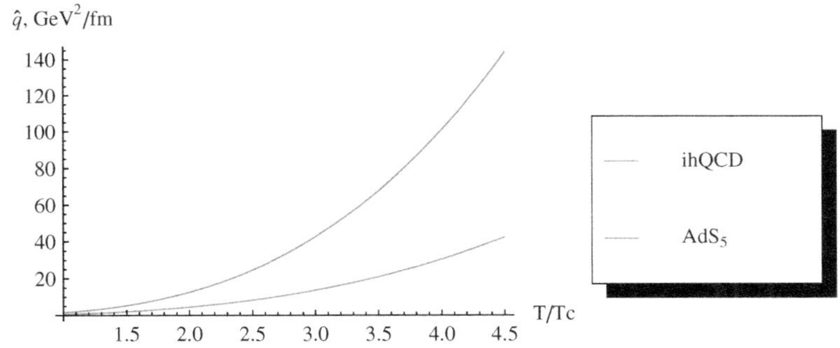

Fig. 4.28 The jet quenching parameter $\hat{q}$ for the Improved Holographic QCD model (*lower curve*) and $\mathcal{N} = 4$ SYM (*upper curve*) are shown in units of GeV^2/fm for temperatures up to $T = 4T_c$

Table 4.6 The jet quenching parameter $\hat{q}$ computed with different cutoffs for the different temperatures shown in the first column

T_{QGP}; MeV	$\hat{q}$(GeV^2/fm) (direct)	$\hat{q}_1$(GeV^2/fm) (direct)
220	–	–
250	0.5	0.6
280	0.8	0.8
310	1.1	1.1
340	1.4	1.4
370	1.8	1.8
400	2.2	2.2

The computation is done in the direct scheme. The second column shows $\hat{q}$ with a cutoff at $r_{\text{cutoff}} = r_h/1000$, where r_h is the location of the horizon. In accordance with the conclusions of appendix $\hat{q}$ does not change significantly as we vary the cutoff from $r_h/1000$ to $r_h/100$

Table 4.7 The jet quenching parameter $\hat{q}$ using the three different comparison schemes

T_{QGP}, MeV	$\hat{q}(\text{GeV}^2/\text{fm})$ (direct)	$\hat{q}(\text{GeV}^2/\text{fm})$ (energy)	$\hat{q}(\text{GeV}^2/\text{fm})$ (entropy)
220	–	0.9	1.0
250	0.5	1.2	1.3
280	0.8	1.6	1.7
310	1.1	2.1	2.2
340	1.4	2.7	2.8
370	1.8	3.4	3.4
400	2.2	4.2	4.2

For lower temperatures the "entropy scheme" gives higher values. As energy is increased the energy and entropy schemes temperatures start to coincide and there is little difference in the jet quenching parameter as well

Table 4.8 The jet quenching parameter $\hat{q}$ using the three different comparison schemes with an effective cutoff provided by the mass of the Charm quark

T_{QGP}, MeV	$\hat{q}_{\text{charm}}$ (GeV^2/fm) (direct)	$\hat{q}_{\text{charm}}$ (GeV^2/fm) (energy)	$\hat{q}_{\text{charm}}$ (GeV^2/fm) (entropy)
220	–	1.3	1.5
250	0.8	1.8	2.0
280	1.2	2.6	2.8
310	1.7	3.5	3.6
340	2.2	4.6	4.7
370	2.8	5.9	6.0
400	3.6	7.6	7.5

Again, for lower temperatures the "entropy scheme" gives higher values. As energy is increased the energy and entropy schemes temperatures start to coincide and there is little difference in the jet quenching parameter as well. Also when the temperature approaches the quark mass the picture of the heavy quark as a hanging string collapses and results are not reliable

Table 4.9 The jet quenching parameter $\hat{q}$ using the three different comparison schemes with an effective cutoff provided by the mass of the Bottom quark

T_{QGP}, MeV	$\hat{q}_{\text{bottom}}$ (GeV^2/fm) (direct)	$\hat{q}_{\text{bottom}}$ (GeV^2/fm) (energy)	$\hat{q}_{\text{bottom}}$ (GeV^2/fm) (entropy)
220	–	1.0	1.1
250	0.6	1.4	1.5
280	0.9	1.9	2.0
310	1.2	2.5	2.6
340	1.6	3.2	3.2
370	2.0	4.0	4.0
400	2.5	5.0	4.9

The results are close to the $\hat{q}$ results computed in Table 4.7 since the mass of the Bottom quark is much larger than the temperatures we examine

4.9 Discussion and Outlook

The construction presented in this paper offers a holographic description of large-N_c Yang–Mills theory that is both realistic and calculable, and in quite good agreement with a large number of lattice results both at zero and finite temperature.

It is a phenomenological model; as such it is not directly associated to an explicit string theory construction. In this respect it is in the same class as the models based on pure AdS backgrounds (with hard or soft walls) [14–16, 106, 107], or on IR deformations of the AdS metric [108, 109]. In comparison to them however, the IHQCD approach has the advantage that the dynamics responsible for strong coupling phenomena (such as confinement and phase transitions) is made explicit in the bulk description, and it is tied to the fact that the coupling constant depends on the energy scale and becomes large in the IR. This makes the model consistent and calculable, once the 5D effective action is specified: the dynamics can be entirely derived from the bulk Einstein's equation. The emergence of an IR mass scale and the finite temperature phase structure are built-in: they need not be imposed by hand and do not suffer from ambiguities related to IR boundary conditions (as in hard wall models) or from inconsistencies in the laws of thermodynamics (as in non-dynamical soft wall models based on a fixed dilaton profile [16] or on a fixed metric [108, 109]). More specifically, in this approach it is guaranteed that the Bekenstein–Hawking temperature of the black hole matches the entropy computed as the derivative of the free energy with respect to temperature.

With an appropriate choice of the potential, a realistic and quantitatively accurate description of essentially all the static properties (spectrum and equilibrium thermodynamics) of the dynamics of pure Yang–Mills can be provided. The main ingredient responsible for the dynamics (the dilaton potential) is fixed through comparison with both perturbative QCD and lattice results. It is worth stressing that such a matching on the quantitative level was only possible because the class of holographic models we discuss *generically* provides a *qualitatively accurate* description of the strong Yang–Mills dynamics. This is a highly non-trivial fact, that strongly indicates that a realistic holographic description of real-world QCD might be ultimately possible.

Although the asymptotics of our potential is dictated by general principles, we base our choice of parameters by comparing with the lattice results for the thermodynamics. There are other physical parameters in the 5D description that do not appear in the potential: the 5D Planck scale, that was fixed by matching the free field thermodynamics in the limit $T \to \infty$; the coefficients in the axion kinetic term, that were set by matching the axial glueball spectrum and the topological susceptibility (from the lattice). The quantities that we use as input in our fit, as well as the corresponding parameters in the 5D model, are shown in the upper half of Table 4.2.

The fact that our potential has effectively two free parameters depends on our choice of the functional form. This functional form contains some degree of arbitrariness, in that only the UV and IR asymptotics of $V(\lambda)$ are fixed by general considerations (matching the perturbative β-function in the UV, and a discrete linear glueball spectrum for the IR). Therefore the results presented in this paper offer more a *description*, rather than a *prediction* of the thermodynamics.

Nevertheless, there are several quantities that we successfully "postdict" (i.e. they agree with the lattice results) once the potential is fixed: apart from the good agreement of the thermodynamic functions over the whole range of temperature explored by the lattice studies (see Figs. 4.5, 4.7 and 4.9), they are the lowest glueball mass ratios and the value of the critical temperature. The comparison of these quantities with the lattice results is shown in the lower half of Table 4.2, and one can see that the agreement is overall very good. Moreover the model predicts the masses of the full towers of glueball states in the $0^{+-}, 0^{++}, 2^{++}$ families.

The fact that IHQCD is consistent with a large number of lattice results is clearly not the end of the story: its added value, and one of the main reasons for its interest lies in its immediate applicability beyond equilibrium thermodynamics, i.e. in the dynamic regimes tested in heavy-ion collision experiments. This is a generic feature of the holographic approach, in which there are no obstructions (as opposed to the lattice) to perform real-time computations and to calculate hydrodynamics and transport coefficients. IHQCD provides a framework to compute these quantities in a case where the static properties agree with the real-word QCD at the quantitative level. Therefore the bulk viscosity, drag force and jet quenching parameters were computed in IHQCD.

4.9.1 Bulk Viscosity

The bulk viscosity was computed by calculating the low frequency asymptotics of the appropriate stress tensor correlator holographically. The result is that the bulk viscosity rises near the phase transition but stays always below the shear viscosity. It floats somewhat above the Buchel bound, with a coefficient of proportionality varying between 1 and 2. Therefore it is expected to affect the elliptic flow at the small percentage level [44, 98]. Knowledge of the bulk viscosity is important in extracting the shear viscosity from the data. This result is not in agreement with the lattice result near T_c. In particular the lattice result gives a value for the viscosity that is ten times larger.

The bulk viscosity keeps increasing in the black-hole branch below the transition point until the large BH turns into the small BH at a temperature $T_{\min}$. The bulk viscosity on the small BH background is always larger than the respective one in the large BH background. In particular, it can be shown that the T derivative of the quantity ζ/s diverges at $T_{\min}$. This is the holographic reason for the presence of a peak in ζ/s near T_c. On the other hand, as it is shown in [29], the presence of $T_{\min}$ (i.e. a small BH branch) is in one-to-one correspondence with color confinement at zero T. We arrive thus at the suggestion that in a (large N) gauge theory that confines color at zero T, there shall be a rise in ζ/s near T_c.

An important ingredient here is the value of the viscosity asymptotically in the small BH branch. There the asymptotic value correlates to the IR behavior of the potential. Taking also into account the fact that this asymptotic value is very close to the value of the bulk viscosity near T_c, we can derive bounds that suggest that the bulk viscosity cannot increase a lot near T_c.

4.9.2 Drag Force

The drag force calculated from IHQCD has the expected behavior. Although it increases with temperature, it does so slower than in $\mathcal{N} = 4$ SYM, signaling the effects of asymptotic freedom.

4.9.3 Diffusion Time

Based on the drag force calculation the diffusion times can be computed for a heavy external quark. The numerical values obtained are in agreement with phenomenological models [50]. To accommodate for the fact that IHQCD exhibits a phase transition around $T = 247\,\text{MeV}$ (i.e. about 30% higher than in QCD), the results are compared using alternative schemes, as proposed in [19, 20]. For example, for an external Charm quark of momentum $p = 10\,\text{GeV}$ (in the alternative scheme) a diffusion time of $\tau = 2.6\,\text{fm}$ at temperature $T = 280\,\text{MeV}$ is found. Similarly, for a Bottom quark of the same momentum and at the same temperature, $\tau = 6.3\,\text{fm}$. Generally the numbers obtained are close to those obtained by [50] and [54].

4.9.4 Jet Quenching

The jet quenching parameter of this model, has been also calculated, based on the formalism of [18, 52] by computing the appropriate light-like Wilson loop. $\hat{q}$ grows with temperature, but slower than the T^3 growth of $\mathcal{N} = 4$ SYM result. Again this can be attributed to the incorporation of asymptotic freedom in IHQCD. Using the alternative scheme to compare with experiment, the results are close to the lower quoted values of $\hat{q}$. For example, for a temperature of $T = 290\,\text{MeV}$, which in the alternative "energy scheme" corresponds to a temperature of $T = 395\,\text{MeV}$ in our model, we find that $\hat{q} \approx 2\,\text{GeV}^2/\text{fm}$.

However, the numbers obtained for this particular definition of jet quenching parameter seem rather low and indicate that this may not be the most appropriate definition in the holographic context. There are other ways to define $\hat{q}$, in particular using the fluctuations of the trailing string solution. This is gives a direct and more detailed input in the associated Langevin dynamics and captures the asymmetry between longitudinal and transverse fluctuations. It would be interesting to compute this, along the lines set in [57, 59, 60].

Acknowledgements We would like to thank the numerous colleagues that have shared their insights with us via discussions and correspondence since 2007 that this line of research has started: O. Aharony, L. Alvarez-Gaumé, B. Bringoltz, R. Brower, M. Cacciari, J. Casalderrey-Solana, R. Emparan , F. Ferrari, B. Fiol, P. de Forcrand, R. Granier de Cassagnac, L. Giusti, S. Gubser, K. Hashimoto, T. Hertog, U. Heinz, D. K. Hong, G. Horowitz, E. Iancu, K. Intriligator, K. Kajantie, F. Karsch, D. Kharzeev, D. Kutasov, H. Liu, B. Lucini, M. Luscher, J. Mas, D. Mateos, H. B. Meyer, C. Morningstar, V. Niarchos, C. Nunez, Y. Oz, H. Panagopoulos, S. Pal, M. Panero, I. Papadimitriou, A. Paredes, A. Parnachev, G. Policastro, S. Pufu, K. Rajagopal, F.Rocha, P. Romatchke, C. Salgado, F. Sannino, M. Shifman, E. Shuryak, S. J. Sin, C. Skenderis, D. T. Son, J. Sonnenschein, S. Sugimoto, M. Taylor, M. Teper, J. Troost, A. Tseytlin, A. Vainshtein, G. Veneziano, A. Vladikas, L. Yaffe, A. Yarom and U. Wiedemann.This work was partially supported by a European Union grant FP7-REGPOT-2008-1-CreteHEPCosmo-228644, and ANR grant NT05-1-41861. Work of LB has been partly funded by INFN, Ecole Polytechnique (UMR du CNRS 7644), MEC and FEDER under grant FPA2008-01838, by the Spanish Consolider-Ingenio 2010 Programme CPAN (CSD2007-00042) and by Xunta de Galicia (Consellería de Educación and grant PGIDIT06PXIB206185PR).E. K. thanks the organizers and especially E. Papantonopoulos for organizing a very interesting and stimulating school.

References

1. Adams, J., et al. [STAR Collaboration]: Experimental and theoretical challenges in the search for the quark gluon plasma: the STAR collaboration's critical assessment of the evidence from RHIC collisions. Nucl. Phys. A **757**, 102 (2005) [ArXiv:nucl-ex/0501009]
2. Back, B.B., et al.: The PHOBOS perspective on discoveries at RHIC. Nucl. Phys. A **757**, 28 (2005) [ArXiv:nucl-ex/0410022]
3. Arsene, I., et al. [BRAHMS Collaboration]: Quark gluon plasma and color glass condensate at RHIC? The perspective from the BRAHMS experiment. Nucl. Phys. A **757**, 1 (2005) [ArXiv:nucl-ex/0410020]
4. Adcox, K., et al. [PHENIX Collaboration]: Formation of dense partonic matter in relativistic nucleus–nucleus collisions at RHIC: experimental evaluation by the PHENIX collaboration. Nucl. Phys. A **757**, 184 (2005) [ArXiv:nucl-ex/0410003]
5. Luzum, M., Romatschke, P.: Conformal relativistic viscous hydrodynamics: applications to RHIC results at $\sqrt{s_{NN}} = 200\,\mathrm{GeV}$. Phys. Rev. C **78**, 034915 (2008) [ArXiv:0804.4015][nucl-th]
6. Policastro, G., Son, D.T., Starinets, A.O.: The shear viscosity of strongly coupled N = 4 supersymmetric Yang–Mills plasma. Phys. Rev. Lett. **87**, 081601 (2001) [ArXiv:hep-th/0104066]
7. Kovtun, P., Son, D.T., Starinets, A.O.: Viscosity in strongly interacting quantum field theories from black-hole physics. Phys. Rev. Lett. **94**, 111601 (2005) [arXiv:hep-th/0405231]
8. Shuryak, E.: Physics of strongly coupled quark–gluon plasma. Prog. Part. Nucl. Phys. **62**, 48–101 (2009) [ArXiv:0807.3033][hep-ph]
9. Son, D.T., Starinets, A.O.: Viscosity, black holes, and quantum field theory. Ann. Rev. Nucl. Part. Sci. **57**, 95 (2007) [ArXiv:0704.0240][hep-th]
10. Natsuume, M.: String theory and quark–gluon plasma [ArXiv:hep-ph/0701201]

11. Karsch, F., Kharzeev, D., Tuchin, K.: Universal properties of bulk viscosity near the QCD phase transition. Phys. Lett. B **663**, 217 (2008) [ArXiv:0711.0914][hep-ph]
12. Meyer, H.B.: A calculation of the bulk viscosity in SU(3) gluodynamics. Phys. Rev. Lett. **100**, 162001 (2008) [ArXiv:0710.3717][hep-lat]
13. Meyer, H.B.: Energy–momentum tensor correlators and viscosity. PoS LATTICE 2008 017 (2008) [ArXiv:0809.5202][hep-lat]
14. Erlich, J., Katz, E., Son, D.T., Stephanov, M.A.: QCD and a holographic model of hadrons. Phys. Rev. Lett. **95**, 261602 (2005) [arXiv:hep-ph/0501128]
15. Da Rold, L., Pomarol, A.: Chiral symmetry breaking from five dimensional spaces. Nucl. Phys. B **721**, 79 (2005) [ArXiv:hep-ph/0501218]
16. Karch, A., Katz, E., Son, D.T., Stephanov, M.A.: Linear confinement and AdS/QCD. Phys. Rev. D **74**, 015005 (2006) [ArXiv:hep-ph/0602229]
17. Herzog, C.P., Karch, A., Kovtun, P., Kozcaz, C., Yaffe, L.G.: Energy loss of a heavy quark moving through N = 4 supersymmetric Yang–Mills plasma. JHEP **0607**, 013 (2006) [ArXiv:hep-th/0605158]
18. Liu, H., Rajagopal, K., Wiedemann, U.A.: Calculating the jet quenching parameter from AdS/CFT. Phys. Rev. Lett. **97**, 182301 (2006) [ArXiv:hep-ph/0605178]
19. Gubser, S.S.: Drag force in AdS/CFT. Phys. Rev. D **74**, 126005 (2006) [ArXiv:hep-th/0605182]
20. Gubser, S.S.: Comparing the drag force on heavy quarks in N = 4 super-Yang–Mills theory and QCD. Phys. Rev. D **76**, 126003 (2007) [ArXiv:hep-th/0611272]
21. Casalderrey-Solana, J., Teaney, D.: Heavy quark diffusion in strongly coupled N = 4 Yang Mills. Phys. Rev. D **74**, 085012 (2006) [ArXiv:hep-ph/0605199]
22. Polchinski, J., Strassler, M.J.: Hard scattering and gauge/string duality. Phys. Rev. Lett. **88**, 031601 (2002) [ArXiv:hep-th/0109174]
23. Gursoy, U., Kiritsis, E.: Exploring improved holographic theories for QCD: Part I. JHEP **0802**, 032 (2008) [ArXiv:0707.1324][hep-th]
24. Gursoy, U., Kiritsis, E., Nitti, F.: Exploring improved holographic theories for QCD: Part II. JHEP **0802**, 019 (2008) [ArXiv:0707.1349][hep-th]
25. Kiritsis, E.: Dissecting the string theory dual of QCD. Fortsch. Phys. **57**, 396–417 (2009) [ArXiv:0901.1772][hep-th]
26. Gubser, S.S., Nellore, A.: Mimicking the QCD equation of state with a dual black hole. Phys. Rev. D**78**, 086007 (2008) [ArXiv:0804.0434][hep-th]
27. Casero, R., Kiritsis, E., Paredes, A.: Chiral symmetry breaking as open string tachyon condensation. Nucl. Phys. B **787**, 98–134 (2007) [ArXiv:hep-th/0702155]
28. Gursoy, U., Kiritsis, E., Mazzanti, L., Nitti, F.: Deconfinement and gluon plasma dynamics in improved holographic QCD. Phys. Rev. Lett. **101**, 181601 (2008) [ArXiv:0804.0899][hep-th]
29. Gursoy, U., Kiritsis, E., Mazzanti, L., Nitti, F.: Holography and thermodynamics of 5D dilaton-gravity. JHEP **0905**, 033 (2009) [ArXiv:0812.0792][hep-th]
30. DeWolfe, O., Rosen, C.: Robustness of sound speed and jet quenching for gauge/gravity models of hot QCD. JHEP **0907**, 022 (2009) [ArXiv:0903.1458][hep-th]
31. Vicari, E., Panagopoulos, H.: Theta dependence of SU(N) gauge theories in the presence of a topological term. Phys. Rep. **470**, 93–150 (2009) [ArXiv:0803.1593][hep-th]
32. Gursoy, U., Kiritsis, E., Mazzanti, L., Nitti, F.: Improved holographic Yang–Mills at finite temperature: comparison with data. Nucl. Phys. B **820**, 148 (2009) [ArXiv:0903.2859][hep-th]
33. Kraus, P.: Dynamics of anti-de Sitter domain walls. JHEP **9912**, 011 (1999) [ArXiv:hep-th/9910149]
34. Kiritsis, E.: Supergravity, D-brane probes and thermal super Yang–Mills: a comparison. JHEP **9910**, 010 (1999) [ArXiv:hep-th/9906206]
35. Kehagias, A., Kiritsis, E.: Mirage cosmology. JHEP **9911**, 022 (1999) [arXiv:hep-th/9910174]

36. Kiritsis, E.: Holography and brane-bulk energy exchange. JCAP **0510**, 014 (2005) [ArXiv:hep-th/0504219]
37. Baier, R., Romatschke, P., Son, D.T., Starinets, A.O., Stephanov, M.A.: Relativistic viscous hydrodynamics, conformal invariance, and holography. JHEP **0804**, 100 (2008) [ArXiv: 0712.2451][hep-th]
38. Bhattacharyya, S., Hubeny, V.E., Minwalla, S., Rangamani, M.: Nonlinear fluid dynamics from gravity. JHEP **0802**, 045 (2008) [ArXiv:0712.2456][hep-th]
39. Meyer, H.B.: A calculation of the shear viscosity in SU(3) gluodynamics. Phys. Rev. D **76**, 101701 (2007) [ArXiv:0704.1801][hep-lat]
40. Kharzeev, D., Tuchin, K.: Bulk viscosity of QCD matter near the critical temperature. JHEP **0809**, 093 (2008) [ArXiv:0705.4280][hep-ph]
41. Karsch, F., Kharzeev, D., Tuchin, K.: Universal properties of bulk viscosity near the QCD phase transition. Phys. Lett. B **663**, 217 (2008) [ArXiv:0711.0914][hep-ph]
42. Teaney, D.: Finite temperature spectral densities of momentum and R-charge correlators in N = 4 Yang Mills theory. Phys. Rev. D **74**, 045025 (2006) [ArXiv:hep-ph/0602044]
43. Romatschke, P., Son, D.T.: Spectral sum rules for the quark–gluon plasma. Phys. Rev. D **80**, 065021 (2009) [ArXiv:0903.3946][hep-ph]
44. Heinz, U.: Talk at the Extra strong quark gluon plasma (ESQGP), Stony Brook (2008)
45. Adare, A., et al. [PHENIX Collaboration]: Energy loss and flow of heavy quarks in Au+Au collisions at $\sqrt{s_{NN}} = 200$ GeV. Phys. Rev. Lett. **98**, 172301 (2007) [ArXiv:nucl-ex/ 0611018]
46. Borghini, N., Wiedemann, U.A.: Predictions for the LHC heavy ion programme. J. Phys. G **35**, 023001 (2008) [ArXiv:0707.0564][hep-ph]
47. Baier, R., Dokshitzer, Y.L., Mueller, A.H., Peigne, S., Schiff, D.: Radiative energy loss of high energy quarks and gluons in a finite-volume quark–gluon plasma. Nucl. Phys. B **483**, 291 (1997) [ArXiv:hep-ph/9607355]
48. Baier, R., Dokshitzer, Y.L., Mueller, A.H., Peigne, S., Schiff, D.: Radiative energy loss and p(T)-broadening of high energy partons in nuclei. Nucl. Phys. B **484**, 265 (1997) [ArXiv:hep-ph/9608322]
49. Zakharov, B.G.: Radiative energy loss of high energy quarks in finite-size nuclear matter and quark–gluon plasma. JETP Lett. **65**, 615 (1997) [ArXiv:hep-ph/9704255]
50. Akamatsu, Y., Hatsuda, T., Hirano, T.: Heavy quark diffusion with relativistic langevin dynamics in the quark–gluon fluid [ArXiv:0809.1499][hep-ph]
51. Gubser, S.S., Pufu, S.S., Rocha, F.D., Yarom, A.: Energy loss in a strongly coupled thermal medium and the gauge–string duality [ArXiv:0902.4041][hep-th]
52. Liu, H., Rajagopal, K., Wiedemann, U.A.: Wilson loops in heavy ion collisions and their calculation in AdS/CFT. JHEP **0703**, 066 (2007) [ArXiv:hep-ph/0612168]
53. Svetitsky, B.: Diffusion of charmed quark in the quark gluon plasma. Phys. Rev. D **37**, 2484 (1988)
54. van Hees, H., Rapp, R.: Thermalization of heavy quarks in the quark–gluon plasma. Phys. Rev. C **71**, 034907 (2005) [ArXiv:nucl-th/0412015]
55. Moore, G.D., Teaney, D.: How much do heavy quarks thermalize in a heavy ion collision?. Phys. Rev. C **71**, 064904 (2005) [ArXiv:hep-ph/0412346]
56. van Hees, H., Greco, V., Rapp, R.: Heavy-quark probes of the quark–gluon plasma at RHIC. Phys. Rev. C **73**, 034913 (2006) [ArXiv:nucl-th/0508055]
57. Gubser, S.S.: Momentum fluctuations of heavy quarks in the gauge–string duality. Nucl. Phys. B **790**, 175 (2008) [ArXiv:hep-th/0612143]
58. de Boer, J., Hubeny, V.E., Rangamani, M., Shigemori, M.: Brownian motion in AdS/CFT. JHEP **0907**, 094 (2004) [ArXiv:0812.5112][hep-th]
59. Son, D.T., Teaney, D.: Thermal noise and stochastic strings in AdS/CFT. JHEP 0907 021, (2009), [ArXiv:0901.2338 [hep-th]
60. Giecold, G.C., Iancu, E., Mueller, A.H.: Stochastic trailing string and Langevin dynamics from AdS/CFT. JHEP 0907 033, (2009), [ArXiv:0903.1840] [hep-th]

61. Debbasch, F., Mallick, K., Rivet, K.P.: Relativistic Ornstein–Uhlenbeck process. J. Stat. Phys. **88**, 945 (1997)
62. Chevalier, C., Debbasch, F.: Relativistic diffusions: a unifying approach. J. Math. Phys. **49**, 043303 (2008)
63. Caron-Huot, S., Kovtun, P., Moore, G.D., Starinets, A., Yaffe, L.G.: Photon and dilepton production in supersymmetric Yang–Mills plasma. JHEP **0612**, 015 (2006) [ArXiv:hep-th/0607237]
64. Sin, S.J., Zahed, I.: Holography of radiation and jet quenching. Phys. Lett. B **608**, 265 (2005) [ArXiv:hep-th/0407215]
65. Sin, S.J., Zahed, I.: Ampere's law and energy loss in AdS/CFT duality. Phys. Lett. B **648**, 318 (2007) [ArXiv:hep-ph/0606049]
66. Witten, E.: Anti-de Sitter space, thermal phase transition, and confinement in gauge theories. Adv. Theor. Math. Phys. **2**, 505 (1998) [ArXiv:hep-th/9803131]
67. Kinar, Y., Schreiber, E., Sonnenschein, J.: Q anti–Q potential from strings in curved spacetime: classical results. Nucl. Phys. B **566**, 103 (2000) [ArXiv:hep-th/9811192]
68. Heitger, J., Simma, H., Sommer, R., Wolff, U. [ALPHA collaboration]: The Schroedinger functional coupling in quenched QCD at low energies. Nucl. Phys. Proc. Suppl. **106**, 859 (2002) [ArXiv:hep-lat/0110201]
69. Alanen, J., Kajantie, K., Suur-Uski, V.: A gauge/gravity duality model for gauge theory thermodynamics. Phys. Rev. D **80**, 126008 (2009) [ArXiv:0911.2114][hep-ph]
70. Boyd, G., Engels, J., Karsch, F., Laermann, E., Legeland, C., Lutgemeier, M., Petersson, B.: Thermodynamics of SU(3) lattice gauge theory. Nucl. Phys. B **469**, 419 (1996) [ArXiv:hep-lat/9602007]
71. Lucini, B., Teper, M., Wenger, U.: The deconfinement transition in SU(N) gauge theories. Phys. Lett. B **545**, 197 (2002) [ArXiv:hep-lat/0206029]
72. Lucini, B., Teper, M., Wenger, U.: The high temperature phase transition in SU(N) gauge theories. JHEP **0401**, 061 (2004) [ArXiv:hep-lat/0307017]
73. Bringoltz, B., Teper, M.: The pressure of the SU(N) lattice gauge theory at large-N. Phys. Lett. B **628**, 113 (2005) [ArXiv:hep-lat/0506034]
74. Panero, M.: Thermodynamics of the QCD plasma and the large-N limit. Phys. Rev. Lett. **103**, 232001 (2009) [ArXiv:0907.3719][hep-lat]
75. Panero, M.: Thermodynamics of the strongly interacting gluon plasma in the large-N limit. PoS LAT2009 172, (2009), [ArXiv:0912.2448][hep-lat]
76. Lucini, B., Teper, M., Wenger, U.: Properties of the deconfining phase transition in SU(N) gauge theories. JHEP **0502**, 033 (2005) [ArXiv:hep-lat/0502003]
77. Kiritsis, E., Nitti, F.: On massless 4D gravitons from 5D asymptotically AdS space-times. Nucl. Phys. B **772**, 67 (2007) [ArXiv:hep-th/0611344]
78. Morningstar, C.J., Peardon, M.J.: The glueball spectrum from an anisotropic lattice study. Phys. Rev. D **60**, 034509 (1999) [ArXiv:hep-lat/9901004]
79. Chen, Y. et al.: Glueball spectrum and matrix elements on anisotropic lattices. Phys. Rev. D **73**, 014516 (2006) [ArXiv:hep-lat/0510074]
80. Lucini, B., Teper, M.: SU(N) gauge theories in four dimensions: Exploring the approach to $N = \infty$. JHEP **0106**, 050 (2001) [ArXiv:hep-lat/0103027]
81. Meyer, H.B.: Glueball Regge trajectories [ArXiv:hep-lat/0508002]
82. Cheng, M. et al.: The spatial string tension and dimensional reduction in QCD. Phys. Rev. D **78**, 034506 (2008) [ArXiv:0806.3264][hep-lat]
83. Schroder, Y., Laine, M.: Spatial string tension revisited. PoS LAT **LAT2005**, 180 (2006) [ArXiv:hep-lat/0509104]
84. Alanen, J., Kajantie, K., Suur-Uski, V.: Spatial string tension of finite temperature QCD matter in gauge/gravity duality. Phys. Rev. D **80**, 075017 (2009) [ArXiv:0905.2032][hep-ph]
85. Zeng, D.f.: Heavy quark potentials in some renormalization group revised AdS/QCD models. Phys. Rev. D **78**, 126006 (2008) [ArXiv:0805.2733][hep-th]

86. Galow, B., Megias, E., Nian, J., Pirner, H.J.: Phenomenology of AdS/QCD and its gravity dual. Nucl.Phys.B **834** 330–362, (2010), [arXiv:0911.0627]
87. Noronha, J.: Connecting polyakov loops to the thermodynamics of $SU(N_c)$ gauge theories using the gauge–string duality. Nucl.Phys.B **834** 330–362, (2010) [ArXiv:0911.0627] [hep-ph]
88. Noronha, J.: Polyakov loops in strongly-coupled plasmas with gravity duals. J.Phys.G **37** 094018, (2010) [ArXiv:1001.3155][hep-th]
89. Del Debbio, L., Giusti, L., Pica, C.: Topological susceptibility in the SU(3) gauge theory. Phys. Rev. Lett. **94**, 032003 (2005) [ArXiv:hep-th/0407052]
90. Moore, G.D., Saremi, O.: Bulk viscosity and spectral functions in QCD. JHEP **0809**, 015 (2008) [ArXiv:0805.4201][hep-ph]
91. Springer, T.: Sound mode hydrodynamics from bulk scalar fields. Phys. Rev. D **79**, 046003 (2009) [[ArXiv:arXiv:0810.4354][hep-th]]
92. Gubser, S.S., Pufu, S.S., Rocha, F.D.: Bulk viscosity of strongly coupled plasmas with holographic duals. JHEP **0808**, 085 (2008) [ArXiv:0806.0407][hep-th]
93. Gursoy, U., Kiritsis, E., Michalogiorgakis, G., Nitti, F.: Thermal transport and drag force in improved holographic QCD. JHEP **0912**, 056 (2009) [ArXiv:0906.1890][hep-ph]
94. Arnold, P., Dogan, C., Moore, G.D.: The bulk viscosity of high-temperature QCD. Phys. Rev. D **74**, 085021 (2006) [arXiv:hep-ph/0608012]
95. Chamblin, H.A., Reall, H.S.: Dynamic dilatonic domain walls. Nucl. Phys. B **562**, 133 (1999) [arXiv:hep-th/9903225]
96. Buchel, A.: Bulk viscosity of gauge theory plasma at strong coupling. Phys. Lett. B **663**, 286 (2008) [ArXiv:0708.3459][hep-th]
97. Kanitscheider, I., Skenderis, K.: Universal hydrodynamics of non-conformal branes. JHEP **0904**, 062 (2009) [ArXiv:0901.1487][hep-th]
98. Song, H., Heinz, U.W.: Extracting the QGP viscosity from RHIC data—a status report from viscous hydrodynamics. J.Phys.G**36** 064033, (2009) [ArXiv:0812.4274][nucl-th]
99. Yarom, A.: Notes on the bulk viscosity of holographic gauge theory plasmas, JHEP**1004** 024, (2010) [ArXiv:0912.2100][hep-th]
100. Cherman, A., Cohen, T.D., Nellore, A.: A bound on the speed of sound from holography. Phys. Rev. D **80**, 066003 (2009) [ArXiv:0905.0903][hep-th]
101. Cherman, A., Nellore, A.: Universal relations of transport coefficients from holography. Phys. Rev. D **80**, 066006 (2009) [ArXiv:0905.2969][hep-th]
102. Hohler, P.M., Stephanov, M.A.: Holography and the speed of sound at high temperatures. Phys. Rev. D **80**, 066002 (2009) [ArXiv:0905.0900][hep-th]
103. Beuf, G., Marquet, C., Xiao, B.W.: Heavy-quark energy loss and thermalization in a strongly coupled SYM plasma. Phys.Rev.D**80** 085001, (2009) [ArXiv:0812.1051][hep-ph]
104. Datta, S., Karsch, F., Petreczky, P., Wetzorke, I.: Behavior of charmonium systems after deconfinement. Phys. Rev. D **69**, 094507 (2004) [ArXiv:hep-lat/0312037]
105. Luzum, M., Romatschke, P.: Viscous hydrodynamic predictions for nuclear collisions at the LHC. Phys.Rev.Lett.**103** 262302, (2009) [ArXiv:arXiv:0901.4588][nucl-th]
106. Herzog, C.P.: A holographic prediction of the deconfinement temperature. Phys. Rev. Lett. **98**, 091601 (2007) [ArXiv:hep-th/0608151]
107. Ballon Bayona, C.A., Boschi-Filho, H., Braga, N.R.F., Pando Zayas, L.A.: On a holographic model for confinement/deconfinement. Phys. Rev. D **77**, 046002 (2008) [ArXiv:0705.1529] [hep-th]
108. Andreev, O.: Some thermodynamic aspects of pure glue, fuzzy bags and gauge/string duality. Phys. Rev. D **76**, 087702 (2007) [ArXiv:0706.3120][hep-ph]
109. Kajantie, K., Tahkokallio, T., Yee, J.T.: Thermodynamics of AdS/QCD. JHEP **0701**, 019 (2007) [ArXiv:hep-ph/0609254]

Chapter 5
The Dynamics of Quark-Gluon Plasma and AdS/CFT

Romuald A. Janik

Abstract In these pedagogical lectures, we present the techniques of the AdS/CFT correspondence which can be applied to the study of real time dynamics of a strongly coupled plasma system. These methods are based on solving gravitational Einstein's equations on the string/gravity side of the AdS/CFT correspondence. We illustrate these techniques with applications to the boost-invariant expansion of a plasma system. We emphasize the common underlying AdS/CFT description both in the large proper time regime where hydrodynamic dynamics dominates, and in the small proper time regime where the dynamics is far from equilibrium. These AdS/CFT methods provide a fascinating arena interrelating General Relativity phenomenae with strongly coupled gauge theory physics.

5.1 Introduction

The current experimental program of heavy-ion collisions at RHIC and the forthcoming experiments at LHC open an interesting window onto properties of QCD matter at high temperatures, where it appears in the guise of a new phase—the quark-gluon plasma. At asymptotically high temperatures it should be a free gas of quarks and gluons, however, at the experimentally accessible energies there are strong indications that the quark-gluon plasma is indeed a strongly coupled system (see e.g. [1]).

R. A. Janik (✉)
Institute of Physics, Jagiellonian University, ul. Reymonta 4, 30-059 Kraków, Poland
e-mail: romuald@th.if.uj.edu.pl

E. Papantonopoulos (ed.), *From Gravity to Thermal Gauge Theories: The Ads/CFT Correspondence,* Lecture Notes in Physics, 828,
DOI: 10.1007/978-3-642-04864-7_5,

This poses numerous problems for its theoretical description, yet at the same time makes its study theoretically interesting. One can roughly differentiate the physical properties of the quark-gluon plasma system into two broad categories—static and dynamic (real-time) properties.

The first of these, the static properties, typically involve the study of equilibrium thermodynamics, the entropy, energy density as a function of temperature and more generally properties which can be directly deduced from the Euclidean formulation of finite temperature gauge theory. In this case lattice QCD is an effective tool for accessing these properties in the nonperturbative, strongly coupled regime. It deals directly with QCD and yields quantitative results directly applicable for the QCD quark-gluon plasma.

The second class, the real time dynamic properties of strongly coupled plasma are much more difficult to access. They have to be formulated directly in Minkowski space and since lattice QCD methods are inherently Euclidean, it is very difficult to extrapolate numerical results to Minkowski signature. Moreover, it is exactly these kind of properties which are particularly relevant for the quark-gluon plasma produced in heavy-ion collisions.

To this end let us recall schematically the basic stages of a heavy-ion collision. First the two ultrarelativistic nuclei collide and the plasma is produced in a state very far from equilibrium. Then in a relatively very short time, it becomes thermalized (or at least the pressure becomes isotropic with the residual anisotropy wholly due to flow). From that point on, hydrodynamic phenomenological models seem to describe the properties of the expanding plasma quite well [2, 3]. In particular the plasma expands and cools, and when the temperature reaches the confinement/deconfinement phase transition one expects hadronization to occur.

It would be interesting to understand these various stages of the plasma dynamics directly from first principles. For example, one would like to *derive* the hydrodynamic behaviour from some theoretical framework and not only use it as a phenomenological model. But what is even more interesting is the understanding of the process of thermalization and what governs the short thermalization time necessary for the applicability of hydrodynamic models.

Unfortunately, for the case of QCD we lack appropriate theoretical methods which would be applicable to these kinds of problems at strong coupling. A possible route that one may take is to consider an analogous set of problems in a different theory for which appropriate real-time nonperturbative tools exist.

The new method for studying nonperturbatively various gauge theories (although not directly QCD) is the AdS/CFT correspondence [4–7], which translates dynamical problems in strongly coupled gauge theory into (usually) gravitational ones in higher number of dimensions. Its main advantage is that it works equally well in Minkowski as well as in Euclidean signature. In these lectures we will consider the AdS/CFT correspondence in its simplest original setting for the maximally supersymmetric $\mathcal{N} = 4$ Super-Yang–Mills theory. Of course, one has to be aware that the cost of switching the theory of interest from QCD to $\mathcal{N} = 4$

SYM is that we may most probably loose direct quantitative applicability of our results to realistic heavy-ion collisions. Moreover, there are certain marked differences between the theories which will cause a huge difference for certain physical phenomenae, at the same time being unimportant for other questions. We will discuss some of these points in these lectures.

Nevertheless, we would like to point out that currently we do not have *any* gauge theory in which we would have a theoretical understanding of the issues described earlier. Therefore it is very interesting to study these issues for the case of $\mathcal{N} = 4$ SYM and use the results as a point of reference for analyzing the situation in QCD. Later, one could try to generalize these results to more complicated versions of the AdS/CFT correspondence for theories closer to QCD. In essence, this motivation is in line with the statement that *$\mathcal{N} = 4$ SYM is the harmonic oscillator of four dimensional gauge theories*. If one tries to develop some theoretical tools, one should better first apply them to the 'harmonic oscillator'.

We have tried to make these lectures very pedagogical and self-contained. Our main emphasis in the presentation is to show how one can use the AdS/CFT correspondence as a tool even for far from equilibrium configurations without presupposing any kind of dynamics (which are in fact not known for nonlinear far from equilibrium systems). Therefore our presentation of hydrodynamics is subordinate to this goal, especially as a very general discussion focused on hydrodynamics *per se* is contained in the lectures by Hubeny at this school [8].

The plan of these lectures is as follows. In Sect. 5.2, we introduce the AdS/CFT correspondence, in Sect. 5.3 we compare some properties of plasma in the $\mathcal{N} = 4$ theory and in QCD. Then we proceed to present the AdS/CFT setup specialized to the study of time-dependent dynamics of strongly coupled plasma. We then illustrate these methods in Sect. 5.5 by discussing two important examples—the appearance from this setup of the standard planar AdS black hole, and a planar shock wave. In this section we also discuss some subtleties arising with different choices of coordinate systems which will be relevant later. Then in Sect. 5.6, we introduce the main physical example of a time-dependent plasma configuration—the boost invariant flow. In Sect. 5.7, we analyze its large proper time asymptotics and show how nonlinear perfect fluid dynamics arises from the AdS/CFT methods. In Sect. 5.8, we show how one can see first corrections coming from shear viscosity, and in Sect. 5.9, for completeness, we will summarize briefly the current status of hydrodynamics in AdS/CFT. Then in Sect. 5.10, we introduce physical situations, where the plasma dynamics is not describable by hydrodynamics, and finally, in Sect. 5.11, we apply the AdS/CFT methods to study boost invariant flow in the far from equilibrium small proper time regime. We close the lectures with conclusions and an appendix with a short guide to the literature regarding work done on related topics which were not covered here.

5.2 The AdS/CFT Correspondence

The AdS/CFT correspondence [4–7], in its original form states the equivalence of two apparently completely different theories: the $\mathcal{N} = 4$ supersymmetric Yang–Mills theory (SYM) in four dimensions and type IIB superstrings in a ten-dimensional curved $AdS_5 \times S^5$ background. Since then, it has been generalized in various directions, extending it to a wider class of gauge theories, at the cost of making the dual string backgrounds more complicated. Throughout these lectures we will stay within the context of the AdS/CFT correspondence for $\mathcal{N} = 4$ SYM theory.

The reason why the AdS/CFT correspondence is so interesting is that the nonperturbative strong coupling regime of the $\mathcal{N} = 4$ gauge theory is mapped to the (semi-)classical strings or just (super)gravity which, in contrast to the gauge theory side, is at least theoretically tractable. Therefore one can use the AdS/CFT correspondence as a new method for accessing the very difficult nonperturbative gauge theory physics.

The AdS/CFT correspondence is an equivalence, so in principle any state/phenomenon on the gauge theory side should have its direct counterpart on the string side and vice-versa. However one should keep in mind that the correspondence is an equivalence of gauge and *string* theory, so the dual counterpart does not have to be in the well understood (super)gravity sector. Fortunately, it will turn out that for the questions considered in these lectures namely the study of the dynamics of plasma expansion, the dual description will be purely gravitational.

Apart from the direct physical interest, the AdS/CFT correspondence is also theoretically very interesting as it translates various dynamical gauge theory questions into a geometrical language described by higher-dimensional General Relativity (GR). This leads to quite fascinating links between the two fields, providing a whole range of physically motivated interesting questions which could be addressed by GR methods. In the other direction, various notions introduced by the GR community like dynamical apparent horizons find their application and new interpretation on the gauge theory side.

5.2.1 *Effective Degrees of Freedom at Strong Coupling*

As an illustration of the use of the AdS/CFT correspondence, and as a justification for the gravitational methods let us consider the question of finding effective degrees of freedom for strongly coupled $\mathcal{N} = 4$ SYM. By the AdS/CFT equivalence, it amounts to asking the same question for superstrings in $AdS_5 \times S^5$.

Let us first recall the case of closed strings in flat space. The string worldsheet action is characterized by a dimensionfull parameter α' (related to the string tension). The various vibrational modes of the string correspond to particles (fields) with distinct masses

$$m_n^2 = \frac{n^2}{\alpha'} \tag{5.1}$$

The massless modes correspond to the graviton (and its whole supergravity multiplet). There is also an infinite tower of massive modes. In the limit of $\alpha' \to 0$, the massive modes become very heavy and effectively decouple at fixed energies leaving as the governing dynamics just (super)gravity.

In the case of strings in $AdS_5 \times S^5$, the α' parameter becomes proportional to $1/\sqrt{\lambda}$ where $\lambda = g_{YM}^2 N_c$ is the 't Hooft coupling of the dual $\mathcal{N} = 4$ gauge theory. The vibrational modes again split into a massless (super)graviton multiplet and a set of massive modes. The formula (5.1) is no longer exact, but the parametric behaviour with α' still holds. So in the strong coupling limit, those massive string modes become very heavy and effectively decouple leaving essentially supergravity modes as the effective degrees of freedom. Since the AdS/CFT correspondence postulates an equivalence with gauge theory, these should also correspond to the effective degrees of freedom of the gauge theory at strong coupling.

Once we lower the coupling, the massive modes become lighter and their effects will no longer be negligible. Initially, their effects may be absorbed into corrections to the gravitational Einstein–Hilbert action (so-called α' corrections), but at low coupling corresponding to the perturbative regime the spacetime description is not known.

Finally, let us note that the $\mathcal{N} = 4$ SYM theory is quite special in that it allows for such a clean separation between gravity modes and massive string modes. Presumably a dual description of real QCD (or even of large N_c pure YM) would not have such a property.

5.3 Why Study $\mathcal{N} = 4$ Plasma?

Since we will be using the AdS/CFT correspondence for $\mathcal{N} = 4$ SYM as a calculational tool for analyzing strongly coupled dynamics of gauge theory plasma, we will be essentially considering plasma in the supersymmetric $\mathcal{N} = 4$ gauge theory. This theory is completely different from QCD at zero temperature. It is supersymmetric, exactly conformal, does not have confinement. However once we turn on some nonzero temperature (or consider not the vacuum but some appropriate state), supersymmetry is broken and temperature (or energy density) introduces a scale. So qualitatively, we may expect to have similarities with QCD plasma in a certain window of temperatures where it is strongly coupled, approximately conformal and (by definition) deconfined.

However we have to keep in mind some definite differences w.r.t. QCD plasma. Firstly, the $\mathcal{N} = 4$ theory has no running coupling so, in contrast to QCD, even at very high temperatures/energy densities the coupling may remain large. Secondly, the equation of state of the $\mathcal{N} = 4$ plasma is exactly conformal ($E = 3p$) which is only an approximation for a certain range of temperatures for QCD plasma. Indeed we know, from lattice QCD, that deviations from a conformal equation of state are important close to T_c. In addition, we have other consequences of conformality like that the bulk viscosity for the $\mathcal{N} = 4$ theory is exactly zero. Thirdly, for the $\mathcal{N} = 4$ theory (on Minkowski spacetime) there is no phase transition—no analog of the confinement/deconfinement phase transition of QCD. Therefore as the plasma expands and cools, in the $\mathcal{N} = 4$ theory it will expand indefinitely, while in QCD it will cool down to the phase transition temperature and hadronize.

From the above discussion we see that the applicability of using $\mathcal{N} = 4$ plasma to model real world phenomena depends on the questions asked. It may give a good qualitative picture for the range of temperatures where we have similarities. However, let us note that in this theory we may compute the dynamics from 'first principles' (using the AdS/CFT correspondence). For QCD, unfortunately, we do not have any similar calculational technique, even numerical, which would enable us to study real-time dynamics of the strongly coupled quark-gluon plasma. Therefore it is interesting to build up results on strong coupling properties of $\mathcal{N} = 4$ plasma and use them as a point of departure for analyzing or describing QCD plasma. Eventually, one might consider more realistic theories with AdS/CFT duals which are closer to QCD. In those cases generically the dual gravitational backgrounds are much more complicated so it is advantageous to start from the simplest setting for the $\mathcal{N} = 4$ SYM theory.

Another motivation for studying the dynamics of $\mathcal{N} = 4$ plasma is that the natural language of the AdS/CFT correspondence is quite new w.r.t. conventional gauge theory methods. So by studying relatively simple examples we may build up some new physical intuitions within this novel language. Also the interrelations with General Relativity physics are fascinating from the purely theoretical point of view. Last but not least, there may be some unexpected discoveries like the celebrated universality of the shear viscosity to entropy ratio η/s [9], which, at strong coupling, remains equal to $1/4\pi$ for *any* theory with a gravitational dual [10, 11] (see the lectures by Starinets at this school [12]).

5.4 The AdS/CFT Setup for Studying Real-Time Dynamics of Plasma

Let us now briefly review the AdS/CFT correspondence on a more technical level, concentrating on the features relevant to the study of the time evolution of a plasma system.

The S^5 factor in the $AdS_5 \times S^5$ background is associated with a global $SO(6) = SU(4)$ symmetry of the $\mathcal{N} = 4$ theory. In the following, we will only consider systems which do not break this symmetry, so the S^5 factor will be irrelevant and the whole dynamics will be concentrated in the AdS_5 factor.

The five-dimensional Anti-de-Sitter spacetime AdS_5 can be given by the following metric

$$ds^2 = \frac{\eta_{\mu\nu} dx^\mu dx^\nu + dz^2}{z^2} \tag{5.2}$$

with $z \geq 0$. $z = 0$ is the boundary of AdS_5, while the region $z > 0$ is often called 'the bulk'. This choice of coordinates covers the Poincare patch of global AdS_5 and is relevant for the case when the dual gauge theory lives in $\mathbb{R}^{1,3}$ Minkowski spacetime. The above geometry can be understood to correspond to the gauge theory vacuum state. In particular gauge theory operators like the energy-momentum tensor $T_{\mu\nu}$ all have vanishing expectation values in this state

$$\langle T_{\mu\nu} \rangle = 0 \tag{5.3}$$

Let us recall that the AdS/CFT correspondence states the equivalence with superstrings in $AdS_5(\times S^5)$. So, on the string side, we may excite any normalizable mode, in particular we may excite gravitons. This will correspond to some states in $\mathcal{N} = 4$ SYM with $\langle T_{\mu\nu} \rangle \neq 0$. We expect a configuration of $\mathcal{N} = 4$ plasma to be a very complicated state which would correspond to exciting very many gravitons. Then it is better to interpret it instead as a change of the background:

$$ds^2 = g^{5D}_{\alpha\beta} dx^\alpha dx^\beta = \frac{g_{\mu\nu}(x^\rho, z) dx^\mu dx^\nu + dz^2}{z^2} \tag{5.4}$$

with the metric coefficients being now generic functions of all the five coordinates. We therefore seek to describe a plasma configuration in terms of the geometry $g_{\mu\nu}(x^\rho, z)$. Let us note that the above choice of the metric (5.4) is always possible after a suitable change of coordinates. Such coordinates, in which the metric has the form (5.4) are called Fefferman–Graham coordinates.

The geometry (5.4) cannot be, however, completely arbitrary. It must form a consistent background for strings, so it must satisfy five-dimensional Einstein's equations with a negative cosmological constant[1]:

$$R_{\alpha\beta} - \frac{1}{2} g^{5D}_{\alpha\beta} R - 6\, g^{5D}_{\alpha\beta} = 0 \tag{5.5}$$

Furthermore, for a physical state in the gauge theory this geometry should not have a naked singularity. This turns out to be a crucial requirement with far reaching consequences for the resulting dynamics of the $\mathcal{N} = 4$ plasma, as we will see in the following.

[1] These equations are equivalent to the original ten-dimensional type IIB supergravity equations when we preserve full $SO(6)$ symmetry of S^5 and no other fields are turned on. The negative cosmological constant is a remnant of the RR five-form under this dimensional reduction.

5.4.1 The Gravity $\longrightarrow \langle T_{\mu\nu} \rangle$ Dictionary

Once one has the geometry (5.4) corresponding to some plasma configuration, the key question is what is the energy momentum tensor of that gauge theory system. The way to derive the answer, *holographic renormalization*, has been explained in the lectures by Skenderis [13]. Here we just summarize the outcome derived in [14], which in the Fefferman–Graham coordinates defined by (5.4) takes a particularly simple form.

Suppose that the metric coefficients $g_{\mu\nu}(x^\rho, z)$ have the following Taylor expansion near the boundary[2]

$$g_{\mu\nu}(x^\rho, z) = \eta_{\mu\nu} + z^4 g^{(4)}_{\mu\nu}(x^\rho) + \cdots \tag{5.6}$$

Then the expectation value of the energy-momentum tensor is

$$\langle T_{\mu\nu}(x^\rho) \rangle = \frac{N_c^2}{2\pi^2} \cdot g^{(4)}_{\mu\nu}(x^\rho) \tag{5.7}$$

The spacetime dependence of the energy-momentum tensor carries a lot of information about the dynamics of the plasma—its energy density, momentum flow, stress tensor. Indeed, it is just this information which is exactly the direct outcome of hydrodynamic simulations of realistic heavy-ion collisions. Finally, let us emphasize, to avoid any chance of confusion, that the $T_{\mu\nu}$ is the energy-momentum tensor of the dual gauge theory. On the gravity side we are always dealing with *vacuum* Einstein's equations.

The construction outlined above leads to the following scenario of investigating a plasma system in strongly coupled $\mathcal{N} = 4$ SYM. One starts from some initial conditions for the five-dimensional Einstein's equations. Then the geometry is evolved forward in time by solving Einstein's equations. Finally using the above formula (5.7), one extracts the $\langle T_{\mu\nu} \rangle$ of the corresponding plasma system. The details of the evolution of $T_{\mu\nu}$ are very interesting from the point of view of physics, especially in the far from equilibrium region, where very little is known about thermalization/isotropisation and transition to a hydrodynamic expansion. We will follow this route in Sect. 5.11.

5.4.2 The $\langle T_{\mu\nu} \rangle \longrightarrow$ Gravity Dictionary

It turns out to be very fruitful to ask also the opposite question. Suppose that we are given a certain spacetime profile of the energy momentum tensor $\langle T_{\mu\nu} \rangle$—how

---

[2] Here we always assume that the gauge theory lives in flat Minkowski space, hence the leading $\eta_{\mu\nu}$ and the absence of a z^2 term (see [14] for details).

to construct the dual five-dimensional geometry? The prescription is really just running the preceding recipe backwards. One has to solve Einstein's equations

$$R_{\alpha\beta} - \frac{1}{2} g^{5D}_{\alpha\beta} R - 6\, g^{5D}_{\alpha\beta} = 0 \tag{5.8}$$

with the boundary condition

$$g_{\mu\nu}(x^\rho, z) = \eta_{\mu\nu} + z^4 g^{(4)}_{\mu\nu}(x^\rho) + \cdots \tag{5.9}$$

where $g^{(4)}_{\mu\nu}(x^\rho)$ is related to $\langle T_{\mu\nu}(x^\rho)\rangle$ through

$$g^{(4)}_{\mu\nu}(x^\rho) = \frac{2\pi^2}{N_c^2} \langle T_{\mu\nu}(x^\rho)\rangle \tag{5.10}$$

It turns out that for a solution to exist, $g^{(4)}_{\mu\nu}(x^\rho)$ has to be traceless and conserved, which is of course physically expected for an energy-momentum tensor in a conformal theory. However here it is just an independent consequence of five-dimensional Einstein's equations.

Once such a $g^{(4)}_{\mu\nu}(x^\rho)$ is chosen, Einstein's equations determine uniquely the solution in the bulk (at least locally i.e. all higher coefficients of the Taylor expansion of $g_{\mu\nu}(x^\rho, z)$ around $z = 0$ are uniquely determined).

In this way we see that for every energy momentum profile which does not violate the standard requirements of energy-momentum conservation and tracelessness we may construct a dual gravity background. However generically, such a geometry will be highly singular with naked singularities in the bulk. The requirement of the absence of naked singularities will very strongly constrain the admissible bulk geometries and hence also the possible spacetime profiles of the energy momentum tensor. This is a nontrivial constraint on the dynamics of the gauge theory as the spacetime profile includes of course the time evolution of $T_{\mu\nu}$.

This line of reasoning was introduced in [15] as a way of determining the possible evolution of the energy momentum tensor. One first picks a family of profiles $T_{\mu\nu}(x^\rho)$, then one constructs for each of them a dual geometry by solving Einstein's equations with appropriate boundary conditions. Finally, one picks the allowed dynamics by requiring that the corresponding dual geometry is nonsingular. We will describe this procedure in detail in the first part of these lectures.

5.5 Exact Analytical Examples

In this section we will illustrate the AdS/CFT methods by analyzing two simple examples of plasma configurations for which the dual gravitational background can be computed in closed form [15]. Both of these examples have also a clear physical interpretation.

5.5.1 A Case Study: Static Uniform Plasma

Let us first consider the simplest configuration of plasma, namely with a uniform and static distribution of energy density filling up the whole spacetime. The energy momentum tensor is just a constant diagonal one:

$$T_{\mu\nu} = \begin{pmatrix} E & 0 & 0 & 0 \\ 0 & p & 0 & 0 \\ 0 & 0 & p & 0 \\ 0 & 0 & 0 & p \end{pmatrix} \tag{5.11}$$

with $E = 3p$. In order to find the dual gravity background, we have to solve Einstein's equations with the boundary conditions given by (5.9) and (5.10). Due to the fact that the energy momentum tensor is constant, the metric will only depend on the z coordinate and the Einstein's equations reduce to ordinary differential equations which can be solved explicitly. The result is

$$ds^2 = -\frac{(1 - z^4/z_0^4)^2}{(1 + z^4/z_0^4)z^2}\, dt^2 + (1 + z^4/z_0^4)\frac{dx_i^2}{z^2} + \frac{dz^2}{z^2} \tag{5.12}$$

where the parameter z_0 is related to E through

$$E = \frac{3N_c^2}{2\pi^2 z_0^4} \tag{5.13}$$

Although it is not evident at first glance, the geometry (5.12) is exactly the standard *AdS planar black hole* [16], but written in the Fefferman–Graham system of coordinates. We will give the explicit form of the coordinate transformation to the standard AdS Schwarzschild form shortly.

The fact that the dual geometry turns out to be a black hole has significant implications for the physics. Let us note that we did not *assume* that a black hole would appear. It came, in a unique way, from solving Einstein's equations with appropriate boundary conditions.

The parameter z_0 appearing in (5.12) is the location of the black hole horizon. The Hawking temperature T_H which is given by

$$T_H = \frac{\sqrt{2}}{\pi z_0} \tag{5.14}$$

is identified with the gauge theory temperature. This may be most easily seen by computing the Hawking temperature through a Wick rotation of the metric (5.12) and requiring the absence of a conical singularity at $z = z_0$. This requirement leads to a specific periodicity condition for the Euclidean time coordinate which is inversely proportional to the Hawking temperature. But, according to the AdS/CFT correspondence the metric induced on the boundary $z = 0$ (up to an overall rescaling by z^2) is exactly the metric of the (now Euclidean) gauge theory.

Thus the gauge theory also has a compactified Euclidean time with the radius given by the same temperature [16].

Another gravitational concept which carries over to the dual gauge theory is the Bekenstein–Hawking entropy which is identified with the entropy of the dual gauge theory plasma system. In this case, the entropy per spatial three-volume is

$$s = \frac{1}{4G_N}\text{Area} = \frac{N_c^2}{2\pi}\left(\frac{\sqrt{2}}{z_0}\right)^3 \tag{5.15}$$

Now we can use the relation between the horizon parameter z_0 and temperature to express the result completely in terms of gauge theoretical quantities

$$s = \frac{1}{2}N_c^2\pi^2T^3 \tag{5.16}$$

Finally, we may use the relation between the energy density E and z_0, and the link with temperature to express the energy density as a function of T. We get

$$E = \frac{3}{8}N_c^2\pi^2T^4 \tag{5.17}$$

The nontrivial factor here is the numerical coefficient, which is *different* from the corresponding one for the free massless gas (Stefan–Boltzmann). Similarly, the entropy density derived earlier is $3/4$ of the free gas answer [17]. This mismatch is quite natural since here we are dealing with a strongly coupled plasma. In fact similar deviations from the Stefan–Boltzmann answer have been observed in lattice studies of QCD thermodynamics above the confinement/deconfinement phase transition.

Before we move on to discuss various systems of coordinates for this geometry, let us note that it is exactly this geometry which is used to describe $\mathcal{N} = 4$ SYM at fixed nonzero temperature T. This interpretation is obvious from the above mentioned Euliclidean continuation, but can also be understood directly in Minkowski signature, where a link with the real-time formalism of finite-temperature QFT appears [18–23].

5.5.1.1 Various Coordinate Systems

The geometry (5.12) has been presented in the Fefferman–Graham coordinates, in which the connection to the gauge theory energy-momentum tensor is simplest. However these coordinates have also some significant drawbacks, of which one has to be aware.

Let us first perform a coordinate transformation to bring the metric (5.12) into the standard AdS Schwarzschild form. To this end set

$$z_{\text{std}} = \frac{z}{\sqrt{1 + z^4/z_0^4}} \tag{5.18}$$

Then, the metric becomes

$$ds^2 = -\frac{1 - z_{\mathrm{std}}^4/\tilde{z}_0^4}{z_{\mathrm{std}}^2} dt^2 + \frac{dx_i^2}{z_{\mathrm{std}}^2} + \frac{1}{1 - z_{\mathrm{std}}^4/\tilde{z}_0^4} \frac{dz^2}{z_{\mathrm{std}}^2} \tag{5.19}$$

with $\tilde{z}_0 = z_0/\sqrt{2}$. Looking at the transformation of coordinates (5.18), we see that the Fefferman–Graham coordinates cover only the region between the boundary and the horizon. Even if one would analytically continue the metric for $z > z_0$ one does not go beyond the horizon but rather returns back to the boundary.

The standard Schwarzschild coordinates also break down at the horizon and, in order to have explicit regularity at the horizon, it is convenient to introduce yet another system of coordinates—the (ingoing) Eddington–Finkelstein coordinates.

These coordinates may be obtained from the standard Schwarzschild ones by redefining the time coordinate:

$$t_{EF} = t - \frac{1}{4}\tilde{z}_0 \left(2 \arctan \frac{z_{\mathrm{std}}}{\tilde{z}_0} + \log \frac{\tilde{z}_0 + z_{\mathrm{std}}}{\tilde{z}_0 - z_{\mathrm{std}}} \right) \tag{5.20}$$

The metric becomes then

$$ds^2 = -\frac{1 - z_{\mathrm{std}}^4/\tilde{z}_0^4}{z_{\mathrm{std}}^2} dt_{EF}^2 + 2\frac{dt_{EF} dz_{\mathrm{std}}}{z_{\mathrm{std}}^2} + \frac{dx_i^2}{z_{\mathrm{std}}^2} \tag{5.21}$$

The crucial advantage of these coordinates is that the horizon is a perfectly regular point and one can enter the region inside the horizon. The lines $x^\mu = \mathrm{const}^\mu$ are null geodesics falling into the black hole and reaching the singularity at $z_{\mathrm{std}} = \infty$. These coordinates were used extensively in V. Hubeny's lectures at this school with z_{std} substituted by $r = 1/z_{\mathrm{std}}$, which brings them to the canonical form.

Finally let us note that in the formula (5.20), the time coordinate gets an infinite shift at the horizon. In the case of the static black hole geometry this is completely harmless as the metric is time-independent, however for the time dependent geometries which will be the focus of these lectures this shift will give rise to some spurious singularities in the Fefferman–Graham treatment (fortunately appearing only at NNNLO in the large proper time expansion of the geometry).

Some comments are in order here. Of course in General Relativity nothing depends on the choice of coordinate system. This is true if we are dealing with an exact solution of Einstein's equations—we may analyze it in any coordinate system we like. However if we perform an expansion in time in some coordinate system and deal with approximate solutions truncated at some order, we may get spurious singularities like in Fefferman–Graham at third order.

In these lectures we will nevertheless present the analysis in Fefferman–Graham coordinates (apart from a brief review of the Eddington–Finkelstein approach of Bhattacharyya et al. [24] in Sect. 5.9) The general formulation in Eddington–Finkelstein is considered in detail in the lectures of V. Hubeny, and has been applied to the boost invariant setting in [25–27]. One motivation for this choice of presentation is that our main focus is in reaching the small proper time regime, where we deal with *exact* solutions of the Einstein's equations and hence do not need to worry about these subtleties. Also there, the analysis of initial conditions in Fefferman–Graham coordinates is simpler.

5.5.2 A Case Study: A Planar Shock Wave

Another case of a gauge theory energy-momentum tensor for which the dual geometry is exactly solvable is a planar shockwave concentrated on the boundary. The only nonvanishing component of $T_{\mu\nu}$ is

$$T_{--} = \mu f(x^-) \tag{5.22}$$

Such a configuration, for $f(x^-) = \mu\delta(x^-)$ represents a planar shock wave of gauge theory matter moving at the speed of light along one light cone direction. It may be understood to represent an analog of an ultrarelativistic nuclei. Then the dual metric is found to be

$$ds^2 = \frac{dx^+dx^- + f(x^-)z^4 dx^{-2} + dx_\perp^2 + dz^2}{z^2} \tag{5.23}$$

This configuration was proposed in [15], being the simplest member of a family of shock wave solutions with $x_\perp$ dependence derived in [28] (see also [29]). A natural question to consider is a collision of two such shock waves, one propagating along the x^- light cone direction, the other along x^+. There have been some preliminary investigation along these lines in [30–34]. However, a complete analysis remains to be done. Finally, let us note that this kind of shock wave is *sourceless* in the bulk, in contrast to the shock waves considered recently in [35] and in the lectures of Yarom [36]. For references on work done on these other kinds of shock waves consult [36].

5.6 Boost-Invariant Flow

Let us now concentrate on a concrete evolving plasma system. Since eventually we would like to solve exactly Einstein's equations, one has to introduce as much symmetries as possible to reduce the complexity of the task, at the same time allowing for nontrivial physics to intervene. A natural choice in the context of heavy ion collisions is the requirement of longitudinal boost invariance. This assumption was introduced by Bjorken [37] back in 1983 to model ultrarelativistic collisions. Basically, the motivation is that at infinite energy, a finite boost along the collision axis would not modify the physics. This is certainly not an ideal approximation, however it is used in basically all hydrodynamic codes for modelling relativistic heavy-ion collisions at RHIC [2, 3]. We will make here an additional assumption that there is no dependence on the transverse coordinates, which corresponds to the limit of infinitely large nuclei. This is not really necessary for discussing the hydrodynamic limit (see [8, 24]) but will be essential in the far from equilibrium regime of small proper times.

When assuming boost invariance, it is natural to pass to proper-time/spacetime rapidity coordinates (τ, y, x_1, x_2)

$$t = \tau \cosh y \quad x_3 = \tau \sinh y \tag{5.24}$$

Then it turns out that the only non-vanishing components of the energy-momentum tensor are $T_{\tau\tau}$, T_{yy} and $T_{xx} \equiv T_{x_1x_1} = T_{x_2x_2}$. Moreover, these components become functions of τ alone.

We should now impose tracelessness $T^\mu_\mu = 0$ and conservation of energy momentum $T^{\mu\nu}_{;\nu} = 0$ conditions, which take the form

$$-T_{\tau\tau} + \frac{1}{\tau^2} T_{yy} + 2T_{xx} = 0$$

$$\tau \frac{d}{d\tau} T_{\tau\tau} + T_{\tau\tau} + \frac{1}{\tau^2} T_{yy} = 0$$

These equations determine $T_{\mu\nu}$ uniquely in terms of a single function $\varepsilon(\tau)$

$$T_{\mu\nu} = \begin{pmatrix} \varepsilon(\tau) & 0 & 0 & 0 \\ 0 & -\tau^3 \frac{d}{d\tau}\varepsilon(\tau) - \tau^2 \varepsilon(\tau) & 0 & 0 \\ 0 & 0 & \varepsilon(\tau) + \frac{1}{2}\tau \frac{d}{d\tau}\varepsilon(\tau) & 0 \\ 0 & 0 & 0 & \varepsilon(\tau) + \frac{1}{2}\tau \frac{d}{d\tau}\varepsilon(\tau) \end{pmatrix} \tag{5.25}$$

The remaining function $\varepsilon(\tau)$ can be interpreted as the energy density of the plasma at mid-rapidity (i.e. at $x_3 = 0$) as a function of (proper-) time.

Let us note that the above decomposition was purely 'kinematical'—valid in *any* conformal 4D theory at any coupling. The determination of $\varepsilon(\tau)$ will be an issue of understanding the dynamics of the theory of interest—here $\mathcal{N} = 4$ SYM. In particular, weak coupling perturbative considerations [38] lead to the free streaming behaviour

$$\varepsilon(\tau) \sim \frac{1}{\tau} \tag{5.26}$$

If, on the other hand, we would suppose that the plasma system behaves as a perfect fluid, then on top of the decomposition (5.25) we would impose

$$T^{\mu\nu} = (\varepsilon + p)u^\mu u^\nu - p\eta^{\mu\nu} \tag{5.27}$$

with $\varepsilon = 3p$. By our symmetry assumptions $u^\mu = (1, 0, 0, 0)$ and we get in particular $p = \frac{1}{\tau^2} T_{yy} = T_{xx}$, which gives a differential equation for $\varepsilon(\tau)$

$$-\tau\frac{d}{d\tau}\varepsilon(\tau)-\varepsilon(\tau)=\varepsilon(\tau)+\frac{1}{2}\tau\frac{d}{d\tau}\varepsilon(\tau) \tag{5.28}$$

with the celebrated Bjorken solution

$$\varepsilon(\tau)=\frac{\text{const.}}{\tau^{\frac{4}{3}}} \tag{5.29}$$

Other dynamical assumptions would modify the functional form of $\varepsilon(\tau)$. E.g. if the fluid would not be a perfect fluid but would have a nonzero viscosity (proportional to $T^3 \sim \varepsilon^{3/4}$ as should be the case for a conformal theory), then (5.29) would no longer be exact but would have corrections starting with

$$\varepsilon(\tau)=\frac{1}{\tau^{\frac{4}{3}}}\left(1-\frac{2\eta_0}{\tau^{\frac{2}{3}}}+\cdots\right) \tag{5.30}$$

with η_0 related to the shear viscosity. Here we set a single dimensional scale to unity. It can be easily reinstated in all terms by dimensional analysis. Further $1/\tau^{4/3}$ corrections are uniquely determined in terms of η_0. If the dynamics would follow 2nd order viscous hydrodynamics, these $1/\tau^{4/3}$ corrections would be *different* and would involve additional, second order transport coefficients.

So we see that the knowledge of $\varepsilon(\tau)$ contains a lot of information on the dynamics of plasma. In the rest of these lectures our goal will be to deduce what $\varepsilon(\tau)$ is singled out by the AdS/CFT correspondence. Initially we will concentrate on the large τ asymptotics of $\varepsilon(\tau)$ and then, in Sect. 5.11, move to consider the behaviour of $\varepsilon(\tau)$ for small τ.

5.7 Large Proper Time Behaviour

Let us first concentrate on the large τ asymptotics of $\varepsilon(\tau)$ and consider determining the exponent s in

$$\varepsilon(\tau)\sim\frac{1}{\tau^s} \quad \text{for } \tau\to\infty \tag{5.31}$$

We will follow here the strategy outlined in Sect. 5.4, and construct, for each s, the dual geometry. Then we will check for which s the dual geometry is nonsingular. This condition will determine s. This approach was proposed in [15], where more details may be found. But first, let us narrow down the range of possible s. We will demand that the energy density in any reference frame is non-negative i.e.

$$T_{\mu\nu}t^{\mu}t^{\nu}\geq 0 \tag{5.32}$$

for any time like four-vector t^{μ}. This leads to the inequalities

$$\varepsilon(\tau) \geq 0 \quad \varepsilon'(\tau) \leq 0 \quad \tau\varepsilon'(\tau) \geq -4\varepsilon(\tau) \tag{5.33}$$

In particular, for (5.31), we obtain that $0 \leq s \leq 4$.

5.7.1 The AdS/CFT Analysis

Now we have to construct the dual geometry to a plasma configuration with the energy density behaving like (5.31). Since the geometry will have the same symmetries as assumed for the plasma system, we are led to the following ansatz

$$ds^2 = \frac{1}{z^2}\left(-e^{a(z,\tau)}d\tau^2 + e^{b(z,\tau)}\tau^2 dy^2 + e^{c(z,\tau)}dx_\perp^2\right) + \frac{dz^2}{z^2} \tag{5.34}$$

Again, let us reiterate that the above ansatz is completely general. At this stage the choice of the Fefferman–Graham system of coordinates is perfectly legitimate.

According to the approach explained in Sect. 5.4, we have to solve Einstein's equations

$$R_{\alpha\beta} - \frac{1}{2}g^{5D}_{\alpha\beta}R - 6\,g^{5D}_{\alpha\beta} = 0 \tag{5.35}$$

with the boundary conditions

$$a(z,\tau) = -z^4\varepsilon(\tau) + z^6 a_6(\tau) + z^8 a_8(\tau) + \cdots \tag{5.36}$$

It is instructive to find the explicit form of the first few coefficients in the above Taylor series.[3] Using a computer algebra system we obtain

$$\begin{aligned} a(\tau,z) = -\varepsilon(\tau)\cdot z^4 + \left\{-\frac{\varepsilon'(\tau)}{4\tau} - \frac{\varepsilon''(\tau)}{12}\right\}\cdot z^6 + \left\{\frac{1}{6}\varepsilon(\tau)^2 + \frac{1}{6}\tau\varepsilon'(\tau)\varepsilon(\tau) + \right. \\ \left. + \frac{1}{16}\tau^2\varepsilon'(\tau)^2 + \frac{\varepsilon'(\tau)}{128\tau^3} - \frac{\varepsilon''(\tau)}{128\tau^2} - \frac{\varepsilon^{(3)}(\tau)}{64\tau} - \frac{1}{384}\varepsilon^{(4)}(\tau)\right\}\cdot z^8 + \cdots \end{aligned} \tag{5.37}$$

Let us now specialize to the case of interest $\varepsilon(\tau) = 1/\tau^s$. We get

$$\begin{aligned} & -z^4\tau^{-s} + z^6\left(\frac{1}{6}\tau^{-s-2}s - \frac{1}{12}\tau^{-s-2}s^2\right) \\ & + z^8\left(-\mathbf{\frac{1}{16}\tau^{-2s}s^2} - \mathbf{\frac{1}{6}\tau^{-2s}} + \mathbf{\frac{1}{6}\tau^{-2s}s} + \frac{1}{96}\tau^{-s-4}s^2 - \frac{1}{384}\tau^{-s-4}s^4\right) + \cdots \end{aligned}$$

where the terms dominant for large τ are outlined in bold. Looking at the above formula and analyzing a couple of higher order terms we may convince ourselves that the dominant terms in $a_n(\tau)$ for large τ will be of the form

[3] This is an *exact* result without any approximation.

$$z^n a_n(\tau) \sim \frac{z^n}{\tau^{\frac{ns}{4}}} = \left(\frac{z}{\tau^{\frac{s}{4}}}\right)^n \quad \text{for large } \tau \tag{5.38}$$

This shows that it is natural to introduce a scaling variable

$$v \equiv \frac{z}{\tau^{\frac{s}{4}}} \tag{5.39}$$

Consequently, the metric coefficients will have an expansion of the form

$$a(z,\tau) = a_0(v) + \frac{1}{\tau^{\#}} a_1(v) + \cdots \tag{5.40}$$

Several comments are in order here. Firstly, the appearance of the scaling variable at late times is a dynamical consequence of the structure of Einstein's equations. We will find later, that for small proper times a similar structure will *not* appear.[4] Secondly, the separation of dynamics into a scaling variable and an expansion in inverse powers of τ corresponds to a gradient expansion (cf. [24] and the lecture by V. Hubeny [8]). Finally, the appearance of a scaling variable reduces the very complicated Einstein's equations to a system of nonlinear *ordinary* differential equations:

$$\begin{aligned}
v(2a'(v)c'(v) + a'(v)b'(v) + 2b'(v)c'(v)) - 6a'(v) - 6b'(v) - 12c'(v) + vc'(v)^2 &= 0 \\
3vc'(v)^2 + vb'(v)^2 + 2vb''(v) + 4vc''(v) - 6b'(v) - 12c'(v) + 2vb'(v)c'(v) &= 0 \\
2vsb''(v) + 2sb'(v) + 8a'(v) - vsa'(v)b'(v) - 8b'(v) + vsb'(v)^2 & \\
+ 4vsc''(v) + 4sc'(v) - 2vsa'(v)c'(v) + 2vsc'(v)^2 &= 0
\end{aligned}$$

which can be solved exactly. The solution is

$$\begin{aligned}
a(v) &= A(v) - 2m(v) \\
b(v) &= A(v) + (2s-2)m(v) \\
c(v) &= A(v) + (2-s)m(v)
\end{aligned}$$

where

$$\begin{aligned}
A(v) &= \frac{1}{2}\left(\log(1+\Delta(s)\,v^4) + \log(1-\Delta(s)\,v^4)\right) \\
m(v) &= \frac{1}{4\Delta(s)}\left(\log(1+\Delta(s)\,v^4) - \log(1-\Delta(s)\,v^4)\right)
\end{aligned}$$

with

$$\Delta(s) = \sqrt{\frac{3s^2 - 8s + 8}{24}}$$

[4] For the case $\varepsilon(\tau) \to \text{const.}$ as $\tau \to 0$.

Now we can analyze the singularities of the above geometry. We see that there is a potential singularity where the argument of the logarithm vanishes. Of course, it may well be a coordinate singularity, so we have to evaluate a curvature invariant like

$$\mathfrak{R} = R^{\mu\nu\alpha\beta} R_{\mu\nu\alpha\beta} \tag{5.41}$$

Moreover, since our geometry is only exact in the scaling limit, we have to evaluate $\mathfrak{R}$ in the same limit i.e. $\tau \to \infty$, $z \to \infty$, keeping the ratio $v = \frac{z}{\tau^{s/4}}$ fixed. The resulting expression is quite complicated and can be found in [15]. Its general structure is

$$\mathfrak{R} = \frac{\text{Numerator}(v; s)}{\left(1 - \Delta(s)^2 v^8\right)^4} \tag{5.42}$$

We see that for generic s, there is a 4th order pole singularity in the curvature. It turns out that this singularity gets cancelled by the numerator *only* for a single value of s:

$$s = \frac{4}{3} \tag{5.43}$$

which is just the asymptotic scaling characteristic of perfect fluid hydrodynamics. In this way we see that nonlinear perfect fluid hydrodynamics arises at late stages of plasma expansion as a consequence of the AdS/CFT correspondence.

Note that in this way we do not approach the 'horizon' directly, but approach it asymptotically along a 'parallel' trajectory. An analogous analysis using Eddington–Finkelstein coordinates enables us to pass through the 'horizon' and require directly the nonsingularity of the metric there. The difference between the two procedures is summarized in Fig. 5.1. Both methods give equivalent results up to two subleading orders in the large proper-time expansion. In order to go beyond that, however, one has to use Eddington–Finkelstein coordinates (currently the only result in this direction beyond 2nd order is in [39] in the boost-invariant setting).

5.7.2 Perfect Fluid Geometry

Let us now examine more closely the dual geometry corresponding to the perfect fluid value of $s = 4/3$. Then the complicated formulas for the metric coefficients involving generically irrational powers and square roots obtained above simplify drastically and we obtain[5]

---

[5] We reinstated here a trivial dimensional scale e_0.

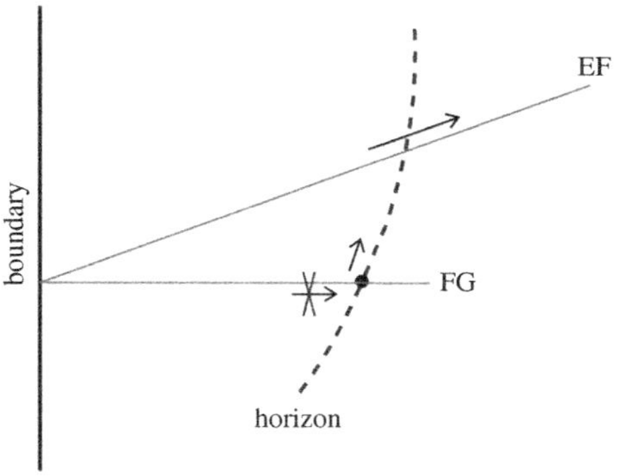

Fig. 5.1 The difference between using Fefferman–Graham and Eddington–Finkelstein coordinates for checking nonsingularity represented by *arrows*

$$ds^2 = \frac{1}{z^2}\left[-\frac{\left(1-\frac{e_0}{3}\frac{z^4}{\tau^{4/3}}\right)^2}{1+\frac{e_0}{3}\frac{z^4}{\tau^{4/3}}}d\tau^2 + \left(1+\frac{e_0}{3}\frac{z^4}{\tau^{4/3}}\right)(\tau^2 dy^2 + dx_\perp^2)\right] + \frac{dz^2}{z^2} \quad (5.44)$$

This geometry is analogous to a black hole (cf. (5.12)) with the position of the horizon moving into the bulk as

$$z_0 = \sqrt[4]{\frac{3}{e_0}} \cdot \tau^{\frac{1}{3}} \quad (5.45)$$

This has a clear physical interpretation. Recall that for a static black hole, the position of the horizon in the bulk is inversely proportional to the temperature. Thus here we have a dual counterpart of the plasma undergoing cooling during expansion.[6] Indeed naively generalizing the static formulas leads to

$$T = \frac{\sqrt{2}}{\pi z_0} = \frac{2^{\frac{1}{2}} e_0^{\frac{1}{4}}}{\pi 3^{\frac{1}{4}}} \tau^{-\frac{1}{3}} \quad (5.46)$$

A more indepth analysis of these phenomenae using the framework of event [41] or dynamical [26, 39, 42] horizons has been made, although a complete understanding of the notions of temperature and entropy in the fully dynamical case is still lacking.

5.8 Plasma Dynamics Beyond Perfect Fluid

One of the key discoveries of the AdS/CFT correspondence was the derivation of the universal value of shear viscosity for plasma at finite nonzero temperature.

[6] The dual counterpart of cooling was first suggested qualitatively in [40].

This was done using linear response theory in [9]. It is thus interesting to examine how do viscous effects manifest themselves in the current nonlinear setting.

Let us first examine the question whether we can see if the perfect fluid dynamics is violated from the dual gravitational point of view. Suppose that it is not and that consequently

$$\varepsilon(\tau) = \frac{1}{\tau^{\frac{4}{3}}} \tag{5.47}$$

is valid at all proper times without any corrections. Then one can find the next orders in the large τ scaling expansion of the metric coefficients

$$a(z,\tau) = a_0(v) + \frac{1}{\tau^{\frac{4}{3}}} a_2(v) + \cdots \tag{5.48}$$

This can be done explicitly since, fortunately, the equations for the corrections are linear albeit still quite complicated. Performing this calculation, and computing the curvature $\mathfrak{R}$ up to this order yields

$$\mathfrak{R} = R_{\alpha\beta\gamma\delta}R^{\alpha\beta\gamma\delta} = \underbrace{R_0(v)}_{\text{nonsingular}} + \frac{1}{\tau^{\frac{4}{3}}} \underbrace{R_2(v)}_{\text{singular!}} + \cdots \tag{5.49}$$

where the indicated singularity is of the very strong 4th order pole type. This means that the perfect fluid behaviour (5.47) has to be violated.

Let us now be completely agnostic about viscous hydrodynamics and assume a completely generic type of corrections:

$$\varepsilon(\tau) = \frac{1}{\tau^{\frac{4}{3}}} \left(1 - \frac{2A}{\tau^r}\right) \tag{5.50}$$

Solving the Einstein's equations with the appropriate boundary conditions set by (5.50), computing the curvature[7] yields

$$\mathfrak{R} = R_{\alpha\beta\gamma\delta}R^{\alpha\beta\gamma\delta} = \underbrace{R_0(v)}_{\text{nonsingular}} + \frac{1}{\tau^r} \underbrace{R_1(v)}_{\text{nonsingular}} + \frac{1}{\tau^{2r}} \underbrace{\tilde{R}_2(v)}_{\text{singular!}} + \frac{1}{\tau^{\frac{4}{3}}} \underbrace{R_2(v)}_{\text{singular!}} + \cdots \tag{5.51}$$

The last two terms are always singular. The only way that we may obtain bounded curvature is to make those two terms cancel between themselves. This requires

$$r = \frac{2}{3} \tag{5.52}$$

which is exactly the correct scaling for a correction coming from shear viscosity. Moreover, we have to fine tune the coefficient A to [44]

---

[7] Various steps of this calculation were done in [43, 44] and [45].

$$A = 2^{-\frac{1}{2}}3^{-\frac{3}{4}} \tag{5.53}$$

which is the value corresponding to the value of the shear viscosity to entropy ratio

$$\frac{\eta}{s} = \frac{1}{4\pi} \tag{5.54}$$

In this way we reconfirmed, in a fully nonlinear setting [44], the value of shear viscosity computed at fixed temperature [9].

The above analysis can be repeated for the NNLO correction [46] with the final result for $\varepsilon(\tau)$:

$$\varepsilon(\tau) = \frac{1}{\tau^{\frac{4}{3}}} - \frac{2}{2^{\frac{1}{2}}3^{\frac{3}{4}}}\frac{1}{\tau^2} + \frac{1 + 2\log 2}{12\sqrt{3}}\frac{1}{\tau^{\frac{8}{3}}} + \cdots \tag{5.55}$$

The coefficient of the NNLO term involves 2nd order transport coefficients. It is at this stage that the pathologies of Fefferman–Graham coordinates surface, leading to a leftover logarithmic singularity in the scaling limit of the curvature (appearing at NNNLO in the metric, which is necessary for obtaining the coefficients of $\varepsilon(\tau)$ at one order lower). The singularity was found to be persistent and not associated with truncating other fields of ten dimensional supergravity [47]. Its origin has been explained in detail in [25] (see also [26, 27]) and can be associated with the singular transformation between coordinates regular at the horizon (Eddington–Finkelstein) and the Fefferman–Graham ones, coupled with performing an expansion of the geometry w.r.t. those coordinates. In order to proceed further, which is however rather impractical analytically, one would have to perform the analysis in Eddington–Finkelstein coordinates (a result in this direction, the NNNLO term in $\varepsilon(\tau)$ is given in [39]).

5.9 Interlude: Hydrodynamics Redux

In the above, we have adopted a very agnostic attitude towards the expected dynamics ruling the time evolution of the energy-momentum tensor of the $\mathcal{N} = 4$ plasma system. We did not assume that one could parametrize the $T_{\mu\nu}$ in terms of such quantities as flow velocity and energy/pressure. We started off from the most general $T_{\mu\nu}$ consistent with the assumed symmetries. By proceeding in this way we have an option of describing dynamics which does not fit at all into a hydrodynamical language. This is in fact the main reason for presenting such an approach here, as in the remaining part of the lectures we would like to address the dynamics of boost-invariant plasma at small proper times where we do not expect hydrodynamic description to be a good starting point.

On the other hand such flexibility has also significant drawbacks. The determination of the transport coefficients of hydrodynamics presented above followed by first deriving from AdS/CFT the explicit form of $\varepsilon(\tau)$. Given that $\varepsilon(\tau)$, one could ascertain that the leading term is a solution of perfect fluid equations of motion, and together with the subleading term is a solution of viscous hydrodynamic equations with a specific value of the shear viscosity (we leave this as an exercise for the reader).

It is thus very interesting to obtain directly the hydrodynamic equations from AdS/CFT without making the passage through explicit solutions. This task was performed in [24] and is presented in detail in the lectures of V. Hubeny at this school [8]. For completeness, let us just summarize here the main idea.

The static black hole geometry presented in Sect. 5.4 is dual to a plasma at rest, which can be described by a flow vector $u^\mu = (1,0,0,0)$ and an energy density E (or equivalently temperature T). By performing a boost one can obtain a dual geometry to a uniformly moving plasma with four-velocity u^μ.

In Eddington–Finkelstein coordinates it is given explicitly as

$$ds^2 = -2u_\mu dx^\mu dr - r^2\left(1 - \frac{T^4}{\pi^4 r^4}\right) u_\mu u_\nu dx^\mu dx^\nu + r^2(\eta_{\mu\nu} + u_\mu u_\nu) dx^\mu dx^\nu \qquad (5.56)$$

where $r = \infty$ corresponds to the boundary, $r = T/\pi$ is the horizon while $r = 0$ is the position of the singularity ($r = 1/z_{EF}$ cf. Sect. 5.4) This is an exact solution of Einstein's equations. Now promote T and u^μ to slowly varying functions of the boundary Minkowski coordinates. The geometry (5.56) ceases to be a solution of Einstein's equations and has to be corrected by terms proportional to gradients. These correction terms can be determined with the integration constants fixed by the requirement of nonsingularity at the horizon.

Now from the corrected geometry one can read off the $T_{\mu\nu}$ which is explicitly expressed (similarly to the metric) in terms of T, u^μ and the gradients of u^μ. The numerical constants coming from nonsingularity become exactly the transport coefficients. In this way one obtains

$$\begin{aligned} T^{\mu\nu}_{\text{rescaled}} = & \underbrace{(\pi T)^4(\eta^{\mu\nu} + 4u^\mu u^\nu)}_{\text{perfect fluid}} - \underbrace{2(\pi T)^3 \sigma^{\mu\nu}}_{\text{viscosity}} + \\ & + \underbrace{(\pi T^2)\left(\log 2 T^{\mu\nu}_{2a} + 2T^{\mu\nu}_{2b} + (2 - \log 2)\left(\frac{1}{3}T^{\mu\nu}_{2c} + T^{\mu\nu}_{2d} + T^{\mu\nu}_{2e}\right)\right)}_{\text{second order hydrodynamics}} \end{aligned}$$

The energy-momentum conservation $\partial_\mu T^{\mu\nu} = 0$ of such a $T^{\mu\nu}$ is by definition the hydrodynamic relativistic Navier–Stokes equation.

From the above construction we see that the appearance of hydrodynamics in the AdS/CFT correspondence is now completely understood. For any solution of the hydrodynamic equations $T(x^\rho)$, $u^\mu(x^\rho)$, the formula (5.56) and its correction terms give an explicit dual metric valid to the same order of the derivative expansion. A similar analysis was performed later in the Fefferman–Graham coordinates [48].

One final thing to note is that the starting point in the above construction is the boosted black hole, which means that one assumes that approximately one is dealing with an energy-momentum tensor of a hydrodynamic type (i.e. parametrizable by a flow velocity and energy density). This does not always need to be the case, as we shall see shortly, and then one has to return to an ab-initio analysis of Einstein's equations of the type presented in Sect. 5.7.

5.10 Plasma Dynamics Beyond Hydrodynamics

The appearance of hydrodynamic behaviour of strongly coupled plasma as a consequence of the AdS/CFT correspondence is certainly very interesting and satisfying theoretically, however perhaps the most fascinating feature of AdS/CFT is its ability to address the behaviour of a plasma system very far from equilibrium, where in QCD we do not even have a well motivated phenomenological model.

As an example of a configuration which cannot be described, even in any approximation, by hydrodynamic methods consider the problem of plasma isotropisation, extensively studied in various variations within QCD [49–52]

Suppose we have a plasma system uniform in space with anisotropic pressures. In weakly coupled gauge theory one could consider a gas of gluons with non isotropic momentum distributions, like gaussians with different widths for the different momentum components. Then one expects that the pressures would isotropise in time. The energy-momentum tensor of such a system would have the form

$$T_{\mu\nu} = \begin{pmatrix} \varepsilon & 0 & 0 & 0 \\ 0 & p_{\|}(t) & 0 & 0 \\ 0 & 0 & p_{\perp}(t) & 0 \\ 0 & 0 & 0 & p_{\perp}(t) \end{pmatrix} \tag{5.57}$$

Note that such a system cannot be described by any form of (even all-order resummed) viscous hydrodynamics, since by symmetry $u^\mu = (1,0,0,0)$ and thus has vanishing derivatives. So all viscous terms vanish, while the leading term is clearly of a different form. However nothing stops us from applying the AdS/CFT analysis using Einstein's equations to such a system. This has been first proposed in [53]. Subsequently, a numerical study of this system was performed in [54].

Another interesting problem, which we will discuss in the remaining part of these lectures, is the behaviour of the boost-invariant plasma system considered before but now at small proper times. Since the hydrodynamic expansion (5.55)

$$\varepsilon(\tau) = \frac{1}{\tau^{\frac{4}{3}}} - \frac{2}{2^{\frac{1}{2}}3^{\frac{3}{4}}}\frac{1}{\tau^2} + \frac{1+2\log 2}{12\sqrt{3}}\frac{1}{\tau^{\frac{8}{3}}} + \cdots \tag{5.58}$$

is an expansion in inverse powers of τ, it completely breaks down as we approach $\tau = 0$. Here the situation is more complicated than in the case of uniform isotropisation (5.57) mentioned above as we expect a mixture of non-equilibrium and hydrodynamic behaviour. In fact it is exactly the question of the transition to hydrodynamics, and what factors are important in setting the scale of this transition, that is very interesting in the context of heavy-ion collisions. A related more general issue is the observation of thermalization (proposed in [55] to be related to a formation of a black hole in the bulk) and an analysis of the concrete way in which this scenario is realised.

5.11 Dynamics at Small Proper Time

For the reasons described above, we will have to deal with the full Einstein's equations. From the point of view of hydrodynamics treated as a gradient expansion these encompass *all orders* of viscous hydrodynamics together with an infinite set of higher transport coefficients. But apart from these there is additional information contained in the Einstein's equations. Taking the case of a planar black hole as an example, all order hydrodynamics may be identified, on the linearized level, with the lowest quasinormal mode and its exact dependence on spatial momentum. But apart from this lowest mode there is an infinite set of higher quasinormal modes which decay exponentially (in the AdS/CFT context, see in particular [56]). And all of these become important in a far from equilibrium situation such as the early time dynamics of the boost-invariant flow.

Here we will again go to the boost invariant setting and repeat the analysis of Sect. 5.7, but now concentrating on the small τ regime. We follow the analysis of [57]. We will adopt the same ansatz for the metric (5.34) and solve Einstein's equations with the boundary conditions (5.36).

Before going into the details of this construction, let us comment why we are using the Fefferman–Graham system of coordinates. In contrast to the late time expansion, here we will aim at solving the Einstein's equations to an arbitrary accuracy—without performing any kind of scaling limit. Therefore any choice of coordinates works equally well. Moreover the constraint equations for initial data in the Fefferman–Graham system of coordinates are particularly transparent.

Let us first determine the qualitative behaviour of $\varepsilon(\tau)$ at small τ. We will do it in two ways.

5.11.1 The Absence of a Scaling Variable

In the large proper time regime, the structure of Einstein's equations naturally led to the introduction of a scaling variable, which reduced the problem to solving ordinary differential equations and a subsequent expansion in inverse powers of τ. Let us now analyze the solution of Einstein's equations at small τ from this point of view.

We again start from the *exact* power series solution

$$a(\tau,z) = -\varepsilon(\tau)\cdot z^4 + \left\{-\frac{\varepsilon'(\tau)}{4\tau} - \frac{\varepsilon''(\tau)}{12}\right\}\cdot z^6 + \left\{\frac{1}{6}\varepsilon(\tau)^2 + \frac{1}{6}\tau\varepsilon'(\tau)\varepsilon(\tau) + \frac{1}{16}\tau^2\varepsilon'(\tau)^2 + \frac{\varepsilon'(\tau)}{128\tau^3} - \frac{\varepsilon''(\tau)}{128\tau^2} - \frac{\varepsilon^{(3)}(\tau)}{64\tau} - \frac{1}{384}\varepsilon^{(4)}(\tau)\right\}\cdot z^8 + \cdots \tag{5.59}$$

and substitute the asymptotics

$$\varepsilon(\tau) \sim \frac{1}{\tau^s} \quad \text{for } \tau \to 0 \tag{5.60}$$

In this way we obtain

$$-z^4\,\tau^{-s} + z^6\left(\frac{1}{6}\,\tau^{-s-2}s - \frac{1}{12}\,\tau^{-s-2}s^2\right) + z^8\left(-\frac{1}{16}\,\tau^{-2s}s^2 - \frac{1}{6}\,\tau^{-2s} + 1/6\,\tau^{-2s}s + \frac{1}{96}\tau^{-s-4}s^2 - \frac{1}{384}\tau^{-s-4}s^4\right) + \cdots \tag{5.61}$$

where the terms dominating for small τ are rendered in bold. This analysis was first done by Kovchegov and Taliotis [58], who deduced that for *generic* s the dominant terms at small τ are of the form

$$\frac{z^4}{\tau^s}\cdot f\left(w \equiv \frac{z}{\tau}\right) \tag{5.62}$$

In [58], the scaling solution was found, but due to its complex branch cut structure,[8] Kovchegov and Taliotis argued that the only acceptable solution had $s = 0$, which is a very interesting result.

But if we again look at (5.61), we see that the terms resummed by the scaling variable *vanish* for $s = 0$ and are no longer dominant. Hence there is no place for a scaling variable at small τ and for $s = 0$ one has to reanalyze the Einstein equations in order to describe the full solution at $\tau \sim 0$.

5.11.2 The Existence of a Regular Initial Condition

One can reach the same conclusion, as well as some more detailed information, on the small τ dependence of $\varepsilon(\tau)$ *assuming* that at $\tau = 0$ we have a regular initial condition.[9] Recall the expression (5.59) and substitute $\tau = 0$. Firstly, we see that

---

[8] And an additional physical bound on s, see [58].

[9] This is an assumption which may, or may not be realistic for ultraenergetic collisions. We prefer to keep the options open and analyze boost invariant flow with regular initial conditions as an interesting nonequilibrium dynamical system for its own sake.

the assumption that the metric coefficients are finite leads to a finite limit of $\varepsilon(\tau)$ as $\tau \to 0$, consistent with $s = 0$. Secondly, the inverse powers of τ appearing in the higher order terms do not lead to a singularity if and only if $\varepsilon(\tau)$ has an expansion only in *even* powers of τ:

$$\varepsilon(\tau) = \varepsilon_0 + \varepsilon_2\tau^2 + \varepsilon_4\tau^4 + \cdots \tag{5.63}$$

A closer analysis reveals that the coefficients ε_{2n} are uniquely determined, through the Einstein's equations, in terms of the coefficients of the initial condition for the metric:

$$a(\tau = 0, z) = a_0z^4 + a_2z^6 + a_4z^* + \cdots \tag{5.64}$$

In order to complete the analysis of the gravitational setup, we have to analyze what are the admissible initial conditions (5.64). Once this is done, one can set up the analysis of the system by evolving the geometry from (5.64) using Einstein's equations and read off $\varepsilon(\tau)$ from the metric. This can be done either numerically solving Einstein's equations [59], or analytically by expressing the coefficients ε_{2n} directly in terms of the coefficients of the initial condition a_{2n} [57].

5.11.3 The Classification of Possible Initial Conditions

As is well known, in General Relativity, the initial conditions cannot be arbitrary but have to satisfy some nonlinear constraint equations. This causes the gravitational initial value problem to be quite nontrivial in general. Fortunately, for the case of the $\tau = 0$ hypersurface in the Fefferman–Graham coordinates, the constraints can be solved exactly.

Let us denote by $E_{\alpha\beta}$, the components of Einstein's equations written in the form

$$E_{\alpha\beta} \equiv R_{\alpha\beta} + 4g_{\alpha\beta} = 0 \tag{5.65}$$

Then the constraints are contained in equations $E_{\tau z}$ and E_{zz}. Denoting $a_0(z) \equiv a(\tau = 0, z)$ etc. we get at once

$$a_0(z) = b_0(z) \quad \dot{a}_0 = \dot{b}_0 = \dot{c}_0 = 0 \tag{5.66}$$

and the only remaining constraint is the single nonlinear equation

$$a_0'' + c_0'' + \frac{1}{2}(a_0')^2 + \frac{1}{2}(c_0')^2 - \frac{1}{z}(a_0' + c_0') = 0 \tag{5.67}$$

Let us note an extremely surprising feature of the above equation. At the linearized level, it has a trivial solution $a_0(z) = -c_0(z)$. So one may expect that for infinitesimal a_0, the function c_0 would be also very small and only slightly deformed from $-a_0$ by taking into account the effect of the nonlinear terms. It turns out, however, that this is never true, and the nonlinearity always causes a blowup of the solution for some finite z. To see this introduce $v(z^2) = \frac{1}{4z} a_0'(z)$ and similarly $w(z^2)$ for c_0. Then the constraint Eq. 5.67 takes the simple form

$$v' + w' + v^2 + w^2 = 0 \tag{5.68}$$

Now it is easy to see, that there does not exist an everywhere bounded ($v = w = 0$ at infinity) solutions of the constraint equations. To this end it is enough to integrate (5.68) to get

$$0 = \int_0^\infty (v' + w') + \int_0^\infty (v^2 + w^2) = \int_0^\infty (v^2 + w^2) \tag{5.69}$$

Therefore, at $\tau = 0$ with boost invariant symmetry, gravity leads to inherently nonlinear dynamics—a linearized regime does not exist at all!

It is not difficult to impose conditions on v and w so that the blowup is not a curvature singularity—at least $\mathfrak{R}$ stays finite there (see [57] for details). Moreover one can solve the constraint (5.68) analytically. Indeed, defining $v_+ = -w - v$, $v_- = w - v$,

$$v_- = \sqrt{2v_+' - v_+^2} \tag{5.70}$$

solves (5.68) for any v_+. Let us conclude with an example of a simple particular solution of the initial value constraints:

$$a_0(z) = b_0(z) = 2 \log \cos az^2 \quad c_0(z) = 2 \log \cosh az^2 \tag{5.71}$$

The huge range of initial conditions that can be imposed is in fact quite natural. On the gauge theory side we may also expect to be able to prepare an initial state with the same energy density in a multitude of ways. Let us contrast this with the large τ expansion (5.58) of $\varepsilon(\tau)$, which only depends on a *single* scale.[10] In other words, once the dominant asymptotics of $\varepsilon(\tau)$ is known, all subleading power like terms are uniquely determined.

The physical interpretation of this difference is quite clear. We expect dissipative effects to wash out differences in initial conditions leaving only a single scale (under the present symmetry assumptions) governing the large proper time expansion of $\varepsilon(\tau)$. On the gravitational side, these effects can be understood as nonlinear generalizations of higher quasinormal modes which die off exponentially (see [60] for some analysis along these lines in the boost invariant setting).

---

[10] In (5.58) this scale has been set to unity, but may be reinstated unambiguously by dimensional analysis.

5.11.4 An Analysis of Some Aspects of the Small Proper Time Behaviour of $\varepsilon(\tau)$

Once we have the allowed initial conditions at $\tau = 0$, we need to solve Einstein's equations with these initial data. Then, as explained in Sect. 5.4, we may read off $\varepsilon(\tau)$ from the solution of Einstein's equations. Since we do not have a scaling variable at our disposal we have to do it exactly. The ideal way to proceed would be to solve Einstein's equations numerically. This study is currently under way [59]. The route followed in [57] was to solve these equations for the metric in a power series in z and τ, obtaining a power series expression for $\varepsilon(\tau)$ up to the order τ^{100} for some initial conditions. A drawback of the above method is that the power series in question has a finite radius of convergence, necessitating the use of Pade approximations as an extrapolation method. Below we will present some analysis of these extrapolated profiles [57]. We have to emphasize, however, that one would require a real numerical solution of Einstein's equation to be sure of all the details.

First let us discuss the transition to hydrodynamics. One possibility of quantifying this is by considering an 'effective exponent' of the power law dependence of $\varepsilon(\tau)$ defined as

$$-\tau \frac{d}{d\tau} \log \varepsilon(\tau) \tag{5.72}$$

Initially it is zero, and it should rise up to $4/3$ for late time expansion. In order to evaluate it we plot a Pade approximant (of constant large τ asymptotics) of the expression (5.72). The result, for the initial condition (5.71) is shown in Fig. 5.2.

We see that it definitely crosses $s = 1$ of free streaming and moves upwards. However, to be sure that it would reach $4/3$ a numerical solution for $\varepsilon(\tau)$ would be needed.

Now assuming the late time exponent $4/3$, we may perform a Pade approximation of $\varepsilon(\tau)$ with this asymptotics to see the profiles of $\varepsilon(\tau)$ for a set of initial conditions. Example plots are shown in Fig. 5.3.

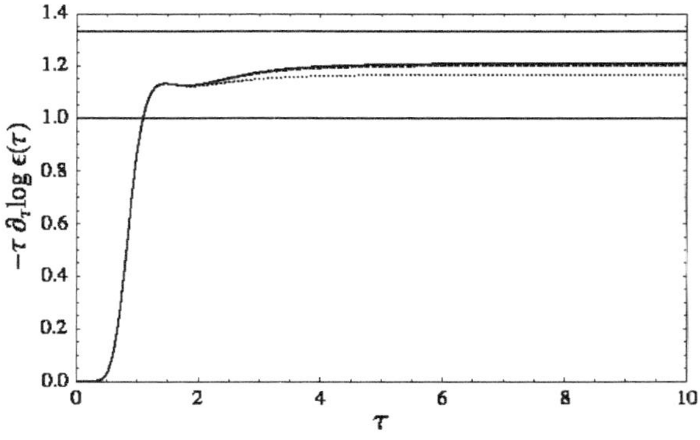

Fig. 5.2 The effective power (5.72) of $\varepsilon(\tau)$ corresponding to the initial condition (5.71)

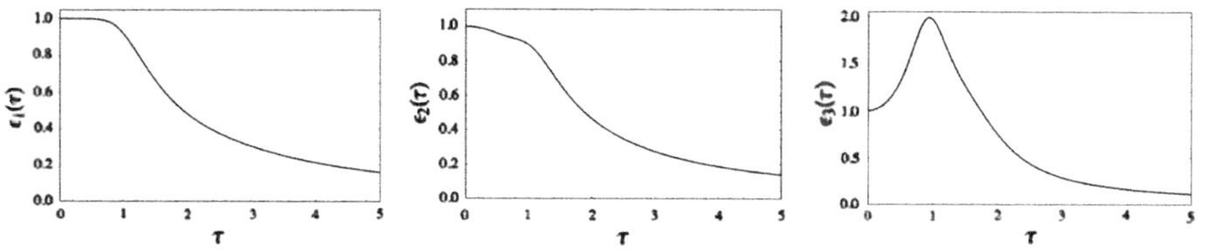

Fig. 5.3 Pade resummed profiles of $\varepsilon(\tau)$ for a set of initial conditions [57]

The main interest of the knowledge of the exact profile of $\varepsilon(\tau)$ for various initial conditions is that then we would be able to study the transition to hydrodynamics and its dependence on various features of the initial conditions. Since the Pade extrapolation introduces some serious systematic uncertainties, we refrain from doing so until we will have at our disposal a direct numerical solution of Einstein equations for these various initial conditions. Apart from the reasons mentioned above, the numerical solution would also allow to analyze the nature of the apparent singularity in the initial data and, more importantly, analyze the geometry for the presence of apparent (dynamical) horizons, relevant for the thermodynamic interpretation. Some of these issues are currently under investigation [59].

Before we close this last part of the lectures let us note a complementary numerical investigation of boost invariant plasma in [61]. The setup of [61] was different from the one presented in these lectures in that the *gauge theory metric* was perturbed in a boost invariant way at some $\tau \sim \tau_0 > 0$ and then set again to flat Minkowski. The metric perturbation produced a change in the geometry which induced a boost invariant flow on the boundary. The main results observed in [61] were a transition to hydrodynamics and a formation of an apparent horizon which moved in from infinity.

Another related work[11] was [62], were a perturbation by a boundary scalar source induced a black hole formation in the bulk.

Looking at all these examples, one sees that Einstein's equations, through the AdS/CFT correspondence, have the potential of describing a multitude of far from equilibrium strongly coupled gauge theory phenomenae. It is however clear that the majority of problems remain still unsolved.

5.12 Conclusions

In these lectures we have described an approach using the AdS/CFT correspondence as a tool for studying real time dynamics of strongly coupled gauge theory plasma. The basic theoretical tool is the possibility of translating, in a completely explicit and constructive way, between the spacetime energy-momentum tensor

[11] This time not in the boost-invariant setting discussed here.

characterizing the gauge theory configuration in question and the five-dimensional metric of the dual geometry. Then one uses the fact, that at strong coupling, the dynamics of the dual geometry is governed by Einstein's equations (with a cosmological constant following from the full ten dimensional supergravity solution). A further dynamical input is the requirement of the absence of naked singularities in the gravity background. This restricts very strongly the allowed spacetime profiles of the gauge theory energy momentum tensor, and consequently the possible gauge theory dynamics.

Using these methods, one may establish the appearance of nonlinear hydrodynamics, as exhibited here in the form of near perfect fluid dynamics in the large proper time asymptotics of boost invariant plasma expansion. Moreover, Einstein equations together with the nonsingularity criterion unambiguously predict viscous first- and higher-order deviations from perfect fluid dynamics with specific values of the transport coefficients appropriate to the case of the $\mathcal{N} = 4$ SYM theory studied here.

Let us note, that we arrived at these results without presupposing anything even about the general form of the gauge theory energy momentum tensor like the presence of something like a flow velocity u^μ etc. This flexibility comes from the fact that Einstein's equations on the string side of the AdS/CFT correspondence require only that the gauge theory energy-momentum tensor is conserved and traceless (throughout these lectures we are dealing exclusively with the conformal case of $\mathcal{N} = 4$ SYM). Therefore one can use the same techniques to address the question of far from equilibrium dynamics where hydrodynamics cannot be used as a starting point of approximation. We exhibited an example of such a study by describing some aspects of the behaviour of the boost invariant plasma expansion in the region of small proper time.

It should be obvious that the majority of questions concerning far from equilibrium dynamics of strongly coupled plasma remain still unanswered even for the case of $\mathcal{N} = 4$ plasma. Once more details are understood, it would be very interesting to address similar problems in more complicated versions of the AdS/CFT correspondence involving gauge theories which might be closer to QCD. Apart from this direct 'application driven' motivation, the study of such time-dependent systems leads to fascinating interrelations with General Relativity. On the one hand, it provides a new setting for investigating such GR concepts as dynamical apparent horizons, black hole formation, providing these GR phenomenae with new interpretation. On the other hand, the well developed technology of numerical relativity might be applied to learn more about the properties of far from equilibrium strongly coupled gauge theory systems.

Last but not least, let us note that the gravity backgrounds obtained as dual descriptions of evolving plasma systems can very well by themselves form a scene for the AdS/CFT correspondence enabling one to study various kind of physics questions. In this way one may study how the expanding plasma system influences properties of mesons, Wilson loops, various correlation functions. Of course, due to the time-dependent nature of those backgrounds this may be quite difficult to do in practice, but the possibility of doing so is certainly very interesting and the results may be rewarding.

Acknowledgement The approach presented in these lectures was introduced with Robi Peschanski, and developed further in collaboration with Micha, Heller, Dongsu Bak, Alex Buchel, Paolo Benincasa, Guillaume Beuf. To all of whom I am grateful for enjoyable collaboration and numerous discussions. I would like to thank the organizers and participants of the *Fifth Aegean Summer School "From Gravity to Thermal Gauge Theory: The AdS/CFT Correspondence"* for a very interesting school. I would also like to thank the *New Frontiers in QCD 2010* program and the Yukawa Institute for Theoretical Physics, Kyoto for hospitality when these lectures were written up. This work was supported in part by Polish science funds as a research project N N202 105136 (2009–2011) and the Marie Curie ToK KraGeoMP (SPB 189/6.PRUE/2007/7).

Appendix

A.1 Topics Not Covered in the Main Text: A Brief Guide to the Literature

In this appendix we would like to give pointers to the literature on other developments related to the approach presented in these lectures.

- Leading α' corrections (i.e. beyond strong coupling in $\mathcal{N} = 4$ SYM theory) to the transport coefficients were computed using the boost invariant flow [63, 64]. This task involved using α' corrected Einstein's equations.
- Beyond $\mathcal{N} = 4$ SYM. A class of general conformal field theories parametrized by higher curvature terms in the dual gravitational action was considered in [65].
- Extension to hydrodynamics with conserved charge(s) was considered in [66–68]. Electric-magnetic equilibration at large proper times was found in [69]. In addition, dilaton driven hydrodynamics was considered in a general way in [68] and the onset of turbulence was observed [70].
- Lower dimensional examples. The case of a 1+1 dimensional conformal field theory allows for an explicit exact dual gravitational solutions [71, 72]. Other investigations in different number of dimensions were performed in [73].
- Various (exact) solutions for $\mathcal{N} = 4$ gauge theory in curved and possibly time-dependent backgrounds were obtained [74–76]. The exact solutions which were found do not involve viscosity effects. An exact gravitational description af a shearless flow in $\mathcal{N} = 4$ in flat space was obtained in [77].
- Further properties of solutions with boost invariant symmetries were studied in [78–82].
- Physics in the expanding plasma. Fundamental flavours were introduced (through D7 brane embeddings) into the late proper time boost invariant geometry [83], diffusion constant was evaluated [84], drag force was computed [85]. In addition, various physical questions were addressed in the case of the

shock wave solutions described in Sect. 5.4 Deep Inelastic Scattering (DIS) was analyzed [86–89] as well as heavy quark energy loss [90, 91].

References

1. Shuryak, E.V.: What RHIC experiments and theory tell us about properties of quark-gluon plasma?. Nucl. Phys. A **750**, 64 (2005)
2. Kolb, P.F., Heinz, U.W.: Hydrodynamic description of ultrarelativistic heavy-ion collisions [arXiv:nucl-th/0305084]
3. Huovinen, P., Ruuskanen, P.V.: Hydrodynamic models for heavy ion collisions. Ann. Rev. Nucl. Part. Sci. **56**, 163 (2006)
4. Maldacena, J.M.: The large N limit of superconformal field theories and supergravity. Adv. Theor. Math. Phys. **2**, 231 (1998)
5. Maldacena, J.M.: The large N limit of superconformal field theories and supergravity. Int. J. Theor. Phys. **38**, 1113 (1999)
6. Gubser, S.S., Klebanov, I.R., Polyakov, A.M.: Gauge theory correlators from non-critical string theory. Phys. Lett. B **428**, 105 (1998)
7. Witten, E.: Anti-de Sitter space and holography. Adv. Theor. Math. Phys. **2**, 253 (1998)
8. Hubeny, V.: V. Hubeny's lectures at this school
9. Policastro, G., Son, D.T., Starinets, A.O.: The shear viscosity of strongly coupled N = 4 supersymmetric Yang–Mills plasma. Phys. Rev. Lett. **87**, 081601 (2001)
10. Buchel, A., Liu, J.T.: Universality of the shear viscosity in supergravity. Phys. Rev. Lett. **93**, 090602 (2004)
11. Kovtun, P., Son, D.T., Starinets, A.O.: Viscosity in strongly interacting quantum field theories from black hole physics. Phys. Rev. Lett. **94**, 111601 (2005)
12. Starinets, A.: A. Starinets's lectures at this school
13. Skenderis, K.: K. Skenderis's lectures at this school
14. de Haro, S., Solodukhin, S.N., Skenderis, K.: Holographic reconstruction of spacetime and renormalization in the AdS/CFT correspondence. Commun. Math. Phys. **217**, 595 (2001)
15. Janik, R.A., Peschanski, R.B.: Asymptotic perfect fluid dynamics as a consequence of AdS/CFT. Phys. Rev. D **73**, 045013 (2006)
16. Witten, E.: Anti-de Sitter space, thermal phase transition, and confinement in gauge theories. Adv. Theor. Math. Phys. **2**, 505 (1998)
17. Gubser, S.S., Klebanov, I.R., Peet, A.W.: Entropy and temperature of black 3-branes. Phys. Rev. D **54**, 3915 (1996)
18. Maldacena, J.M.: Eternal black holes in Anti-de-Sitter. JHEP **0304**, 021 (2003)
19. Kraus, P., Ooguri, H., Shenker, S.: Inside the horizon with AdS/CFT. Phys. Rev. D **67**, 124022 (2003)
20. Fidkowski, L., Hubeny, V., Kleban, M., Shenker, S.: The black hole singularity in AdS/CFT. JHEP **0402**, 014 (2004)
21. Herzog, C.P., Son, D.T.: Schwinger–Keldysh propagators from AdS/CFT correspondence. JHEP **0303**, 046 (2003)
22. Skenderis, K., van Rees, B.C.: Real-time gauge/gravity duality. Phys. Rev. Lett. **101**, 081601 (2008)
23. Skenderis, K., van Rees, B.C.: Real-time gauge/gravity duality: prescription, renormalization and examples. JHEP **0905**, 085 (2009)
24. Bhattacharyya, S., Hubeny, V.E., Minwalla, S., Rangamani, M.: Nonlinear fluid dynamics from gravity. JHEP **0802**, 045 (2008)
25. Heller, M.P., Surowka, P., Loganayagam, R., Spalinski, M., Vazquez, S.E.: Consistent holographic description of boost-invariant plasma. Phys. Rev. Lett. **102**, 041601 (2009)

26. Kinoshita, S., Mukohyama, S., Nakamura, S., Oda, K.y.: A holographic dual of Bjorken flow. Prog. Theor. Phys. **121**, 121 (2009)
27. Kinoshita, S., Mukohyama, S., Nakamura, S., Oda, K.y.: Consistent anti-de Sitter-space/conformal-field-theory dual for a time-dependent finite temperature system. Phys. Rev. Lett. **102**, 031601 (2009)
28. Emparan, R.: Exact gravitational shockwaves and Planckian scattering on branes. Phys. Rev. D **64**, 024025 (2001)
29. Beuf, G.: Gravity dual of $N = 4$ SYM theory with fast moving sources. Phys. Lett. B **686**, 55 (2010)
30. Grumiller, D., Romatschke, P.: On the collision of two shock waves in AdS5. JHEP **0808**, 027 (2008)
31. Albacete, J.L., Kovchegov, Y.V., Taliotis, A.: Modeling heavy ion collisions in AdS/CFT. JHEP **0807**, 100 (2008)
32. Albacete, J.L., Kovchegov, Y.V., Taliotis, A.: Asymmetric collision of two shock waves in AdS_5. JHEP **0905**, 060 (2009)
33. Kovchegov, Y.V., Lin, S.: Toward thermalization in heavy ion collisions at strong coupling. JHEP **1003**, 057 (2010) [arXiv:0911.4707 [hep-th]]
34. Lin, S., Shuryak, E.: Grazing collisions of gravitational shock waves and entropy production in heavy ion collision. Phys. Rev. D **79**, 124015 (2009)
35. Gubser, S.S., Pufu, S.S., Yarom, A.: Entropy production in collisions of gravitational shock waves and of heavy ions. Phys. Rev. D **78**, 066014 (2008)
36. Yarom, A.: A. Yarom's lectures at this school
37. Bjorken, J.D.: Highly relativistic nucleus–nucleus collisions: the central rapidity region. Phys. Rev. D **27**, 140 (1983)
38. Kovchegov, Y.V.: Can thermalization in heavy ion collisions be described by QCD diagrams?. Nucl. Phys. A **762**, 298 (2005)
39. Booth, I., Heller, M.P., Spalinski, M.: Black brane entropy and hydrodynamics: the boost-invariant case. Phys. Rev. D **80**, 126013 (2009)
40. Shuryak, E., Sin, S.J., Zahed, I.: A gravity dual of RHIC collisions. J. Korean Phys. Soc. **50**, 384 (2007)
41. Bhattacharyya, S. et al.: Local fluid dynamical entropy from gravity. JHEP **0806**, 055 (2008)
42. Figueras, P., Hubeny, V.E., Rangamani, M., Ross, S.F.: Dynamical black holes and expanding plasmas. JHEP **0904**, 137 (2009)
43. Nakamura, S., Sin, S.J.: A holographic dual of hydrodynamics. JHEP **0609**, 020 (2006)
44. Janik, R.A.: Viscous plasma evolution from gravity using AdS/CFT. Phys. Rev. Lett. **98**, 022302 (2007)
45. Heller, M.P.: MSc thesis (2009) (Unpublished)
46. Heller, M.P., Janik, R.A.: Viscous hydrodynamics relaxation time from AdS/CFT. Phys. Rev. D **76**, 025027 (2007)
47. Benincasa, P., Buchel, A., Heller, M.P., Janik, R.A.: On the supergravity description of boost invariant conformal plasma at strong coupling. Phys. Rev. D **77**, 046006 (2008)
48. Gupta, R.K., Mukhopadhyay, A.: On the universal hydrodynamics of strongly coupled CFTs with gravity duals. JHEP **0903**, 067 (2009)
49. Mrowczynski, S.: Plasma instability at the initial stage of ultrarelativistic heavy ion collisions. Phys. Lett. B **314**, 118 (1993)
50. Arnold, P., Lenaghan, J., Moore, G.D., Yaffe, L.G.: Apparent thermalization due to plasma instabilities in quark gluon. Phys. Rev. Lett. **94**, 072302 (2005)
51. Arnold, P., Moore, G.D., Yaffe, L.G.: The fate of non-abelian plasma instabilities in 3+1 dimensions. Phys. Rev. D **72**, 054003 (2005)
52. Romatschke, P., Strickland, M.: Collective modes of an anisotropic quark gluon plasma. Phys. Rev. D **68**, 036004 (2003)
53. Janik, R.A., Witaszczyk, P.: Towards the description of anisotropic plasma at strong coupling. JHEP **0809**, 026 (2008)

54. Chesler, P.M., Yaffe, L.G.: Horizon formation and far-from-equilibrium isotropization in supersymmetric Yang–Mills plasma. Phys. Rev. Lett. **102**, 211601 (2009)
55. Nastase, H.: The RHIC fireball as a dual black hole [arXiv:hep-th/0501068]
56. Horowitz, G.T., Hubeny, V.E.: Quasinormal modes of AdS black holes and the approach to thermal equilibrium. Phys. Rev. D **62**, 024027 (2000)
57. Beuf, G., Heller, M.P., Janik, R.A., Peschanski, R.: Boost-invariant early time dynamics from AdS/CFT. JHEP **0910**, 043 (2009)
58. Kovchegov, Y.V., Taliotis, A.: Early time dynamics in heavy ion collisions from AdS/CFT correspondence. Phys. Rev. C **76**, 014905 (2007)
59. Work in progress
60. Janik, R.A., Peschanski, R.B.: Gauge/gravity duality and thermalization of a boost-invariant perfect fluid. Phys. Rev. D **74**, 046007 (2006)
61. Chesler, P.M., Yaffe, L.G.: Boost invariant flow, black hole formation, and far-from-equilibrium dynamics in N = 4 supersymmetric Yang–Mills theory. Phys. Rev. D **82**, 026006 (2010) [arXiv:0906.4426 [hep-th]]
62. Bhattacharyya, S., Minwalla, S.: Weak field black hole formation in asymptotically AdS spacetimes. JHEP **0909**, 034 (2009)
63. Buchel, A.: Shear viscosity of boost invariant plasma at finite coupling. Nucl. Phys. B **802**, 281 (2008)
64. Buchel, A., Paulos, M.: Second order hydrodynamics of a CFT plasma from boost invariant expansion. Nucl. Phys. B **810**, 40 (2009)
65. Buchel, A., Heller, M.P., Myers, R.C.: sQGP as hCFT. Phys. Lett. B **680**, 521 (2009)
66. Banerjee, N., Bhattacharya, J., Bhattacharyya, S., Dutta, S., Loganayagam, R., Surowka, P.: Hydrodynamics from charged black branes. JHEP **1101**, 094 (2011) [arXiv:0809.2596 [hep-th]]
67. Torabian, M., Yee, H.U.: Holographic nonlinear hydrodynamics from AdS/CFT with multiple/non-Abelian symmetries. JHEP **0908**, 020 (2009)
68. Bhattacharyya, S., Loganayagam, R., Minwalla, S., Nampuri, S., Trivedi, S.P., Wadia, S.R.: Forced fluid dynamics from gravity. JHEP **0902**, 018 (2009)
69. Bak, D., Janik, R.A.: From static to evolving geometries: R-charged hydrodynamics from supergravity. Phys. Lett. B **645**, 303 (2007)
70. Bhattacharyya, S., Minwalla, S., Wadia, S.R.: The incompressible non-relativistic Navier–Stokes equation from gravity. JHEP **0908**, 059 (2009)
71. Kajantie, K., Louko, J., Tahkokallio, T.: Gravity dual of 1+1 dimensional Bjorken expansion. Phys. Rev. D **76**, 106006 (2007)
72. Kajantie, K., Louko, J., Tahkokallio, T.: Gravity dual of conformal matter collisions in 1+1 dimensions. Phys. Rev. D **77**, 066001 (2008)
73. Bhattacharyya, S., Loganayagam, R., Mandal, I., Minwalla, S., Sharma, A.: Conformal nonlinear fluid dynamics from gravity in arbitrary dimensions. JHEP **0812**, 116 (2008)
74. Sin, S.J., Nakamura, S., Kim, S.P.: Elliptic flow, Kasner universe and holographic dual of RHIC fireball. JHEP **0612**, 075 (2006)
75. Kajantie, K., Tahkokallio, T.: Spherically expanding matter in AdS/CFT. Phys. Rev. D **75**, 066003 (2007)
76. Kajantie, K., Louko, J., Tahkokallio, T.: Isotropic AdS/CFT fireball. Phys. Rev. D **78**, 126011 (2008)
77. Friess, J.J., Gubser, S.S., Michalogiorgakis, G., Pufu, S.S.: Expanding plasmas and quasinormal modes of anti-de Sitter black holes. JHEP **0704**, 080 (2007)
78. Alsup, J., Middleton, C., Siopsis, G.: AdS/CFT correspondence with heat conduction. Phys. Lett. B **654**, 35 (2007)
79. Alsup, J., Siopsis, G.: Bjorken flow from an AdS Schwarzschild black hole. Phys. Rev. Lett. **101**, 031602 (2008)
80. Kajantie, K., Louko, J., Tahkokallio, T.: Casimir energy in the gauge/gravity description of Bjorken flow?. Phys. Rev. D **78**, 126012 (2008)
81. Alsup, J., Siopsis, G.: Dissipative Bjorken hydrodynamics from an AdS Schwarzschild black hole. Phys. Rev. D **79**, 066011 (2009)

82. Yee, H.U.: Parity asymmetric boost invariant plasma in AdS/CFT correspondence. Phys. Lett. B **680**, 286 (2009)
83. Grosse, J., Janik, R.A., Surowka, P.: Flavors in an expanding plasma. Phys. Rev. D **77**, 066010 (2008)
84. Kim, K.Y., Sin, S.J., Zahed, I.: Diffusion in an expanding plasma using AdS/CFT. JHEP **0804**, 047 (2008)
85. Giecold, G.C.: Heavy quark in an expanding plasma in AdS/CFT. JHEP **0906**, 002 (2009)
86. Albacete, J.L., Kovchegov, Y.V., Taliotis, A.: DIS on a large nucleus in AdS/CFT. JHEP **0807**, 074 (2008)
87. Albacete, J.L., Kovchegov, Y.V., Taliotis, A.: DIS in Ads. AIP Conf. Proc. **1105**, 356 (2009)
88. Taliotis, A.: DIS from the AdS/CFT correspondence. Nucl. Phys. A **830**, 299C (2009)
89. Dominguez, F.: Particle production in DIS off a shockwave in AdS. JHEP **1009**, 007 (2010) [arXiv:0912.1641 [hep-th]]
90. Horowitz, W.A., Kovchegov, Y.V.: Shock treatment: heavy quark drag in a novel AdS geometry. Phys. Lett. B **680**, 56 (2009)
91. Horowitz, W.A.: Shock treatment: heavy quark energy loss in a novel AdS/CFT geometry. Nucl. Phys. A **830**, 773C (2009)

Chapter 6
Fluid Dynamics from Gravity

Veronika E. Hubeny

Abstract We give a brief review of the recently-formulated fluid/gravity correspondence. Originating from the AdS/CFT correspondence, this constitutes a one-to-one map between configurations of a conformal fluid dynamics in d dimensions and solutions to Einstein's equations in $d + 1$ dimensions. The map is fully constructive; for a given fluid configuration, it is completely algorithmic to write down the spacetime geometry and deduce its causal properties. In particular, the bulk solutions describe a regular generic, non-uniform and dynamical, black hole which at late times settles down to a stationary planar black hole. We briefly indicate the iterative construction of such solutions, extract the key physical properties, and discuss further generalizations and open questions.

6.1 Introduction

In these lectures I will review the recent developments in formulating a relation between fluid dynamics and gravity, which came to be called the fluid/gravity correspondence. This correspondence is rooted in the well-known AdS/CFT (or more generally the gauge/gravity) correspondence, reviewed elsewhere in these proceedings.

As is often the case with important developments, there are several diverse and complementary takes on the underlying motivation for formulating and using such a correspondence. Iconically speaking, we were motivated by the need to:

V. E. Hubeny (✉)
Centre for Particle Theory & Department of Mathematical Sciences, Science Laboratories, South Road, Durham, DH1 3LE, UK
e-mail: veronika.hubeny@durham.ac.uk

E. Papantonopoulos (ed.), *From Gravity to Thermal Gauge Theories: The Ads/CFT Correspondence,* Lecture Notes in Physics, 828,
DOI: 10.1007/978-3-642-04864-7_6, © Springer-Verlag Berlin Heidelberg 2011

- probe nature of spacetime in quantum gravity,
- extract universal dynamics of wide class of gravitational theories,
- explore properties of strongly-coupled gauge theories (with particular view to exploring QCD), and
- elucidate long-standing questions in fluid dynamics.

I will now briefly elaborate on each of these in turn, simultaneously reminding the reader of the key ingredients of the AdS/CFT correspondence to be used later.

6.1.1 Quantum Gravity

One of the most important long-standing goals of the past few decades has been to unravel the underlying quantum theory of gravity. Since the classical theory of gravity, Einstein's general relativity, identifies gravity with the curvature of spacetime, the key question of quantum gravity is *what is the fundamental nature of spacetime?* Serendipitously, one invaluable tool that came to our aid in tackling this question during the last decade is the celebrated AdS/CFT correspondence, for the simple reason that it allows us, in principle, to recast the long-standing quantum gravitational questions in terms of non-gravitational quantum field theory.

In its simplest formulation, string theory (which in particular contains gravity while being a consistent quantum theory), on asymptotically five-dimensional Anti-de Sitter (AdS) spacetime times a five-sphere, is dual to Super Yang-Mills gauge theory which is a four-dimensional conformal field theory (CFT). Such a correspondence is holographic; the CFT may be thought of as living on the boundary of AdS. We will therefore refer to the AdS side as the "bulk" and the CFT side as the "boundary" theory. According to the AdS/CFT dictionary, different asymptotically AdS spacetimes manifest themselves by different states in the boundary theory. For example, the vacuum state in the CFT corresponds to pure AdS. Metric perturbations from AdS are related to the stress-energy-momentum tensor expectation value in the CFT.

More relevantly, putting a black hole in the bulk has the effect of heating up the boundary theory. Specifically, a large[1] Schwarzschild-AdS black hole corresponds to (approximately) thermal state in the gauge theory. Conceptually, this can be understood as the late-time configuration that a generic state evolves to: in the bulk, gravity, as well as the negative curvature, tends to make a generic large-energy configuration collapse and form a black hole which then quickly settles down to the Schwarzschild-AdS geometry. Correspondingly, in the field theory, a generic large-energy excitation will eventually thermalize.

[1] AdS is a space of constant negative curvature, which introduces a length scale, called the AdS scale R_{AdS}, corresponding to the radius of curvature. The black hole size is then measured in terms of this AdS scale; large black holes have horizon radius $r_+ > R_{\text{AdS}}$.

While appealingly simple, this level of understanding is far too coarse to allow us to extract the more interesting aspects. We need to probe the AdS/CFT dictionary further to uncover what happens in regions where the classical description of the black hole breaks down, such as near the curvature singularity, or in the more general, dynamical, situations. Ultimately, we would like to answer such questions as: Which CFT configurations admit a dual spacetime description? or What types of spacetime singularities are physically allowed? More generally, we wish to probe spacetime dynamics. Although the fluid/gravity correspondence will not enable us to answer these intriguing questions fully, it will take the first step in this direction.

6.1.2 Universal Implications

While it is heartening to have a framework wherein we can explore the questions of interest, it is much more gratifying for that framework to exhibit certain degree of universality. The fluid/gravity framework does indeed achieve this. Specifically, the setup which we will consider in the bulk,[2] namely Einstein's gravity with negative cosmological constant (and no other bulk matter content), is in fact relevant for a much wider class of theories. This is because Einstein's equations (with negative cosmological constant) constitute a consistent truncation of all two-derivative gravitational theories interacting with other fields (with spin <2) which have AdS as a solution. This means that for every CFT having SUGRA bulk dual description, we have a decoupled sector exhibiting universal dynamics for the stress tensor. In particular, the stress tensor correlators (from which we can extract the transport coefficients) are in this sense universal. More importantly, since uncharged planar black holes solve our theory, they likewise lie in this universal sector, so this universality holds at any temperature.

6.1.3 QCD

A complementary long-standing goal of theoretical physics is to solve QCD. As observed elsewhere, AdS/CFT provides a convenient window of opportunity even for this programme, since it allows us to get a handle on certain strongly coupled field theories. Unfortunately, QCD is in many respects markedly different from the super Yang-Mills CFT, so it is only for certain universal features that the AdS/CFT correspondence has any direct relevance. Nevertheless, as remarked elsewhere in these proceedings, many quantities of interest in QCD, such as its

---

[2] In what follows, the five-sphere of $AdS_5 \times S^5$ will play a passive role, so we will ignore it and consider the five-dimensional (asymptotically) AdS bulk spacetime only.

transport coefficients, do fall into this category, so that while not quantitatively accurate, the predictions from fluid/gravity do give a reasonable indication of the qualitative features. This is especially useful since presently we do not have other means, apart from possibly lattice calculations, to calculate in the strong coupling regime. This makes the AdS/CFT, and specifically the fluid/gravity, correspondence relevant to high-energy experiments. Indeed, the predictions which will be described later in these lectures have been used in analysing data from the Relativistic Heavy Ion Collider.

6.1.4 Fluid Dynamics

Although many detailed questions about the fundamental degrees of freedom in a given strongly coupled field theory remain well beyond our reach, a more macroscopic effective description is not only more easily accessible, but also is interesting in its own right. The key observation is that any interacting field theory admits an effective description in terms of fluid dynamics. Fortuitously, fluid dynamics is a subject that has been well-studied for more than a century, theoretically as well as observationally/experimentally and numerically. Nevertheless, there are important questions in fluid dynamics which have yet to be understood. While we have a good handle on near-equilibrium physics, dynamics away from thermodynamic equilibrium is more difficult to tackle. For example, one of the famous Clay Millenium Prize Problems concerns the global regularity (existence and smoothness) of the Navier–Stokes equations. Intriguingly, the solutions often include turbulence, which, despite its importance in science and engineering, still remains one of the greatest unsolved problems in physics. Furthermore, there are causality issues in the conventional Navier–Stokes equations which plague numerical simulations. As already suggested by its name, the fluid/gravity correspondence is indeed relevant for addressing such important questions in fluid dynamics.

The structure of these lectures is as follows. We will first review the necessary background for constructing the fluid/gravity map in Sect. 6.2. In particular, we will briefly discuss the two sides of the correspondence, conformal fluid dynamics on the boundary and gravity in the bulk, and then indicate the main idea of constructing the map between them. In Sect. 6.3 we will sketch the actual construction, which involves expanding in boundary derivatives; in particular, we will explicitly correct the zeroth order configuration to first order. The second order solution, calculated in [5], is discussed in Sect. 6.4; in particular, we will present the boundary stress tensor and transport coefficients for our conformal fluid, followed by a discussion of the bulk geometry from which we extract the location and area of the event horizon. Finally, Sect. 6.5 summarizes the salient points and indicates several open questions.

These lectures are based primarily on [5] which first formulated the nonlinear fluid/gravity correspondence and constructed the second order solution, and on [6] which analysed the causal structure of this solution, focussing on the event horizon area and corresponding entropy current in the boundary fluid. This work builds on earlier derivations of linearized fluid dynamics from linearized gravity [18] (see [22] for an excellent review), and on earlier examples of the duality between non-linear fluid dynamics and gravity [14, 4]. Since its inception, the fluid/gravity correspondence has flourished into an active line or research, with large number of extensions and generalizations. Some of the key developments and reviews of further progress include [7, 8, 19], among many others. Since these lectures are intended as a brief overview serving rather as an invitation into the subject than as a full, self-contained review (excellent samples of which already exist in the literature), I am purposefully keeping the list of references to the bare minimum; please see the above-cited works for a more complete set of background references.

6.2 Background

We begin with a brief review of conformal fluid dynamics, proceed to discuss the dual gravitational solutions, and then motivate the construction of the explicit mapping between them. To keep the discussion and formulas as clean as possible, we will restrict attention to the simplest case of uncharged conformal fluid on four-dimensional Minkowski spacetime $\mathbf{R}^{3,1}$ (though initially we will quote the d-dimensional results). Several generalizations to this setup will be mentioned in Sect. 6.5.

Before proceeding, we make few remarks on notation: the bulk metric will be denoted by g_{MN} with the capital Latin indices taking values over the $d+1$ bulk dimensions; we will separate the coordinates into the radial coordinate r and the remaining "boundary coordinates" x^{μ}, where the μ index ranges over the d boundary directions (which includes time). The stress tensor in the boundary theory is denoted $T^{\mu\nu}$, and in writing its conservation ($\nabla_{\mu}T^{\mu\nu}=0$), the ∇_{μ} is the covariant derivative with respect to the boundary metric $\eta_{\mu\nu}$, which in this case simply reduces to the partial derivative ∂_{μ}.

6.2.1 *Conformal Fluid Dynamics*

Fluid dynamics is the continuum effective description of any (interacting) microscopic quantum field theory. In order to meaningfully describe the system in terms of the fluid variables, the fluid description assumes that the system achieves local thermodynamic equilibrium. This means that the regime of validity where such a description is valid requires that the scale of variation of the dynamical degrees of freedom, L, be much larger than the microscopic scale ℓ_{mfp}, typically set by the temperature, T. In this *long-wavelength approximation*, local equilibrium then demands that $LT \equiv \frac{1}{\epsilon} \gg 1$.

A conformal fluid is characterized by a traceless symmetric stress tensor, which in d spacetime dimensions has $\frac{d(d+1)}{2} - 1$ degrees of freedom, along with a collection of charge currents (which for simplicity we have set to zero). In a fluid dynamical characterization of the same system, the number of basic degrees of freedom is drastically reduced. The conformal invariance fixes the equation of state, thereby determining the pressure in terms of the energy density, which can in turn be expressed in terms of the temperature. Hence the basic variables are the local temperature $T(x)$ and velocity $u_\mu(x)$ (unit-normalized so that $u_\mu u^\mu = -1$), which constitute just d degrees of freedom.

The equations of fluid dynamics are then simply the equations of local conservation of the stress tensor (as well as the charge currents in more general situations), supplemented by constitutive relations that express these currents as functions of the fluid dynamical variables. As fluid dynamics is a long wavelength effective theory, such constitutive relations are usually specified in a derivative expansion. At any given order, thermodynamics plus symmetries determine the form of this expansion up to a finite number of undetermined coefficients. In general, the coefficients can be obtained either from measurements or from microscopic computations. However, as we will see, in the present framework these coefficients are fully determined by the gravity side (which in a sense knows about the microscopics of the boundary field theory).

Purely based on the symmetries, we can then write down an expression for the stress tensor of a d-dimensional conformal fluid:

$$T^{\mu\nu} = \alpha T^d (\eta^{\mu\nu} + d u^\mu u^\nu) + \pi^{\mu\nu}_{\text{dissipative}}.$$

The first two terms describe the ideal conformal fluid stress tensor, while $\pi^{\mu\nu}_{\text{dissipative}}$ incorporates all the dissipative terms, and α is a constant giving the overall normalization. As variations of $T(x)$ and $u_\mu(x)$ are small, we can expand $\pi^{\mu\nu}$ in a derivative expansion $\partial_\mu \equiv \frac{\partial}{\partial x^\mu}$ in the boundary directions; the leading term will turn out to be proportional to the shear viscosity. The dynamical content of the fluid equations is encoded in the conservation of the stress tensor

$$\partial_\mu T^{\mu\nu} = 0. \tag{6.1}$$

Fluid dynamics viewed in this derivative expansion constructs an effective field theory for the slowly varying modes $T(x)$ and $u_\mu(x)$, analogously to the chiral Lagrangian for pions.

6.2.2 Gravity in the Bulk

We now turn to the gravitational solutions in asymptotically AdS spacetime. Motivated by the AdS/CFT correspondence, we will consider two-derivative theories of gravity with an AdS_{d+1} "vacuum", such as the IIB SUGRA on $AdS_5 \times S^5$. As mentioned above, the solution space has a universal sub-sector, pure gravity with negative cosmological constant, for which the bulk field equations are simply Einstein's equations,

$$E_{MN} \equiv R_{MN} - \frac{1}{2} R g_{MN} + \Lambda g_{MN} = 0. \tag{6.2}$$

(Note that without loss of generality we can measure distances in AdS units, which amounts to taking $R_{\text{AdS}} = 1$ and thereby sets $\Lambda = -6$ in five dimensions.) We will focus on this sub-sector in the long-wavelength limit. Apart from the pure AdS_5 solution, there is a 4-parameter family of solutions representing asymptotically-AdS_5 boosted planar black holes. We will use these solutions to construct general dynamical spacetimes characterized by fluid-dynamical configurations.

Roughly speaking, our construction may be thought of as a "collective coordinate method" for black hole horizons. Recall that the isometry group of AdS_5 is $SO(4,2)$. The Poincare algebra plus dilatations form a distinguished subalgebra of this group (one that preserves the boundary). Out of these, the $SO(3)$ rotations and translations in world-volume $\mathbf{R}^{3,1}$ leave the static planar AdS black hole invariant, but the remaining symmetry generators, dilatations and boosts, act nontrivially on this solution, generating a 4-parameter family of boosted planar black holes, parameterized by the temperature T and the velocity u^μ of the brane. Our construction effectively promotes these parameters to Goldstone fields (or collective coordinate fields) $T(x)$ and $u^\mu(x)$, and determines their dynamics, order by order in the boundary derivative expansion. Note that this is distinct from linearization: we make no assumptions about the amplitudes of these slow variations.

6.2.3 Fluid/Gravity Map

Before proceeding to sketch the construction in more detail, we pause to stress an important point in mapping these long-wavelength gravity solutions to corresponding fluid configurations. A well-known procedure of holographic renormalization (see e.g. [3, 11]) links the boundary stress tensor to the behaviour of the bulk metric near the AdS boundary. Given any asymptotically AdS spacetime, we can read-off the induced stress tensor on the boundary, since the latter is related to the normalizable modes of the gravitational field in AdS. In particular, expanding the bulk metric in the Fefferman–Graham form near the boundary $z = 0$,

$$ds^2 = \frac{dz^2 + (\eta_{\mu\nu} + \alpha z^d T_{\mu\nu}) dw^\mu dw^\nu}{z^2},$$

the stress tensor is simply given by $T_{\mu\nu}$. Conversely, given a boundary stress tensor, there is a procedure to holographically reconstruct the bulk metric in a radial expansion around the boundary.

Naively, this might seem puzzling: as mentioned above, a conformally invariant stress tensor in d dimensions has $\frac{d(d+1)}{2} - 1$ degrees of freedom. If any such stress tensor yielded a regular bulk spacetime, we would have a discrepancy between the fluid side which has only d degrees of freedom and whose dynamics is correspondingly specified by only d equations, and the gravity side that would seemingly allow more degrees of freedom. In other words, passing from a generic quantum conformal field theory stress tensor to the stress tensor of its effective description in terms of fluid dynamics constitutes a drastic reduction in the number of degrees of freedom required to specify the spacetime. How is this manifested in the bulk? The answer lies in regularity. As a series expansion around the boundary, the holographic reconstruction cannot guarantee that the metric does not become nakedly singular at some finite radial value in the bulk. In fact, for a generic stress tensor it will. The fluid/gravity construction demonstrates that the regular solutions are given precisely by such stress tensors which are fluid dynamical. Moreover, we claim that the gravity solutions thus constructed are the most general regular long-wavelength[3] solutions to Einstein's equations with negative cosmological constant. They typically correspond to deformed and dynamical black holes; i.e. the solutions admit a regular event horizon which shields a curvature singularity.

The heuristic picture of a generic evolution, on the two sides of the fluid/gravity correspondence, is as follows. Suppose we start with some generic high energy initial conditions. On the CFT side, the system quickly settles down to local thermodynamic equilibrium, whose bulk dual is described by a dynamical, non-uniform (planar) black hole. On both sides, such configuration is described by local velocity and temperature fields which exhibit slow variation in the boundary directions. The subsequent evolution is described by equations of fluid dynamics on the boundary, which originate from Einstein's equations governing the bulk evolution. Finally, at late times, the system relaxes to global thermal equilibrium, given by a stationary state parameterized by a constant temperature and velocity. In the bulk, this is one of the well-known stationary solutions describing a planar black hole in AdS mentioned in the previous subsection and given explicitly below.

Our construction utilizes the long-wavelength regime of fluid dynamics: we write Einstein's equations as a perturbative expansion in boundary derivatives (however keeping the exact radial dependence). This allows us to solve the equations order by order in this boundary derivative expansion. In turns out that Einstein's equations at a given order implement the fluid stress tensor conservation equations at lower order. Therefore, order by order, we can use the lower-order fluid dynamical solution to construct the bulk metric, and then read off the corrected fluid dynamical stress tensor. In [5], we constructed the boundary stress tensor $T^{\mu\nu}$ and corresponding bulk metric g_{MN} to second order in boundary

---

[3] Note, however, that there are regular solutions which do not fall into the long-wavelength category, such as small black holes in AdS, which correspondingly are not described by fluid configurations.

derivative expansion. This yields a map between fluid dynamics and gravity, which we now proceed to sketch in more detail.

6.3 Construction of Bulk Metric and Boundary Stress Tensor

To explain our iterative procedure, we will first review the zeroth order configuration, and then discuss the deformation of this solution to first order.

6.3.1 0th Order

We start with the stationary solution to Einstein's equations with negative cosmological constant corresponding to the boosted planar Schwarzschild-AdS black hole:

$$ds^2 = -2u_\mu \, dx^\mu \, dr + r^2(\eta_{\mu\nu} + [1 - f(r/\pi T)u_\mu u_\nu) \, dx^\mu \, dx^\nu, \tag{6.3}$$

where $f(r) \equiv 1 - \frac{1}{r^4}$. Albeit perhaps less familiar, this form of the solution is very convenient since the metric is manifestly regular on the event horizon at $r = \pi T$ as well as being boundary-covariant. To obtain it from a more familiar form, we can start with the static Schwarzschild-AdS black hole in the planar limit:

$$ds^2 = r^2\left(-f(r)\,dt^2 + \sum_i (dx^i)^2\right) + \frac{dr^2}{r^2 f(r)},$$

change to ingoing Eddington coordinates to avoid the coordinate singularity on the horizon: $v = t + r_*$ where $dr_* = \frac{dr}{r^2 f(r)}$, and finally "covariantize" by boosting: $v \to u_\mu x^\mu, x_i \to P_{i\mu} x^\mu$, where $P_{\mu\nu}$ is the spatial projector, $P_{\mu\nu} = \eta_{\mu\nu} + u_\mu u_\nu$.

This solution is now parameterized by four parameters: the temperature T and the boosts u^μ (which have three independent components in four-dimensions due to normalization). The metric given in Eq. 6.3 constitutes the 0th order solution in our iterative procedure. The causal structure of this solution is identical to that of the static Schwarzschild-AdS black hole: there is a spacelike curvature singularity at $r = 0$, cloaked by a regular event horizon, and a timelike boundary. The AdS/CFT correspondence maps this bulk solution to an ideal fluid characterized by temperature T and fluid velocity u^μ. In particular, the induced stress tensor on the boundary for $d = 4$ is

$$T^{\mu\nu} = \pi^4 T^4 (\eta^{\mu\nu} + 4u^\mu u^\nu). \tag{6.4}$$

Note that this is an exact solution since the conservation equation (6.1) is automatically satisfied for this constant stress tensor, and correspondingly the metric (6.3) satisfies the bulk Einstein's equations (6.2) for any constant T and u^μ.

6.3.2 1st Order

A general fluid configuration in local, but not global, equilibrium can be described by promoting the four parameters T and u^μ to physical fields dependent on the boundary coordinates x^μ, i.e., to $T(x)$ and $u^\mu(x)$. If these fields vary slowly compared to the microscopic scale ℓ_{mfp}, i.e. if

$$\frac{\partial_\mu \log T}{T} \sim \mathcal{O}(\epsilon), \quad \frac{\partial_\mu u}{T} \sim \mathcal{O}(\epsilon)$$

for small ϵ, the fluid configuration still satisfies the conditions of local equilibrium. In each local domain of slow variation, which we refer to as tube, the bulk gravitational solution is approximately that of a uniform black brane. Remarkably, the bulk solution can be constructed by patching together these tubular domains! Of course, if we just replace u_μ and T in the metric (6.3) by $T(x)$ and $u^\mu(x)$, the resulting metric (call it $\overset{(0)}{g}_{MN}$) will no longer solve Einstein's equations (6.2). Instead, the metric $\overset{(0)}{g}_{MN}$ will need to be corrected by higher-order piece ($\overset{(1)}{g}_{MN}$, etc.), which we can obtain iteratively as an expansion in ϵ. We will find that the resulting corrected metric can be constructed systematically to any desired order, and is valid well inside the event horizon, thus allowing verification of its regularity. It is worthwhile to stress that the success of such a procedure rests on the fact the our seed metric $\overset{(0)}{g}_{MN}$ is manifestly regular on the horizon, since otherwise the expansion would break down near the coordinate singularity at horizon.

To implement the construction algebraically, we express the line element in a boundary derivative expansion of the fields $u_\mu(x)$ and $T(x)$, and use ϵ as a book-keeping parameter (counting the number of x^μ derivatives):

$$g_{MN} = \sum_{k=0}^{\infty} \epsilon^k \overset{(k)}{g}_{MN}, \quad T = \sum_{k=0}^{\infty} \epsilon^k T^{(k)}, \quad u_\mu = \sum_{k=0}^{\infty} \epsilon^k u_\mu^{(k)}. \tag{6.5}$$

The term $\overset{(k)}{g}_{MN}$ corrects the metric at the kth order, such that Einstein's equations will be satisfied to $\mathcal{O}(\epsilon^k)$ provided the functions $T(x)$ and $u^\mu(x)$ obey a certain set of equations of motion, which turn out to be precisely the stress tensor conservation equations of boundary fluid dynamics at $\mathcal{O}(\epsilon^{k-1})$.

Specifically, we can obtain the equations for $\overset{(k)}{g}_{MN}$ by substituting the expansion (6.5) into Einstein's equations (6.2), and extracting the coefficient of $\mathcal{O}(\epsilon^k)$. Schematically, these take the form

$$\mathcal{H}[g^{(0)}(u_\mu^{(0)}, T^{(0)})]g^{(k)}(x^\mu) = s_k \tag{6.6}$$

where $\mathcal{H}$ is a second-order linear differential operator in the variable r alone and s_k are regular source terms which are built out of $g^{(k)}$ with $n < k - 1$. Since $g^{(k)}(x^\mu)$ is already of $\mathcal{O}(\epsilon^k)$, and since every boundary derivative appears with an additional power of ϵ, $\mathcal{H}$ is an ultralocal operator in the field theory directions. Moreover, at a given x^μ, the precise form of this operator $\mathcal{H}$ depends only on the local values of T and u^μ but not on their derivatives at x^μ. Furthermore, the operator $\mathcal{H}$ is independent of k; we have the same homogeneous operator at every order in perturbation theory. This allows us to find an explicit solution of (6.6) systematically at any order. The source term s_k however gets more complicated with each order, and reflects the nonlinear nature of the theory.

Bit more explicitly, the equations of motion split up into two kinds: Constraint equations, $E_{r\mu} = 0$ which implement stress-tensor conservation (at one lower order), and Dynamical equations $E_{\mu\nu} = 0$ and $E_{rr} = 0$ which allow determination of $g^{(k)}$. We solve the dynamical equations

$$g^{(k)} = \text{particular}(s_k) + \text{homogeneous}(\mathcal{H})$$

subject to regularity in the interior and asymptotically AdS boundary conditions. Using the rotational symmetry group of the seed solution (6.3) it turns out to be possible to make a judicious choice of variables such that the operator $\mathcal{H}$ is converted into a decoupled system of first order differential operators. It is then simple to solve the (6.6) for an arbitrary source s_k by direct integration. For the details of the procedure, as well as discussion of convenient gauge choice, etc., we refer the reader to the original work [5] or the review [19].

Instead, here we simply quote the result for the bulk metric and boundary stress tensor, corrected to first order in ϵ. To first order the bulk metric takes the form

$$\begin{aligned} ds^2 = &- 2u_\mu\, dx^\mu\, dr + r^2(\eta_{\mu\nu} + [1 - f(r/\pi T)]u_\mu u_\nu)\, dx^\mu\, dx^\nu \\ &+ 2r\left[\frac{r}{\pi T}F(r/\pi T)\sigma_{\mu\nu} + \frac{1}{3}u_\mu u_\nu \partial_\lambda u^\lambda - \frac{1}{2}u^\lambda \partial_\lambda (u_\nu u_\mu)\right] dx^\mu\, dx^\nu, \end{aligned} \tag{6.7}$$

where $T(x)$ and $u_\mu(x)$ are any slowly-varying functions which satisfy the conservation equation (6.1) for the zeroth order ideal fluid stress tensor (6.4),

$$F(r) \equiv \int_r^\infty dx\, \frac{x^2 + x + 1}{x(x+1)(x^2+1)} = \frac{1}{4}\left[\ln\left(\frac{(1+r)^2(1+r^2)}{r^4}\right) - 2\arctan(r) + \pi\right] \tag{6.8}$$

and $\sigma^{\mu\nu}$ is the transverse traceless symmetric part of $\partial^\mu u^\nu$ called shear, i.e.

$$\sigma^{\mu\nu} = P^{\mu\alpha} P^{\nu\beta} \partial_{(\alpha} u_{\beta)} - \frac{1}{3} P^{\mu\nu} \partial_\alpha u^\alpha.$$

Note that the first line of (6.7) is simply the zeroth order (boundary-derivative-free) solution (6.3), whereas each of the terms in the second line have exactly one boundary derivative.[4]

The induced fluid stress tensor on the boundary, which can be easily obtained[5] from the bulk metric (6.7), is given by

$$T^{\mu\nu} = \pi^4 T^4 (4u^\mu u^\nu + \eta^{\mu\nu}) - 2\pi^3 T^3 \sigma^{\mu\nu}. \tag{6.9}$$

Here the first two (derivative-free) terms describe a perfect fluid with pressure (or negative free energy density) $\pi^4 T^4$, and correspondingly (using thermodynamics) entropy density $s = 4\pi^4 T^3$. The shear viscosity η of this fluid may be read off from the coefficient of $\sigma^{\mu\nu}$ and is given by $\pi^3 T^3$. Notice that $\eta/s = 1/(4\pi)$, in agreement with the well-known result of [18].

6.4 Solution to Second Order

In the previous section we have illustrated at first order how our iterative procedure can be implemented (in principle systematically to any order) to construct a generic long-wavelength solution. Such a procedure was carried out in [5], where the bulk metric and boundary stress tensor was calculated explicitly to second order in the boundary derivative expansion. In this section we will discuss the new physics which can be extracted from such a construction.

Note that already at first order, the bulk metric (6.7) was a much lengthier expression than the boundary stress tensor (6.9). This remains true in general; in fact, already at second order the expression for the metric is far too unwieldy to write down here. In the following subsections, we will therefore only write the second order boundary stress tensor explicitly but indicate the bulk metric only schematically.

---

[4] Note that (6.7) does not have any $\partial_\mu T$ terms appearing explicitly, since by implementing the zeroth order stress tensor conservation, we have expressed the temperature derivatives in terms of the velocity derivatives.

[5] We use the standard prescription (cf. [3]),

$$T^\mu_\nu = -2 \lim_{r\to\infty} r^4 (K^\mu_\nu - \delta^\mu_\nu)$$

where $K_{\mu\nu}$ is the extrinsic curvature tensor on a constant-r surface.

6.4.1 The Four-Dimensional Conformal Fluid from AdS_5

The second order stress tensor can be written as[6]

$$T^{\mu\nu} = (\pi T)^4(\eta^{\mu\nu} + 4u^\mu u^\nu) - 2(\pi T)^3\sigma^{\mu\nu} \\ + (\pi T)^2\left((\ln 2)T_{2a}^{\mu\nu} + 2T_{2b}^{\mu\nu} + (2 - \ln 2)\left[\frac{1}{3}T_{2c}^{\mu\nu} + T_{2d}^{\mu\nu} + T_{2e}^{\mu\nu}\right]\right) \tag{6.10}$$

where the expressions in the second line are given in terms of the fluid velocity and derived quantities,

$$\begin{aligned} T_{2a}^{\mu\nu} &= \epsilon^{\alpha\beta\gamma(\mu}\sigma_\gamma^{\nu)}u_\alpha\lambda_\beta, \quad T_{2b}^{\mu\nu} = \sigma^{\mu\alpha}\sigma_\alpha^\nu - \frac{1}{3}P^{\mu\nu}\sigma^{\alpha\beta}\sigma_{\alpha\beta}, \\ T_{2c}^{\mu\nu} &= \partial_\alpha u^\alpha\sigma^{\mu\nu}, \quad T_{2d}^{\mu\nu} = \mathcal{D}u^\mu\mathcal{D}u^\nu - \frac{1}{3}P^{\mu\nu}\mathcal{D}u^\alpha\mathcal{D}u_\alpha, \\ T_{2e}^{\mu\nu} &= P^{\mu\alpha}P^{\nu\beta}\mathcal{D}(\partial_{(\alpha}u_{\beta)}) - \frac{1}{3}P^{\mu\nu}P^{\alpha\beta}\mathcal{D}(\partial_\alpha u_\beta) \end{aligned} \tag{6.11}$$

with $\mathcal{D} \equiv u^\mu\partial_\mu$, $\lambda_\mu \equiv \epsilon_{\alpha\beta\gamma\mu}u^\alpha\partial^\beta u^\gamma$ where $\epsilon_{\alpha\beta\gamma\mu}$ is the Levi-Civita tensor, and as before $P^{\mu\nu}$ is the spatial projector and $\sigma^{\mu\nu}$ is the shear. Note that all of these expressions are second order in boundary derivatives (denoted by the 2 in the subscript).

In fact, one can package the expressions in (6.11) more compactly by using the decomposition of the 4-velocity gradient $\partial^\nu u^\mu$ into transverse, traceless and trace parts,

$$\partial^\nu u^\mu = -a^\mu u^\nu + \sigma^{\mu\nu} + \omega^{\mu\nu} + \frac{1}{3}\theta P^{\mu\nu},$$

where expansion, acceleration, and vorticity, are respectively defined as:

$$\theta = \partial_\mu u^\mu, \quad a^\mu = \mathcal{D}u^\mu, \quad \omega^{\nu\mu} = P^{\mu\alpha}P^{\nu\beta}\partial_{[\alpha}u_{\beta]}.$$

We can then rewrite the expressions (6.11) as:

$$\begin{aligned} T_{2a}^{\mu\nu} &= -2\omega^{\rho\langle\mu}\sigma_\rho^{\nu\rangle} \\ T_{2b}^{\mu\nu} &= \sigma^{\rho\langle\mu}\sigma_\rho^{\nu\rangle} \\ T_{2c}^{\mu\nu} &= \theta\sigma^{\mu\nu} \\ T_{2d}^{\mu\nu} &= a^{\langle\mu}a^{\nu\rangle} \\ \frac{1}{3}T_{2c}^{\mu\nu} + T_{2d}^{\mu\nu} + T_{2e}^{\mu\nu} &= \langle\mathcal{D}\sigma^{\mu\nu}\rangle + \frac{1}{3}\theta\sigma^{\mu\nu} \end{aligned} \tag{6.12}$$

[6] Here we quote the stress tensor as written originally in [5]; subsequently, slightly simpler and more general expressions were found; cf. e.g. [19] for a review. (Also, for simplicity we have absorbed an overall factor of $16\pi G_N^{(5)}$ which would appear in the conventionally defined stress tensor as conjugate to boundary time translations.)

Here we have adopted the notation employed in e.g. [1], using the angular brackets around the indices $A^{\langle\mu\nu\rangle}$ to denote the symmetric, transverse, traceless part of $A^{\mu\nu}$.

It however turns out to be most useful to write the second order stress tensor in a different basis of operators, $\mathcal{T}_k^{\mu\nu}$, which are manifestly conformally covariant, cf. [1, 16]:

$$T_{(2)}^{\mu\nu} = \tau_\pi \eta \mathcal{T}_1^{\mu\nu} + \kappa \mathcal{T}_2^{\mu\nu} + \lambda_1 \mathcal{T}_3^{\mu\nu} + \lambda_2 \mathcal{T}_4^{\mu\nu} + \lambda_3 \mathcal{T}_5^{\mu\nu}$$

and whose coefficients are the physically meaningful transport coefficients.

At first order the only nontrivial transport coefficient is the shear viscosity,

$$\eta = \frac{N^2}{8\pi}(\pi T)^3$$

whereas at second order the fluid parameters (relaxation timescales, etc.) extracted from the stress tensor (6.10) are

$$\tau_\pi = \frac{2 - \ln 2}{\pi T}, \qquad \lambda_1 = \frac{2\eta}{\pi T}, \qquad \lambda_2 = \frac{2\eta \ln 2}{\pi T}, \qquad \lambda_3 = 0.$$

These coefficients agree with the independent results of [1], who in addition derive the curvature coupling term, $\kappa = \frac{\eta}{\pi T}$. These coefficients correspond to the various relaxation timescales discussed in the literature (see e.g. [17]) in context of high energy collider physics. As indicated earlier, the fact that we are dealing with a particular conformal fluid, namely one that is dual to gravitational dynamics in asymptotically AdS spacetimes, leads to these coefficients being determined as fixed numbers.

6.4.2 The Spacetime Geometry Dual to Fluids

Let us now turn to discuss the bulk geometry obtained at the second order in boundary derivative expansion (whose first order part is given by (6.7)). As mentioned previously, this bulk solution is "tubewise" approximated by a planar black hole. This means that in each tube, defined by a small neighbourhood of given x^μ, but fully extended in the radial direction r, the radial dependence of the metric is approximately that of a boosted planar black hole at some temperature T and horizon velocity u^μ. These parameters vary from one position x^μ to another in a manner consistent with fluid dynamics. Our choice of coordinates is such that each tube extends along an ingoing radial null geodesic. Apart from technical advantages, this is conceptually rather pleasing, since it suggests a mapping between the boundary and the bulk which is natural from causality considerations.

It is worth stressing that although we refer to the metric written in (6.7) and its second order extension presented in [5] as "a solution" in the singular form rather than the plural, these expressions actually correspond to not just a single solution or finite family of solutions, but rather a continuously infinite family of solutions, specified by the four functions $T(x)$ and $u^\mu(x)$ of four variables. The flip side of the coin is that while very general, such a metric is not fully explicit: in order to be so, we need to use a given solution to fluid dynamics as input.

However, even in the absence of the explicit functional dependence of $T(x)$ and $u^\mu(x)$, it is possible to extract certain salient features of any such geometry. The most important feature of our geometry is the presence of an event horizon. In [6] we have demonstrated explicitly that the event horizon is regular, and determined its location in terms of the functions $T(x)$ and $u^\mu(x)$. Here we only schematically motivate these results. Intriguingly, it turns out that the location of the event horizon $r_H(x^\mu)$ in the bulk is determined locally by the fluid dynamical data at a point x^μ (within the derivative expansion), rather than globally as usual in general relativity.

To motivate this physically, within each tube characterized by a given x^μ, the position of the horizon is approximately at $r_H \approx \pi T$ corresponding to that tube. Since T varies as a function of x^μ, so will the horizon position $r_H(x^\mu)$. In the corrected solution, the surface $r = \pi T(x)$ is not the event horizon (for example, it is not a null surface in general), but if the variation $T(x)$ is slow, the deviation from the true event horizon is likewise small.

One can determine the position of the event horizon within our perturbation scheme using the fact that the solution settles down at late times to a uniformly boosted planar black hole. In particular, if we expand the horizon location as a series in boundary derivatives, tagged as before by ϵ,

$$r = r_H(x) = \pi T(x) + \sum_{k=1}^{\infty} \epsilon^k r_{(k)}(x),$$

then the coefficient functions $r_{(k)}(x)$ are determined algebraically by demanding that the surface given by $r = r_H(x)$ be null.

A very simple toy model which captures the gist of this argument is given by a time-dependent but spherically symmetric black hole, the Vaidya spacetime:

$$ds^2 = -\left(1 - \frac{2m(v)}{r}\right) dv^2 + 2\, dv\, dr + r^2\, d\Omega^2 .$$

This metric describes a four-dimensional asymptotically flat black hole accreting null dust, so that the mass $m(v)$ increases with time. Assuming that at late times the black hole settles down to Schwarzschild, $m(v) \to m_f$ as $v \to \infty$, and denoting the location of event horizon by $r = r_H(v)$, we can find its position by demanding that it describes the null surface which at late times approaches the correct event horizon $r_H \to 2m_F$ as $v \to \infty$. Note that the normal n to a null surface will be simultaneously tangent to that surface and likewise null. For Vaidya, the normal 1-form $n = dr - \dot{r}\, dv$ (where $\dot{} \equiv \frac{d}{dv}$) is null when

$$r_H(v) = 2m(v) + 2r_H(v)\dot{r}_H(v).$$

Of course, the exact solution to this equation yields the horizon $r_H(v)$ non-locally in terms of $m(v)$, requiring the knowledge of $m(v)$ for all $v < \infty$. However, when

$m(v)$ varies slowly, so that $\dot{m} = \mathcal{O}(\epsilon), m\ddot{m} = \mathcal{O}(\epsilon^2)$, etc., we can determine this location in terms of an ϵ expansion. For the ansatz

$$r_H(v) = 2m(v) + am(v)\dot{m}(v) + bm(v)\dot{m}(v)^2 + cm(v)^2\ddot{m}(v) + \cdots$$

this expansion gives $a = 8, b = 64, c = 32, \ldots$ This toy model illustrates that in spite of the event horizon being defined globally (as the boundary of the causal past of the future null infinity) and therefore requiring knowledge of the mass $m(v)$ for all time $v < \infty$, for slowly varying $m(v)$ we can nevertheless express $r_H(v_0)$ solely in terms of m its derivatives at $v = v_0$.

Returning to the problem of interest, we can similarly locate the event horizon $r_H(x^\mu)$ in our dynamical non-uniform planar black hole geometry in terms of T, u^μ, and all their derivatives, at the given point x^μ. At first order, the position of the horizon is unchanged, whereas at second order it is corrected by terms which scale with square of the shear and vorticity (see [6, 19] for explicit expressions.)

Once we identify the position of the event horizon in our geometry, it is easy to check that this horizon is regular. In fact, our construction manifestly guarantees regularity: the only curvature singularity of the seed metric is at $r = 0$, and the source terms which appear in correcting the metric order by order do not introduce any additional singularities. The final issue to check is the regime of validity of our expansion, and this can be seen to extend well inside the event horizon.

Therefore we have an explicit one-to-one map relating conformal fluid configurations on $\mathbf{R}^{3,1}$ to asymptotically AdS_5 inhomogeneous black brane solutions having regular event horizons. This is a remarkable statement about gravity, suggesting that fluid dynamical configurations within the AdS/CFT correspondence naturally uphold Cosmic Censorship.

To extract further physics from the position of the event horizon, let us consider the proper area of its spatial slices. By the second law of black hole mechanics, the horizon area cannot decrease with time; or equivalently, the expansion of the horizon generators must be non-negative. A well-known identification with thermodynamics translates this statement to that of the entropy increasing, or more locally, the entropy current having non-negative divergence. Having obtained the event horizon for our geometry explicitly in terms of the metric functions $u^\mu(x)$ and $T(x)$, we can verify these statements, and identify the entropy current naturally induced on the boundary.

To obtain the boundary entropy current J_S^μ from the bulk geometry, we can pull-back the area form A on the event horizon to the boundary. We perform this pull-back along a tube of constant x^μ, i.e. along ingoing radial null geodesics. This yields the expression

$$\begin{aligned}(4\pi\eta)^{-1}J_S^\mu = {} & [1 + b^2(A_1\sigma_{\alpha\beta}\sigma^{\alpha\beta} + A_2\omega_{\alpha\beta}\omega^{\alpha\beta} + A_3\mathcal{R})]u^\mu \\ & + b^2[B_1\mathcal{D}_\lambda\sigma^{\mu\lambda} + B_2\mathcal{D}_\lambda\omega^{\mu\lambda}] \\ & + C_1 b\lambda^\mu + C_2 b^2 u^\lambda\mathcal{D}_\lambda\lambda^\mu + \cdots \end{aligned} \tag{6.13}$$

with $b \equiv \frac{1}{\pi T}$ and

$$A_1 = \frac{1}{4} + \frac{\pi}{16} + \frac{\ln 2}{4}, \quad A_2 = -\frac{1}{8}, \quad A_3 = \frac{1}{8}, \\ B_1 = \frac{1}{4}, \quad B_2 = \frac{1}{2}, \quad C_1 = C_2 = 0 \tag{6.14}$$

which can be easily verified to have non-negative divergence. At leading order, the divergence is proportional to $\sigma_{\mu\nu}\sigma^{\mu\nu}$ which is manifestly positive for any non-zero shear. In case of zero shear, the relations $B_1 = 2A_3$ and $C_1 + C_2 = 0$ guarantee non-negativity. We can readily see that these relations are indeed satisfied for the entropy current given by our gravity dual construction (6.14).

The expression (6.13) was written suggestively in a most general form yielding a Weyl-covariant entropy current for any set of constant coefficients A_i, B_i, C_i. In general, to the order we work, the seven coefficients get reduced to five independent ones allowing for Weyl-covariant entropy current with non-negative divergence. From the field theory side, this would therefore suggest a 5-parameter ambiguity in constructing a sensible entropy current, purely based on the symmetries and the requirement that it correctly reproduces equilibrium physics. This is at first sight puzzling, since our gravity construction seemingly fixed all these parameters. In fact, this was not quite the case: there is still an ambiguity even on the gravity side, corresponding to the freedom to add total derivative terms without changing the horizon area, and the pull-back being ambiguous to boundary diffeomorphisms. However, at second order this results in a 2-parameter ambiguity for Weyl covariant current with positive divergence, so that some mismatch remains.[7]

6.5 Summary

One of the most intriguing features of the fluid/gravity correspondence is that it provides us with a window into the *generic* behavior of gravity in a nonlinear regime, mapping long-wavelength (but arbitrary amplitude) perturbations of AdS black holes to the more familiar physics of fluid dynamics. Apart from the obvious conceptual advantages, one has a tremendous computational simplification for numerical studies of gravitational solutions since the fluid dynamics lives in one lower dimension.

The bulk spacetime solutions discussed here describe a generic (time-dependent and non-uniform) planar black hole. We have demonstrated that each of these solutions (with regular fluid data $T(x)$ and $u_\mu(x)$) has its singularities hidden from the boundary by a regular event horizon. In this sense, all gravitational solutions dual to regular solutions of fluid dynamics uphold the cosmic censorship

[7] Although including higher orders may remedy this discrepancy [21].

conjecture. Moreover, our construction seems to imply a new variant of Uniqueness theorem: for a given boundary fluid configuration, there corresponds a unique regular bulk metric.

At the technical level, the key to our construction lies in utilizing the long-wavelength regime of fluid dynamics. This allows us to write the bulk metric, and corresponding boundary stress tensor, to any order in a boundary derivative expansion in a completely procedural manner. Furthermore, once having obtained the bulk geometry, this expansion likewise enables us to determine the radial position of its event horizon, remarkably expressed locally in terms of the fluid dynamical data. We can then easily verify that the area of this event horizon is necessarily non-decreasing, in accordance with the Area Theorem. The corresponding entropy current induced from the bulk solution is then guaranteed to satisfy the Second Law of thermodynamics.

The boundary fluid stress tensor contains new quantities of interest, namely the various transport coefficients which characterise the fluid. At first order we have recovered the previously-known value of viscosity and verified that $\frac{\eta}{s} = \frac{1}{4\pi}$. More importantly, we have predicted the second order fluid parameters $(\tau_\pi, \lambda_1, \lambda_2, \lambda_3)$. This has been of interest in QCD phenomenology, especially in understanding certain characteristic features of the quark-gluon plasma.

Although (in interest of familiarity and brevity of presentation) we have sketched the fluid/gravity correspondence in the simplest setting, many useful generalizations have already been carried out. One of the earliest such generalizations involved extending the correspondence to other dimensions, relating a d-dimensional conformal fluid to asymptotically AdS_{d+1} black hole (see [23] for the interesting case of $d = 3$ and [8, 13] for general d). More ambitiously, one may also consider fluids on curved manifolds (rather than just the Minkowski spacetime $\mathbf{R}^{d-1,1}$), as has been initiated in [7]. In addition, one can include matter in the bulk. This allows for richer dynamics, but typically at the expense of losing universality. Early examples of such extensions include considering the dilaton (which corresponds to forcing of the fluid) in [7], Maxwell $U(1)$ field [2, 12], multiple Maxwell fields and scalars, magnetic and dyonic charges, as well as more exotic models (see e.g. [19] for further references). Moreover, one can even extend the correspondence to con-conformal fluids [10, 15] as well as to non-relativistic fluids [9, 20], which allows us to make closer contact with familiar everyday systems.

Nevertheless, many future directions and puzzles remain, as well the need for further generalizations. For example, of particular current interest is to understand the fluid/gravity correspondence for extremal fluids (and in particular superfluids) which are presently attracting much attention. Also, to mimic many of the familiar aspects of fluid flows, we need to understand how to confine the fluid within walls in the gravity dual. Still more ambitiously, to understand the rich phenomena rooted in quantum processes, we would like to get a better handle on finite-N effects.

More prosaically, even within the original context, several intriguing issues deserve further investigation. We have established one-to-one map between long-wavelength gravitational solutions and solutions of fluid dynamics. Naively, one might think of this as merely a low-energy effective description of the AdS/CFT correspondence. However is it not true that any gravitational solution in AdS admits a fluid description. In particular, it would be useful to understand the role of non-long-wavelength bulk classical or semiclassical solutions, such as small AdS black holes. Likewise, there is room for more detailed bulk analysis, such as probing the allowed horizon topology and dynamics in more general situations, the nature of curvature singularity, or Cosmic Censorship.

Another intriguing aspect is the striking difference between the phenomenology of turbulent flows in 3 + 1 and 2 + 1 dimensions, as pointed out in [23]. In the 3 + 1 dimensional turbulent energy cascade, large scale eddies give rise to smaller scale eddies, eventually transferring energy down to scales where viscosity becomes important and energy is dissipated. In contrast, the 2 + 1 dimensional turbulent flows are characterized by an "inverse cascade," in which smaller scale eddies merge into large scale eddies, creating large long-lived vortical structures. If these qualitative differences extend to relativistic fluids, they would suggest a profound difference in gravitational dynamics between four and five dimensional gravity. In particular, we might predict that black holes in AdS_4 would take much longer to equilibrate owing to the fact that the fluctuations on the horizon could coalesce in macroscopic structures. From the gravitational standpoint, this would certainly seem very surprising and counter-intuitive.

More generally, our correspondence offers new insight into the black hole membrane paradigm. The conventional membrane paradigm provides a simple picture of black hole dynamics in terms of classical physics of fluid living on a "membrane" (or stretched horizon) just outside the event horizon. Taking a more general view of trying to encode the black hole dynamics by fluid dynamics localized on a membrane in the spacetime, the immediate natural question is: where should such a membrane live? Perhaps the most obvious candidate is the event horizon; but this is problematic due to its null nature, and more importantly, because it is globally defined so we cannot fix its position without knowing the full future evolution of the spacetime. Alternately, several (quasi)local notions of a black hole have been proposed, such as the so-called dynamical horizon, which however are spacelike surfaces inside the event horizon, and therefore do not admit the standard notion of evolution. A more popular suggestion is the stretched horizon, which is the formulation given by the membrane paradigm. However, there likewise remain ambiguities in localizing stretched horizon. Within the fluid/gravity correspondence, the full spacetime dynamics is mapped to the dynamics of a conformal fluid, which albeit reminiscent of the membrane paradigm, has one important twist: the membrane lives on the boundary of the spacetime (which is unambiguously defined and admits a fluid description with well-defined dynamics), and gives a perfect mirror of the bulk physics. This "membrane at the end of the universe" picture is a natural consequence of the holographic nature of the fluid/gravity correspondence.

Finally, we have mentioned in the Introduction that one of the outstanding problems in fluid dynamics is understanding turbulence. We might therefore ask how relevant is the fluid/gravity correspondence (appropriately generalized) to tackling such a problem. We argue that the correspondence is indeed relevant and worth vigorously pursuing further (with a handful of proposals already appearing in the literature). The important point is that the regime where our derivative expansion is valid corresponds to large Reynolds number (where the viscous terms are small relative to the leading order terms), so the phenomenology of turbulence should be directly relevant to the study of near-equilibrium AdS black hole dynamics. At least for the widely studied case of non-relativistic fluids described by the Navier–Stokes equations, turbulence has many striking qualitative features, including the tendency to dynamically break symmetries as Reynolds number is increased, the sharp onset of turbulence at critical Reynolds numbers, and an "energy cascade" in fully developed turbulence in which energy is transferred in a predictable way between eddies at different scales. Less is known about turbulence for the microscopically relativistic fluids relevant in our context, but it would be fascinating to understand the gravitational interpretation for those features that do generalize to the relativistic case.

Acknowledgements It is a pleasure to thank my collaborators, Sayantani Bhattacharyya, R. Loganayagam, Gautam Mandal, Shiraz Minwalla, Takeshi Morita, Harvey Reall, Mukund Rangamani, and Mark Van Raamsdonk for marvelous collaborations on various aspects of fluid dynamics. I would also like to thank the participants of the Fifth Aegean summer school for their enthusiasm and questions, and the organizers, especially Eleftherios Papantonopoulos, for putting together an excellent summer school. This work was supported in part by STFC Rolling grant and by the National Science Foundation under the Grant No. NSF PHY05-51164.

References

1. Baier, R., Romatschke, P., Son, D.T., Starinets, A.O., Stephanov, M.A.: Relativistic viscous hydrodynamics, conformal invariance,and holography. JHEP **0804**, 100 (2008)
2. Banerjee, N., Bhattacharya, J., Bhattacharyya, S., Dutta, S., Loganayagam, R., Surowka, P.: Hydrodynamics from charged black branes, arXiv:0809.2596 [hep-th]
3. Balasubramanian, V., Kraus, P.: A stress tensor for anti-de sitter gravity. Commun. Math. Phys. **208**, 413–428 (1999)
4. Bhattacharyya, S., Lahiri, S., Loganayagam, R., Minwalla, S.: Large rotating AdS black holes from fluid mechanics. JHEP **0809**, 054 (2008)
5. Bhattacharyya, S., Hubeny, V.E., Minwalla, S., Rangamani, M.: Nonlinear fluid dynamics from gravity. JHEP **0802**, 045 (2008)
6. Bhattacharyya, S., Hubeny, V.E., Loganayagam, R., Mandal, G., Minwalla, S., Morita, T., Rangamani, M., Reall, H.S.: Local fluid dynamical entropy from gravity. JHEP **0806**, 055 (2008)
7. Bhattacharyya, S., Loganayagam, R., Minwalla, S., Nampuri, S., Trivedi, S.P., Wadia, S.R.: Forced fluid dynamics from gravity. JHEP **0902**, 018 (2009)
8. Bhattacharyya, S., Loganayagam, R., Mandal, I., Minwalla, S., Sharma, A.: Conformal nonlinear fluid dynamics from gravity in arbitrary dimensions. JHEP **0812**, 116 (2008)
9. Bhattacharyya, S., Minwalla, S., Wadia, S.R.: The incompressible non-relativistic Navier–Stokes equation from gravity. JHEP **0908**, 059 (2009)
10. David, J.R., Mahato, M., Wadia, S.R.: Hydrodynamics from the D1-brane. JHEP **04**, 042 (2009)

11. de Haro, S., Solodukhin, S.N., Skenderis, K.: Holographic reconstruction of spacetime and renormalization in the AdS/CFT correspondence. Commun. Math. Phys. **217**, 595–622 (2001)
12. Erdmenger, J., Haack, M., Kaminski, M., Yarom, A.: Fluid dynamics of R-charged black holes. JHEP **0901**, 055 (2009)
13. Haack, M., Yarom, A.: Nonlinear viscous hydrodynamics in various dimensions using AdS/CFT. JHEP **10**, 063 (2008)
14. Janik, R.A., Peschanski, R.: Asymptotic perfect fluid dynamics as a consequence of ads/cft. Phys. Rev. **D73**, 045013 (2006)
15. Kanitscheider, I., Skenderis, K.: Universal hydrodynamics of non-conformal branes. JHEP **04**, 062 (2009)
16. Loganayagam, R.: Entropy current in conformal hydrodynamics. JHEP **0805**, 087 (2008)
17. Muronga, A.: Causal theories of dissipative relativistic fluid dynamics for nuclear collisions. Phys. Rev. **C69**, 034903 (2004)
18. Policastro, G., Son, D.T., Starinets, A.O.: The shear viscosity of strongly coupled $N = 4$ supersymmetric yang-mills plasma. Phys. Rev. Lett. **87**, 081601 (2001)
19. Rangamani, M.: Gravity and hydrodynamics: lectures on the fluid–gravity correspondence. Class. Quant. Grav. **26**, 224003 (2009)
20. Rangamani, M., Ross, S.F., Son, D.T., Thompson, E.G.: Conformal non-relativistic hydrodynamics from gravity. JHEP **0901**, 075 (2009)
21. Romatschke, P.: Relativistic viscous fluid dynamics and non-equilibrium entropy, arXiv:0906.4787 [hep-th]
22. Son, D.T., Starinets, A.O.: Viscosity, black holes, and quantum field theory. Ann. Rev. Nucl. Part. Sci. **57**, 95 (2007)
23. Van Raamsdonk, M.: Black hole dynamics from atmospheric science. JHEP **05**, 106 (2008)

Chapter 7
The Gauge-Gravity Duality and Heavy Ion Collisions

Amos Yarom

Abstract This chapter provides a review of two particular applications of the gauge-gravity duality to heavy ion collisions. The first involves a study of the wake of a quark as it travels through the quark gluon plasma and its possible connection to measurements of jet correlations carried out at the relativistic heavy ion collider at Brookhaven. The second section provides, via the gauge/gravity duality, a lower bound on the entropy produced in a collision of two energetic distributions. This is then compared to particle multiplicity in gold–gold collisions.

In its simplest form, the gauge-gravity duality [21, 47, 61] relates the planar limit of $\mathcal{N}=4$ super-Yang–Mills (SYM) at large 't Hooft coupling to classical gravity in an asymptotically anti-de-Sitter (AdS) geometry. The particle content of $\mathcal{N}=4$ super-Yang–Mills is completely determined by super-symmetry and includes gluons, four Majorana fermions in the adjoint representation of the gauge group and six real scalars also in the adjoint. While some deformations of the $\mathcal{N}=4$ SYM theory and their dual are well understood, the string theory dual of quantum chromodynamics (QCD) is currently unavailable. Nevertheless, there have been numerous attempts to extract information about the the strongly coupled deconfined phase of QCD using AdS/CFT.

A critic may argue that any attempt to compare QCD with $\mathcal{N}=4$ SYM or its variants is destined to fail since there is no control parameter for such a comparison. However, even though there is no direct means of comparing the two, one may be able to address open problems in QCD using the gauge-gravity duality. This can be achieved either by uncovering universal properties of strongly coupled gauge theories—ones that are shared by all gauge theories with a holographic dual,

A. Yarom (✉)
Joseph Henry Laboratories, Princeton University Princeton, Princeton NJ 08540, USA
e-mail: ayarom@princeton.edu

E. Papantonopoulos (ed.), *From Gravity to Thermal Gauge Theories: The Ads/CFT Correspondence*, Lecture Notes in Physics, 828,
DOI: 10.1007/978-3-642-04864-7_7,

or by providing a precise answer to a question for which even a qualitative solution is lacking.

In this chapter, two particular applications of AdS/CFT to heavy ion collisions will be reviewed. The first involves a study of the wake of a quark as it travels through the quark gluon plasma and its possible connection to measurements of jet correlations carried out at the relativistic heavy ion collider at Brookhaven. The second section provides, via the gauge/gravity duality, a lower bound on the entropy produced in a collision of two energetic distributions. This is then compared to particle multiplicity in gold–gold collisions.

7.1 The Wake of a Quark

7.1.1 Jet Correlations at RHIC

At the relativistic heavy ion collider at brookhaven (RHIC) two gold ions collide with a total center of mass energy of about 39 TeV. Each gold ion has 197 nucleons. So, in the center of mass frame each nucleon has an energy of roughly

$$E_{\mathrm{n}} \sim 39\,\mathrm{TeV}/(2 \times 197) \sim 100\,\mathrm{GeV}. \tag{7.1}$$

Given that the mass of a nucleon is roughly 1 GeV, the Lorentz factor for the ions is roughly a hundred,

$$\gamma \sim E_{\mathrm{n}}/M_{\mathrm{n}} \sim 100, \tag{7.2}$$

which implies that the ions will be Lorentz contracted. For this reason the moving ions are usually depicted as flat "pancakes". See Fig. 7.1. Shortly after the collision thousands of hadrons reach the detector and it is an experimental challenge to extract information about the intermediate stages of the collision from the final particle distribution at the detector. Presumably, at the late stages of the collision a

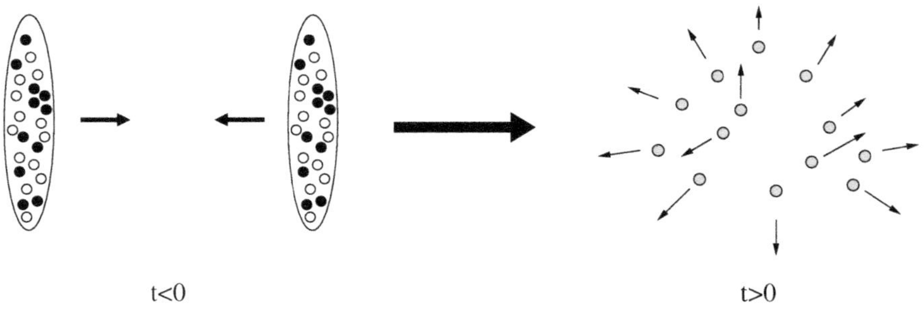

Fig. 7.1 A head on collision of two gold ions. Before the collision there are roughly 400 nucleons, each one has a center of mass energy of approximately 100 GeV. This is why the heavy ions are usually depicted as flattened 'pancakes'. After the collision thousands of hadrons reach the detector. Presumably a quark-gluon plasma is formed in the intermediate stages of the collision

strongly coupled quark-gluon plasma (QGP) is formed—a configuration in which the quarks and gluons are strongly coupled but deconfined. Some indication for the existence of a QGP is the Boltzman distribution of hadrons reaching the detector at a temperature which is roughly the deconfinement temperature [37].

One method to quantify the resulting distribution of hadrons is a measurement of the degree of correlation between the emerging hadrons. In proton–proton collisions it is expected that correlated hadrons are a result of a pair being created in the collision region: from momentum conservation, the jets reaching the detector must come out back to back, as shown in Fig. 7.2. In gold–gold collisions, at mid-rapidity, and when considering trigger jets with transverse momentum in the range $2.5\,\mathrm{GeV} < p_T < 4\,\mathrm{GeV}$ and away-side jets of transverse momentum $1\,\mathrm{GeV} < p_T < 2.5\,\mathrm{GeV}$, instead of back-to-back jets one finds a local minimum in the correlation function at angles of π, and a maximum at an angle of, roughly $\pi - 1$ [2–4, 32, 36, 58, 59]. As depicted in Fig. 7.3, this implies that instead of back-to-back jets, the jets are "split".

A possible explanation of this phenomenon involves the hydrodynamic excitations of the parton pair as it moves through the quark-gluon plasma [14, 15]. Imagine a pair produced very close to the surface of the plasma such that one of the quarks/gluons immediately leaves the QGP while its partner has to plow through the plasma. As the partner particle traverses the plasma it will loose energy and eventually slow to a halt. But, if its initial velocity is higher than the speed of sound of the plasma it will create a shock wave. Once hadronization occurs the energy contained in the shock wave will reach the detector as particles and the observed signal will be that of a split jet. While such a scenario is consistent with observations, its drawback is that point like excitations such as quarks also generate a laminar wake far behind them. In the setup we are considering, such a wake, if it exists, will upon hadronization generate a broad-shoulder like signal which will be observed, at the detector, at an angle of π instead of the observed minimum. This is depicted in Fig. 7.4.

While it is true that a point-like source will generate a laminar wake behind it, it is possible that the collective excitation of the away-side quark, which is what sources the hydrodynamic modes, is not point-like. It is possible that very close to the quark, the dynamics of the plasma is not governed by hydrodynamics. On the contrary,

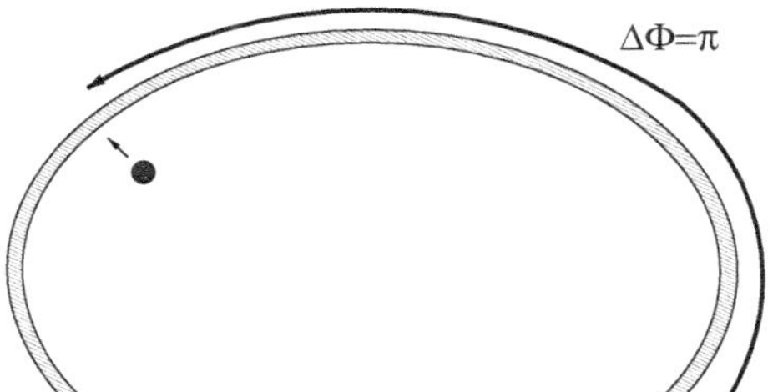

Fig. 7.2 A schematic diagram of the detector (*shaded*) and two correlated hadrons reaching it at an angle of π relative to one another. One should keep in mind that this is a highly oversimplified picture of the detector and that in practice the experimental situation is much more complex

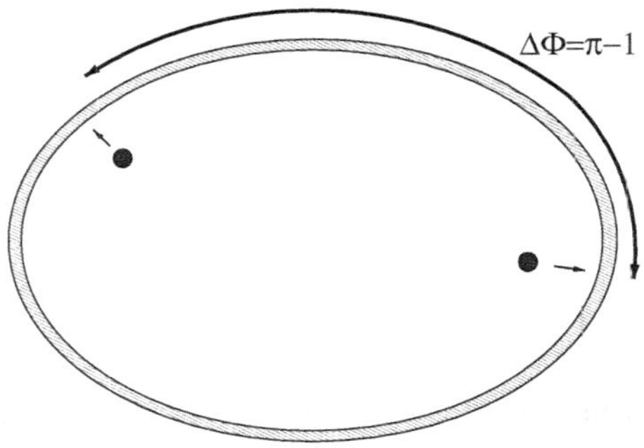

Fig. 7.3 A schematic diagram of the detector (*shaded*) and two correlated hadrons reaching it at an angle of $\pi - 1$ relative to one another. One should keep in mind that this is a highly oversimplified picture of the detector and that in practice the experimental situation is much more complex

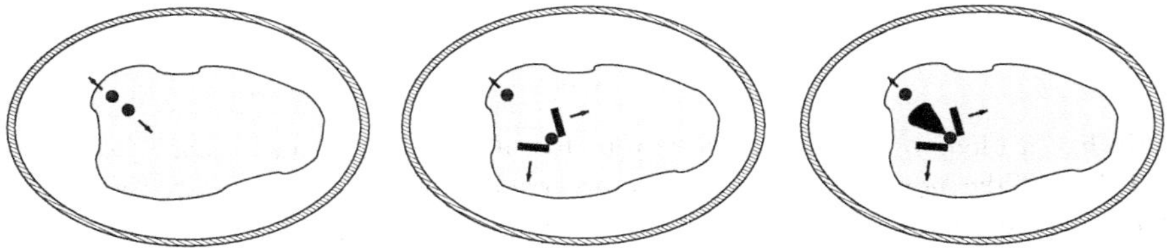

Fig. 7.4 A simplified picture of a proposed mechanism for the generation of the jet-splitting effect. The detector is depicted by the *gray* ellipsoid, and the quark gluon plasma is the *light gray* region in the center. A quark pair is created very close to the surface of the plasma (far *left*). One quark exits the QGP and appears as a trigger jet at the detector. The other quark traverses the plasma generating a shock wave (*middle*), or a shock wave and wake (far *right*). In the first case, the signal arriving at the detector will have a local minimum at an angle of π relative to the trigger whereas in the second case the signal will have a broad shoulder-like structure

hydrodynamics describes the long wavelength excitations of the QGP far away from the moving source. What sources the hydrodynamic modes is not the moving quark (or parton) but rather the quark plus near-field excitation. As was shown in [14, 15], it is possible that such an effective source for hydrodynamics will not behave like a point-like particle and therefore will not generate a wake behind it.

While the phenomenological model in [14, 15] seems to explain some of the data, it is unsatisfying that one needs to assume that the effective excitation of a quark moving in a strongly coupled QGP is tuned precisely so that a wake will not be generated behind the quark. Without some handle over the interaction between the quark and the QGP it is impossible to know whether or not such a model is viable. This is where the AdS/CFT correspondence comes in. Some attempts at treating this problem by more conventional methods can be found in, for example [10, 35, 46–49, 55, 56].

7.1.2 A Holographic Computation

As explained in the previous section, it is difficult to compute properties of the wake of a quark moving in a QCD plasma at temperatures above the deconfinement temperature where QCD is presumably strongly coupled. However, it is possible to carry out an exact computation of the linear response of the QGP to a heavy quark in the $\mathcal{N} =$ SYM theory. As emphasized earlier, there is no control parameter which allows us to go from $\mathcal{N} = 4$ SYM to QCD. Nevertheless, we will see that the results of the $\mathcal{N} = 4$ SYM theory computation will provide important information regarding the possible validity of the scenario described above where the quark presumably does not create a wake behind it.

The planar limit of strongly coupled $\mathcal{N} = 4$ SYM is equivalent to the supergravity limit of type *IIB* superstring theory in an asymptotically $\mathrm{AdS}_5 \times S^5$ geometry. The string tension α' is related to the 't Hooft coupling, λ, in the gauge theory, and the number of colors in the gauge theory, N, is related to the gravitational constant in the bulk G_N,

$$G_N \sim 1/N^2 \quad \alpha' \sim \sqrt{\lambda}\,. \tag{7.3}$$

The duality between $\mathcal{N} = 4$ SYM and type *IIB* string theory can be succinctly stated as an equality between the partition function of string theory and the generating function of the gauge theory. In the supergravity limit this implies that the on-shell supergravity action of *IIB* supergravity gives the generating function of the planar limit of $\mathcal{N} = 4$ SYM. Explaining the duality in detail will take us far afield, and is beyond the scope of this work. There is an immense body of literature which explains the internal working of the duality to which the interested reader is referred to. Some canonical works are [5, 21, 47, 61].

Since the duality provides the generating function for a strongly coupled gauge theory, we can use it to compute a particular one-point function: that of the stress tensor $\langle T_{\mu\nu}\rangle$. The AdS/CFT prescription for carrying out this computation is as follows [12, 13, 21, 41, 61]. We work in a gauge where the line element takes the form:

$$ds^2 = \frac{L^2}{z^2} g_{\mu\nu} dx^\mu dx^\nu + \frac{L^2}{z^2} dz^2 \tag{7.4}$$

where we have restricted our attention to the five non-compact dimensions and have omitted an S^5 part of the line element. The S^5 will play a minor role in the current analysis. The energy momentum tensor of the boundary theory can be read off a near boundary ($z \to 0$) expansion of the metric

$$g_{\mu\nu} = \eta_{\mu\nu} + \cdots + \frac{3}{16\pi G_N} \langle T_{\mu\nu}\rangle z^4 + \mathcal{O}(z^5) \tag{7.5}$$

where η is the Minkowski metric and by ... we mean intermediate terms in the series expansion.

Let us see how the prescription (7.5) works in practice. Consider first AdS_5 space which, according to the AdS/CFT dictionary is dual to the vacuum of the SYM gauge theory. The line element for a Poincaré patch of AdS_5 can be written in the form

$$ds^2 = \frac{L^2}{z^2}\left(-dt^2 + \sum_i (dx^i)^2\right) + \frac{L^2}{z^2}dz^2. \tag{7.6}$$

where z runs from 0 to ∞ and the other coordinates run from $-\infty$ to ∞. Note that apart from the overall warp factor $\frac{L^2}{z^2}$ the line element of AdS_5 space is identical to the Minkowski space line element; if we look at a surface of constant $z = z_0$ then the induced metric on this surface is precisely that of Minkowski space stretched by an overall warp factor $\frac{L^2}{z_0^2}$. See Fig. 7.5. Going back to the prescription in (7.5) we obtain immediately

$$\langle T_{\mu\nu}\rangle = \begin{pmatrix} 0 & 0 & 0 & 0 \\ 0 & 0 & 0 & 0 \\ 0 & 0 & 0 & 0 \\ 0 & 0 & 0 & 0 \end{pmatrix} \tag{7.7}$$

as expected for the vacuum state of $\mathcal{N} = 4$ SYM.

Moving on to a more complicated example, we consider the state dual to a Schwarzschild black hole in AdS which is, presumably, a thermal state with temperature T [61]. Before carrying out such a computation let us see what sort of stress-tensor we expect for such a state. From a field theory point of view, we expect that the stress energy tensor for a large N $SU(N)$ gauge theory be proportional to N^2 the number of degrees of freedom, and in a conformal theory, in the absence of any other scales, it should also be proportional to T^4, from dimensional arguments. Finally, by studying the variation of the action by an infinitesimal dilatation one finds that the stress energy tensor is traceless. Hence, in a conformal theory on expects that

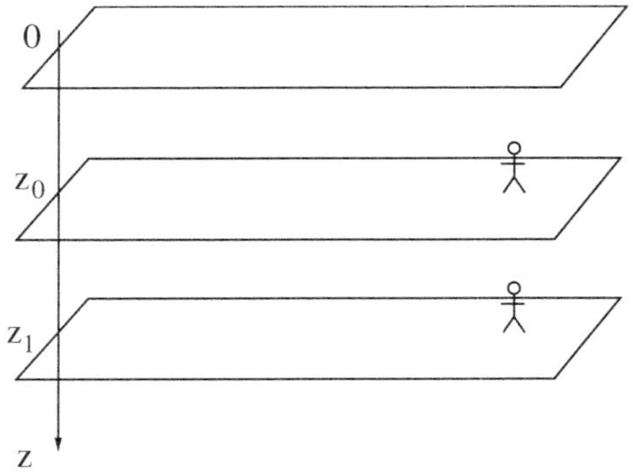

Fig. 7.5 A cartoon of AdS space. Using the parameterization in (7.6), an observer confined to $z = z_0$ would not be able to distinguish the surface she lives on from Minkowski space. However, when moving in the z direction the observer will notice that distances transverse to the z direction are warped. The boundary of AdS is located at $z = 0$

Fig. 7.6 A cartoon of an AdS black hole. The patch covered by the coordinate system in (7.9) covers the region of space from the boundary to the event horizon which is located at $z = z_0$

$$\langle T_{\mu\nu} \rangle = CN^2T^4 \begin{pmatrix} 3 & 0 & 0 & 0 \\ 0 & 1 & 0 & 0 \\ 0 & 0 & 1 & 0 \\ 0 & 0 & 0 & 1 \end{pmatrix} \tag{7.8}$$

where C is a number which depends on the details of the theory.

A thermal state in the gauge theory is dual to a black hole [61]. The line element of an AdS_5 black hole is given by

$$ds^2 = \frac{L^2}{z^2} \left(-f(z)dt^2 + \sum_{i=1}^{3} (dx^i)^2 + \frac{dz^2}{f(z)} dz^2 \right) \tag{7.9}$$

where

$$f(z) = 1 - \left(\frac{z}{z_0} \right)^4 \tag{7.10}$$

The Hawking temperature of this black hole is $T = 1/(4\pi z_0)$ and is equal to the temperature of the thermal state of the gauge theory. If $f(z) = 1$, we obtain the AdS geometry given in (7.6). When $f(z)$ is as given in (7.10), the patch of spacetime we are interested in ends at $z = z_0$, where the event horizon of the black hole is located. See Fig. 7.6.

To use (7.5) we need to bring the metric in (7.9) to the form (7.4).[1] This can be achieved by using a coordinate ρ such that

$$\frac{L^2}{z^2}\frac{dz^2}{f} = \frac{L^2}{\rho^2} d\rho^2 \tag{7.11}$$

in which case the line element takes the form

$$ds^2 = \frac{L^2}{\rho^2} \left(\frac{\rho^4 - 4}{4(\rho^4 + 4)z_0^2} \right) dt^2 + \left(1 + \frac{\rho^4}{4z_0^2} \right) \sum_i (dx^i)^2 + \frac{L^2 d\rho^2}{\rho^2} \tag{7.12}$$

[1] Alternately, we can work in a covariant formalism such as the one discussed in [41].

Taking the appropriate $\rho \to 0$ limit of the metric components, we find, in the notation of (7.4), that

$$g_{tt} = -1 + \frac{3\rho^4}{4z_0^2} + \mathcal{O}(\rho^5) \quad g_{x^ix^i} = 1 + \frac{\rho^4}{4z_0^2}. \tag{7.13}$$

Thus, the energy momentum tensor is given by

$$\langle T_{\mu\nu} \rangle = \frac{\pi^2}{8} N^2 T^4 \begin{pmatrix} 3 & 0 & 0 & 0 \\ 0 & 1 & 0 & 0 \\ 0 & 0 & 1 & 0 \\ 0 & 0 & 0 & 1 \end{pmatrix} \tag{7.14}$$

which is precisely the form given in (7.14) with $C = \pi^2/8$.

Recall that our goal in this section is to compute the response of a thermal medium to a moving quark and to check whether the laminar wake is strong or weak. We know what the dual of a thermal state is—it is given by the black hole geometry in (7.9). What we need to do next is place a quark in the thermal medium. According to [21, 31, 41], a heavy (infinitely massive) quark is dual to the endpoint of an open string located at the asymptotically AdS boundary. A stationary quark, for example, would be dual to a string whose endpoint is standing still at the boundary of AdS. By symmetry, the string must hang straight down into the bulk of AdS. To formally derive the profile of the string we need to solve the string equation of motion, with appropriate boundary conditions.

The dynamics of the string are governed by the Nambu–Goto action:

$$S_{\mathrm{NG}} = -\frac{1}{2\pi\alpha'} \int \sqrt{-|G_{\mu\nu}\partial_\alpha X^\mu \partial^\alpha X^\nu|}\, d\tau d\sigma \tag{7.15}$$

where G is the spacetime metric and X is the string embedding function. The boundary conditions we impose on the string is that its endpoint on the boundary of AdS is stationary. Just from symmetry, it should be clear that the only solution to the equations of motion is one where the string hangs straight down into AdS (see Fig. 7.7) . Thus, if we use a gauge where $\tau = X^0$ and $\sigma = X^5$, we find that the solution to the string equations of motion must be:

$$X^\mu = (\tau \quad 0 \quad 0 \quad 0 \quad \sigma). \tag{7.16}$$

Of more relevance to the problem at hand is a moving quark. i.e., a string whose endpoint is not stationary. The simplest case to consider is a configuration where the string endpoint is moving at constant velocity. In this case, if we insert the ansatz

$$X^\mu = (t \quad vt + \xi(z) \quad 0 \quad 0 \quad z) \tag{7.17}$$

into the Nambu–Goto action (7.15), we find

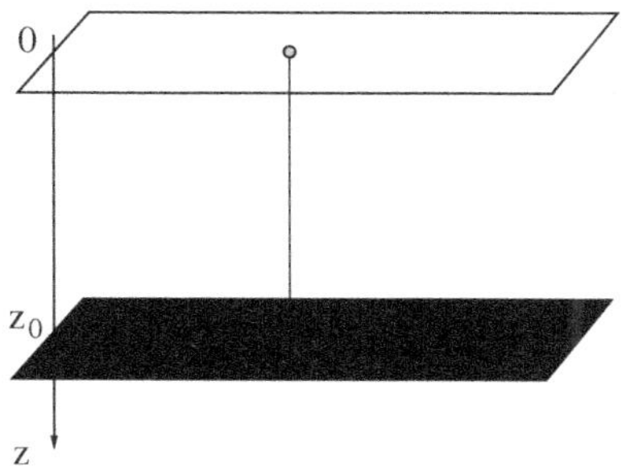

Fig. 7.7 A sketch of an open string hanging into AdS space. The string endpoint located at the AdS boundary is stationary and its other end stretches into the black hole. Since the boundary conditions and the equations of motion are rotationally invariant then the solution should have the same symmetry. Thus, we conclude that the string hangs from its endpoint straight down into AdS space

$$S_{\mathrm{NG}} = -\frac{1}{2\pi\alpha'}\int \frac{L^2}{z^2}\sqrt{f(z)(\xi'(z))^2 + 1 - \frac{v^2}{f(z)}}\,dz dt \tag{7.18}$$

$$= -\frac{1}{2\pi\alpha'}\int \mathcal{L}_{\mathrm{NG}}\,dz dt. \tag{7.19}$$

Since the action (7.15) does not depend explicitly on ξ but only on ξ', then there is a conserved quantity Π_ξ given by

$$\Pi_\xi \equiv \frac{\partial \mathcal{L}_{\mathrm{NG}}}{\partial \xi'} = \frac{L^2 f(z)\xi'(z)}{z^2\sqrt{f(z)(\xi')^2 + 1 - v^2/f(z)}}. \tag{7.20}$$

Inverting (7.20) we obtain the following equation of motion for ξ',

$$(\xi')^2 = \frac{\Pi_\xi^2(v^2 - f(z))}{f(z)^2\left(\Pi_\xi^2 - \frac{L^4}{z^4}f(z)\right)} \tag{7.21}$$

Unless the solution is trivial, $\Pi_\xi = 0$, the only way in which $(\xi')^2$ can be real is that the numerator and denominator on the right hand side of (7.21) flip sign simultaneously. Thus,

$$\Pi_\xi = \frac{L^4}{z_*^4}v^2 \tag{7.22}$$

where z_* is defined through

$$f(z_*) = v^2. \tag{7.23}$$

With this value of Π_ξ the solution to (7.21) is

$$\xi(z) = \frac{v z_0}{4}\left(\ln\frac{1-\frac{z}{z_0}}{1+\frac{z}{z_0}} + 2\arctan\frac{z}{z_0}\right) \tag{7.24}$$

Geometrically what we find is that the string is trailing behind its endpoint. See Fig. 7.8.

At this point we have all the ingredients to complete our computation: we know how to compute the stress tensor of the boundary theory QGP and we know how to represent a moving quark in this background. What we are looking for is the response of the QGP to the moving quark. From the previous discussion this maps into the response of the metric to the moving string: like a point particle, the string warps space–time as it moves through it. In other words, as the string moves through AdS-space it will source the metric. The metric fluctuations near the boundary map into the response of the medium to the moving quark. See Fig. 7.9.

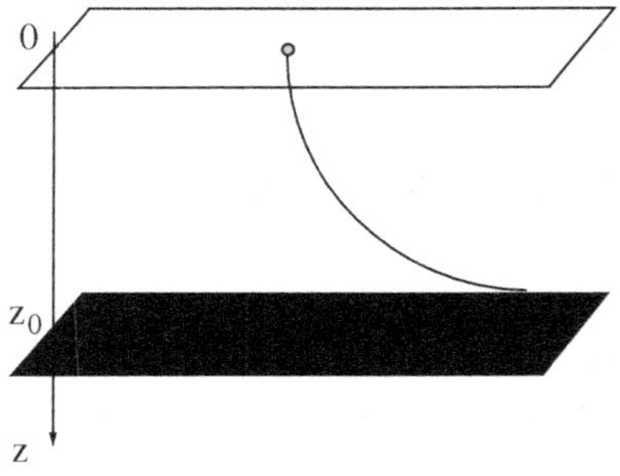

Fig. 7.8 A sketch of an open string hanging into AdS space. The string endpoint located at the AdS boundary and is moving at constant velocity v to the left. The causal solution to the equations of motion is that the rest of the string trails behind its endpoint

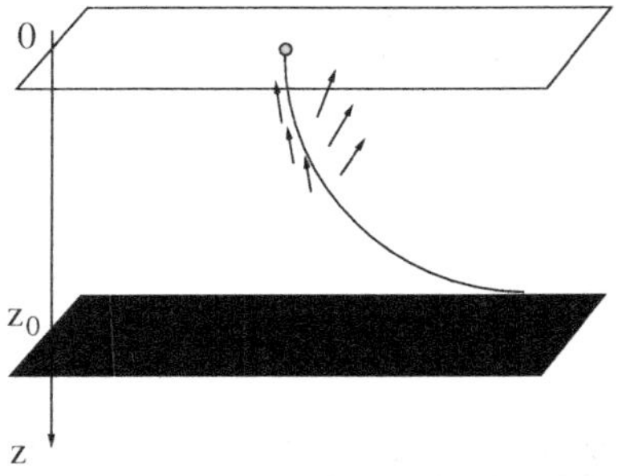

Fig. 7.9 A sketch of an open string sourcing the metric as it moves through AdS space. The near boundary value of the response of the metric to the moving string translates into the response of the medium to the moving quark

To solve the coupled, string + gravity, problem, consider the total action

$$S = \frac{1}{16\pi G_N}\int \sqrt{-G}\left(R+\frac{12}{L^2}\right)d^5x - \frac{1}{2\pi\alpha'}\int \sqrt{-G\partial X\partial X}d\tau d\sigma$$
$$= \frac{1}{16\pi G_N}\int \left[\sqrt{-G}\left(R+\frac{12}{L^2}\right)\right.$$
$$\left. - \frac{16\pi G_N}{2\pi\alpha'}\int \sqrt{-G\partial X\partial X}\delta^{(5)}(x^\mu - X^\mu)d\tau d\sigma\right]d^5x. \tag{7.25}$$

The Einstein equations are

$$R^{\mu\nu} - \frac{1}{2}G^{\mu\nu}R - \frac{6}{L^2}G_{\mu\nu} = \mathcal{T}^{\mu\nu} \tag{7.26}$$

where

$$\mathcal{T}^{\mu\nu} = -\frac{8\pi G_N}{2\pi\alpha'}\int \frac{\sqrt{-|G\partial X\partial X|}}{\sqrt{-G}}\delta^{(5)}(x^\mu - X^\mu)\partial_\alpha X^\mu \partial_\alpha X^\nu d\tau d\sigma\,. \tag{7.27}$$

We will not write out explicitly the equations of motion for the string. The AdS/CFT dictionary tells us that

$$G_N = \frac{\pi L^3}{2N^2} \quad \alpha' = \frac{L^2}{\sqrt{\lambda}}, \tag{7.28}$$

so

$$\frac{G_N}{\alpha'} \sim \frac{\sqrt{\lambda}}{N^2}. \tag{7.29}$$

In what follows we will work in the limit where

$$\frac{\sqrt{\lambda}}{N^2} \ll 1. \tag{7.30}$$

The limit (7.30) is useful since it implies that the energy momentum tensor of the string $\mathcal{T}^{\mu\nu}$ decouples from gravity and we can treat it as a linear perturbation. From the boundary theory point of view we will be looking at the linear response of the medium to the moving quark.

To proceed consider the expansion

$$G_{\mu\nu} = G^{(0)}_{\mu\nu} + \frac{\sqrt{\lambda}}{N^2}G^{(1)}_{\mu\nu} + \mathcal{O}\left(\frac{\lambda}{N^4}\right). \tag{7.31}$$

To leading order, the Einstein equation reduces to

$$R^{(0)\mu\nu} - \frac{1}{2}G^{(0)\mu\nu}R^{(0)} - \frac{6}{L^2}G^{(0)\mu\nu} = 0 \tag{7.32}$$

whose solution is the AdS black hole we have encountered previously,

$$G^{(0)}_{\mu\nu}dx^{\mu}dx^{\nu} = \frac{L^2}{z^2}\left(-fdt^2 + \sum_{i=1}^{3}(dx^i)^2 + \frac{dz^2}{f}\right). \tag{7.33}$$

Since at this order the metric does not back-react on the string, the solution to the string equations of motion is the same as before,

$$X^{(0)\mu} = (t, vt + \xi(z), 0, 0, z) \tag{7.34}$$

where we have expanded the embedding function X^{μ} in the same way we have expanded the metric, c.f., Eq. 7.33. The function $\xi(z)$ is given in (7.24).

At the next order in $\sqrt{\lambda}/N^2$ the equation of motion for the metric takes the form

$$\begin{aligned}&\frac{\sqrt{\lambda}}{N^2}\left(R^{(1)\mu\nu} - \frac{1}{2}G^{(1)\mu\nu}R^{(0)} - \frac{1}{2}G^{(0)\mu\nu}R^{(1)} - \frac{6}{L^2}G^{(1)\mu\nu}\right)\\ &= -\frac{4\pi G_N}{\pi\alpha'}\int\frac{\sqrt{-|G^{(0)}\partial X^{(0)}\partial X^{(0)}|}}{\sqrt{-|G^{(0)}|}}\delta^{(5)}(X^{\mu}-x^{\mu})\partial_{\alpha}X^{(0)\mu}\partial^{\alpha}X^{(0)\nu}d\tau d\sigma. \end{aligned} \tag{7.35}$$

Written in components, the resulting equation is a rather complicated second order linear differential equation. There are many good references that explain how to decouple the various equations [16, 20, 23, 24]. One straightforward way to simplify them is to

- Work in Fourier space,

$$\widehat{G}^{(1)}_{\mu\nu} = \int G^{(1)}_{\mu\nu}e^{-i\left(k_1(x^1-vt)+k_2x^2+k_3x^3\right)}, \tag{7.36}$$

  so that the partial differential equation reduces to an ordinary differential equation.
- Fix a gauge. A useful gauge is $G^{(1)}_{z\mu} = 0$ which one might call the axial gauge.
- Use the rotational symmetry in the x^2, x^3 plane to set $G^{(1)}_{\mu 3} = 0$ ($\mu \neq 3$).

Even after implementing these points (or using the fancier methods described in [24]) the equations of motion are still analytically intractable. (Though solutions can be obtained for very small or very large momentum see [20, 23, 62, 63].) To proceed, one must resort to numerics.

In most of what remains of this section we will present the final result of the numerical analysis. Equation 7.35 was solved numerically and the near boundary asymptotics of the metric (7.5),

$$\langle T_{mn}\rangle = \frac{L^3}{4\pi G_N}g^{(4)}_{mn}, \tag{7.37}$$

where used in order to translate the numerical solution into a boundary theory stress tensor of the form

$$\langle T_{mn} \rangle = \langle T^{(0)}_{mn} \rangle + \frac{\sqrt{\lambda}}{N^2} \langle T^{(1)}_{mn} \rangle + \cdots . \tag{7.38}$$

In Fig. 7.10 we have plotted the numerical expression for the energy density $\langle T^{(1)}_{00} \rangle$ for $v = 3/4 > 1/\sqrt{3}$. In this contour plot dark regions signify a positive energy density relative to the background value $\langle T^{(0)}_{00} \rangle$ which has been subtracted. Light regions correspond to negative energy. The location of the Mach cone is signified by the dashed line. Clearly, the large distance asymptotics match onto the expected hydrodynamic behavior, a property which can be confirmed analytically by carrying out a large momentum expansion. Since we have good control over the asymptotic behavior of the solution, we can zoom in on the region near the quark. At scales which are roughly three orders of magnitude smaller than the ones in Fig. 7.10 we find that a new structure emerges. This can structure is depicted in Fig. 7.11. The dashed line in this figure signifies the location of the Mach cone. The fact that the features of this plot are not associated with the Mach cone provides evidence that the physical mechanism responsible for the energy distribution close to the quark does not have a hydrodynamical origin. Further evidence for non hydrodynamical behavior is given by the extra lobes, appearing at velocities above the speed of sound. In fact, there is an apparent transition from a configuration where there is a region of energy depletion behind the quark to a configuration where there is a region of energy depletion in front of the quark. The physical mechanism responsible for this behavior is unclear at the moment. But, as we will see shortly, it might play an important role in the jet-splitting effect.

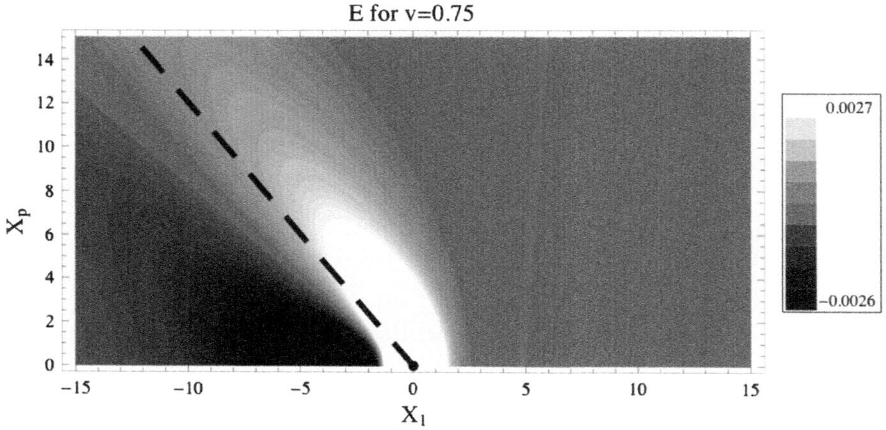

Fig. 7.10 (Reproduced from [24].) A contour plot of the energy density $\langle T^{00} \rangle$ surrounding a massive quark moving in the $\mathcal{N} = 4$ SYM plasma. The quark is located at the origin. *Dark regions* correspond to a higher energy density the surrounding plasma and light regions to a lower energy density. The *dashed line* signifies the location of the Mach cone

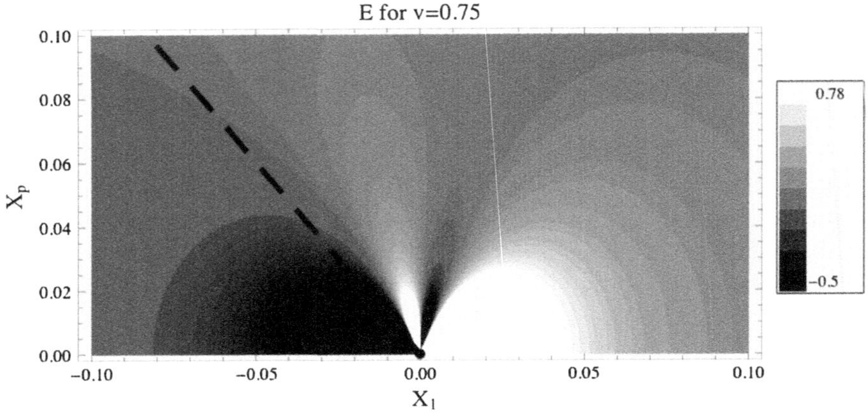

Fig. 7.11 (Reproduced from [24].) Energy density for a massive quark moving through an $\mathcal{N} = 4$ SYM plasma. The quark is located at the origin. The extra lobes which can be observed in the figure and the mismatch between the location of the Mach cone and the regions of over-energy indicate that the physical mechanism responsible for the near field is not hydrodynamical in origin

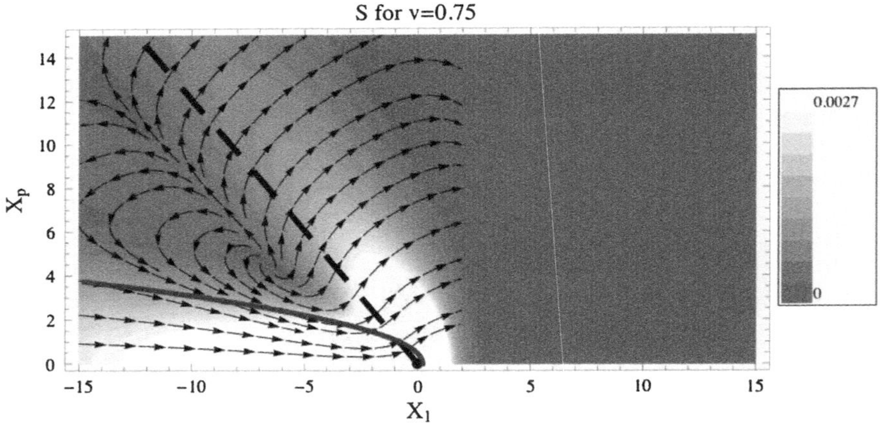

Fig. 7.12 (Reproduced from [24].) The Poynting vector $S^i = \langle T^{0i} \rangle$ for a massive quark moving through an $\mathcal{N} = 4$ SYM plasma. The quark is located at the origin. The magnitude of the Poynting vector is color coded from light to dark. The *arrows* signify its direction. The *dashed* and *thick lines* specify the location of the Mach cone and laminar wake—according to a hydrodynamic analysis

We had set out to find the strength of the wake behind the moving quark. This wake manifests itself as a flux of energy far behind it. It can be observed by computing the Poynting vector $\langle T^{0i} \rangle$. From Fig. 7.12, we see that there is a strong wake behind the quark. A closer study shows that the far-field behavior of the wake is precisely the one expected of a point particle [25, 27, 29, 30]. Thus, a heavy quark in $\mathcal{N} = 4$ SYM generates a strong wake behind it—so strong that once the QGP hadronizes, instead of a minimum at $\Delta\Phi = \pi$ a broad-shoulder like structure should appear. At this point it should be mentioned that this sort of behavior is not unique to the $\mathcal{N} = 4$ theory and that a strong wake will appear for any theory with a holographic dual [29, 30].

In [11, 51], the AdS/CFT analysis described so far was taken one step further. It is possible to estimate the final distribution of hadrons reaching the detector from the energy distribution of the QGP using a mathematical prescription which goes under the name of 'Cooper–Frye hadronization' [17]. Of course, the $\mathcal{N} = 4$ SYM theory does not have a confined phase, but nevertheless, it is interesting to ask what the hadron distribution would look like had the $\mathcal{N} = 4$ SYM plasma hadronized. Following the discussion in this section, one would expect that the appearance of a wake in these theories will result in a high degree of correlation between events reaching the detector at an angle of π (and with appropriate transverse momenta) and the trigger jet. It turns out that this is not the case. Surprisingly, even though there's a strong wake, jet splitting does seem to occur. As discussed in [50], this effect comes about in an unexpected way: it is not the far field behavior that is responsible for the effect but rather the near field which was observed in [23, 62]. The relation between this observation and the actual data at RHIC is unclear. One might treat it as an indication that the correct physics responsible for the jet-splitting effect lies in the near field energy distribution around the quark and not in the far-field hydrodynamic behavior. Certainly, further study is warranted.

7.2 Entropy Production

Information about the entropy produced during the collision process can be obtained my measuring the total particle multiplicity. Recall that before the collision we have two pancake shaped gold ions, and after the collision thousands of particles are produced (see Fig. 7.1). Certainly, a large amount of entropy is created in this process. One method of estimating the amount of entropy produced is to compute the entropy per hadron for a gas of hadrons at the deconfinement temperature, which according to [26, 52] is

$$s/N \sim 7.5, \tag{7.39}$$

and multiply this ratio by the total amount of charged hadrons reaching the detector. Figure 7.13 shows a plot of the total number of charged particles produced in the collision process as a function of the center of mass energy.

In this section we would like to construct a process dual to a collision and then use the AdS/CFT correspondence to estimate the amount of entropy produced. As is well known, the $\mathcal{N} = 4$ SYM theory is deconfined, so it would be difficult to

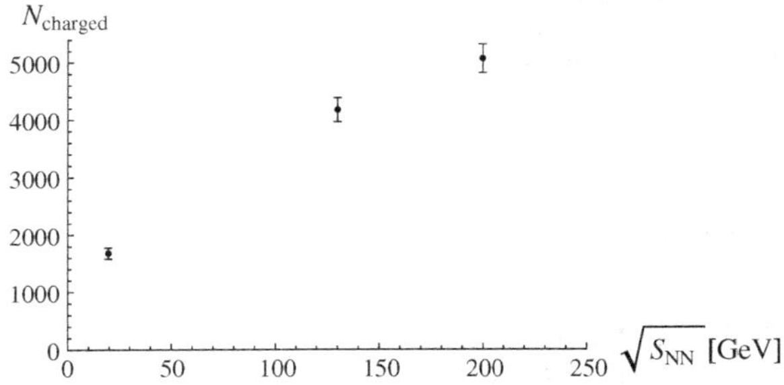

Fig. 7.13 Total number of charged particles reaching the detector as a function of center of mass energy. The total multiplicity is proportional to the amount of entropy produced during the collision process. Data taken from [7]

generate a collision of ions. But, one can consider a collision of two energetic distributions. If these distributions thermalize and generate a thermal state then entropy will be produced.

There are many setups one can use to generate a boundary theory configuration where a thermal state is produced. In this section we use the simplest setup we could think of: the collision of two light-like particles in the bulk of AdS (see Fig. 7.14). This is the simplest case to consider since, as we will see shortly, one has some control of the back-reacted geometry.

Consider the action of gravity coupled to a light-like particle:

$$S = \int \frac{1}{16\pi G_N} \int \sqrt{-|G|}\left(R + \frac{12}{L^2}\right) d^5x + \int \frac{1}{2e} G_{\mu\nu}\partial_\eta X^\mu X^\nu d\eta \tag{7.40}$$

where the term on the right hand side is the action for a point particle: e is the einbein on the particle worldline and η parameterizes it. To start, we wish to describe a single light like particle moving in the bulk of AdS at some constant depth $z = z_*$ (see Fig. 7.15).

The equation of motion for a light-like particle in AdS admits a solution in which the particle moves along a line of constant $z = z_*$. The equations of motion for the metric are

$$R_{\mu\nu} - \frac{1}{2}G_{\mu\nu}R - \frac{6}{L^2}G_{\mu\nu} = 8\pi G_N \mathcal{T}_{\mu\nu}. \tag{7.41}$$

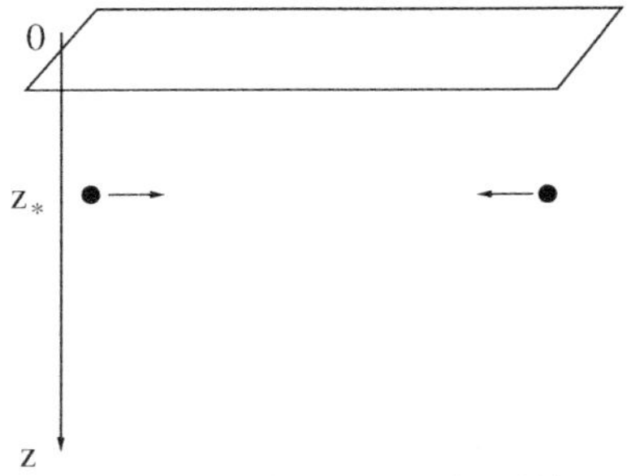

Fig. 7.14 A head-on collision of two light-like particles moving along the light-like geodesic $z = z_*$ in an AdS background

Fig. 7.15 A single lightlike particle moving along the lightlike geodesic $z = z_*$ in an AdS background

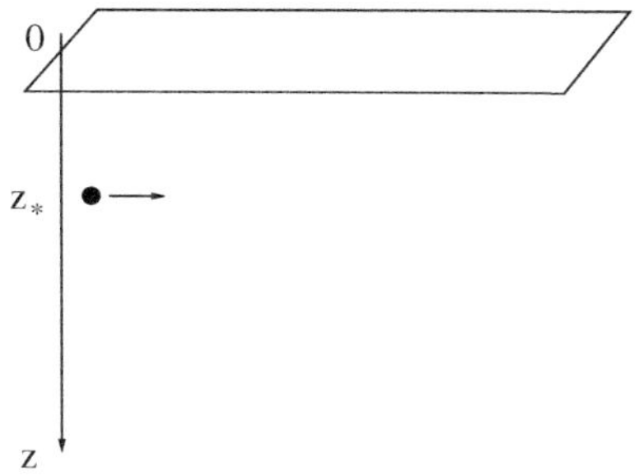

Working in light-like coordinates:

$$u = t - x^3, \quad v = t + x^3 \tag{7.42}$$

we find that

$$\mathcal{T}_{uu} = \frac{Ez^3}{L^3}\delta(u)\delta(z - z_*)\delta(x^1)\delta(x^2) \tag{7.43}$$

and all other components of the energy momentum tensor vanish. Plugging the ansatz

$$ds^2 = \frac{L^2}{z^2}\left(-dudv + \sum_{i=1}^{2}(dx^i)^2 + dz^2 + \frac{z}{L}\Phi(x^1, x^2, z)\delta(u)du^2\right) \tag{7.44}$$

into the equations of motion (7.41), one finds a single non trivial equation:

$$\left(\frac{z^3}{L^3}\partial_z\frac{1}{z}\partial_z + \frac{z^2}{L^2}\left(\partial_1^2 + \partial_2^2\right) - \frac{3}{L^2}\right)\Phi = -\frac{16\pi G_N E z^4}{L^4}\delta(z - z_*)\delta(x^1)\delta(x^2). \tag{7.45}$$

The solution is

$$\Phi(x^1, x^2, z) = \frac{2G_N}{L}\left(\frac{1 + 8q(1+q) - 4\sqrt{q}1 + q(1+2q)}{\sqrt{q(1+q)}}\right) \tag{7.46}$$

where

$$q = \frac{(x^1)^2 + (x^2)^2 + (z - z_*)^2}{4z_* z}. \tag{7.47}$$

Equation 7.46 gives the response of the metric to a moving light like particle. Moreover, it is a full solution to the fully back-reacted gravity plus lightlike particle system. One way to see this is to take a black hole solution and boost it to the speed of light [26]. The first study of such configurations in flat space–time can be found in [6]. Shock-waves in the context of AdS/CFT were discussed in [19, 33, 34, 54, 57].

With the solution for a single point-like particle at hand we can easily construct a solution involving two lightlike particles, as depicted in Fig. 7.14—at least for spacetime points outside the future light cone of the collision. See Fig. 7.16. Outside this region causality tells us that we can add the solutions for the two lightlike particles. In other words the line element

$$ds^2 = \frac{L^2}{z^2}\left(-dudv + \sum_{i=1}^{2}(dx^i)^2 + dz^2 + \frac{z}{L}\Phi\delta(v)dv^2 + \frac{z}{L}\Phi\delta(u)du^2\right) \quad (7.48)$$

with Φ given in (7.46) is a solution to the Einstein equations in the presence of two lightlike particles in the region outside the future light-cone of the collision point.

We would eventually like to see a black hole form, and to compute its area which is dual to the entropy produced in the collision. To that end we need control over the metric in the future light cone of the collision. This is because the definition of an event horizon is non local and requires information on geodesics reaching future null infinity. Heuristically, the event horizon can be defined as the boundary of all the causal curves reaching future null infinity. Unfortunately, it is a difficult problem to solve for the metric in the future light cone of the collision point. So, instead of carrying out a direct computation, we follow a different strategy. There is another notion of a horizon called an apparent horizon or a marginally trapped surface. A marginally trapped surface (in a five dimensional spacetime) can be heuristically defined as a three dimensional surface for which the outward pointing null vector propagates neither inward nor outward and the other null vector propagates inward. A more formal definition of a marginally

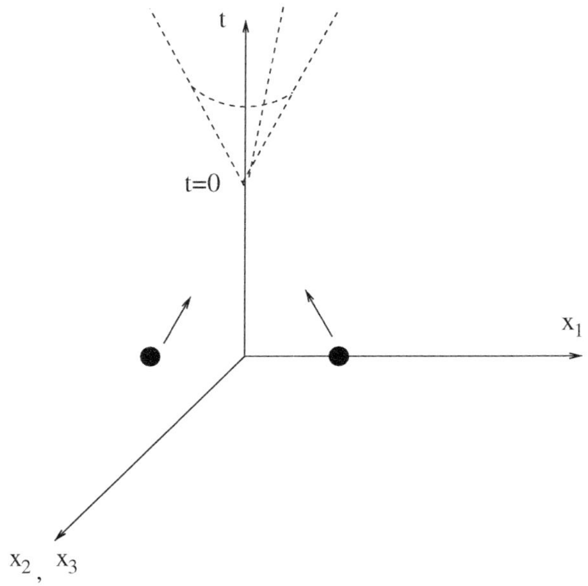

Fig. 7.16 A spacetime diagram of a constant z slice of AdS space. Causality dictates that the two lightlike particles affect each other only on or inside the future light cone of the collision point

trapped surface is the following: let ℓ_μ and n_μ be the null normal vectors to the surface. Then, a marginally trapped surface satisfies:

$$\Theta = h^{\mu\nu} D_\nu \ell_\mu = 0. \tag{7.49}$$

Θ is called the expansion, and $h^{\mu\nu}$ is the induced metric on the surface. The virtue of a trapped surface is that its definition is local, so one can compute it even without knowing the full geometry. Barring some issues regarding cosmic censorship, one can argue that a trapped surface will also always be on or inside an event horizon. The latter property turns out to be very useful; the area increase theorems of general relativity tell us that the area of an event horizon can only increase. Thus, we can use the following strategy to obtain a lower bound on the area (and thus the entropy) of the black hole which forms in a collision of two lightlike particles: We find a marginally trapped surface and compute its area. Since the marginally trapped surface is located inside or on the event horizon of the black hole, the area of the black hole should be bigger or equal to the area of the trapped surface. Following the area increase theorem, this area provides a lower bound on the final area of the black hole and so on the entropy of the configuration.

Thus, our main challenge reduces to finding a marginally trapped surface. In what follows we use an ansatz first suggested by Penrose [53], and elaborated on in [18, 64]. The ansatz is that the trapped surface is composed of two parts. Part I is the surface $u = 0$ and $v = -\psi(q)$, where q is as given in (7.47) and ψ is a function which will be determined shortly. Part II of the surface is given by $v = 0$ and $u = -\psi(q)$. We will find ψ by requiring that the expansion, defined in (7.49), vanishes on the surface. A null normal to part I of the surface is given by

$$\ell_\mu^{(I)} dx^\mu = A du + B(dv + d\psi). \tag{7.50}$$

Requiring that this vector is lightlike and outward pointing implies that[2]

$$A = -\frac{1}{4}(\partial\psi)^2 \quad B = -\frac{z^2}{L^2}. \tag{7.51}$$

The inward pointing null vector is given by

$$n_\mu^{(I)} dx^\mu = -\frac{1}{4}(\partial\psi)^2 du + \frac{z^2}{L^2}(dv + d\psi). \tag{7.52}$$

From symmetry the normals in part II of the surface will take the form

$$n_\mu^{(II)} dx^\mu = -\frac{1}{4}(\partial\psi)^2 dv + \frac{z^2}{L^2}(du + d\psi). \tag{7.53}$$

---

[2] Note that the metric is singular at $u = 0$ and $v < 0$. In order for the metric to be finite we used the coordinate transformation $v \longrightarrow v + \frac{z}{L}\phi(q)\Theta(u)$ with Θ the Heaviside step function which is unity when its argument is positive.

$$\ell_{\mu}^{(II)}dx^{\mu} = -\frac{1}{4}(\partial\psi)^2 dv - \frac{z^2}{L^2}(du + d\psi). \tag{7.54}$$

Once we have the null normals, to compute the expansion we need the induced metric $h_{\mu\nu}$ which, by definition, is orthogonal to both normals. Using an ansatz

$$h_{\mu}^{\nu} = \delta_{\mu}^{\nu} + A\ell_{\mu}\ell^{\nu} + B(\ell_{\mu}n^{\nu} + n_{\mu}\ell^{\nu}) + Cn_{\mu}n^{\nu} \tag{7.55}$$

for the induced metric, the unknown functions A, B and C can be determined through the condition that

$$h_{\mu}^{\nu}\ell_{\nu} = 0 \quad \text{and} \quad h_{\mu}^{\nu}\ell_{\nu} = 0. \tag{7.56}$$

With ℓ and h at hand we can compute the expansion and find a function ψ for which the expansion vanishes. Defining $\Psi = \frac{L}{z}\psi$ the vanishing of the expansion, $\Theta = 0$, is equivalent to

$$\left(q(1+q)\partial_q^2 + \frac{3}{2}(1+2q)\partial_q - 3\right)(\Psi - \Phi) = 0 \tag{7.57}$$

where Φ is given by (7.46) and q is as in (7.47).[3] The boundary conditions implied by continuity of the normals are

$$\Psi\big|_C = 0 \quad \text{and} \quad (\partial\Psi)^2\big|_C = 4. \tag{7.58}$$

where C is the mutual boundary of regions I and II. Note (7.57) and (7.58) constitute a rather unusual boundary value problem since we are imposing both Dirichlet and Neumann boundary conditions. The equivalent electrostatic problem would be to solve the Laplace equation on a surface C where the potential vanishes and the electric field has a specified non zero value. The problem isn't over-determined since to solve such a problem we need to find both the surface C and the electric potential. Going back to the problem at hand, we note that the homogenous equation (7.57) admits two solutions only one of which is finite at the origin:

$$\Psi - \Phi = (1+2q)K \tag{7.59}$$

where K is an undetermined integration constant. From symmetry, the boundary C should be a surface of constant q, $q = q_C$. Plugging the general solution (7.59) into the boundary conditions (7.58) we find that the latter turn into algebraic constraints,

[3] It is no coincidence that the resulting equation depends only on q. The variable q can be thought of as a radial variable of the AdS hyperboloid which one finds when embedding AdS_5 in $\mathbf{R}^{4,2}$.

$$\frac{EG_N}{L} = 2q_c(1+q_c)(1+2q_c) \quad K = -\frac{\Phi(q_C)}{1+2q_C}. \tag{7.60}$$

Solving (7.60) we can find both Ψ and the surface C.

Recall that our goal was to find a marginally trapped surface, compute its area, and use that as a lower bound on the entropy of the black hole produced. The area of the trapped surface is given by

$$A = \int \sqrt{h_S} d^3x = \cdots = 4\pi L^3 \int_0^{x_c} \frac{x^2 dx}{\sqrt{1+x^2}} \tag{7.61}$$

where $x_c = 2\sqrt{q_C(1+q_C)}$. This integral can be computed explicitly, but we will only need its asymptotic form:

$$\frac{2A}{4G_N} = \pi\left(\frac{L^3}{G_N}\right)(2Ez_*)^{2/3}\left(1 + \mathcal{O}\left((Ez_*)^{-1}\right)\right). \tag{7.62}$$

Using the Bekenstein–Hawking law for the black hole entropy we find that a lower bound on the entropy of the resulting black hole is given by (7.62),

$$S_b = \frac{A}{2G_N} \tag{7.63}$$

We now have an exact solution to a well defined problem in classical gravity. The next step in our analysis is to map this into a boundary quantities. First, consider the early stages of the collision. Recall from (7.5) that the boundary theory stress tensor is related to the metric through

$$\langle T_{mn} \rangle = \frac{L^3}{4\pi G_N} g^{(4)}_{mn}, \tag{7.64}$$

Expanding the metric in (7.44) in a near boundary Taylor series and extracting the appropriate coefficient in the series expansion we find that

$$\langle T_{uu} \rangle = \frac{2Ez_*^4}{\pi\left(x_\perp^2 + z_*^2\right)^3}\delta(u) \quad \langle T_{vv} \rangle = \frac{2Ez_*^4}{\pi(x_\perp^2 + z_*^2)^3}\delta(v). \tag{7.65}$$

Thus, from the boundary point of view we are looking at the collision of two energetic objects moving at the speed of light and with an energy distribution which is spread in the transverse directions. The total energy of such a configuration is given by E and the energy averaged RMS size of the distribution is given by $z_*^2 = \int \langle T_{uu} \rangle x_\perp^2 d^3x$. It is natural to identify the energy of the lightlike particle E with the beam energy,

$$E = 19.7\,\text{TeV} \tag{7.66}$$

and to identify z_* with its QCD counterpart which can be estimated using a standard exponential-like distribution of the nucleus, see, for example [1, 40]. We find

$$z_* = 4.3\,\mathrm{fm} \tag{7.67}$$

To fully convert the entropy bound (7.63) to a boundary quantity we need to identify G_N with N^2. For the $\mathcal{N} = 4$ SYM theory one has

$$G_N = \frac{\pi L^3}{2N^2}. \tag{7.68}$$

A more general version of (7.68) is

$$G_N = \frac{V_5}{2\pi N^2} \tag{7.69}$$

FIX! where V_5 is the volume of a five dimensional compact space. Assuming V_5 is tunable, we have some freedom in choosing G_5/L^3.[4] Recall from (7.5) that

$$\langle T_{00} \rangle = \frac{3L^3}{16\pi G_N z_0^4}. \tag{7.70}$$

This can be compared with the lattice prediction,

$$\langle T_{00} \rangle_{\text{lattice}} \sim 11T^4 \tag{7.71}$$

for QCD. Comparing (7.70) with (7.71) we obtain

$$\frac{L^3}{G_5} \sim 2 \tag{7.72}$$

(which is unfortunately small.)

Using (7.66), (7.67) and (7.72) in (7.62), we find

$$S_b = 35{,}000 \left(\frac{\sqrt{s_{NN}}}{200\,\mathrm{GeV}} \right)^{2/3}. \tag{7.73}$$

Figure 7.17 shows a comparison of the data obtained at RHIC with the lower bound on the entropy [converted to total multiplicity using (7.39)]. While the bound is in surprisingly good agreement with the data, there is some cause for worry. More traditional models relating the total entropy produced to the beam energy predict a scaling behavior of the form $S_b \sim E^{1/2}$ instead of the $E^{2/3}$ scaling given in (7.73). While the current data agrees with both the 2/3 scaling and the 1/2 scaling, at energies as high as the LHC energies, the prediction of (7.73) is

---

[4] Until now we have ignored the compact five dimensional manifold attached to every point in AdS_5 space.

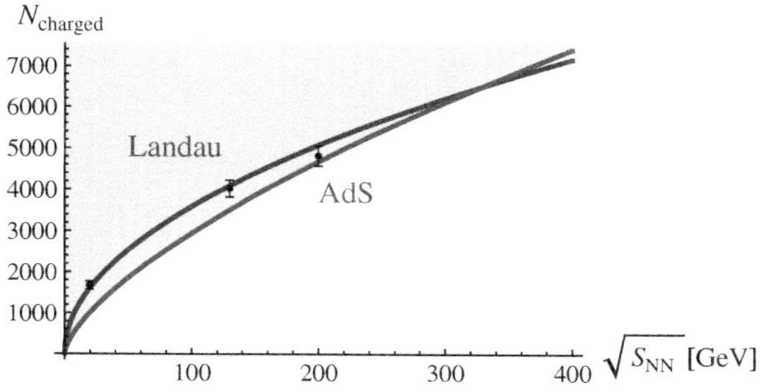

Fig. 7.17 (Reproduced from [26].) Various models which predict total multiplicity as a function of center of mass energy. The bound on the *shaded region* marks the allowed region according to the AdS/CFT model described in the main text. The *thick blue* line follows from Landau's model [41] which predicts an $S_b \sim E^{1/2}$ behavior. The data points were taken from [7]

150% higher than the traditional predictions. It is tempting to speculate that this is not a problem but a prediction of AdS/CFT. However, a closer look at the approximation being made shows us that, in the particular problem we are dealing with, the relation between the gauge-gravity duality and QCD is, perhaps, being pushed too hard. There are two important features of QCD that the AdS/CFT duality does not capture. One is asymptotic freedom and the other is confinement. Both these features of QCD may play an important role in the early stages of the collision.

So how can we estimate what effect these features have on the entropy bound? As a general rule of thumb the infrared physics is captured by the geometry far from the boundary (large values of z) while the ultraviolet physics are captured by the near boundary geometry. Thus, as a very (very) coarse approximation of UV and IR effects we can simply chop off the parts of the trapped surface below a certain IR cutoff and above a UV cutoff. As depicted in Fig. 7.18, the result is that

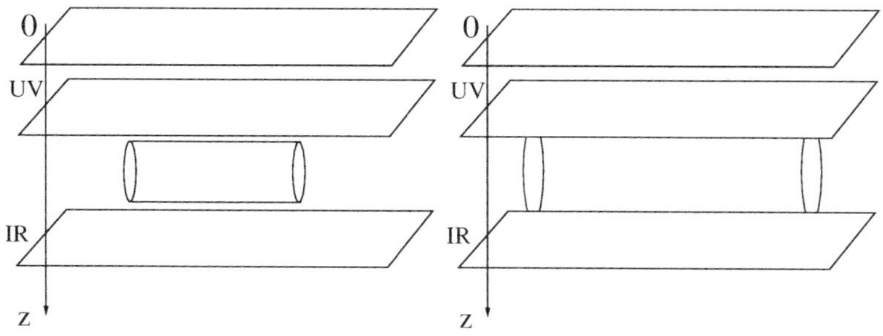

Fig. 7.18 The effects of the UV and IR cutoffs on the area of the trapped surface. The surface is depicted as a cylinder. Once the energy of the collision is large enough, the trapped surface enters the UV and IR regions where its area gets modified

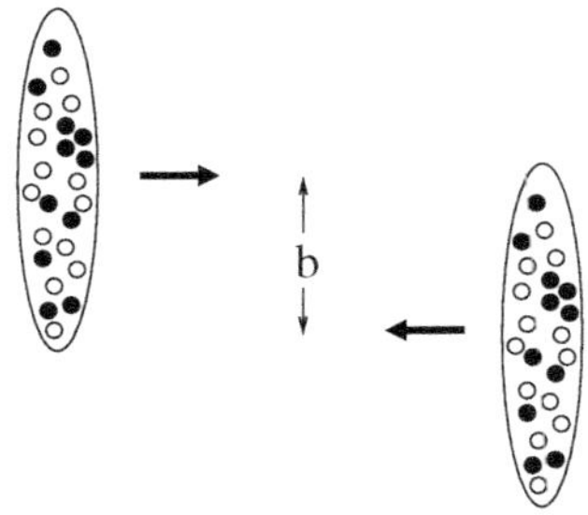

Fig. 7.19 A caricature of a collision of two gold ions with a non trivial impact parameter b

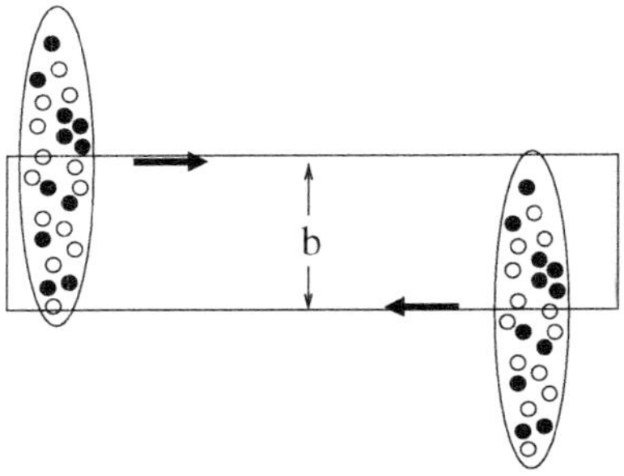

Fig. 7.20 A cartoon of an off center collision of two heavy ions. The nucleons outside the interaction region (*boxed*) are "spectators" and do not participate in the collision process

the area of the trapped surface will decrease as it enters the UV and IR region.[5] With $\Lambda_{\text{UV}} = 2\,\text{GeV}$ and $\Lambda_{\text{IR}} = 0.2\,\text{GeV}$, we find that at RHIC energies $S_b \sim E^{2/3}$ as before, but at LHC energies the power law dependence of S_b on E decreases appreciably until it reaches $S_b \sim E^{1/3}$ [28].

Until now we have focused on central collisions. It is also possible to consider off center ones, i.e., ones where the impact parameter is non zero (see Fig. 7.19). Ideally one would like to measure the total particle multiplicity as a function of the impact parameter. In practice the impact parameter is difficult to measure. What is measured instead is the number of nucleons participating in the collision N_{part}. Due to confinement, only those nucleons that are in the interaction region will participate in the collision. This is depicted in Fig. 7.20. The other nucleons are "spectators" and are detected by the paddle detectors located along the beam axis. In practice, the data is usually presented in a plot such as the one in 7.21 where the dependance of the ratio of the total multiplicity to the number of participating nucleons is presented as a function of the number of participating nucleons.

From the point of view of the gauge-gravity duality, generating an off center collision of light-like particles is straightforward: One considers colliding light-like particles with a non trivial impact parameter. From the bulk point of view the impact parameter can be both in the radial AdS direction z, and in the

[5] Actually, a closer analysis shows that the UV region is more important than the IR region. This is because the AdS warp factor L/z diverges near the boundary.

transverse x^2, or x^3 directions. The former would correspond to a head-on collision of different sized energy distributions and the latter to an off center collision of identical nuclei. It is interesting that in a conformal theory these two setups are related by a conformal transformation. In what follows we will focus on an impact parameter in the x^2, x^3 plane.

When dealing with an off center collision, one complication is that the initial configuration, and the shape of the trapped surface have less symmetry. One way around this is to solve the problem numerically [42]. Another is to work perturbatively in the impact parameter [28]. In the latter case, one finds that

$$S_b = 35{,}000 \left(\frac{\sqrt{s_{NN}}}{200\,\text{GeV}} \right)^{2/3} \frac{\sinh^{-1} \beta}{\beta \sqrt{1 + \beta^2}} \tag{7.74}$$

where β is related to the impact parameter b in the boundary theory through

$$\beta = 0.12 b/\text{fm}. \tag{7.75}$$

To compare equation (7.74) with the data, we need to somehow convert the value of the impact parameter to the number of participating nucleons. This can be done using a standard optical-Glauber method. We once again refer the reader to the literature [28, 38, 39, 45] for details.

In Fig. 7.21 we have plotted the lower bound on the entropy given in (7.74) together with the data. As can be clearly seen, once the impact parameter becomes large the dependance of N on N_{part} deviates from a linear one. Even if we include a UV and IR cutoff the situation does not improve by much. One possible reason for the deviation from linearity is that in the strongly interacting conformal theory there are no spectator particles: as opposed to a collision of two ions, after the collision, spatially separated distributions will still interact. To somehow take the "spectators effect" into account we can identify the photon energy E not with

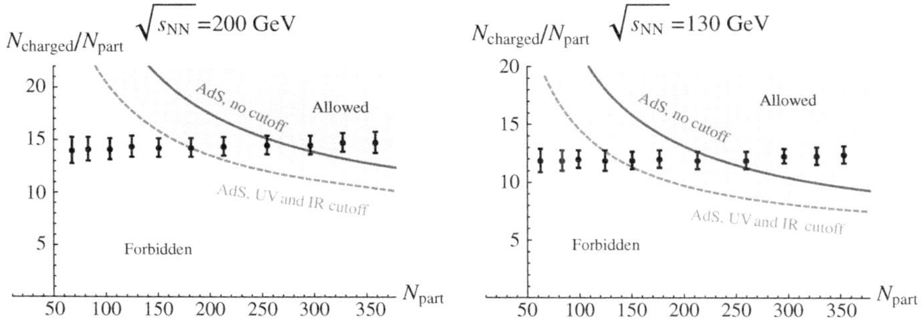

Fig. 7.21 (Reproduced from [28].) Ratio of the total multiplicity N_{charged} to the number of participating nucleons N_{part} as a function of N_{part}. The inner red region marks the allowed region according to the AdS/CFT model without imposing a UV and IR cutoff. The outer green region marks the region which is allowed once a cutoff is introduced. The data was taken from [8]

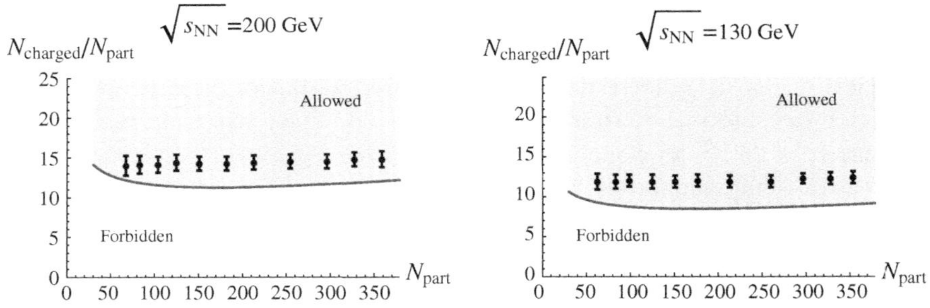

Fig. 7.22 (Reproduced from [28].) Ratio of the total multiplicity N_{charged} to the number of participating nucleons N_{part} as a function of N_{part}. The red region marks the allowed region according to the AdS/CFT model once one makes the replacement (7.76) which takes into account the non-participating nucleons, or spectators. See the main text for details. The data was taken from [8]

the beam energy but with the energy of the nucleons participating in the collision. Thus in (7.74) we should make the substitution

$$\sqrt{s_{NN}} \to \sqrt{s_{NN}} N_{\mathrm{part}}/394. \tag{7.76}$$

With this substitution the fit to the data seems much better. See Fig. 7.22. However, one can not escape the feeling that the manipulation that lead to this sort of fit is somewhat contrived and that one can do better. It would be interesting to carry out this analysis in a background which does capture effects such as confinement and asymptotic freedom in a controlled way.

References

1. Adams, J. et al.: Production of e+ e− pairs accompanied by nuclear dissociation in ultra-peripheral heavy ion collision. Phys. Rev. **C70**, 031,902 (2004). doi:10.1103/PhysRevC.70.031902
2. Adams, J. et al.: Distributions of charged hadrons associated with high transverse momentum particles in p p and Au + Au collisions at s(NN)**(1/2) = 200-GeV. Phys. Rev. Lett. **95**, 152,301 (2005). doi:10.1103/PhysRevLett.95.152301
3. Adare, A. et al.: System size and energy dependence of jet-induced hadron pair correlation shapes in Cu + Cu and Au + Au collisions at s(NN)**(1/2) = 200-GeV and 62.4-GeV. Phys. Rev. Lett. **98**, 232,302 (2007). doi:10.1103/PhysRevLett.98.232302
4. Adler, S.S. et al.: Modifications to di-jet hadron pair correlations in Au + Au collisions at s(NN)**(1/2) = 200-GeV. Phys. Rev. Lett. **97**, 052,301 (2006). doi:10.1103/PhysRevLett.97.052301
5. Aharony, O., Gubser, S.S., Maldacena, J.M., Ooguri, H., Oz, Y.: Large N field theories, string theory and gravity. Phys. Rept. **323**, 183–386 (2000). doi:10.1016/S0370-2693(98)00377-3
6. Aichelburg, P.C., Sexl, R.U.: On the Gravitational field of a massless particle. Gen. Relativ. Gravit. **2**, 303–312 (1971). doi:10.1007/BF00758149

7. Back, B.B. et al.: The significance of the fragmentation region in ultrarelativistic heavy ion collisions. Phys. Rev. Lett. **91**, 052,303 (2003). doi:10.1103/PhysRevLett.91.052303
8. Back, B.B. et al.: Charged-particle pseudorapidity distributions in Au + Au collisions at $s(NN)^{1/2}$ = 62.4-GeV. Phys. Rev. **C74**, 021,901 (2006). doi:10.1103/PhysRevC.74.021901
9. Balasubramanian, V., Kraus, P.: A stress tensor for anti-de Sitter gravity. Commun. Math. Phys. **208**, 413–428 (1999). doi:10.1007/s002200050764
10. Betz, B.: Jet propagation and Mach-cone formation in (3 + 1)-dimensional ideal hydrodynamics. arXiv:0910.4114 [nucl-th]
11. Betz, B., Gyulassy, M., Noronha, J., Torrieri, G.: Anomalous conical di-jet correlations in pQCD vs AdS/CFT. Phys. Lett. **B675**, 340 (2009)
12. Bianchi, M., Freedman, D.Z., Skenderis, K.: How to go with an RG flow. JHEP **08**, 041 (2001)
13. Bianchi, M., Freedman, D.Z., Skenderis, K.: Holographic renormalization. Nucl. Phys. **B631**, 159–194 (2002)
14. Casalderrey-Solana, J., Shuryak, E.V., Teaney, D.: Conical flow induced by quenched QCD jets. J. Phys. Conf. Ser. **27**, 22–31 (2005). doi:10.1016/j.nuclphysa.2006.06.091
15. Casalderrey-Solana, J., Shuryak, E.V., Teaney, D.: Hydrodynamic flow from fast particles. arXiv:hep-ph/0602183
16. Chesler, P.M., Yaffe, L.G.: The wake of a quark moving through a strongly-coupled $\mathcal{N} = 4$ supersymmetric Yang–Mills plasma. Phys. Rev. Lett. **99**, 152,001 (2007). doi: 10.1103/PhysRevLett.99.152001
17. Cooper, F., Frye, G.: Comment on the single particle distribution in the hydrodynamic and statistical thermodynamic models of multiparticle production. Phys. Rev. **D10**, 186 (1974). doi:10.1103/PhysRevD.10.186
18. Eardley, D.M., Giddings, S.B.: Classical black hole production in high-energy collisions. Phys. Rev. **D66**, 044,011 (2002). doi:10.1103/PhysRevD.66.044011
19. Emparan, R.: Exact gravitational shockwaves and Planckian scattering on branes. Phys. Rev. **D64**, 024,025 (2001). doi:10.1103/PhysRevD.64.024025
20. Friess, J.J., Gubser, S.S., Michalogiorgakis, G., Pufu, S.S.: The stress tensor of a quark moving through N = 4 thermal plasma. Phys. Rev. **D75**, 106,003 (2007). doi:10.1103/PhysRevD.75.106003
21. Gubser, S.S.: Drag force in AdS/CFT. Phys. Rev. **D74**, 126,005 (2006). doi: 10.1103/PhysRevD.74.126005
22. Gubser, S.S., Klebanov, I.R., Polyakov, A.M.: Gauge theory correlators from non-critical string theory. Phys. Lett. **B428**, 105–114. doi:10.1016/S0370-2693(98)00377-3 (1998)
23. Gubser, S.S., Pufu, S.S.: Master field treatment of metric perturbations sourced by the trailing string. Nucl. Phys. **B790**, 42–71 (2008). doi:10.1016/j.nuclphysb.2007.08.015
24. Gubser, S.S., Pufu, S.S., Rocha, F.D., Yarom, A.: Energy loss in a strongly coupled thermal medium and the gauge-string duality. arXiv:0902.4041 [hep-th]
25. Gubser, S.S., Pufu, S.S., Yarom, A.: Energy disturbances due to a moving quark from gauge-string duality. JHEP **09**, 108 (2007). doi:10.1088/1126-6708/2007/09/108
26. Gubser, S.S., Pufu, S.S., Yarom, A.: Entropy production in collisions of gravitational shock waves and of heavy ions. Phys. Rev. **D78**, 066,014 (2008). doi:10.1103/PhysRevD.78.066014
27. Gubser, S.S., Pufu, S.S., Yarom, A.: Sonic booms and diffusion wakes generated by a heavy quark in thermal AdS/CFT. Phys. Rev. Lett. **100**, 012,301 (2008). doi:10.1103/PhysRevLett.100.012301
28. Gubser, S.S., Pufu, S.S., Yarom, A.: Off-center collisions in AdS5 with applications to multiplicity estimates in heavy-ion collisions. JHEP **11**, 050 (2009). doi:10.1088/1126-6708/2009/11/050
29. Gubser, S.S., Yarom, A.: Universality of the diffusion wake in the gauge-string duality. Phys. Rev. **D77**, 066,007 (2008). doi:10.1103/PhysRevD.77.066007
30. Gubser, S.S., Yarom, A.: Linearized hydrodynamics from probe-sources in the gauge-string duality. Nucl. Phys. **B813**, 188–219 (2009). doi:10.1016/j.nuclphysb.2008.12.032

31. Herzog, C.P., Karch, A., Kovtun, P., Kozcaz, C., Yaffe, L.G.: Energy loss of a heavy quark moving through N = 4 supersymmetric Yang–Mills plasma. JHEP **07**, 013 (2006)
32. Horner, M.J.: Low- and intermediate-p(T) di-hadron distributions in Au + Au collisions at s(NN)**(1/2) = 200-GeV from STAR. J. Phys. **G34**, S995–S998 (2007). doi:10.1088/0954-3899/34/8/S142
33. Horowitz, G.T., Itzhaki, N.: Black holes, shock waves, and causality in the AdS/CFT correspondence. JHEP **02**, 010 (1999)
34. Hotta, M., Tanaka, M.: Shock wave geometry with nonvanishing cosmological constant. Class. Quant. Grav. **10**, 307–314 (1993)
35. Hwa, R.C.: Hadron correlations in jets and ridges through parton recombination. arXiv:0904.2159 [nucl-th]
36. Jia, J.: [for the PHENIX Collaboration], Mapping out the jet correlation landscape: jet quenching and medium response. arXiv:0705.3060 [nucl-ex]
37. Kaneta, M., Xu, N.: Centrality dependence of chemical freeze-out in Au + Au collisions at RHIC. arXiv:nucl-th/0405068
38. Kharzeev, D., Lourenco, C., Nardi, M., Satz, H.: A quantitative analysis of charmonium suppression in nuclear collisions. Z. Phys. **C74**, 307–318 (1997). doi:10.1007/s002880050392
39. Kharzeev, D., Nardi, M.: Hadron production in nuclear collisions at RHIC and high density QCD. Phys. Lett. **B507**, 121–128 (2001). doi:10.1016/S0370-2693(01)00457-9
40. Klein, S., Nystrand, J.: Exclusive vector meson production in relativistic heavy ion collisions. Phys. Rev. **C60**, 014,903 (1999). doi:10.1103/PhysRevC.60.014903
41. Landau, L.D.: On the multiparticle production in high-energy collisions. Izv. Akad. Nauk SSSR Ser. Fiz. **17**, 51–64 (1953)
42. Lin, S., Shuryak, E.: Grazing collisions of gravitational shock waves and entropy production in heavy ion collision. Phys. Rev. **D79**, 124,015 (2009). doi:10.1103/PhysRevD.79.124015
43. Maldacena, J.M.: The large N limit of superconformal field theories and supergravity. Adv. Theor. Math. Phys. **2**, 231–252 (1998)
44. Maldacena, J.M.: Wilson loops in large N field theories. Phys. Rev. Lett. **80**, 4859–4862 (1998). doi:10.1103/PhysRevLett.80.4859
45. Miller, M.L., Reygers, K., Sanders, S.J., Steinberg, P.: Glauber modeling in high energy nuclear collisions. Ann. Rev. Nucl. Part. Sci. **57**, 205–243. doi:10.1146/annurev.nucl.57.090506.123020 (2007)
46. Neufeld, R.B.: Mach cones in the quark-gluon plasma: viscosity, speed of sound, and effects of finite source structure. Phys. Rev. **C79**, 054,909 (2009). doi:10.1103/PhysRevC.79.054909
47. Neufeld, R.B., Muller, B.: The sound produced by a fast parton in the quark-gluon plasma is a 'crescendo'. Phys. Rev. Lett. **103**, 042,301 (2009). doi:10.1103/PhysRevLett.103.042301
48. Neufeld, R.B., Muller, B., Ruppert, J.: Sonic Mach cones induced by fast partons in a perturbative quark-gluon plasma. Phys. Rev. **C78**, 041,901 (2008). doi:10.1103/PhysRevC.78.041901
49. Neufeld, R.B., Renk, T.: The Mach cone signal and energy deposition scenarios in linearized hydrodynamics. arXiv:1001.5068 [nucl-th]
50. Noronha, J., Gyulassy, M., Torrieri, G.: Near zone hydrodynamics of AdS/CFT jet wakes. J. Phys. **G35**, 104,061 (2008). doi:10.1088/0954-3899/35/10/104061
51. Noronha, J., Gyulassy, M., Torrieri, G.: Di-jet conical correlations associated with heavey quark jets in anti-de sitter space/conformal field theory correspondence. Phys. Rev. Lett **102**, 102,301 (2009). doi:10.1103/PhysRevLett.102.102301
52. Pal, S., Pratt, S.: Entropy production at RHIC. Phys. Lett. **B578**, 310–317 (2004). doi:10.1016/j.physletb.2003.10.054
53. Penrose, R.: (1974). Unpublished
54. Podolsky, J., Griffiths, J.B.: Impulsive waves in de Sitter and anti-de Sitter space-times generated by null particles with an arbitrary multipole structure. Class. Quant. Grav. **15**, 453–463 (1998). doi:10.1088/0264-9381/15/2/018

55. Renk, T., Ruppert, J.: Mach cones in an evolving medium. Phys. Rev. **C73**, 011,901 (2006). doi:10.1103/PhysRevC.73.011901
56. Renk, T., Ruppert, J.: The rapidity structure of Mach cones and other large angle correlations in heavy-ion collisions. Phys. Lett. **B646**, 19–23 (2007). doi:10.1016/j.physletb.2007.01.008
57. Sfetsos, K.: On gravitational shock waves in curved space-times. Nucl. Phys. **B436**, 721–746 (1995). doi:10.1016/0550-3213(94)00573-W
58. Ulery, J.G.: Two- and three-particle jet correlations from STAR. Nucl. Phys. **A774**, 581–584 (2006). doi:10.1016/j.nuclphysa.2006.06.092
59. Ulery, J.G.: Are there Mach cones in heavy ion collisions? Three-particle correlations from STAR. Int. J. Mod. Phys. **E16**, 2005–2010 (2007). doi:10.1142/S0218301307007374
60. Witten, E.: Anti-de Sitter space and holography. Adv. Theor. Math. Phys. **2**, 253–291 (1998)
61. Witten, E.: Anti-de Sitter space, thermal phase transition, and confinement in gauge theories. Adv. Theor. Math. Phys. **2**, 505–532 (1998)
62. Yarom, A.: On the energy deposited by a quark moving in an $N = 4$ SYM plasma. Phys. Rev. **D75**, 105,023 (2007). doi:10.1103/PhysRevD.75.105023
63. Yarom, A.: The high momentum behavior of a quark wake. Phys. Rev. **D75**, 125,010 (2007). doi:10.1103/PhysRevD.75.125010
64. Yoshino, H., Nambu, Y.: Black hole formation in the grazing collision of high- energy particles. Phys. Rev. **D67**, 024,009 (2003). doi:10.1103/PhysRevD.67.024009

Chapter 8
AdS/CFT on the Brane

Jiro Soda

Abstract It is widely recognized that the AdS/CFT correspondence is a useful tool to study strongly coupled field theories. On the other hand, Randall-Sundrum (RS) braneworld models have been actively discussed as a novel cosmological framework. Interestingly, the geometrical set up of braneworlds is quite similar to that in the AdS/CFT correspondence. Hence, it is legitimate to seek a precise relation between these two different frameworks. In this lecture, I will explain how the AdS/CFT correspondence is related to the RS braneworld models. There are two different versions of RS braneworlds, namely, the single-brane model and the two-brane model. In the case of the single-brane model, we reveal the relation between the geometrical and the AdS/CFT correspondence approach using the gradient expansion method. It turns out that the high energy and the Weyl term corrections found in the geometrical approach correspond to the CFT matter correction found in the AdS/CFT correspondence approach. In the case of two-brane system, we also show that the AdS/CFT correspondence play an important role in the sense that the low energy effective field theory can be described by the conformally coupled scalar-tensor theory where the radion plays the role of the scalar field. We also discuss dilatonic braneworld models from the point of view of the AdS/CFT correspondence.

8.1 Introduction

It is believed that string theory is a candidate of the unified theory of everything. Remarkably, string theory can be consistently formulated only in 10 dimensions [1].

J. Soda (✉)
Department of Physics, Kyoto University, Kyoto 606-8501, Japan
e-mail: jiro@tap.scphys.kyoto-u.ac.jp

E. Papantonopoulos (ed.), *From Gravity to Thermal Gauge Theories: The Ads/CFT Correspondence,* Lecture Notes in Physics, 828,
DOI: 10.1007/978-3-642-04864-7_8,

This fact requires a mechanism to fill the gap between our real world and the higher dimensions. Conventionally, the extra dimensions are considered to be compactified to a small compact space of the order of the Planck scale. However, recent developments of superstring theory invented a new idea, the so-called braneworld where matter resides on the hypersurface in higher dimensional spacetime [2–6] (see also earlier independent works [7, 8]). This hypersurface is called (mem)brane. This idea originates from D-brane solutions in string theory. Interestingly, the D-brane solution also gives rise to the AdS/CFT correspondence which claims that classical gravity in an anti-de Sitter(AdS) spacetime is equivalent to a strongly coupled conformally invariant field theory (CFT). Since the origin is the same, braneworlds and the AdS/CFT correspondence may be related to each other. In particular, Randall and Sundrum (RS) braneworld models [9, 10] have a similar geometrical setup to that in the AdS/CFT correspondence. Hence, in this lecture, I will try to reveal relations between the AdS/CFT correspondence and RS braneworld models [11].

The method we will use is the gradient expansion method. Physically, it is a low energy expansion method. Historically, the method has been used in the cosmological context [12–15]. In particular, it is known to be useful for analyzing the evolution of cosmological perturbations during inflation. Since the AdS spacetime can be regarded as the inflating universe in the spatial direction, the gradient expansion method is also expected to be useful in the AdS spacetime. First, by utilizing the gradient expansion method, we approximately solve Einstein equations in the bulk. Then, the junction conditions at the brane give the effective equations of motion on the brane. Thus, we can understand the low energy physics in the braneworld. The similar but slightly different method is also used in the AdS/CFT correspondence. We will identify a concrete relation between the geometrical and the AdS/CFT correspondence approach by detailed comparison. The difference shows up when we consider two brane systems. Indeed, we do not have the conventional AdS/CFT correspondence for the two-brane system. Instead, we have a conformally coupled radion on the brane which reflects the conformal symmetry of the theory. This observation is useful for understanding why brane inflation suffers from the eta problem. It is apparent that the gradient expansion method can be applicable to various braneworld models. Moreover, as we will see, the gradient expansion method provides a unified view of branworlds and a useful tool to make cosmological predictions.

The organization of this lecture is as follows: In Sect. 8.2, we introduce RS models and derive the effective Friedman equation on the brane. Here, two important corrections, i.e., the dark radiation and the high energy corrections, are identified. This sets our starting point. In Sect. 8.3, we review two different views from the brane, namely, the AdS/CFT correspondence and the geometrical holography. These two frameworks give us a complementary picture of the braneworlds. In Sect. 8.4, we present key questions to make our concerns manifest. In Sect. 8.5, we review the gradient expansion method. In Sects. 8.6 and 8.7, we apply the gradient expansion method to the single-brane model and to the two-brane model, respectively. We obtain the effective theory for both cases. In Sect. 8.8, we give answers to the key questions. This completes explanation of relations between the AdS/CFT correspondence and RS braneworld models. In Sect. 8.9, we extend our analysis to models with a bulk scalar field. Since the presence of bulk fields would break the conformal invariance, it is interesting to consider dilatonic braneworlds in conjunction with the AdS/CFT correspondence. The final section is devoted to the conclusion.

8.2 Braneworlds in AdS Spacetime

In this section, we will introduce RS braneworld models. We will derive the effective Friedmann equation and identify the effects of extra-dimensions. In this lecture, we will concentrate on cosmology although we can apply the results to black hole physics.

8.2.1 RS Models

Randall and Sundrum proposed a simple model where a four-dimensional brane with the tension σ is embedded in the five-dimensional asymptotically anti-de Sitter (AdS) bulk with a curvature scale ℓ. This single-brane model is described by the action [10]

$$S = \frac{1}{2\kappa^2}\int d^5x\sqrt{-g}\left(\mathcal{R}+\frac{12}{\ell^2}\right) - \sigma\int d^4x\sqrt{-h} + \int d^4x\sqrt{-h}\mathcal{L}_{\rm matter}, \qquad (8.1)$$

where $\mathcal{R}$ and κ^2 are the scalar curvature and gravitational constant in five-dimensions, respectively. We impose Z_2 symmetry on this spacetime, with the brane at the fixed point. The matter $\mathcal{L}_{\rm matter}$ is confined to the brane. Throughout this lecture, $h_{\mu\nu}$ represents the induced metric on the brane. Remarkably, the internal dimension is non-compact in this model. Hence, we do not have to care about the stability problem. The basic equations consist of the equations of motion in the bulk and junction conditions at the brane position due to the presence of the brane. Alternatively, the basic equations can be regarded as the 5-dimensional Einstein equations with singular sources. Let us recall the 4-dimensional components of 5-dimensional Einstein tensor can be expressed by

$$\overset{(5)}{G}_{\mu\nu} = \overset{(4)}{G}_{\mu\nu} + \mathcal{L}_{n^\mu}\left[K_{\mu\nu} - g_{\mu\nu}K\right] + \cdots, \qquad (8.2)$$

where $\mathcal{L}_{n^\mu}$ denotes the Lie derivative along the unit normal vector to the brane, n^μ. Here, we defined the extrinsic curvature by

$$K_{\mu\nu} = -\left(\delta^\rho_\mu - n_\mu n^\rho\right)\nabla_\rho n_\nu. \tag{8.3}$$

By integrating Einstein equations along the normal to the brane, we obtain the jump of the extrinsic curvature $(K^+_{\mu\nu} - g_{\mu\nu}K^+) - (K^-_{\mu\nu} - g_{\mu\nu}K^-)$ from the left hand side and the total energy momentum tensor on the brane from the right hand side due to the delta function sources. Thus, taking into account the Z_2 symmetry $K_{\mu\nu} \equiv K^+_{\mu\nu} = -K^-_{\mu\nu}$, we obtain the junction conditions

$$\left[K^\mu{}_\nu - \delta^\mu_\nu K\right]\Big|_{\text{at the brane}} = \frac{\kappa^2}{2}\left(-\sigma\delta^\mu_\nu + T^\mu{}_\nu\right) \tag{8.4}$$

Here, $T_{\mu\nu}$ represents the energy-momentum tensor of the matter.

Originally, they proposed the two-brane model as a possible solution of the hierarchy problem [9]. The action reads

$$\begin{aligned} S = &\frac{1}{2\kappa^2}\int d^5x\sqrt{-g}\left(\mathcal{R} + \frac{12}{\ell^2}\right) \\ &- \sum_{i=\oplus,\ominus}\sigma_i\int d^4x\sqrt{-h_i} + \sum_{i=\oplus,\ominus}\int d^4x\sqrt{-h}\mathcal{L}^i_{\text{matter}}, \end{aligned} \tag{8.5}$$

where $\oplus$ and $\ominus$ represent the positive and the negative tension branes, respectively. In principle, one can consider multiple-branes although they are not discussed in this lecture.

8.2.2 Cosmology

The homogeneous cosmology of the single-brane model is easy to analyze [16]. Because of the Birkoff theorem due to the symmetry on the brane, it is sufficient to consider AdS black hole spacetime:

$$ds^2 = -h(r)dt^2 + \frac{dr^2}{h(r)} + r^2\left[d\chi^2 + f_k^2(\chi)\left(d\theta^2 + \sin^2\theta d\phi^2\right)\right], \tag{8.6}$$

where

$$f_k = \begin{cases} \sin\chi & \text{for}\, k = 1 \\ \chi & \text{for}\, k = 0 \\ \sinh\chi & \text{for}\, k = -1 \end{cases} \tag{8.7}$$

and

$$h(r) = k - \frac{M}{r^2} + \frac{r^2}{\ell^2}. \tag{8.8}$$

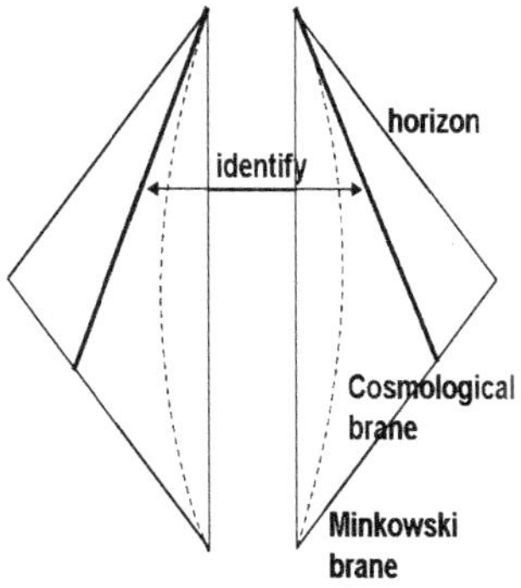

Fig. 8.1 The Minkowski brane represented by the dotted line is a static brane in the Poincare coordinate system of the AdS spacetime. While, the cosmological brane represented by the thick line is moving in the bulk. The motion of the brane induces the expansion of the brane universe. The Cauchy horizon of AdS spacetime corresponds to the big-bang singularity. In the case of AdS-Schwarzschild spacetime, the horizon should be the past horizon of the black hole. The big bang is located beyond the horizon

Note that M is the mass of the black hole, k is the curvature of the horizon and ℓ is the AdS curvature radius. Suppose the brane is moving in this spacetime with a trajectory $t = t(\tau)\,, r = a(\tau)$, where τ is a proper time of the brane (see Fig. 8.1). The induced metric on the brane becomes

$$ds^2_{ind} = -d\tau^2 + a^2(\tau)\left[d\chi^2 + f_k^2(\chi)\left(d\theta^2 + \sin^2\theta d\phi^2\right)\right]. \tag{8.9}$$

This is nothing but the Friedman-Robertson-Walker spacetime where a is the scale factor. The motion of the brane cannot be arbitrary. It is constrained by the junction condition:

$$K^{\chi}{}_{\chi} = -\frac{\kappa^2\sigma}{6} - \frac{\kappa^2}{2}\left[T^{\chi}{}_{\chi} - \frac{1}{3}T\right], \tag{8.10}$$

where $K^{\chi}{}_{\chi}, T^{\chi}{}_{\chi}, T$ are a $\chi\chi$ component of the extrinsic curvature, a $\chi\chi$ component and the trace part of the energy momentum tensor of matter on the brane, respectively. From the normalization condition $n^{\mu}n_{\mu} = 1$ of the unit normal vector $n^{\mu} = (-\dot{a}, \dot{t})$, we obtain

$$\dot{t} = \sqrt{\frac{1}{h(a)} + \frac{\dot{a}^2}{h^2(a)}}. \tag{8.11}$$

Here, the dot is a derivative with respect to the proper time τ. Now, one can calculate $K_{\chi\chi}$ as

$$K_{\chi\chi} = -\nabla_{\chi}n_{\chi} = -\partial_{\chi}n_{\chi} + \Gamma^{r}_{\chi\chi} = -\frac{1}{2}n^{r}\frac{\partial}{\partial\chi}g_{\chi\chi}. \tag{8.12}$$

Hence, from Eqs. 8.10, 8.11 and 8.12, we have an equation

$$\frac{1}{a}\sqrt{h(a)+\dot{a}^2} = \frac{\kappa^2}{6}(\sigma+\rho). \tag{8.13}$$

Thus, we get

$$\frac{k}{a^2} - \frac{M}{a^4} + \ell^2 + \frac{\dot{a}^2}{a^2} = \frac{\kappa^4}{36}(\sigma+\rho)^2, \tag{8.14}$$

or

$$H^2 = \frac{\kappa^4\sigma^2}{36} - \frac{1}{\ell^2} + \frac{\kappa^4\sigma}{18}\rho - \frac{k}{a^2} + \frac{M}{a^4} + \frac{\kappa^4}{36}\rho^2. \tag{8.15}$$

By setting $\kappa^2\sigma\ell = 6$, we finally derived the effective Friedmann equation as [17–21]

$$H^2 = -\frac{k}{a^2} + \frac{\kappa^2}{3\ell}\rho + \frac{M}{a^4} + \frac{\kappa^4}{36}\rho^2, \tag{8.16}$$

where $H = \dot{a}/a$ is the Hubble parameter. The Newton's constant can be identified as $8\pi G_N = \kappa^2/\ell$. The curvature of the horizon k corresponds spatial curvature of the universe. The black hole mass M is referred to as the dark radiation [22] which is not real radiation fluid but a reflection of the bulk geometry. This effect exists even in the low energy regime. The last term represents the high energy effect of the braneworld [23].

As to the two-brane model, the same effective Friedmann equation (8.16) can be expected on each brane because the above Eq. 8.16 has been deduced without referring to the bulk equations of motion.

Given this cosmological background, it is natural to investigate cosmological perturbation in the braneworld [24]. In the case of the single-brane model, it is shown that the gravity in Minkowski brane is localized on the brane in spite of the noncompact extra dimension. Consequently, it turned out that the conventional linearized Einstein equation approximately holds at scales large compared with the curvature scale ℓ. It should be stressed that this result can be attained by imposing the outgoing boundary conditions. It turns out that this is also true in the cosmological background [25].

In the case of the two-brane model, Garriga and Tanaka analyzed linearized gravity and have shown that the gravity on the brane behaves as the Brans-Dicke theory at low energy [26]. Thus, the conventional linearized Einstein equations do not hold even on scales large compared with the curvature scale ℓ in the bulk. Charmousis et al. have clearly identified the Brans-Dicke field as the radion mode [27].

In the end, we would like to know how nonlinear gravity in the braneworld is deviated from the conventional Einstein gravity. A partial answer will be given in the following sections.

8.3 View from the Brane

In the previous section, we have considered an isotropic and homogeneous universe and seen that the effective Friedmann equation on the brane can be regarded as the conventional Friedmann equation with two kind of corrections, i.e., the dark radiation and high-energy corrections. Here, we review two different approaches to extend the above result to more general cases.

8.3.1 AdS/CFT Correspondence

Let us start with the AdS/CFT correspondence [28–31]. After solving the equations of motion in the bulk with the boundary value fixed and substituting the solution $g_{\rm cl}$ into the 5-dimensional Einstein-Hilbert action $S_{\rm 5d}$, we obtain the effective action for the boundary field $h = g_{\rm cl}|_{\rm boundary}$. The statement of the AdS/CFT correspondence is that the resultant effective action can be equated with the partition functional of some conformally invariant field theory (CFT), namely

$$\exp[iS_{\rm 5d}[g_{\rm cl}]] \approx < \exp\left[i\int h\mathcal{O}\right] >_{\rm CFT}, \tag{8.17}$$

where $\mathcal{O}$ is the field in CFT. In the right hand side, h should be interpreted as a source field. This action must be defined at the AdS infinity where the conformal symmetry exists as the asymptotic symmetry. Hence, there exist infrared divergences which must be subtracted by counter terms. Thus, the correct formula becomes

$$\exp[iS_{\rm 5d}[g_{\rm cl}] + iS_{\rm ct}] = < \exp\left[i\int h\mathcal{O}\right] >_{\rm CFT}, \tag{8.18}$$

where we added the counter terms

$$S_{\rm ct} = S_{\rm brane} - S_{\rm 4d} - [R^2\text{terms}], \tag{8.19}$$

where $S_{\rm brane}$ and S_{4d} are the brane action and the 4-dimensional Einstein-Hilbert action, respectively. Here, the higher curvature terms $[R^2\text{terms}]$ should be understood symbolically.

In the case of the braneworld, the brane acts as the cutoff. Therefore, there is no divergences in the above expressions. In other words, no counter term is necessary. We can regard the above relation as the definition of the "cut off" CFT. Thus, we can freely rearrange the terms as follows

$$S_{\rm 5d} + S_{\rm brane} = S_{\rm 4d} + S_{\rm CFT} + [R^2\text{terms}], \tag{8.20}$$

where we have defined

$$\exp iS_{\mathrm{CFT}} \equiv < \exp\left[i \int h\mathcal{O}\right] >_{\mathrm{CFT}}. \tag{8.21}$$

This tells us that the brane models can be described as the conventional Einstein theory with the cutoff CFT and higher order curvature terms [32–35]. In terms of the equations of motion, the AdS/CFT correspondence reads

$$\overset{(4)}{G}_{\mu\nu} = \frac{\kappa^2}{\ell}\left(T_{\mu\nu} + T^{\mathrm{CFT}}_{\mu\nu}\right) + [R^2\text{terms}], \tag{8.22}$$

where the R^2 terms represent the higher order curvature terms and $T^{CFT}_{\mu\nu}$ denotes the energy-momentum tensor of the cutoff version of conformal field theory. When we apply this result to cosmology, we see CFT corresponds to the dark radiation in the braneworld and the higher curvature terms can be reduced to the high-energy corrections.

8.3.2 Geometrical Holography

Here, let us review the geometrical approach [36]. In the Gaussian normal coordinate system:

$$ds^2 = dy^2 + g_{\mu\nu}(y, x^\mu)dx^\mu dx^\nu, \tag{8.23}$$

we can write the 5-dimensional Einstein tensor $\overset{(5)}{G}_{\mu\nu}$ in terms of the 4-dimensional Einstein tensor $\overset{(4)}{G}_{\mu\nu}$ and the extrinsic curvature as

$$\begin{aligned}\overset{(5)}{G}_{\mu\nu} = \overset{(4)}{G}_{\mu\nu} + K_{\mu\nu,y} - g_{\mu\nu}K_{,y} - KK_{\mu\nu} + 2K_{\mu\lambda}K^{\lambda}{}_{\nu} \\ + \frac{1}{2}g_{\mu\nu}\left(K^2 + K^{\alpha}{}_{\beta}K^{\beta}{}_{\alpha}\right) = \frac{6}{\ell^2}g_{\mu\nu},\end{aligned} \tag{8.24}$$

where we have introduced the extrinsic curvature

$$K_{\mu\nu} = -\frac{1}{2}g_{\mu\nu,y}, \tag{8.25}$$

and the last equality comes from the 5-dimensional Einstein equations. On the other hand, the Weyl tensor in the bulk can be expressed as

$$C_{y\mu y\nu} = K_{\mu\nu,y} - g_{\mu\nu}K_{,y} + K_{\mu}{}^{\lambda}K_{\lambda\nu} + g_{\mu\nu}K^{\alpha\beta}K_{\alpha\beta} - \frac{3}{\ell^2}g_{\mu\nu}. \tag{8.26}$$

Now, one can eliminate $K_{\mu\nu,y} - g_{\mu\nu}K_{,y}$ from (8.24) using (8.26) and obtain

$$\overset{(4)}{G}_{\mu\nu} = -C_{y\mu y\nu} + KK_{\mu\nu} - K_{\mu\lambda}K^{\lambda}{}_{\nu}$$
$$- \frac{1}{2} g_{\mu\nu}\left(K^2 - K^{\alpha}{}_{\beta}K^{\beta}{}_{\alpha}\right) + \frac{3}{\ell^2} g_{\mu\nu}. \tag{8.27}$$

Taking into account the Z_2 symmetry, we also obtain the junction conditions

$$\left[K^{\mu}{}_{\nu} - \delta^{\mu}_{\nu}K\right]\Big|_{y=0} = \frac{\kappa^2}{2}\left(-\sigma\delta^{\mu}_{\nu} + T^{\mu}{}_{\nu}\right). \tag{8.28}$$

Here, $T_{\mu\nu}$ represents the energy-momentum tensor of the matter. Evaluating Eq. 8.27 at the brane and substituting the junction condition into it, we have the "effective" equations of motion

$$\overset{(4)}{G}_{\mu\nu} = \frac{\kappa^2}{\ell} T_{\mu\nu} + \kappa^4 \pi_{\mu\nu} - E_{\mu\nu} \tag{8.29}$$

where we have defined the quadratic of the energy momentum tensor

$$\pi_{\mu\nu} = -\frac{1}{4} T_{\mu}{}^{\lambda} T_{\lambda\nu} + \frac{1}{12} T T_{\mu\nu} + \frac{1}{8} g_{\mu\nu}\left(T^{\alpha\beta}T_{\alpha\beta} - \frac{1}{3}T^2\right) \tag{8.30}$$

and the projection of Weyl tensor $C_{y\mu y\nu}$ onto the brane

$$E_{\mu\nu} = C_{y\mu y\nu}\big|_{y=0}.$$

Here, we assumed the relation

$$\kappa^2\sigma = \frac{6}{\ell} \tag{8.31}$$

so that the effective cosmological constant vanishes.

Because of the traceless property of $E_{\mu\nu}$, when we consider an isotropic and homogeneous universe, it is easy to show that this gives the dark radiation component $\propto 1/a^4$. The existence of the high-energy corrections $\propto \rho^2$ is apparent in this approach.

The geometrical approach is useful to classify possible corrections to the conventional Einstein equations. One defect of this approach is the fact that the projected Weyl tensor $E_{\mu\nu}$ can not be determined without solving the equations in the bulk.

8.4 Does AdS/CFT Play Any Role in Braneworld?

To make our concerns explicit, we give a sequence of questions. We treat the single-brane model and two-brane model, separately.

8.4.1 Single-Brane Model

Is the Einstein theory recovered even in the non-linear rregime?

In the case of the linear theory, it is known that the conventional Einstein theory is recovered at low energy. On the other hand, the cosmological consideration suggests the deviation from the conventional Friedmann equation even in the low energy regime. This is due to the dark radiation term. Therefore, we need to clarify when the conventional Einstein theory can be recovered on the brane.

How does the AdS/CFT come into the braneworld?

It was argued that the cutoff CFT comes into the braneworld. However, no one knows what is the cutoff CFT. It is a vague concept at least from the point of view of the classical gravity. Moreover, it should be noted that the AdS/CFT correspondence is a specific conjecture. Indeed, originally, Maldacena conjectured that the supergravity on $AdS_5 \times S^5$ is dual to the four-dimensional $\mathcal{N} = 4$ super Yang-Mills theory [28, 29]. Nevertheless, the AdS/CFT correspondence seems to be related to the brane world model as has been demonstrated by several people [33–42]. Hence, it is important to reveal the role of the AdS/CFT correspondence starting from the
5-dimensional general relativity.

How are the AdS/CFT and geometrical approach related?

The geometrical approach gives the effective equations of motion (8.29)

$$\overset{(4)}{G}_{\mu\nu} = \frac{\kappa^2}{\ell} T_{\mu\nu} + \kappa^4 \pi_{\mu\nu} - E_{\mu\nu}.$$

On the other hand, the AdS/CFT correspondence yields the other effective equations of motion (8.22)

$$\overset{(4)}{G}_{\mu\nu} = \frac{\kappa^2}{\ell} \left(T_{\mu\nu} + T^{\mathrm{CFT}}_{\mu\nu} \right) + [R^2 \mathrm{terms}].$$

An apparent difference is remarkable.

It is an interesting issue to clarify how these two descriptions are related. Shiromizu and Ida tried to understand the AdS/CFT correspondence from the geometrical point of view [43]. They argued that π^μ_μ corresponds to the trace anomaly of the cutoff CFT on the brane. However, this result is rather paradoxical because there exists no trace anomaly in an odd dimensional brane although π^μ_μ exists even in that case. Thus, the more precise relation between the geometrical and the AdS/CFT approaches should be given.

Moreover, since both the geometrical and AdS/CFT approaches seem to have their own merit, it would be beneficial to understand the mutual relationship.

8.4.2 Two-Brane Model

How is the geometrical approach consistent with the Brans-Dicke picture?

Irrespective of the existence of other branes, the geometric approach gives the same effective equations (8.29). The effect of the bulk geometry comes into the brane world only through $E_{\mu\nu}$. In this picture, the two-brane system can be regarded as the Einstein theory with some corrections due to the Weyl tensor in the bulk.

On the other hand, the linearized gravity on the brane behaves as the Brans-Dicke theory on scales large compared with the curvature scale ℓ in the bulk [26]. Therefore, the conventional linearized Einstein equations do not hold at low energy.

In the geometrical approach, no radion appears. While, from the linear analysis, it turns out that the system can be described by the Brans-Dicke theory where the extra scalar field is nothing but the radion. How can we reconcile these seemingly incompatible pictures?

What replaces the AdS/CFT correspondence in the two-brane model?

In the single-brane model, there are continuum Kaluza-Klein (KK)-spectrum around the zero mode. They induce the CFT matter in the 4-dimensional effective action. In the two-brane system, the spectrum become discrete and then a mass gap exists. Hence, we can not expect CFT matter on the brane, although KK-modes exist and affect the physics on the brane. So, it is interesting to know if the AdS/CFT correspondence play a role in the two-brane system.

8.5 Gradient Expansion Method

Our aim in this lecture is to show the gradient expansion method gives the answers to all of the questions presented in the previous section. Here, we review the formalism developed by us [44–47].

We use the Gaussian normal coordinate system (8.23) to describe the geometry of the brane world. Note that the brane is located at $y = 0$ in this coordinate system. Decomposing the extrinsic curvature into the traceless part and the trace part

$$K_{\mu\nu} = \Sigma_{\mu\nu} + \frac{1}{4} h_{\mu\nu} K, \quad K = -\frac{\partial}{\partial y} \log \sqrt{-g}, \tag{8.32}$$

we obtain the basic equations which hold in the bulk;

$$\Sigma^{\mu}{}_{\nu,y} - K \Sigma^{\mu}{}_{\nu} = -\left[\overset{(4)}{R}{}^{\mu}{}_{\nu} - \frac{1}{4} \delta^{\mu}_{\nu} \overset{(4)}{R} \right], \tag{8.33}$$

$$\frac{3}{4} K^2 - \Sigma^{\alpha}{}_{\beta} \Sigma^{\beta}{}_{\alpha} = \left[\overset{(4)}{R} \right] + \frac{12}{\ell^2}, \tag{8.34}$$

$$K_{,y} - \frac{1}{4} K^2 - \Sigma^{\alpha\beta} \Sigma_{\alpha\beta} = -\frac{4}{\ell^2}, \tag{8.35}$$

$$\nabla_\lambda \Sigma_\mu{}^\lambda - \frac{3}{4}\nabla_\mu K = 0, \tag{8.36}$$

where $\overset{(4)}{R}{}^\mu{}_\nu$ is the curvature on the brane and ∇_μ denotes the covariant derivative with respect to the metric $g_{\mu\nu}$. One also have the junction condition

$$\left[K^\mu{}_\nu - \delta^\mu_\nu K\right]\Big|_{y=0} = \frac{\kappa^2}{2}\left(-\sigma\delta^\mu_\nu + T^\mu{}_\nu\right). \tag{8.37}$$

Recall that we are considering the Z_2 symmetric spacetime.

The problem now is separated into two parts. First, we will solve the bulk equations of motion with the Dirichlet boundary condition at the brane,

$$g_{\mu\nu}(y=0, x^\mu) = h_{\mu\nu}(x^\mu). \tag{8.38}$$

After that, the junction condition will be imposed at the brane. As it is the condition for the induced metric $h_{\mu\nu}$, it is naturally interpreted as the effective equations of motion for gravity on the brane.

Along the normal coordinate y, the metric varies with a characteristic length scale ℓ; $g_{\mu\nu,y} \sim g_{\mu\nu}/\ell$. Denote the characteristic length scale of the curvature fluctuation on the brane as L; then we have $R \sim g_{\mu\nu}/L^2$. For the reader's reference, let us take $\ell = 1$ mm, for example. Then, the relation (8.31) give the scale, $\kappa^2 = (10^8\,\mathrm{GeV})^{-3}$ and $\sigma = 1\,\mathrm{TeV}^4$. In this lecture, we will consider the low energy regime in the sense that the energy density of matter, ρ, on the brane is smaller than the brane tension, ie., $\rho/\sigma \ll 1$. In this regime, a simple dimensional analysis

$$\frac{\rho}{\sigma} \sim \frac{\ell\frac{\kappa^2}{\ell}\rho}{\kappa^2\sigma} \sim \left(\frac{\ell}{L}\right)^2 \ll 1 \tag{8.39}$$

implies that the curvature on the brane can be neglected compared with the extrinsic curvature at low energy. Here, we have used the relation (8.31) and Einstein's equation on the brane, $R \sim g_{\mu\nu}/L^2 \sim \kappa^2\rho/\ell$. Thus, the anti-Newtonian or gradient expansion method used in the cosmological context is applicable to our problem [12].

At zeroth order, we can neglect the curvature term. Then we have

$$\overset{(0)}{\Sigma}{}^\mu{}_{\nu,y} - \overset{(0)}{K}\overset{(0)}{\Sigma}{}^\mu{}_\nu = 0, \tag{8.40}$$

$$\frac{3}{4}\overset{(0)}{K}{}^2 - \overset{(0)}{\Sigma}{}^\alpha{}_\beta \overset{(0)}{\Sigma}{}^\beta{}_\alpha = \frac{12}{\ell^2}, \tag{8.41}$$

$$\overset{(0)}{K}{}_{,y} - \frac{1}{4}\overset{(0)}{K}{}^2 - \overset{(0)}{\Sigma}{}^{\alpha\beta}\overset{(0)}{\Sigma}{}_{\alpha\beta} = -\frac{4}{\ell^2}, \tag{8.42}$$

$$\nabla_\lambda \overset{(0)}{\Sigma}{}^\lambda{}_\mu - \frac{3}{4}\nabla_\mu \overset{(0)}{K} = 0. \tag{8.43}$$

Equation 8.40 can be readily integrated into

$$\overset{(0)}{\Sigma}{}^{\mu}{}_{\nu} = \frac{C^{\mu}{}_{\nu}(x^{\mu})}{\sqrt{-g}}, \quad C^{\mu}{}_{\mu} = 0, \tag{8.44}$$

where $C^{\mu}{}_{\nu}$ is the constant of integration. Equation 8.43 also requires $C^{\mu}{}_{\nu|\mu} = 0$. If it could exist, it would represent a radiation like fluid on the brane and hence a strongly anisotropic universe. In fact, as we see soon, this term must vanish in order to satisfy the junction condition. Therefore, we simply put $C^{\mu}{}_{\nu} = 0$, hereafter. Now, it is easy to solve the remaining equations. The result is

$$\overset{(0)}{K} = \frac{4}{\ell}. \tag{8.45}$$

Using the definition of the extrinsic curvature

$$\overset{(0)}{K}_{\mu\nu} = -\frac{1}{2}\frac{\partial}{\partial y}\overset{(0)}{g}_{\mu\nu}, \tag{8.46}$$

we get the zeroth order metric as

$$ds^2 = dy^2 + b^2(y)h_{\mu\nu}(x^{\mu})dx^{\mu}dx^{\nu}, \quad b(y) = e^{-2\frac{y}{\ell}}, \tag{8.47}$$

where the tensor $h_{\mu\nu}$ is the induced metric on the brane, which conforms to the boundary condition (8.38).

From the zeroth order solution, we obtain

$$\left[\overset{(0)}{K}{}^{\mu}{}_{\nu} - \delta^{\mu}_{\nu}\overset{(0)}{K}\right]\Big|_{y=0} = -\frac{3}{\ell}\delta^{\mu}_{\nu} = -\frac{\kappa^2}{2}\sigma\delta^{\mu}_{\nu} \tag{8.48}$$

Then we get the well known relation

$$\kappa^2\sigma = 6/\ell. \tag{8.49}$$

Here, we will assume that this relation holds exactly. It is apparent that $C^{\mu}{}_{\nu}$ is not allowed to exist.

The iteration scheme consists in writing the metric $g_{\mu\nu}$ as a sum of local tensors built out of the induced metric on the brane, the number of gradients increasing with the order. Hence, we will seek the metric as a perturbative series

$$g_{\mu\nu}(y, x^{\mu}) = b^2(y)\left[h_{\mu\nu}(x^{\mu}) + \overset{(1)}{g}_{\mu\nu}(y, x^{\mu}) + \overset{(2)}{g}_{\mu\nu}(y, x^{\mu}) + \cdots\right], \tag{8.50}$$

where $b^2(y)$ is extracted and we put the Dirichlet boundary condition

$$\overset{(i)}{g}_{\mu\nu}(y = 0, x^{\mu}) = 0, \tag{8.51}$$

so that $g_{\mu\nu}(y=0,x) = h_{\mu\nu}(x)$ holds at the brane. Other quantities can be also expanded as

$$K^{\mu}{}_{\nu} = \frac{1}{\ell}\delta^{\mu}_{\nu} + \overset{(1)}{K}{}^{\mu}{}_{\nu} + \overset{(2)}{K}{}^{\mu}{}_{\nu} + \cdots$$
$$\Sigma^{\mu}{}_{\nu} = +\overset{(1)}{\Sigma}{}^{\mu}{}_{\nu} + \overset{(2)}{\Sigma}{}^{\mu}{}_{\nu} + \cdots. \tag{8.52}$$

In our scheme, in contrast to the AdS/CFT correspondence where the Dirichlet boundary condition is imposed at infinity, we impose it at the finite point $y = 0$, the location of the brane. Furthermore, we carefully consider the constants of integration, i.e., homogeneous solutions. These homogeneous solutions are ignored in the calculation of AdS/CFT correspondence. However, they play an important role in the braneworld. Note that the scheme can be applicable to other systems [48–51].

8.6 Single Brane Model (RS2)

Now, we will apply the gradient expansion method to the single-brane models and obtain the effective equations on the brane.

8.6.1 Einstein Gravity at Lowest Order

The next order solutions are obtained by taking into account the terms neglected at zeroth order. At first order, Eqs. 8.33–8.36 become

$$\overset{(1)}{\Sigma}{}^{\mu}{}_{\nu,y} - \frac{4}{\ell}\overset{(1)}{\Sigma}{}^{\mu}{}_{\nu} = -\left[\overset{(4)}{R}{}^{\mu}{}_{\nu} - \frac{1}{4}\delta^{\mu}_{\nu}\overset{(4)}{R}\right]^{(1)}, \tag{8.53}$$

$$\frac{6}{\ell}\overset{(1)}{K} = \left[\overset{(4)}{R}\right]^{(1)}, \tag{8.54}$$

$$\overset{(1)}{K}{}_{,y} - \frac{2}{\ell}\overset{(1)}{K} = 0, \tag{8.55}$$

$$\overset{(1)}{\Sigma}{}_{\mu}{}^{\lambda}{}_{|\lambda} - \frac{3}{4}\overset{(1)}{K}{}_{|\mu} = 0. \tag{8.56}$$

where the superscript (1) represents the order of the derivative expansion and a stroke $|$ denotes the covariant derivative with respect to the metric $h_{\mu\nu}$. Here, $[\overset{(4)}{R}{}^{\mu}{}_{\nu}]^{(1)}$ means that the curvature is approximated by taking the Ricci tensor of $b^2 h_{\mu\nu}$ in place of $\overset{(4)}{R}{}^{\mu}{}_{\nu}$. It is also convenient to write it in terms of the Ricci tensor of $h_{\mu\nu}$, denoted $R^{\mu}{}_{\nu}(h)$.

Substituting the zeroth order metric into $\overset{(4)}{R}$, we obtain

$$\overset{(1)}{K} = \frac{\ell}{6b^2} R(h). \tag{8.57}$$

Hereafter, we omit the argument of the curvature for simplicity. Simple integration of Eq. 8.53 also gives the traceless part of the extrinsic curvature as

$$\overset{(1)}{\Sigma}{}^{\mu}{}_{\nu} = \frac{\ell}{2b^2}(R^{\mu}{}_{\nu} - \frac{1}{4}\delta^{\mu}_{\nu}R) + \frac{\chi^{\mu}{}_{\nu}(x)}{b^4}, \tag{8.58}$$

where the homogeneous solution satisfies the constraints

$$\chi^{\mu}{}_{\mu} = 0, \quad \chi^{\mu}{}_{\nu|\mu} = 0. \tag{8.59}$$

As we see later, this term corresponds to dark radiation at this order. The metric can be obtained as

$$\overset{(1)}{g}_{\mu\nu} = -\frac{\ell^2}{2}\left(\frac{1}{b^2} - 1\right)\left(R_{\mu\nu} - \frac{1}{6}h_{\mu\nu}R\right) - \frac{\ell}{2}\left(\frac{1}{b^4} - 1\right)\chi_{\mu\nu}, \tag{8.60}$$

where we have imposed the boundary condition, $\overset{(1)}{g}_{\mu\nu}(y = 0, x^{\mu}) = 0$.

Let us focus on the role of $\chi^{\mu}{}_{\nu}$ in this part. At this order, the junction condition can be written as

$$\begin{aligned}\left[\overset{(1)}{K}{}^{\mu}{}_{\nu} - \delta^{\mu}_{\nu}\overset{(1)}{K}\right]\Big|_{y=0} &= \frac{\ell}{2}\left(R^{\mu}{}_{\nu} - \frac{1}{2}\delta^{\mu}_{\nu}R\right) + \chi^{\mu}{}_{\nu} \\ &= \frac{\kappa^2}{2}T^{\mu}{}_{\nu}.\end{aligned} \tag{8.61}$$

Using the solutions Eqs. 8.57, 8.58 and the formula

$$E^{\mu}{}_{\nu} = K^{\mu}{}_{\nu,y} - \delta^{\mu}_{\nu}K_{,y} - K^{\mu}{}_{\lambda}K^{\lambda}{}_{\nu} + \delta^{\mu}_{\nu}K^{\alpha}{}_{\beta}K^{\beta}{}_{\alpha} - \frac{3}{\ell^2}\delta^{\mu}_{\nu}, \tag{8.62}$$

we calculate the projective Weyl tensor as

$$\overset{(1)}{E}{}^{\mu}{}_{\nu} = \frac{2}{\ell}\chi^{\mu}{}_{\nu}. \tag{8.63}$$

Then we obtain the effective Einstein equation

$$R^{\mu}{}_{\nu} - \frac{1}{2}\delta^{\mu}_{\nu}R = \frac{\kappa^2}{\ell}T^{\mu}{}_{\nu} - \overset{(1)}{E}{}^{\mu}{}_{\nu}. \tag{8.64}$$

At this order, we do not have the conventional Einstein equations. Recall that the dark radiation exists even in the low energy regime. Indeed, the low energy effective Friedmann equation becomes

$$H^2 = \frac{\kappa^2}{3\ell}\rho + \frac{\mathcal{C}}{a^4}. \tag{8.65}$$

This equation can be obtained from Eq. 8.64 by imposing the maximal symmetry on the spatial part of the brane world and the conditions (8.59). Hence, we observe that $\chi^\mu{}_\nu$ is the generalization of the dark radiation found in the cosmological context.

The nonlocal tensor $\chi_{\mu\nu}$ must be determined by the boundary conditions in the bulk. The natural choice is asymptotically AdS boundary condition. For this boundary condition, we have $\chi_{\mu\nu} = 0$. It is this boundary condition that leads to the conventional Einstein theory in linearized gravity. Assuming this, we have

$$R^\mu{}_\nu - \frac{1}{2}\delta^\mu_\nu R = \frac{\kappa^2}{\ell} T^\mu{}_\nu. \tag{8.66}$$

Thus, the conventional Einstein theory is recovered at the leading order!

8.6.2 AdS/CFT Emerges

In this subsection, we do not include the $\chi_{\mu\nu}$ field because we have adopted the AdS boundary condition. Of course, we have calculated the second order solutions with the contribution of the $\chi_{\mu\nu}$ field. It merely adds extra terms such as $\chi^\mu{}_\nu \chi^\nu{}_\mu$, etc.

At second order, the basic equations can be easily deduced. Substituting the solution up to first order into the Ricci tensor and picking up the second order quantities, we obtain the solutions at second order. The trace part is deduced algebraically as

$$\overset{(2)}{K} = \frac{\ell^3}{8b^4}\left(R^\alpha{}_\beta R^\beta{}_\alpha - \frac{2}{9}R^2\right) - \frac{\ell^3}{12b^2}\left(R^\alpha{}_\beta R^\beta{}_\alpha - \frac{1}{6}R^2\right). \tag{8.67}$$

By integrating the equation for the traceless part, we have

$$\overset{(2)}{\Sigma}{}^\mu{}_\nu = -\frac{\ell^2}{2}\left(\frac{y}{b^4} + \frac{\ell}{2b^2}\right)\mathcal{S}^\mu{}_\nu - \frac{\ell^3}{24b^2}\left(RR^\mu{}_\nu - \frac{1}{4}\delta^\mu_\nu R^2\right) + \frac{\ell^3}{b^4} t^\mu{}_\nu, \tag{8.68}$$

where $\mathcal{S}^\mu{}_\nu$ is defined by

$$\delta \int d^4x\sqrt{-h}\frac{1}{2}\left[R^{\alpha\beta}R_{\alpha\beta} - \frac{1}{3}R^2\right] = \int d^4x\sqrt{-h}\mathcal{S}_{\mu\nu}\delta g^{\mu\nu}. \tag{8.69}$$

The tensor $\mathcal{S}^{\mu}{}_{\nu}$ is transverse and traceless,

$$\mathcal{S}^{\mu}{}_{\nu|\mu} = 0\,, \quad \mathcal{S}^{\mu}{}_{\mu} = 0. \tag{8.70}$$

The homogeneous solution $t^{\mu}{}_{\nu}$ must be traceless. Moreover, it must satisfy the momentum constraint. To be more precise, we must solve the constraint equation

$$t^{\mu}{}_{\nu|\mu} - \frac{1}{16}R^{\alpha}{}_{\beta}R^{\beta}{}_{\alpha|\nu} + \frac{1}{48}RR_{|\nu} - \frac{1}{24}R_{|\lambda}R^{\lambda}{}_{\nu} = 0. \tag{8.71}$$

As one can see immediately, there are ambiguities in integrating this equation. Indeed, there are two types of covariant local tensor whose divergences vanish:

$$\delta \int d^4x\sqrt{-h}\frac{1}{2}R^{\alpha\beta}R_{\alpha\beta} = \int d^4x\sqrt{-h}\mathcal{H}_{\mu\nu}\delta g^{\mu\nu}, \tag{8.72}$$

$$\delta \int d^4x\sqrt{-h}\frac{1}{2}R^2 = \int d^4x\sqrt{-h}\mathcal{K}_{\mu\nu}\delta g^{\mu\nu}. \tag{8.73}$$

Notice that $\mathcal{S}^{\mu}{}_{\nu} = \mathcal{H}^{\mu}{}_{\nu} - \mathcal{K}^{\mu}{}_{\nu}/3$. Hence, only $\mathcal{S}^{\mu}{}_{\nu}$ and $\mathcal{K}^{\mu}{}_{\nu}$ are independent. Thanks to the Gauss-Bonnet topological invariant, we do not need to consider the Riemann squared term. In addition to these local tensors, we have to take into account the nonlocal tensor $\tau^{\mu}{}_{\nu}$ with the property $\tau^{\mu}{}_{\nu|\mu} = 0$. Thus, we get

$$\begin{aligned} t^{\mu}{}_{\nu} = &\frac{1}{32}\delta^{\mu}_{\nu}\left(R^{\alpha}{}_{\beta}R^{\beta}{}_{\alpha} - \frac{1}{3}R^2\right) + \frac{1}{24}\left(RR^{\mu}{}_{\nu} - \frac{1}{4}\delta^{\mu}_{\nu}R^2\right) \\ &+ \tau^{\mu}{}_{\nu} + \left(\alpha + \frac{1}{4}\right)\mathcal{S}^{\mu}{}_{\nu} + \frac{\beta}{3}\mathcal{K}^{\mu}{}_{\nu}, \end{aligned} \tag{8.74}$$

where the constants α and β are free parameters representing the degree of initial conditions in the bulk. Hence, they represents the freedom of gravitational waves in the bulk. The condition $t^{\mu}{}_{\mu} = 0$ leads to

$$\tau^{\mu}{}_{\mu} = -\frac{1}{8}\left(R^{\alpha}{}_{\beta}R^{\beta}{}_{\alpha} - \frac{1}{3}R^2\right) - \beta\Box R. \tag{8.75}$$

This expression is the reminiscent of the trace anomaly of the CFT. It is possible to use the result of CFT at this point. For example, we can choose the $\mathcal{N} = 4$ super Yang-Mills theory as the conformal matter. In that case, we simply put $\beta = 0$. This is the way the AdS/CFT correspondence comes into the brane world scenario.

Up to the second order, the junction condition gives

$$R^{\mu}{}_{\nu} - \frac{1}{2}\delta^{\mu}_{\nu}R + 2\ell^2\left[\tau^{\mu}{}_{\nu} + \alpha\mathcal{S}^{\mu}{}_{\nu} + \frac{\beta}{3}\mathcal{K}^{\mu}{}_{\nu}\right] = \frac{\kappa^2}{\ell}T^{\mu}{}_{\nu}. \tag{8.76}$$

If we define

$$T^{\rm CFT}_{\mu\nu} = -2\frac{\ell^3}{\kappa^2}\tau_{\mu\nu}, \tag{8.77}$$

we can write

$$\overset{(4)}{G}_{\mu\nu} = \frac{\kappa^2}{\ell}\left(T_{\mu\nu} + T^{\rm CFT}_{\mu\nu}\right) - 2\ell^2\alpha\mathcal{S}_{\mu\nu} - \frac{2\ell^2}{3}\beta\mathcal{K}_{\mu\nu}. \tag{8.78}$$

Let us try to arrange the terms so as to reveal the geometrical meaning of the above equation. We can calculate the projective Weyl tensor as

$$\overset{(2)}{E}{}^{\mu}{}_{\nu} = \ell^2\left[P^{\mu}{}_{\nu} + 2\tau^{\mu}{}_{\nu} + 2\alpha\mathcal{S}^{\mu}{}_{\nu} + \frac{2}{3}\beta\mathcal{K}^{\mu}{}_{\nu}\right], \tag{8.79}$$

where

$$P^{\mu}{}_{\nu} = -\frac{1}{4}R^{\mu}{}_{\lambda}R^{\lambda}{}_{\nu} + \frac{1}{6}RR^{\mu}{}_{\nu} + \frac{1}{8}\delta^{\mu}_{\nu}R^{\alpha}{}_{\beta}R^{\beta}{}_{\alpha} - \frac{1}{16}\delta^{\mu}_{\nu}R^2. \tag{8.80}$$

Substituting this expression into Eq. 8.76 yields our main result

$$\overset{(4)}{G}_{\mu\nu} = \frac{\kappa^2}{\ell}T_{\mu\nu} + \ell^2 P_{\mu\nu} - \overset{(2)}{E}_{\mu\nu}. \tag{8.81}$$

Notice that $E^{\mu}{}_{\nu}$ contains the nonlocal part and the free parameters α and β. On the other hand, $P^{\mu}{}_{\nu}$ is determined locally. One can see the relationship in a more transparent way. Within the accuracy we are considering, we can get $P^{\mu}{}_{\nu} = \pi^{\mu}{}_{\nu}$ using the lowest order equation $R^{\mu}{}_{\nu} = \kappa^2/\ell(T^{\mu}{}_{\nu} - 1/2\delta^{\mu}_{\nu}T)$. Hence, we can rewrite Eq. 8.81 as

$$\overset{(4)}{G}_{\mu\nu} = \frac{\kappa^2}{\ell}T_{\mu\nu} + \kappa^4\pi_{\mu\nu} - \overset{(2)}{E}_{\mu\nu}. \tag{8.82}$$

Now, the similarity between Eqs. 8.29 and 8.82 is apparent. Thus we get an explicit relation between the geometrical approach and the AdS/CFT approach. However, we note that our Eq. 8.82 is a closed system of equations provided that the specific conformal field theory is chosen.

Now we can read off the effective action as

$$\begin{aligned} S_{\rm eff} &= \frac{\ell}{2\kappa^2}\int d^4x\sqrt{-h}R + S_{\rm matter} + S_{\rm CFT} \\ &\quad + \frac{\alpha\ell^2}{2\kappa^2}\int d^4x\sqrt{-h}\left[R^{\mu\nu}R_{\mu\nu} - \frac{1}{3}R^2\right] + \frac{\beta\ell^2}{6\kappa^2}\int d^4x\sqrt{-h}R^2, \end{aligned} \tag{8.83}$$

where we have used the relations Eqs. 8.69, 8.72 and 8.73 and we denoted the nonlocal effective action constructed from $\tau^{\mu}{}_{\nu}$ as $S_{\rm CFT}$.

8.7 Two-Brane Model (RS1)

In this section, we will apply the gradient expansion method to the two-brane models and reveal a role of the AdS/CFT correspondence.

8.7.1 Scalar-Tensor Theory Emerges

We consider the two-brane system depicted in Fig. 8.2. Without matter on the branes, we have the relation $g_{\mu\nu}^{\ominus-\mathrm{brane}} = e^{-2d/\ell} g^{\oplus-\mathrm{brane}} \equiv \Omega^2 g^{\oplus-\mathrm{brane}}$ where d is the distance between the two branes. Although Ω is constant for vacuum branes, it becomes the function of the 4-dimensional coordinates if we put the matter on the brane.

Adding the energy momentum tensor to each of the two branes, and allowing deviations from the pure AdS_5 bulk, the effective (non-local) Einstein equations on the branes at low energies take the form [45],

$$G^{\mu}{}_{\nu}(h) = \frac{\kappa^2}{\ell} \overset{\oplus}{T}{}^{\mu}{}_{\nu} - \frac{2}{\ell} \chi^{\mu}{}_{\nu}, \tag{8.84}$$

$$G^{\mu}{}_{\nu}(f) = -\frac{\kappa^2}{\ell} \overset{\ominus}{T}{}^{\mu}{}_{\nu} - \frac{2}{\ell} \frac{\chi^{\mu}{}_{\nu}}{\Omega^4}. \tag{8.85}$$

where $h_{\mu\nu} = g_{\mu\nu}^{\oplus-\mathrm{brane}}$, $f_{\mu\nu} = g_{\mu\nu}^{\ominus-\mathrm{brane}} = \Omega^2 h_{\mu\nu}$ and the terms proportional to $\chi_{\mu\nu}$ are 5-dimensional Weyl tensor contributions which describe the non-local 5-dimensional effect. Although Eqs. 8.84 and 8.85 are non-local individually, with undetermined $\chi_{\mu\nu}$, one can combine both equations to reduce them to

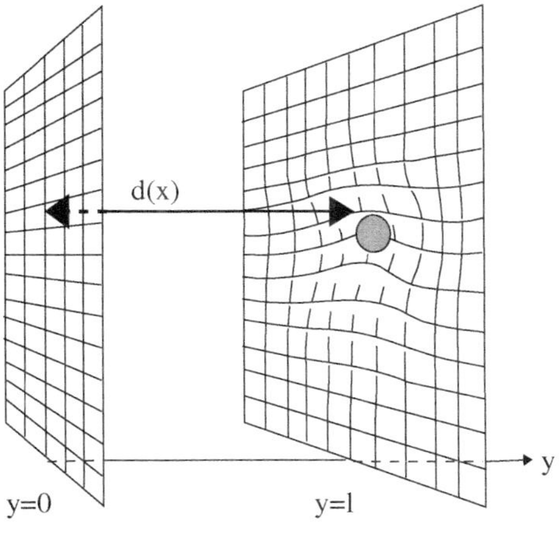

Fig. 8.2 In the two brane system, the radion is defined as a distance between two branes. It could depend on the position because the brane could have bending. Hence, for observers on the brane, it appears as a scalar field

local equations for each brane. Since $\chi_{\mu\nu}$ appears only algebraically, one can easily eliminate $\chi_{\mu\nu}$ from Eqs. 8.84 and 8.85. Defining a new field $\Psi = 1 - \Omega^2$, we find

$$G^{\mu}{}_{\nu}(h) = \frac{\kappa^2}{\ell\Psi}\overset{\oplus}{T}{}^{\mu}{}_{\nu} + \frac{\kappa^2(1-\Psi)^2}{\ell\Psi}\overset{\ominus}{T}{}^{\mu}{}_{\nu} + \frac{1}{\Psi}\left(\Psi^{|\mu}{}_{|\nu} - \delta^{\mu}_{\nu}\Psi^{|\alpha}{}_{|\alpha}\right) + \frac{3}{2\Psi(1-\Psi)}\left(\Psi^{|\mu}\Psi_{|\nu} - \frac{1}{2}\delta^{\mu}_{\nu}\Psi^{|\alpha}\Psi_{|\alpha}\right), \quad (8.86)$$

$$\Box\Psi = \frac{\kappa^2}{3\ell}(1-\Psi)\left\{\overset{\oplus}{T} + (1-\Psi)\overset{\ominus}{T}\right\} - \frac{1}{2(1-\Psi)}\Psi^{|\mu}\Psi_{|\mu}, \quad (8.87)$$

where | denotes the covariant derivative with respect to the metric $h_{\mu\nu}$. Since Ω (or equivalently Ψ) contains the information of the distance between the two branes, we call Ω (or Ψ) the radion.

We can also determine $\chi^{\mu}{}_{\nu}$ by eliminating $G^{\mu}{}_{\nu}$ from Eqs. 8.84 and 8.85. Then,

$$\chi^{\mu}{}_{\nu} = -\frac{\kappa^2(1-\Psi)}{2\Psi}\left(\overset{\oplus}{T}{}^{\mu}{}_{\nu} + (1-\Psi)\overset{\ominus}{T}{}^{\mu}{}_{\nu}\right) - \frac{\ell}{2\Psi}\left[\left(\Psi^{|\mu}{}_{|\nu} - \delta^{\mu}_{\nu}\Psi^{|\alpha}{}_{|\alpha}\right) + \frac{3}{2(1-\Psi)}\left(\Psi^{|\mu}\Psi_{|\nu} - \frac{1}{2}\delta^{\mu}_{\nu}\Psi^{|\alpha}\Psi_{|\alpha}\right)\right]. \quad (8.88)$$

Note that the index of $\overset{\ominus}{T}{}^{\mu}{}_{\nu}$ is to be raised or lowered by the induced metric on the $\ominus$-brane, $f_{\mu\nu}$.

The effective action for the $\oplus$-brane which gives Eqs. 8.86 and 8.87 is

$$S_{\oplus} = \frac{\ell}{2\kappa^2}\int d^4x\sqrt{-h}\left[\Psi R - \frac{3}{2(1-\Psi)}\Psi^{|\alpha}\Psi_{|\alpha}\right] + \int d^4x\sqrt{-h}\mathcal{L}^{\oplus} + \int d^4x\sqrt{-h}(1-\Psi)^2\mathcal{L}^{\ominus}. \quad (8.89)$$

The above action can be used to make cosmological predictions [52]. It should be stressed that the radion has the conformal coupling. In fact, using the transformation $\Psi = 1 - \psi^2$, we obtain

$$S_{\oplus} = \frac{6\ell}{\kappa^2}\int d^4x\sqrt{-h}\left[\frac{1}{12}R - \frac{1}{12}\psi^2 R - \frac{1}{2}\psi^{|\alpha}\psi_{|\alpha}\right] + \cdots. \quad (8.90)$$

This is nothing but Einstein theory with a conformally coupled scalar field ψ.

8.7.2 AdS/CFT in Two-Brane System?

In the two-brane case, it is difficult to proceed to the next order calculations. Hence, we need to invent a new method [53]. For this purpose, we shall start with the effective Einstein equation obtained by Shromizu et al. [36]

$$G_{\mu\nu} = T_{\mu\nu} + \pi_{\mu\nu} - E_{\mu\nu} \tag{8.91}$$

where $\pi_{\mu\nu}$ is the quadratic of energy momentum tensor $T_{\mu\nu}$ and $E_{\mu\nu}$ represents the effect of the bulk geometry. Here we have set $8\pi G = 1$. This geometrical projection approach can not give a concrete prediction, because we do not know $E_{\mu\nu}$ without solving the equations of motion in the bulk. Fortunately, in the case of the homogeneous cosmology, the property $E^{\mu}{}_{\mu} = 0$ determines the dynamics as

$$H^2 = \frac{1}{3}\rho + \rho^2 + \frac{\mathcal{C}}{a^4}. \tag{8.92}$$

This reflects the interplay between the bulk and the brane dynamics on the brane.

What we want to seek is an effective theory which contains the information of the bulk as finite number of constant parameters like $\mathcal{C}$ in the homogeneous universe. When we succeed in obtaining it, the cosmological perturbation theory can be constructed in a usual way. Although the concrete prediction can not be made, qualitative understanding of the evolution of the cosmological fluctuations can be obtained. This must be useful to make observational predictions.

In the two-brane system, the mass spectrum is known from the linear analysis [26]. At low energy, the propagator for the KK mode with the mass m can be expanded as

$$\frac{-1}{\Box - m^2} = \frac{1}{m^2}\left[1 + \frac{\Box}{m^2} + \frac{\Box^2}{m^4} + \cdots\right]. \tag{8.93}$$

However, massless modes can not be expanded in this way, hence we must take into account all of the massless modes to construct braneworld effective action. It seems legitimate to assume this consideration is valid even in the non-linear regime. Thus, at low energy, the action can be expanded by the local terms with increasing orders of derivatives of the metric $g_{\mu\nu}$ and the radion Ψ [45].

Let us illustrate our method using the following action truncated at the second order derivatives:

$$S_{\text{eff}} = \frac{1}{2}\int d^4x\sqrt{-g}\left[\Psi R - 2\Lambda(\Psi) - \frac{\omega(\Psi)}{\Psi}\nabla^{\mu}\Psi\nabla_{\mu}\Psi\right], \tag{8.94}$$

which is nothing but the scalar-tensor theory with coupling function $\omega(\Psi)$ and the potential function $\Lambda(\Psi)$. Note that this is the most general local action which contains up to the second order derivatives and has the general coordinate invariance. It should be stressed that the scalar-tensor theory is, in general, not related to the braneworld. However, we know a special type of scalar-tensor theory corresponds to the low energy braneworld [45, 54–58]. Here, we will present a simple derivation of this known fact.

For the vacuum brane, we can put $T_{\mu\nu} + \pi_{\mu\nu} = -\lambda g_{\mu\nu}$. Hence, the geometrical effective equation reduces to

$$G_{\mu\nu} = -E_{\mu\nu} - \lambda g_{\mu\nu}. \tag{8.95}$$

First, we must find $E_{\mu\nu}$. The above action (8.96) gives the equations of motion for the metric as

$$G_{\mu\nu} = -\frac{\Lambda}{\Psi} g_{\mu\nu} + \frac{1}{\Psi} (\nabla_\mu \nabla_\nu \Psi - g_{\mu\nu} \Box \Psi) + \frac{\omega}{\Psi^2} \left(\nabla_\mu \Psi \nabla_\nu \Psi - \frac{1}{2} g_{\mu\nu} \nabla^\alpha \Psi \nabla_\alpha \Psi \right). \qquad (8.96)$$

The right hand side of this Eq. 8.96 should be identified with $-E_{\mu\nu} - \lambda g_{\mu\nu}$. Hence, the condition $E^\mu{}_\mu = 0$ becomes

$$\Box \Psi = -\frac{\omega}{3\Psi} \nabla^\mu \Psi \nabla_\mu \Psi - \frac{4}{3} (\Lambda - \lambda \Psi). \qquad (8.97)$$

This is the equation for the radion Ψ. However, we also have the equation for Ψ from the action (8.94) as

$$\Box \Psi = \left(\frac{1}{2\Psi} - \frac{\omega'}{2\omega} \right) \nabla^\alpha \Psi \nabla_\alpha \Psi - \frac{\Psi}{2\omega} R + \frac{\Psi}{\omega} \Lambda', \qquad (8.98)$$

where the prime denotes the derivative with respect to Ψ. In order for these two Eqs. 8.97 and 8.98 to be compatible, Λ and ω must satisfy

$$-\frac{\omega}{3\Psi} = \frac{1}{2\Psi} - \frac{\omega'}{2\omega}, \qquad (8.99)$$

$$\frac{4}{3} (\Lambda - \lambda \Psi) = \frac{\Psi}{\omega} (2\lambda - \Lambda'), \qquad (8.100)$$

where we used $R = 4\lambda$ which comes from the trace part of Eq. 8.95. Equations 8.99 and 8.100 can be integrated as

$$\Lambda(\Psi) = \lambda + \lambda\gamma(1 - \Psi)^2, \quad \omega(\Psi) = \frac{3}{2} \frac{\Psi}{1 - \Psi}, \qquad (8.101)$$

where the constant of integration γ represents the ratio of the cosmological constant on the negative tension brane to that on the positive tension brane. Here, one of constants of integration is absorbed by rescaling of Ψ. In doing so, we have assumed the constant of integration is positive. We can also describe the negative tension brane if we take the negative signature.

Thus, we get the effective action

$$S_{\text{eff}} = \int d^4x \sqrt{-g} \left[\frac{1}{2} \Psi R - \frac{3}{4(1 - \Psi)} \nabla^\mu \Psi \nabla_\mu \Psi - \lambda - \lambda\gamma(1 - \Psi)^2 \right]. \qquad (8.102)$$

Surprisingly, this completely agrees with the previous result (8.89). Our simple symmetry principle $E^\mu{}_\mu = 0$ has determined the action completely.

As we have shown in [59], if $\gamma < -1$ there exists a static deSitter two-brane solution which turns out to be unstable. In particular, two inflating branes can collide at $\Psi = 0$. This process is completely smooth for the observer on the brane. This fact led us to the born-again scenario [59, 60]. The similar process occurs also in the ekpyrotic (cyclic) model [61, 62] where the moduli approximation is used. It can be shown that the moduli approximation is nothing but the lowest order truncation of the low energy gradient expansion method developed by us [63–66]. Hence, it is of great interest to see the leading order corrections due to KK modes to this process.

Let us now apply the conformal symmetry method explained above to the higher order case. First, we need to write down the most generic action containing up to fourth order derivatives. Then, we impose the symmetry to determine unknown functionals. The action reads

$$
\begin{aligned}
S_{\text{eff}} = & \frac{1}{2}\int d^4x\sqrt{-g}\left[\Psi R - 2\Lambda(\Psi) - \frac{\omega(\Psi)}{\Psi}\nabla^\mu\Psi\nabla_\mu\Psi\right] \\
& + \int d^4x\sqrt{-g}\Big[A(\Psi)\left(\nabla^\mu\Psi\nabla_\mu\Psi\right)^2 + B(\Psi)(\Box\Psi)^2 \\
& + C(\Psi)\nabla^\mu\Psi\nabla_\mu\Psi\Box\Psi + D(\Psi)R\Box\Psi \\
& + E(\Psi)R\nabla^\mu\Psi\nabla_\mu\Psi + F(\Psi)R^{\mu\nu}\nabla_\mu\Psi\nabla_\nu\Psi \\
& + G(\Psi)R^2 + H(\Psi)R^{\mu\nu}R_{\mu\nu} \\
& + I(\Psi)R^{\mu\nu\lambda\rho}R_{\mu\nu\lambda\rho} + \cdots\Big],
\end{aligned}
\tag{8.103}
$$

where $A, B, \cdots$ denote arbitrary functionals of the radion.

Now we impose the conformal symmetry on the fourth order derivative terms in the action (8.103) as we did in the previous example. Starting from the action (8.103), one can read off the equation for the metric from which $E_{\mu\nu}$ can be identified. The compatibility between the equations of motion for Ψ and the equation $E^\mu{}_\mu = 0$ constrains the coefficient functionals in the action (8.103). Surprisingly, every coefficient functionals are determined up to constants.

Thus, we find the 4-dimensional effective action with KK corrections as

$$
\begin{aligned}
S_{\text{eff}} = & \int d^4x\sqrt{-g}\left[\frac{1}{2}\Psi R - \frac{3}{4(1-\Psi)}\nabla^\mu\Psi\nabla_\mu\Psi - \lambda - \lambda\gamma(1-\Psi)^2\right] \\
& + \ell^2\int d^4x\sqrt{-g}\Bigg[\frac{1}{4(1-\Psi)^4}\left(\nabla^\mu\Psi\nabla_\mu\Psi\right)^2 \\
& + \frac{1}{(1-\Psi)^2}(\Box\Psi)^2 + \frac{1}{(1-\Psi)^3}\nabla^\mu\Psi\nabla_\mu\Psi\Box\Psi \\
& + \frac{2}{3(1-\Psi)}R\Box\Psi + \frac{1}{3(1-\Psi)^2}R\nabla^\mu\Psi\nabla_\mu\Psi \\
& + jR^2 + kR^{\mu\nu}R_{\mu\nu}\Bigg],
\end{aligned}
\tag{8.104}
$$

where constants j and k can be interpreted as the variety of the effects of the bulk gravitational waves. These constants have the same origin as the previous parameters α and β. It should be noted that this action becomes non-local after integrating out the radion field. This fits the fact that KK effects are non-local usually. In principle, we can continue this calculation to any order of derivatives.

8.8 The Answers

We have developed the low energy gradient expansion scheme to give insights into the physics of the braneworld such as the black hole physics and the cosmology. In particular, we have concentrated on the specific questions in this lecture. Here, we summarize our answers obtained by the gradient expansion method. Our understanding of RS braneworlds would be also useful for other brane models.

8.8.1 Single-Brane Model

Is the Einstein theory recovered even in the non-linear regime?

We have obtained the effective theory at the lowest order as

$$\overset{(4)}{G}{}^{\mu}{}_{\nu} = \frac{\kappa^2}{\ell} T^{\mu}{}_{\nu} - \frac{2}{\ell}\chi^{\mu}{}_{\nu}. \tag{8.105}$$

Here we have the correction $\chi_{\mu\nu}$ which can be interpreted as the dark radiation in the cosmological situation.

On the other hand, in the linearized gravity, the conventional Einstein theory is recovered at low energy. This is because the out-going boundary condition is imposed. In other words, the asymptotic AdS boundary condition is imposed. In the nonlinear case, this corresponds to the requirement that the dark radiation term $\chi_{\mu\nu}$ must be zero. For this boundary condition, the conventional Einstein theory is recovered. Hence, the standard Friedmann equation holds.

In this sense, the answer is yes.

How does the AdS/CFT come into the braneworld?

The CFT emerges as the constant of integration which satisfies the trace anomaly relation

$$\tau^{\mu}{}_{\mu} = -\frac{1}{8}\left(R^{\alpha}{}_{\beta}R^{\beta}{}_{\alpha} - \frac{1}{3}R^2\right) - \beta\Box R. \tag{8.106}$$

This constant can not be determined a priori. Here, the AdS/CFT correspondence could come into the braneworld. Namely, if we identify some CFT with $\tau_{\mu\nu}$, then we can determine the boundary condition.

How are the AdS/CFT and geometrical approach related?

The key quantity in the geometric approach is obtained as

$$\overset{(2)}{E}{}^{\mu}{}_{\nu} = \ell^2 \left[P^{\mu}{}_{\nu} + 2\tau^{\mu}{}_{\nu} + 2\alpha S^{\mu}{}_{\nu} + \frac{2}{3}\beta \mathcal{K}^{\mu}{}_{\nu} \right]. \tag{8.107}$$

The above expression contains $\tau_{\mu\nu}$ which can be interpreted as the CFT matter. Hence, once we know $E_{\mu\nu}$, no enigma remains. In particular, $P_{\mu\nu} \approx \pi_{\mu\nu}$ is independent of the $\tau_{\mu\nu}$. In odd dimensions, there exists no trace anomaly, but $P_{\mu\nu}$ exists. In 4-dimensions, $\pi^{\mu}{}_{\mu}$ accidentally coincides with the trace anomaly in CFT.

It is interesting to note that the high energy and the Weyl term corrections found in the geometrical approach merge into the CFT matter correction found in the AdS/CFT approach.

8.8.2 Two-Brane Model

How is the geometrical approach consistent with the Brans-Dicke picture?

In the geometrical approach, no radion seems to appear. On the other hand, the linear theory predicts the radion as the crucial quantity. The resolution can be attained by obtaining $E_{\mu\nu}$ ($\chi_{\mu\nu}$ in our notation). The resultant expression

$$\chi^{\mu}{}_{\nu} = -\frac{\kappa^2(1-\Psi)}{2\Psi}\left(\overset{\oplus}{T}{}^{\mu}{}_{\nu} + (1-\Psi)\overset{\ominus}{T}{}^{\mu}{}_{\nu} \right) - \frac{\ell}{2\Psi}\left[\left(\Psi^{|\mu}{}_{|\nu} - \delta^{\mu}_{\nu}\Psi^{|\alpha}{}_{|\alpha} \right) + \frac{3}{2(1-\Psi)}\left(\Psi^{|\mu}\Psi_{|\nu} - \frac{1}{2}\delta^{\mu}_{\nu}\Psi^{|\alpha}\Psi_{|\alpha} \right) \right]$$

contains the radion in an intriguing way. The dark radiation consists of the radion and the matter.

We have shown that the radion transforms the Einstein theory with Weyl correction into the conformally coupled scalar-tensor theory where the radion plays the role of the scalar field. Thus, it turned out that the radion is hidden by the projected Weyl tensor $E_{\mu\nu}$ in the geometrical approach.

What replaces the AdS/CFT correspondence in the two-brane model?

In the case of the single-brane model, the out-going boundary condition at the Cauchy horizon is assumed. This conforms to AdS/CFT correspondence. Indeed, the continuum KK-spectrum are projected on the brane as CFT matter.

On the other hand, the boundary condition in the two-brane system allows only the discrete KK-spectrum. Hence, we can not expect CFT matter on the brane. Instead, the radion controls the bulk/brane correspondence in two-brane model. In fact, the higher derivative terms of the radion mimics the effect of the bulk geometry (KK-effect) as we have shown explicitly. Hence, the conventional AdS/CFT correspondence does not exist. Instead, there exists the

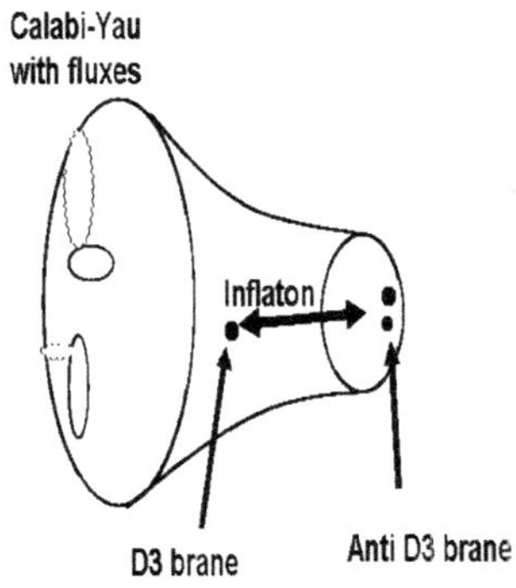

Fig. 8.3 The warped throat is attached to Calabi-Yau manifold. Anti-D3-branes are stacked at the tip of the warped throat. A D3-brane can move in the throat and the radion, i.e. the distance between the D3-brane and anti-D3-branes, plays a role of the inflaton

AdS/CFT correspondence realized by the conformally coupled radion. The conformal coupling can be regarded as a reflection of the symmetry of the bulk geometry.

Here, I would like to mention D-brane inflation models proposed in [67]. There, the inflaton is identified as the radion which is the distance between a D3-brane and anti-D3-branes as is shown in Fig. 8.3. Naively, it seems possible to realize a slow roll inflation due to the warped geometry. However, our result of two-brane system suggests the existence of the conformal coupling of the radion, which ruins the slow role inflation of the model. In fact, the curvature coupling gives a large mass of the inflaton which causes the notorious eta problem. Hence, a fine tuning is unavoidable.

8.9 AdS/CFT in Dilatonic Braneworld

In the previous sections, we have considered RS braneworlds. If we take into account bulk fields, the pure AdS bulk would not be expected. Hence, it is interesting to see the role of AdS/CFT correspondence in those cases. In this section, we will consider a bulk scalar field and call this kind of models dilatonic braneworlds [68].

In addition to the above theoretical interest, there is an phenomenological interest in dilatonic braneworlds. Let us see how the inflationary universe can be realized in the braneworld. The formula for the effective cosmological constant in the braneworld reads

$$\Lambda_{\mathrm{eff}} = \frac{\kappa^4 \sigma^2}{12} - \frac{3}{\ell^2}, \tag{8.108}$$

where σ and ℓ are the tension of the brane and the curvature scale in the bulk which is determined by the bulk vacuum energy, respectively. For $\kappa^2 \sigma = 6/\ell$, we have Minkowski spacetime. In order to obtain the inflationary universe, we need the positive effective cosmological constant. In the braneworld model, there are two possibilities. One is to increase the brane tension and the other is to increase ℓ. The brane tension can be controlled by the scalar field on the brane. The bulk curvature scale ℓ can be controlled by the bulk scalar field. The former case is a natural extension of the 4-dimensional inflationary scenario. The latter possibility is a novel one peculiar to the brane model. Recall that, in the superstring theory, scalar fields are ubiquitous. Indeed, the dilaton and moduli exists in the bulk generically, because they arise as the modes associated with the closed string. Moreover, when the supersymmetry is spontaneously broken, they may have the non-trivial potential. Hence, it is natural to consider the inflationary scenario driven by these fields [69, 70]. Therefore, dilatonic braneworlds are phenomenologically interesting. In this section, we would like to discuss this dilatonic braneworld from the point of view of the AdS/CFT correspondence.

8.9.1 Dilatonic Braneworld

We consider a S_1/Z_2 orbifold spacetime with the two branes as the fixed points. In this first Randall-Sundrum (RS1) model, the two flat 3-branes are embedded in AdS_5 and the brane tensions given by $\overset{\oplus}{\sigma} = 6/(\kappa^2\ell)$ and $\overset{\ominus}{\sigma} = -6/(\kappa^2\ell)$. Our system is described by the action

$$S = \frac{1}{2\kappa^2}\int d^5x\sqrt{-g}\mathcal{R} - \int d^5x\sqrt{-g}\left[\frac{1}{2}g^{AB}\partial_A\varphi\partial_B\varphi + U(\varphi)\right] - \sum_{i=\oplus,\ominus}\overset{i}{\sigma}\int d^4x\sqrt{-g^{i-\mathrm{brane}}} + \sum_{i=\oplus,\ominus}\int d^4x\sqrt{-g^{i-\mathrm{brane}}}\,\mathcal{L}^i_{\mathrm{matter}}, \tag{8.109}$$

where $g^{i-\mathrm{brane}}_{\mu\nu}$ and $\overset{i}{\sigma}$ are the induced metric and the brane tension on the i-brane, respectively. We assume the potential $U(\varphi)$ for the bulk scalar field takes the form $U(\varphi) = -\frac{6}{\kappa^2\ell^2} + V(\varphi)$, where the first term is regarded as a 5-dimensional cosmological constant and the second term is an arbitrary potential function. The brane tension σ is tuned so that the effective cosmological constant on the brane vanishes. The above setup realizes a flat braneworld after inflation ends and the field φ reaches the minimum of its potential.

Inflation in the braneworld can be driven by a scalar field either on the brane or in the bulk. We derive the effective equations of motion which are useful for both models. Here, we begin with the single-brane system. Since we know the effective 4-dimensional equations hold irrespective of the existence of other branes [47], the analysis of the single-brane system is sufficient to derive the effective action for the two-brane system as we see in the next subsection.

We again adopt the Gaussian normal coordinate system to describe the geometry of the brane model; $ds^2 = dy^2 + g_{\mu\nu}(y, x^\mu)dx^\mu dx^\nu$, where the brane is assumed to be located at $y = 0$. Let us decompose the extrinsic curvature into the traceless part $\Sigma_{\mu\nu}$ and the trace part K as $K_{\mu\nu} = -\frac{1}{2}g_{\mu\nu,y} = \Sigma_{\mu\nu} + \frac{1}{4}g_{\mu\nu}K$. Then, we can obtain the basic equations off the brane using these variables. First, the Hamiltonian constraint equation leads to

$$\frac{3}{4}K^2 - \Sigma^\alpha{}_\beta \Sigma^\beta{}_\alpha = \overset{(4)}{R} - \kappa^2 \nabla^\alpha \varphi \nabla_\alpha \varphi + \kappa^2 (\partial_y \varphi)^2 - 2\kappa^2 U(\varphi), \tag{8.110}$$

where $\overset{(4)}{R}$ is the curvature on the brane and ∇_μ denotes the covariant derivative with respect to the metric $g_{\mu\nu}$. Momentum constraint equation becomes

$$\nabla_\lambda \Sigma_\mu{}^\lambda - \frac{3}{4}\nabla_\mu K = -\kappa^2 \partial_y \varphi \partial_\mu \varphi. \tag{8.111}$$

Evolution equation in the direction of y is given by

$$\Sigma^\mu{}_{\nu,y} - K\Sigma^\mu{}_\nu = -\left[\overset{(4)}{R}{}_{\mu\nu} - \kappa^2 \nabla^\mu \varphi \nabla_\nu \varphi\right]_{\text{traceless}}. \tag{8.112}$$

Finally, the equation of motion for the scalar field reads

$$\partial_y^2 \varphi - K\partial_y \varphi + \nabla^\alpha \nabla_\alpha \varphi - U'(\varphi) = 0, \tag{8.113}$$

where the prime denotes derivative with respect to the scalar field φ.

As we have the singular source at the brane position, we must consider the junction conditions. Assuming a Z_2 symmetry of spacetime, we obtain the junction conditions for the metric and the scalar field

$$\left[\Sigma^\mu{}_\nu - \frac{3}{4}\delta^\mu_\nu K\right]\Big|_{y=0} = -\frac{\kappa^2}{2}\sigma\delta^\mu_\nu + \frac{\kappa^2}{2}T^\mu{}_\nu, \tag{8.114}$$

$$\left[\partial_y \varphi\right]\big|_{y=0} = 0, \tag{8.115}$$

where $T^\mu{}_\nu$ is the energy-momentum tensor for the matter fields on the brane.

8.9.2 AdS/Radion Correspondence

We assume the inflation occurs at low energy in the sense that the additional energy due to the bulk scalar field is small, $\kappa^2 \ell^2 V(\varphi) \ll 1$, and the curvature on the brane R is also small, $R\ell^2 \ll 1$. It should be stressed that the low energy does not necessarily implies weak gravity on the brane. Under these circumstances, we can use a gradient expansion scheme to solve the bulk equations of motion.

At zeroth order, we ignore matters on the brane. Then, from the junction condition (8.114), we have

$$\left[\overset{(0)}{\Sigma}{}^{\mu}{}_{\nu} - \frac{3}{4}\delta^{\mu}_{\nu}\overset{(0)}{K}\right]\Big|_{y=0} = -\frac{\kappa^2}{2}\sigma\delta^{\mu}_{\nu}. \tag{8.116}$$

As the right hand side of (8.116) contains no traceless part, we get $\overset{(0)}{\Sigma}{}^{\mu}{}_{\nu} = 0$. We also take the potential for the bulk scalar field $U(\varphi)$ to be $-6/(\kappa^2\ell^2)$. We discard the terms with 4-dimensional derivatives since one can neglect the long wavelength variation in the direction of x^μ at low energies. Thus, the equations to be solved are given by

$$\frac{3}{4}\overset{(0)}{K}{}^2 = \kappa^2(\partial_y\overset{(0)}{\varphi})^2 + \frac{12}{\ell^2}, \tag{8.117}$$

$$\partial_y^2\overset{(0)}{\varphi} - \overset{(0)}{K}\partial_y\overset{(0)}{\varphi} = 0. \tag{8.118}$$

The junction condition (8.115) at this order $\left[\partial_y\overset{(0)}{\varphi}\right]|_{y=0}=0$ tells us that the solution of Eq. 8.118 must be $\overset{(0)}{\varphi} = \eta(x^\mu)$, where $\eta(x^\mu)$ is an arbitrary constant of integration. Now, the solution of Eq. 8.118 yields $\overset{(0)}{K} = 4/\ell$. Other Eqs. 8.111 and 8.112 are trivially satisfied at zeroth order. Using the definition $\overset{(0)}{K}_{\mu\nu} = -\overset{(0)}{g}_{\mu\nu,y}/2$, we have the lowest order metric

$$\overset{(0)}{g}_{\mu\nu}(y,x^\mu) = b^2(y)h_{\mu\nu}(x^\mu), \quad b(y) \equiv e^{-y/\ell}, \tag{8.119}$$

where the induced metric on the brane, $h_{\mu\nu} \equiv g_{\mu\nu}(y=0,x^\mu)$, arises as a constant of integration. The junction condition for the induced metric (8.116) merely implies well known relation $\kappa^2\sigma = 6/\ell$ and that for the scalar field (8.115) is trivially satisfied. At this leading order analysis, we can not determine the constants of integration $h_{\mu\nu}(x^\mu)$ and $\eta(x^\mu)$ which are constant as far as the short length scale ℓ variations are concerned, but are allowed to vary over the long wavelength scale. These constants should be constrained by the next order analysis.

Now, we take into account the effect of both the bulk scalar field and the matter on the brane perturbatively. Our iteration scheme is to write the metric $g_{\mu\nu}$ and the scalar field φ as a sum of local tensors built out of the induced metric and the induced scalar field on the brane, in the order of expansion parameters, that is, $O((R\ell^2)^n)$ and $O(\kappa^2\ell^2V(\varphi))^n$, $n = 0,1,2,\cdots$ [47]. Then, we expand the metric and the scalar field as

$$\begin{aligned} g_{\mu\nu}(y,x^\mu) &= b^2(y)\left[h_{\mu\nu}(x^\mu) + \overset{(1)}{g}_{\mu\nu}(y,x^\mu) + \overset{(2)}{g}_{\mu\nu}(y,x^\mu) + \cdots\right], \\ \varphi(y,x^\mu) &= \eta(x^\mu) + \overset{(1)}{\varphi}(y,x^\mu) + \overset{(2)}{\varphi}(y,x^\mu) + \cdots. \end{aligned} \tag{8.120}$$

Here, we put the boundary conditions $\overset{(i)}{g}_{\mu\nu}(y=0,x^\mu)=0\,,\,\overset{(i)}{\varphi}(y=0,x^\mu)=0\,,$ $i=1,2,3,\ldots$ so that we can interpret $h_{\mu\nu}$ and η as induced quantities. Extrinsic curvatures can be also expanded as

$$K=\frac{4}{\ell}+\overset{(1)}{K}+\overset{(2)}{K}+\cdots\,,\qquad \Sigma^\mu_\nu=\overset{(1)}{\Sigma}{}^\mu_\nu+\overset{(2)}{\Sigma}{}^\mu_\nu+\cdots. \tag{8.121}$$

Using the formula such as $\overset{(4)}{R}(\overset{(0)}{g}_{\mu\nu})=R(h_{\mu\nu})/b^2$, we obtain the solution

$$\overset{(1)}{K}=\frac{\ell}{6b^2}\left(R(h)-\kappa^2\eta^{|\alpha}\eta_{|\alpha}\right)-\frac{\ell}{3}\kappa^2V(\eta), \tag{8.122}$$

where $R(h)$ is the scalar curvature of $h_{\mu\nu}$ and $|$ denotes the covariant derivative with respect to $h_{\mu\nu}$. Substituting the results at zeroth order solutions into Eq. 8.112, we obtain

$$\overset{(1)}{\Sigma}{}^\mu{}_\nu=\frac{\ell}{2b^2}\left[R^\mu{}_\nu(h)-\kappa^2\eta^{|\mu}\eta_{|\nu}\right]_{\text{traceless}}+\frac{\chi^\mu{}_\nu}{b^4}, \tag{8.123}$$

where $R^\mu{}_\nu(h)$ denotes the Ricci tensor of $h_{\mu\nu}$ and $\chi^\mu{}_\nu$ is a constant of integration which satisfies the constraint $\chi^\mu{}_\mu=0$. Hereafter, we omit the argument of the curvature for simplicity. Integrating the scalar field equation (8.113) at first order, we have

$$\partial_y\overset{(1)}{\varphi}=\frac{\ell}{2b^2}\Box\eta-\frac{\ell}{4}V'(\eta)+\frac{C}{b^4}, \tag{8.124}$$

where C is also a constant of integration. At first order in this iteration scheme, we get two kinds of constants of integration, $\chi^\mu{}_\nu$ and C.

Given the matter fields $T_{\mu\nu}$ on the brane, the junction condition (8.114) becomes

$$\left[\overset{(1)}{\Sigma}{}^\mu{}_\nu-\frac{3}{4}\delta^\mu_\nu\overset{(1)}{K}\right]\Big|_{y=0}=\frac{\kappa^2}{2}T^\mu{}_\nu. \tag{8.125}$$

At this order, the junction condition (8.115) yields

$$\left[\partial_y\overset{(1)}{\varphi}\right]\Big|_{y=0}=0. \tag{8.126}$$

These junction conditions give the effective equations of motion on the brane.

Now, we are in a position to discuss the effective equations of motion for the dilatonic two-brane models. The point is the fact that the equations of motion on each brane take the same form if we use the induced metric on each brane [47]. The effective Einstein equations on each positive ($\oplus$) and negative ($\ominus$) tension brane at low-energies yield

$$G^{\mu}{}_{\nu}(h) = \kappa^2\left(\eta^{|\mu}\eta_{|\nu} - \frac{1}{2}\delta^{\mu}_{\nu}\eta^{|\alpha}\eta_{|\alpha} - \frac{1}{2}\delta^{\mu}_{\nu}V\right) - \frac{2}{\ell}\chi^{\mu}{}_{\nu} + \frac{\kappa^2}{\ell}\overset{\oplus}{T}{}^{\mu}{}_{\nu}, \tag{8.127}$$

$$G^{\mu}{}_{\nu}(f) = \kappa^2\left(\eta^{;\mu}\eta_{;\nu} - \frac{1}{2}\delta^{\mu}_{\nu}\eta^{;\alpha}\eta_{;\alpha} - \frac{1}{2}\delta^{\mu}_{\nu}V\right) - \frac{2}{\ell}\frac{\chi^{\mu}{}_{\nu}}{\Omega^4} - \frac{\kappa^2}{\ell}\overset{\ominus}{T}{}^{\mu}{}_{\nu}. \tag{8.128}$$

where $f_{\mu\nu}$ is the induced metric on the negative tension brane and ; denotes the covariant derivative with respect to $f_{\mu\nu}$. When we set the position of the positive tension brane at $y = 0$, that of the negative tension brane $\bar{y}$ in general depends on x^{μ}, i.e. $\bar{y} = \bar{y}(x^{\mu})$. Hence, the warp factor at the negative tension brane $\Omega(x^{\mu}) \equiv b(\bar{y}(x))$ also depends on x^{μ}. Because the metric always comes into equations with derivatives, the zeroth order relation is enough in this first order discussion. Hence, the metric on the positive tension brane is related to the metric on the negative tension brane as $f_{\mu\nu} = \Omega^2 h_{\mu\nu}$.

Although Eqs. 8.127 and 8.128 are non-local individually, with undetermined $\chi^{\mu}{}_{\nu}$, one can combine both equations to reduce them to local equations for each brane. We can therefore easily eliminate $\chi^{\mu}{}_{\nu}$ from Eqs. 8.127 and 8.128, since $\chi^{\mu}{}_{\nu}$ appears only algebraically. Eliminating $\chi^{\mu}{}_{\nu}$ from both Eqs. 8.127 and 8.128, we obtain

$$\begin{aligned} G^{\mu}{}_{\nu} &= \frac{\kappa^2}{\ell\Psi}\overset{\oplus}{T}{}^{\mu}{}_{\nu} + \frac{\kappa^2(1-\Psi)^2}{\ell\Psi}\overset{\ominus}{T}{}^{\mu}{}_{\nu} \\ &+ \frac{1}{\Psi}\left[\Psi^{|\mu}{}_{|\nu} - \delta^{\mu}_{\nu}\Psi^{|\alpha}{}_{|\alpha} + \frac{3}{2}\frac{1}{1-\Psi}\left(\Psi^{|\mu}\Psi_{|\nu} - \frac{1}{2}\delta^{\mu}_{\nu}\Psi^{|\alpha}\Psi_{|\alpha}\right)\right] \\ &+ \kappa^2\left(\eta^{|\mu}\eta_{|\nu} - \frac{1}{2}\delta^{\mu}_{\nu}\eta^{|\alpha}\eta_{|\alpha} - \delta^{\mu}_{\nu}V_{\text{eff}}\right)s, \quad V_{\text{eff}} = \frac{2-\Psi}{2}V, \end{aligned} \tag{8.129}$$

where we defined a new field $\Psi = 1 - \Omega^2$ which we refer to by the name "radion". The bulk scalar field induces the energy-momentum tensor of the conventional 4-dimensional scalar field with the effective potential which depends on the radion.

We can also determine the dark radiation $\chi^{\mu}{}_{\nu}$ by eliminating $G^{\mu}{}_{\nu}(h)$ from Eqs. 8.127 and 8.128,

$$\begin{aligned} \frac{2}{\ell}\chi^{\mu}{}_{\nu} &= -\frac{1}{\Psi}\left[\Psi^{|\mu}{}_{|\nu} - \delta^{\mu}_{\nu}\Psi^{|\alpha}{}_{|\alpha} + \frac{3}{2}\frac{1}{1-\Psi}\left(\Psi^{|\mu}\Psi_{|\nu} - \frac{1}{2}\delta^{\mu}_{\nu}\Psi^{|\alpha}\Psi_{|\alpha}\right)\right] \\ &+ \frac{\kappa^2}{2}(1-\Psi)\delta^{\mu}_{\nu}V - \frac{\kappa^2}{\ell}\frac{1-\Psi}{\Psi}\left[\overset{\oplus}{T}{}^{\mu}{}_{\nu} + (1-\Psi)\overset{\ominus}{T}{}^{\mu}{}_{\nu}\right]. \end{aligned} \tag{8.130}$$

Due to the property $\chi^{\mu}{}_{\mu} = 0$, we have

$$\Box\Psi = \frac{\kappa^2}{3\ell}(1-\Psi)\left[\overset{\oplus}{T} + (1-\Psi)\overset{\ominus}{T}\right] - \frac{1}{2(1-\Psi)}\Psi^{|\alpha}\Psi_{|\alpha} - \frac{2\kappa^2}{3}\Psi(1-\Psi)V. \tag{8.131}$$

Note that Eqs. 8.129 and 8.131 are derived from a scalar-tensor type theory coupled to the additional scalar field.

Similarly, the equations for the scalar field on branes become

$$\Box_h \eta - \frac{V'}{2} + \frac{2}{\ell} C = 0, \tag{8.132}$$

$$\Box_f \eta - \frac{V'}{2} + \frac{2}{\ell} \frac{C}{\Omega^4} = 0, \tag{8.133}$$

where the subscripts refer to the induced metric on each brane. Notice that the scalar field takes the same value for both branes at this order. Eliminating the dark source C from these Eqs. 8.132 and 8.133, we find the equation for the scalar field takes the form

$$\Box_h \eta - V'_{\text{eff}} = -\frac{\Psi^{|\mu}}{\Psi} \eta_{|\mu}. \tag{8.134}$$

Notice that the radion acts as a source for η. And we can also get the dark source as

$$\frac{2}{\ell} C = -\frac{V'}{2}(1 - \Psi) + \frac{\Psi^{|\mu}}{\Psi} \eta_{|\mu}. \tag{8.135}$$

Now the effective action for the positive tension brane which gives Eqs. 8.129, 8.131 and 8.134 can be read off as

$$\begin{aligned} S = \frac{\ell}{2\kappa^2} \int d^4x \sqrt{-h} \left[\Psi R - \frac{3}{2(1-\Psi)} \Psi^{|\alpha} \Psi_{|\alpha} - \kappa^2 \Psi \left(\eta^{|\alpha} \eta_{|\alpha} + 2 V_{\text{eff}}(\eta, \Psi) \right) \right] \\ + \int d^4x \sqrt{-h} \overset{\oplus}{\mathcal{L}} + \int d^4x \sqrt{-h} (1-\Psi)^2 \overset{\ominus}{\mathcal{L}}, \end{aligned} \tag{8.136}$$

where the last two terms represent actions for the matter on each brane. Thus, we found the radion field couples with the induced metric and the induced scalar field on the brane non-trivially. Surprisingly, at this order, the nonlocality of $\chi_{\mu\nu}$ and C are eliminated by the radion. We see the radion has a conformal coupling. However, in the present case, the radion couples to the dilaton field which breaks a conformal invariance. Hence, this gives non-conformal holography.

As this is a closed system, we can analyze a primordial spectrum to predict the cosmic background fluctuation spectrum [71]. Interestingly, χ^{μ}_{ν} and C vanishes in the single brane limit, $\Psi \to 1$, as can be seen from (8.130) and (8.135). The dynamics is simply governed by Einstein theory with the single scalar field. Therefore, we can conclude that the bulk inflaton can drive inflation when the slow role conditions are satisfied.

8.9.3 AdS/CFT and KK Corrections: Single-Brane Cases

It would be important to take into account the KK effects as corrections to the leading order result. It can be accomplished in the single-brane models. Using our approach, in the single brane limit, we can deduce the effective action with KK corrections as [47, 72] (see also [73, 74])

$$S = \frac{\ell}{2\kappa^2}\int d^4x\sqrt{-h}\left[\left(1+\frac{\ell^2}{12}\kappa^2 V\right)R - \kappa^2\left(1+\frac{\ell^2}{12}\kappa^2 V - \frac{\ell^2}{4}V''\right)\eta^{|\alpha}\eta_{|\alpha}\right.$$
$$\left. -2\kappa^2 V_{\text{eff}} - \frac{\ell^2}{4}\left(R^{\alpha\beta}R_{\alpha\beta} - \frac{1}{3}R^2\right)\right] + \int d^4x\sqrt{-h}\mathcal{L}_{\text{matter}} + S_{\text{CFT}}, \tag{8.137}$$

where the last term comes from the energy-momentum tensor of CFT matter $\tau_{\mu\nu}$ and the effective potential at this order is defined by

$$V_{\text{eff}} = \frac{1}{2}V + \frac{\ell^2\kappa^2}{48}V^2 - \frac{\ell^2}{64}V'^2. \tag{8.138}$$

It is interesting to note that the effective potential contains the terms which looks like F-terms in supersymmetric models.

Thus, even in the dilatonic braneworld, the AdS/CFT correspondence seems to play an important role in the single-brane case.

8.10 Conclusion

In this lecture, I have reviewed the gradient expansion method in the context of braneworlds. Using the formalism, I have tried to explain how the AdS/CFT correspondence is related to the braneworld models.

In the case of the RS single-brane model, we clarified when the conventional Einstein equations hold at low energy. Moreover, we revealed the relation between the geometrical and the AdS/CFT correspondence approach using the gradient expansion method. We have shown that the high energy and the Weyl term corrections found in the geometrical approach correspond to the CFT matter corrections found in the AdS/CFT approach.

In the case of the RS two-brane sysytem, we showed that the AdS/CFT correspondence plays an important role in the sense that the low energy effective field theory can be described by the conformally coupled scalar-tensor theory where the radion plays the role of the scalar field. We also presented the symmetry method to derive KK corrections in the two-brane system.

These effective theories for RS braneworlds can be used to make cosmological predictions. More importantly, it turned out that the gradient expansion method provides a unified view of RS braneworlds.

We have also considered the bulk scalar field with a nontrivial potential and derived the non-linear low energy effective action for the dilatonic two-brane model using the gradient expansion method. As a result, we have shown that the effective theory reduces to the scalar-tensor theory with the non-trivial coupling between the radion and the bulk scalar field. Since the radion has a conformal coupling, the conformal symmetry is relevant even for the dilatonic braneworlds. In this sense, the AdS/CFT correspondence is related to dilatonic braneworlds. However, the radion couples to the scalar field which is non-conformal. Hence, the conformal invariance is violated.

Our phenomenological motivation to consider dilatonic braneworlds was a possibility of the bulk inflaton. Concerning to this issue, taking into account the fact that $\chi_{\mu\nu}$ and C becomes zero when two branes get separated infinitely, one can conclude that the bulk inflaton can drive the inflation on the brane as far as the slow roll conditions are satisfied. We also obtained KK corrections in the single brane limit which contain CFT corrections.

These results tell us that there exist profound relations between braneworlds and the AdS/CFT correspondence, although the correspondence is slightly deformed in the dilatonic cases.

In this lecture, we have considered only codimesion-one braneworlds. It is important to extend the analysis to higher codimension models [75–84]. It is intriguing to study a role of the AdS/CFT correspondence in these higher codimension braneworlds.

Acknowledgments I would like to thank Sugumi Kanno for collaborations on which this lecture is based and useful comments on the manuscript. I am grateful to the organizers of the 5th Aegean Summer school, especially Eleftherios Papantonopoulos, for inviting me to give a lecture and their kind hospitality during the summer school. The present work is supported by the Japan-U.K. Research Cooperative Program, Grant-in-Aid for Scientific Research Fund of the Ministry of Education, Science and Culture of Japan No. 18540262, Grant-in-Aid for Scientific Research on Innovative Area No. 21111006 and the Grant-in-Aid for the Global COE Program "The Next Generation of Physics, Spun from Universality and Emergence".

References

1. Polchinski, J.: String Theory I and II. Cambridge University Press, Cambridge (1998)
2. Horava, P., Witten, E.: Nucl. Phys. B **460**, 506 [arXiv:hep-th/9510209] (1996)
3. Horava, P., Witten, E.: Nucl. Phys. B **475**, 94 [arXiv:hep-th/9603142] (1996)
4. Lukas, A., Ovrut, B.A., Stelle, K.S., Waldram, D.: Phys. Rev. D **59**, 086001 [arXiv:hep-th/9803235] (1999)
5. Antoniadis, I., Arkani-Hamed, N., Dimopoulos, S., Dvali, G.R.: Phys. Lett. B **436**, 257 [arXiv:hep-ph/9804398] (1998)
6. Arkani-Hamed, N., Dimopoulos, S., Dvali, G.R.: Phys. Rev. D **59**, 086004 [arXiv:hep-ph/9807344] (1999)
7. Akama, K.: Lect. Notes Phys. **176**, 267 [arXiv:hep-th/0001113] (1982)
8. Rubakov, V.A., Shaposhnikov, M.E.: Phys. Lett. B **125**, 136 (1983)
9. Randall, L., Sundrum, R.: Phys. Rev. Lett. **83**, 3370 [arXiv:hep-ph/9905221] (1999)

10. Randall, L., Sundrum, R.: Phys. Rev. Lett **83**, 4690 [arXiv:hep-th/9906064] (1999)
11. Kanno, S., Soda, J.: TSPU Vestnik **44N7**, 15 [arXiv:hep-th/0407184] (2004)
12. Tomita, K.: Prog. Theor. Phys. **54**, 730 (1975)
13. Salopek, D.S., Stewart, J.M.: Phys. Rev. D **47**, 3235 (1993)
14. Comer, G.L., Deruelle, N., Langlois, D., Parry, J.: Phys. Rev. D **49**, 2759 (1994)
15. Soda, J., Ishihara, H., Iguchi, O.: Prog. Theor. Phys. **94**, 781 [arXiv:gr-qc/9509008] (1995)
16. Brax, P., van de Bruck, C., Davis, A.C.: . Prog. Phys. **67**, 2183 [arXiv:hep-th/0404011] (2004)
17. Binetruy, P., Deffayet, C., Ellwanger, U., Langlois, D.: Phys. Lett. B **477**, 285 [arXiv:hep-th/9910219] (2000)
18. Kraus, P.: JHEP **9912**, 011 [arXiv:hep-th/9910149] (1999)
19. Flanagan, E.E., Tye, S.H.H., Wasserman, I.: Phys. Rev. D **62**, 044039 [arXiv:hep-ph/9910498] (2000)
20. Ida, D.: JHEP **0009**, 014 [arXiv:gr-qc/9912002] (2000)
21. Kaloper, N.: Phys. Rev. D **60**, 123506 [arXiv:hep-th/9905210] (1999)
22. Mukohyama, S.: Phys. Lett. B **473**, 241 [arXiv:hep-th/9911165] (2000)
23. Maartens, R.: Living Rev. Rel. **7**, 7 [arXiv:gr-qc/0312059] (2004)
24. Maartens, R.: [arXiv:astro-ph/0402485]
25. Koyama, K., Soda, J.: Phys. Rev. D **62**, 123502 [arXiv:hep-th/0005239] (2000)
26. Garriga, J., Tanaka, T.: Phys. Rev. Lett. **84**, 2778 [arXiv:hep-th/9911055] (2000)
27. Charmousis, C., Gregory, R., Rubakov, V.A.: Phys. Rev. D **62**, 067505 [arXiv:hep-th/9912160] (2000)
28. Maldacena, J.M.: Adv. Theor. Math. Phys. **2**, 231 (1998)
29. Maldacena, J.M.: Int. J. Theor. Phys. **38**, 1113 [arXiv:hep-th/9711200] (1999)
30. Gubser, S.S., Klebanov, I.R., Polyakov, A.M.: Phys. Lett. B **428**, 105 [arXiv:hep-th/9802109] (1998)
31. Witten, E.: Adv. Theor. Math. Phys. **2**, 253 [arXiv:hep-th/9802150] (1998)
32. de Haro, S., Skenderis, K., Solodukhin, S.N.: Class. Quant. Grav. **18**, 3171 [arXiv:hep-th/0011230] (2001)
33. Gubser, S.S.: Phys. Rev. D **63**, 084017 [arXiv:hep-th/9912001] (2001)
34. Anchordoqui, L., Nunez, C., Olsen, K.: JHEP **0010**, 050 [arXiv:hep-th/0007064] (2000)
35. de Haro, S., Solodukhin, S.N., Skenderis, K.: Commun. Math. Phys. **217**, 595 [arXiv:hep-th/0002230] (2001)
36. Shiromizu, T., Maeda, K., Sasaki, M.: Phys. Rev. D **62**, 024012 [arXiv:gr-qc/9910076] (2000)
37. Koyama, K., Soda, J.: JHEP **0105**, 027 [arXiv:hep-th/0101164] (2001)
38. Nojiri, S., Odintsov, S.D.: Phys. Lett. B **484**, 119 [arXiv:hep-th/0004097] (2000)
39. Nojiri, S., Odintsov, S.D.: [arXiv:hep-th/0105068]
40. Nojiri, S., Odintsov, S.D., Zerbini, S.: Phys. Rev. D **62**, 064006 [arXiv:hep-th/0001192] (2000)
41. Kiritsis, E.: JCAP **0510**, 014 [arXiv:hep-th/0504219] (2005)
42. Tanaka T.: [arXiv:gr-qc/0402068]
43. Shiromizu, T., Ida, D.: Phys. Rev. D **64**, 044015 [arXiv:hep-th/0102035] (2001)
44. Kanno, S., Soda, J.: Phys. Rev. D **66**, 043526 [arXiv:hep-th/0205188] (2002)
45. Kanno, S., Soda, J.: Phys. Rev. D **66**, 083506 [arXiv:hep-th/0207029] (2002)
46. Kanno, S., Soda, J.: Astrophys. Space Sci. **283**, 481 [arXiv:gr-qc/0209087] (2003)
47. Kanno, S., Soda, J.: Gen. Rel. Grav. **36**, 689 [arXiv:hep-th/0303203] (2004)
48. Shiromizu, T., Koyama, K., Onda, S., Torii, T.: Phys. Rev. D **68**, 063506 [arXiv:hep-th/0305253] (2003)
49. Onda, S., Shiromizu, T., Koyama, K., Hayakawa, S.: Phys. Rev. D **69**, 123503 [arXiv:hep-th/0311262] (2004)
50. Shiromizu, T., Torii, T., Uesugi, T.: Phys. Rev. D **67**, 123517 [arXiv:hep-th/0302223] (2003)
51. Takahashi, K., Shiromizu, T.: Phys. Rev. D **70**, 103507 [arXiv:hep-th/0408043] (2004)
52. Koyama, K.: Phys. Rev. Lett. **91**, 221301 [arXiv:astro-ph/0303108] (2003)

53. Kanno, S., Soda, J.: Phys. Lett. B **588**, 203 [arXiv:hep-th/0312106] (2004)
54. Wiseman, T.: Class. Quant. Grav. **19**, 3083 [arXiv:hep-th/0201127] (2002)
55. Shiromizu, T., Koyama, K.: Phys. Rev. D **67**, 084022 [arXiv:hep-th/0210066] (2003)
56. Brax, P., van de Bruck, C., Davis, A.C., Rhodes, C.S.: [arXiv:hep-th/0209158]
57. Chiba, T.: Phys. Rev. D **62**, 021502 [arXiv:gr-qc/0001029] (2000)
58. Kobayashi, S., Koyama, K.: JHEP **0212**, 056 [arXiv:hep-th/0210029] (2002)
59. Kanno, S., Sasaki, M., Soda, J.: Prog. Theor. Phys. **109**, 357 [arXiv:hep-th/0210250] (2003)
60. Kanno, S., Soda, J., Wands D.: JCAP **0508**, 002 [arXiv:hep-th/0506167] (2005)
61. Khoury, J., Ovrut, B.A., Steinhardt, P.J., Turok, N.: Phys. Rev. D **64**, 123522 [arXiv:hep-th/0103239] (2001)
62. Steinhardt, P.J., Turok, N.: Science **296**: 1436 (2002)
63. Kanno, S., Soda, J.: Phys. Rev. D **71**, 044031 [arXiv:hep-th/0410061] (2005)
64. Kanno, S.:Phys. Rev. D **72**, 024009 [arXiv:hep-th/0504087] (2005)
65. Brax, P., van de Bruck, C., Davis, A.C., Rhodes, C.S.: Phys. Rev. D **67**, 023512 [arXiv:hep-th/0209158] (2003)
66. McFadden, P.L., Turok, N.G.: Phys. Rev. D **71**, 086004 [arXiv:hep-th/0412109] (2005)
67. Kachru, S., Kallosh, R., Linde, A.D., Maldacena, J.M., McAllister, L.P., Trivedi S.P.: JCAP **0310**, 013 [arXiv:hep-th/0308055] (2003)
68. Maeda, K.i., Wands, D.: Phys. Rev. D **62**, 124009 [arXiv:hep-th/0008188] (2000)
69. Kobayashi, S., Koyama, K., Soda, J.: Phys. Lett. B **501**, 157 [arXiv:hep-th/0009160] (2001)
70. Himemoto, Y., Sasaki, M.: Phys. Rev. D **63**, 044015 [arXiv:gr-qc/0010035] (2001)
71. Soda, J., Kanno, S.: Gen. Rel. Grav. **37**, 1621 [arXiv:gr-qc/0508083] (2005)
72. Kanno, S., Soda, J.: Gen. Rel. Grav. **36**, 689 [arXiv:hep-th/0303203] (2004)
73. Soda, J., Kanno, S.: Astrophys. Space Sci. **283**, 639 [arXiv:gr-qc/0209086] (2003)
74. Soda, J., Kanno, S.: [arXiv:gr-qc/0410066]
75. Cline, J.M., Descheneau, J., Giovannini, M., Vinet, J.: JHEP **0306**, 048 [arXiv:hep-th/0304147] (2003)
76. Burgess, C.P., Quevedo, F., Tasinato, G., Zavala, I.: JHEP **0411**, 069 [arXiv:hep-th/0408109] (2004)
77. Vinet, J., Cline, J.M.: Phys. Rev. D **71**, 064011 [arXiv:hep-th/0501098] (2005)
78. Bostock, P., Gregory, R., Navarro, I., Santiago, J.: Phys. Rev. Lett. **92**, 221601 [arXiv:hep-th/0311074] (2004)
79. Kanno, S., Soda, J.: JCAP **0407**, 002 [arXiv:hep-th/0404207] (2004)
80. Charmousis, C., Kofinas, G., Papazoglou, A.: [arXiv:0907.1640 [hep-th]]
81. Papantonopoulos, E., Papazoglou, A., Zamarias, V.: Nucl. Phys. B **797**, 520 [arXiv:0707.1396 [hep-th]] (2008)
82. Papantonopoulos, E.: [arXiv:gr-qc/0601011]
83. Kanno, S., Langlois, D., Sasaki, M., Soda, J.: Prog. Theor. Phys. **118**, 701 [arXiv:0707.4510 [hep-th]] (2007)
84. Chen, F., Cline, J.M., Kanno, S.: Phys. Rev. D **77**, 063531 [arXiv:0801.0226 [hep-th]] (2008)

Part III
Condensed Matter and the AdS/CFT Correspondence

Chapter 9
Condensed Matter and AdS/CFT

Subir Sachdev

Abstract I review two classes of strong coupling problems in condensed matter physics, and describe insights gained by application of the AdS/CFT correspondence. The first class concerns non-zero temperature dynamics and transport in the vicinity of quantum critical points described by relativistic field theories. I describe how relativistic structures arise in models of physical interest, present results for their quantum critical crossover functions and magneto-thermoelectric hydrodynamics. The second class concerns symmetry breaking transitions of two-dimensional systems in the presence of gapless electronic excitations at isolated points or along lines (i.e. Fermi surfaces) in the Brillouin zone. I describe the scaling structure of a recent theory of the Ising-nematic transition in metals, and discuss its possible connection to theories of Fermi surfaces obtained from simple AdS duals.

9.1 Introduction

The past couple of decades have seen vigorous theoretical activity on the quantum phases and phase transitions of correlated electron systems in two spatial dimensions. Much of this work has been motivated by the cuprate superconductors, but the list of interesting materials continues to increase unabated [1].

Methods from field theory have had a strong impact on much of this work. Indeed, they have become part of the standard toolkit of condensed matter physicists. In these lectures, I focus on two classes of strong-coupling problems which have not yielded accurate solutions via the usual arsenal of field-theoretic methods. I will also discuss how the AdS/CFT correspondence, discovered by string

S. Sachdev (✉)
Department of Physics, Harvard University, Cambridge, MA 02138, USA
e-mail: sachdev@physics.harvard.edu

E. Papantonopoulos (ed.), *From Gravity to Thermal Gauge Theories: The Ads/CFT Correspondence,* Lecture Notes in Physics, 828,
DOI: 10.1007/978-3-642-04864-7_9,

theorists, has already allowed substantial progress on some of these problems, and offers encouraging prospects for future progress.

The first class of strong-coupling problems are associated with the real-time, finite temperature behavior of strongly interacting quantum systems, especially those near quantum critical points. Field-theoretic or numerical methods often allow accurate determination of the *zero* temperature or of *imaginary time* correlations at non-zero temperatures. However, these methods usual fail in the real-time domain at non-zero temperatures, particularly at times greater than $\hbar/k_BT$, where T is the absolute temperature. In systems near quantum critical points the natural scale for correlations is $\hbar/k_BT$ itself, and so lowering the temperature in a numerical study does not improve the situation.

The second class of strong-coupling problems arise near two-dimensional quantum critical points with fermionic excitations. When the fermions have a massless Dirac spectrum, with zero excitation energy at a finite number of points in the Brillouin zone, conventional field-theoretic methods do allow significant progress. However, in metallic systems, the fermionic excitations have zeros along a line in the Brillouin zone (the Fermi surface), allowing a plethora of different low energy modes. Metallic quantum critical points play a central role in many experimental systems, but the interplay between the critical modes and the Fermi surface has not been fully understood (even at zero temperature). Readers interested only in this second class of problems can jump ahead to Sect. 9.7.

These lectures will start with a focus on the first class of strong-coupling problems. We will begin in Sect. 9.2 by introducing a variety of model systems and their quantum critical points; these are motivated by recent experimental and theoretical developments. We will use these systems to introduce basic ideas on the finite temperature crossovers near quantum critical points in Sect. 9.3. In Sect. 9.4, we will focus on the important *quantum critical region* and present a general discussion of its transport properties. An important recent development has been the complete exact solution, via the AdS/CFT correspondence, of the dynamic and transport properties in the quantum critical region of a variety of (supersymmetric) model systems in two and higher dimensions: this will be described in Sect. 9.5. The exact solutions are found to agree with the earlier general ideas discussed here in Sect. 9.4. As has often been the case in the history of physics, the existence of a new class of solvable models leads to new and general insights which apply to a much wider class of systems, almost all of which are not exactly solvable. This has also been the case here, as we will review in Sect. 9.6: a hydrodynamic theory of the low frequency transport properties has been developed, and has led to new relations between a variety of thermo-electric transport co-efficients.

The latter part of these lectures will turn to the second class of strong coupling problems, by describing the role of fermions near quantum critical points. In Sect. 9.7 we will consider some simple symmetry breaking transitions in d-wave superconductors. Such superconductors have fermionic excitations with a massless Dirac spectrum, and we will show how they become critical near the quantum phase transition. We will review how the field-theoretic $1/N$ expansion does allow solution of a large class of such problems. Finally, in Sect. 9.8 we will consider phase transitions of metallic systems with Fermi surfaces. We will discuss how the $1/N$ expansion fails here, and review the results of recent work involving the AdS/CFT correspondence.

Some portions of the discussions below have been adapted from other review articles by the author [2, 3].

9.2 Model Systems and Their Critical Theories

9.2.1 Coupled Dimer Antiferromagnets

Some of the best studied examples of quantum phase transitions arise in insulators with unpaired $S = 1/2$ electronic spins residing on the sites, i, of a regular lattice. Using S_i^a $(a = x, y, z)$ to represent the spin $S = 1/2$ operator on site i, the low energy spin excitations are described by the Heisenberg exchange Hamiltonian

$$H_J = \sum_{i<j} J_{ij} S_i^a \cdot S_j^a + \cdots \tag{9.1}$$

where $J_{ij} > 0$ is the antiferromagnetic exchange interaction. We will begin with a simple realization of this model is illustrated in Fig. 9.1. The $S = 1/2$ spins reside on the sites of a square lattice, and have nearest neighbor exchange equal to either J or J/λ. Here $\lambda \geq 1$ is a tuning parameter which induces a quantum phase transition in the ground state of this model.

At $\lambda = 1$, the model has full square lattice symmetry, and this case is known to have a Néel ground state which breaks spin rotation symmetry. This state has a checkerboard polarization of the spins, just as found in the classical ground state, and as illustrated on the left side of Fig. 9.1. It can be characterized by a vector order parameter φ^a which measures the staggered spin polarization

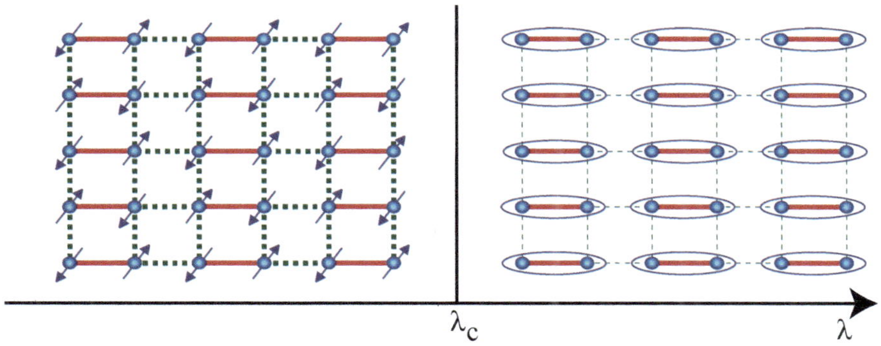

Fig. 9.1 The coupled dimer antiferromagnet. The full red lines represent an exchange interaction J, while the dashed green lines have exchange J/λ. The ellipses represent a singlet valence bond of spins $(|\uparrow\downarrow\rangle - |\downarrow\uparrow\rangle)/\sqrt{2}$

$$\varphi^a = \eta_i S_i^a \tag{9.2}$$

where $\eta_i = \pm 1$ on the two sublattices of the square lattice. In the Néel state we have $\langle \varphi^a \rangle \neq 0$, and we expect that the low energy excitations can be described by long wavelength fluctuations of a field $\varphi^a(x, \tau)$ over space, x, and imaginary time τ.

On the other hand, for $\lambda \gg 1$ it is evident from Fig. 9.1 that the ground state preserves all symmetries of the Hamiltonian: it has total spin $S = 0$ and can be considered to be a product of nearest neighbor singlet valence bonds on the J links. It is clear that this state cannot be smoothly connected to the Néel state, and so there must at least one quantum phase transition as a function λ.

Extensive quantum Monte Carlo simulations [4–6] on this model have shown there is a direct phase transition between these states at a critical λ_c, as in Fig. 9.1. The value of λ_c is known accurately, as are the critical exponents characterizing a second-order quantum phase transition. These critical exponents are in excellent agreement with the simplest proposal for the critical field theory, [6] which can be obtained via conventional Landau-Ginzburg arguments. Given the vector order parameter φ^a, we write down the action in d spatial and one time dimension,

$$\mathcal{S}_{LG} = \int d^d r d\tau \left[\frac{1}{2}\left[(\partial_\tau \varphi^a)^2 + v^2 (\nabla \varphi^a)^2 + s(\varphi^a)^2 \right] + \frac{u}{4}\left[(\varphi^a)^2 \right]^2 \right], \tag{9.3}$$

as the simplest action expanded in gradients and powers of φ^a which is consistent will all the symmetries of the lattice antiferromagnet. The transition is now tuned by varying $s \sim (\lambda - \lambda_c)$. Notice that this model is identical to the Landau-Ginzburg theory for the thermal phase transition in a $d + 1$ dimensional ferromagnet, because time appears as just another dimension. As an example of the agreement: the critical exponent of the correlation length, ν, has the same value, $\nu = 0.711\ldots$, to three significant digits in a quantum Monte Carlo study of the coupled dimer antiferromagnet, [6] and in a 5-loop analysis [7] of the renormalization group fixed point of $\mathcal{S}_{LG}$ in $d = 2$. Similar excellent agreement is obtained for the double-layer antiferromagnet [8, 9] and the coupled-plaquette antiferromagnet [10].

In experiments, the best studied realization of the coupled-dimer antiferromagnet is $TlCuCl_3$. In this crystal, the dimers are coupled in all three spatial dimensions, and the transition from the dimerized state to the Néel state can be induced by application of pressure. Neutron scattering experiments by Ruegg and collaborators [11] have clearly observed the transformation in the excitation spectrum across the transition, and these observations are in good quantitative agreement with theory [1].

9.2.2 Deconfined Criticality

We now consider an analog of transition discussed in Sect. 9.2.1, but for a Hamiltonian $H = H_0 + \lambda H_1$ which has full square lattice symmetry at all λ. For H_0, we choose a form of H_J, with $J_{ij} = J$ for all nearest neighbor links. Thus at

$\lambda = 0$ the ground state has Néel order, as in the left panel of Fig. 9.1. We now want to choose H_1 so that increasing λ leads to a spin singlet state with spin rotation symmetry restored. A large number of choices have been made in the literature, and the resulting ground state invariably [12] has valence bond solid (VBS) order; a VBS state has been observed in the organic antiferromagnet $\mathrm{EtMe_3P[Pd(dmit)_2]_2}$ [13, 14]. The VBS state is superficially similar to the dimer singlet state in the right panel of Fig. 9.1: the spins primarily form valence bonds with near-neighbor sites. However, because of the square lattice symmetry of the Hamiltonian, a columnar arrangement of the valence bonds as in Fig. 9.1, breaks the square lattice rotation symmetry; there are 4 equivalent columnar states, with the valence bond columns running along different directions. More generally, a VBS state is a spin singlet state, with a non-zero degeneracy due to a spontaneously broken lattice symmetry. Thus a direct transition between the Néel and VBS states involves two distinct broken symmetries: spin rotation symmetry, which is broken only in the Néel state, and a lattice rotation symmetry, which is broken only in the VBS state. The rules of Landau-Ginzburg-Wilson theory imply that there can be no generic second-order transition between such states.

It has been argued that a second-order Néel-VBS transition can indeed occur [15], but the critical theory is not expressed directly in terms of either order parameter. It involves a fractionalized bosonic spinor $z_\alpha (\alpha = \uparrow, \downarrow)$, and an emergent gauge field A_μ. The key step is to express the vector field φ^a in terms of z_α by

$$\varphi^a = z_\alpha^* \sigma^a_{\alpha\beta} z_\beta \tag{9.4}$$

where σ^a are the 2×2 Pauli matrices. Note that this mapping from φ^a to z_α is redundant. We can make a spacetime-dependent change in the phase of the z_α by the field $\theta(x, \tau)$

$$z_\alpha \to e^{i\theta} z_\alpha \tag{9.5}$$

and leave φ^a unchanged. All physical properties must therefore also be invariant under Eq. 9.5, and so the quantum field theory for z_α has a U(1) gauge invariance, much like that found in quantum electrodynamics. The effective action for the z_α therefore requires introduction of an ‘emergent’ U(1) gauge field A_μ (where $\mu = x, \tau$ is a three-component spacetime index). The field A_μ is unrelated the electromagnetic field, but is an internal field which conveniently describes the couplings between the spin excitations of the antiferromagnet. As we did for $\mathcal{S}_{LG}$, we can write down the quantum field theory for z_α and A_μ by the constraints of symmetry and gauge invariance, which now yields

$$\mathcal{S}_z = \int d^2 r d\tau \left[|(\partial_\mu - iA_\mu) z_\alpha|^2 + s|z_\alpha|^2 + u(|z_\alpha|^2)^2 + \frac{1}{2g^2} (\epsilon_{\mu\nu\lambda} \partial_\nu A_\lambda)^2 \right] \tag{9.6}$$

For brevity, we have now used a “relativistically” invariant notation, and scaled away the spin-wave velocity v; the values of the couplings s, u are different from, but related to, those in $\mathcal{S}_{LG}$. The Maxwell action for A_μ is generated from short distance z_α fluctuations, and it makes A_μ a dynamical field; its coupling g is unrelated to the electron charge. The action $\mathcal{S}_z$ is a valid description of the Néel state for $s<0$ (the critical upper value of s will have fluctuation corrections away from 0), where the gauge theory enters a Higgs phase with $\langle z_\alpha \rangle \neq 0$. This description of the Néel state as a Higgs phase has an analogy with the Weinberg-Salam theory of the weak interactions—in the latter case it is hypothesized that the condensation of a Higgs boson gives a mass to the W and Z gauge bosons, whereas here the condensation of z_α quenches the A_μ gauge boson. As written, the $s > 0$ phase of $\mathcal{S}_z$ is a 'spin liquid' state with a $S = 0$ collective gapless excitation associated with the A_μ photon. Non-perturbative effects [12] associated with the monopoles in A_μ (not discussed here), show that this spin liquid is ultimately unstable to the appearance of VBS order.

Numerical studies of the Néel-VBS transition have focussed on a specific lattice antiferromagnet proposed by Sandvik [16–19]. There is strong evidence for VBS order proximate to the Néel state, along with persuasive evidence of a second-order transition. However, some studies [20, 21] support a very weak first order transition.

9.2.3 Graphene

The last few years have seen an explosion in experimental and theoretical studies [22] of graphene: a single hexagonal layer of carbon atoms. At the currently observed temperatures, there is no evident broken symmetry in the electronic excitations, and so it is not conventional to think of graphene as being in the vicinity of a quantum critical point. However, graphene does indeed undergo a bona fide quantum phase transition, but one without any order parameters or broken symmetry. This transition may be viewed as being 'topological' in character, and is associated with a change in nature of the Fermi surface as a function of carrier density.

Pure, undoped graphene has a conical electronic dispersion spectrum at two points in the Brillouin zone, with the Fermi energy at the particle-hole symmetric point at the apex of the cone. So there is no Fermi surface, just a Fermi point, where the electronic energy vanishes, and pure graphene is a 'semi-metal'. By applying a gate voltage, the Fermi energy can move away from this symmetric point, and a circular Fermi surface develops, as illustrated in Fig. 9.2. The Fermi surface is electron-like for one sign of the bias, and hole-like for the other sign. This change from electron to hole character as a function of gate voltage constitutes the quantum phase transition in graphene. As we will see below, with regard to its dynamic properties near zero bias, graphene behaves in almost all respects like a canonical quantum critical system.

The field theory for graphene involves fermionic degrees of freedom. Representing the electronic orbitals near one of the Dirac points by the two-component

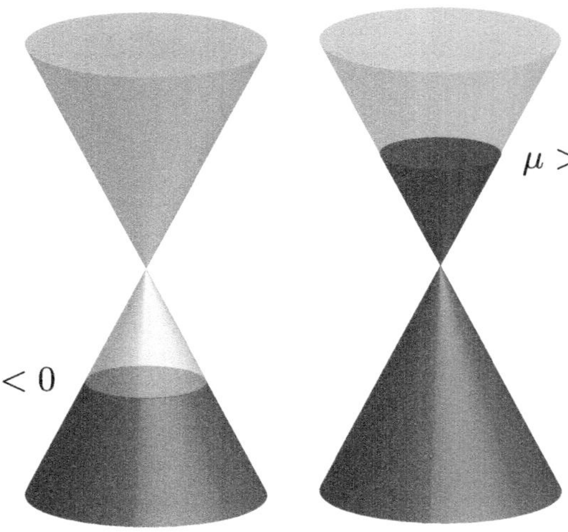

Fig. 9.2 Dirac dispersion spectrum for graphene showing a 'topological' quantum phase transition from a hole Fermi surface for $\mu < 0$ to a electron Fermi surface for $\mu > 0$

fermionic spinor Ψ_s, where s is a sublattice index (we suppress spin and 'valley' indices), we have the effective electronic action

$$\begin{aligned} \mathcal{S}_\Psi = & \int d^2r \int d\tau\, \Psi_s^\dagger \left[(\partial_\tau + iA_\tau - \mu)\delta_{ss'} + iv_F \tau^x_{ss'} \partial_x + iv_F \tau^y_{ss'} \partial_y \right] \Psi_{s'} \\ & + \frac{1}{2g^2} \int \frac{d^2q}{4\pi^2} \int d\tau\, \frac{q}{2\pi} |A_\tau(\mathbf{q}, \tau)|^2, \end{aligned} \qquad (9.7)$$

where $\tau^i_{ss'}$ are Pauli matrices in the sublattice space, μ is the chemical potential, v_F is the Fermi velocity, and A_τ is the scalar potential mediating the Coulomb interaction with coupling $g^2 = e^2/\epsilon$ (ϵ is a dielectric constant). This theory undergoes a quantum phase transition as a function of μ, at $\mu = 0$, similar in many ways to that of $\mathcal{S}_{LG}$ as a function of s. The interaction between the fermionic excitations here has coupling g^2, which is the analog of the non-linearity u in $\mathcal{S}_{LG}$. The strength of the interactions is determined by the dimensionless 'fine structure constant' $\alpha = g^2/(\hbar v_F)$ which is of order unity in graphene. While u flows to a non-zero fixed point value under the renormalization group, α flows logarithmically slowly to zero. For many purposes, it is safe to ignore this flow, and to set α equal to a fixed value.

9.3 Finite Temperature Crossovers

The previous section has described four model systems at $T = 0$: we examined the change in the nature of the ground state as a function of some tuning parameter, and motivated a quantum field theory which describes the low energy excitations on both sides of the quantum critical point.

We now turn to the important question of the physics at non-zero temperatures. All of the models share some common features, which we will first explore for the coupled dimer antiferromagnet. For $\lambda > \lambda_c$ (or $s > 0$ in $\mathcal{S}_{LG}$), the excitations consist of a triplet of $S = 1$ particles (the 'triplons'), which can be understood perturbatively in the large λ expansion as an excited $S = 1$ state on a dimer, hopping between dimers (see Fig. 9.3). The mean field theory tells us that the excitation energy of this dimer vanishes as $\sqrt{s}$ upon approaching the quantum critical point. Fluctuations beyond mean field, described by $\mathcal{S}_{LG}$, show that the exponent is modified to $s^{z\nu}$, where $z = 1$ is the dynamic critical exponent, and ν is the correlation length exponent. Now imagine turning on a non-zero temperature. As long as T is smaller than the triplon gap, i.e. $T < s^{z\nu}$, we expect a description in terms of a dilute gas of thermally excited triplon particles. This leads to the behavior shown on the right-hand-side of Fig. 9.3, delimited by the crossover indicted by the dashed line. Note that the crossover line approaches $T = 0$ only at the quantum critical point.

Now let us look a the complementary behavior at $T > 0$ on the Néel-ordered side of the transition, with $s < 0$. In two spatial dimensions, thermal fluctuations prohibit the breaking of a non-Abelian symmetry at all $T > 0$, and so spin rotation symmetry is immediately restored. Nevertheless, there is an exponentially large spin correlation length, ξ, and at distances shorter than ξ we can use the ordered ground state to understand the nature of the excitations. Along with the spin-waves, we also found the longitudinal 'Higgs' mode with energy $\sqrt{-2s}$ in mean field theory. Thus, just as was this case for $s > 0$, we expect this spin-wave+Higgs picture to apply at all temperatures lower than the natural energy scale; i.e. for $T < (-s)^{z\nu}$. This leads to the crossover boundary shown on the left-hand-side of Fig. 9.3.

Having delineated the physics on the two sides of the transition, we are left with the crucial *quantum critical* region in the center of Fig. 9.3. This is present for

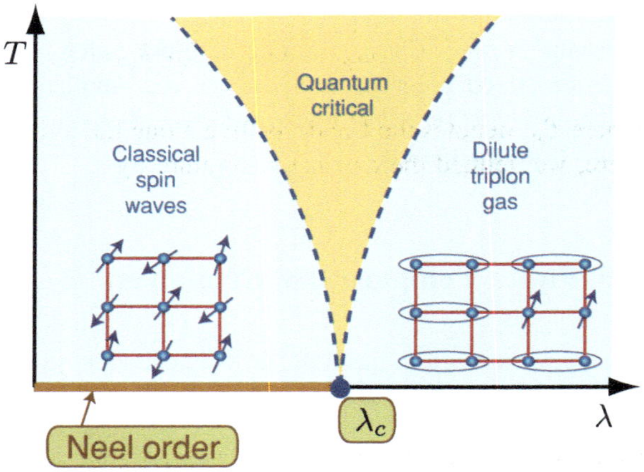

Fig. 9.3 Finite temperature crossovers of the coupled dimer antiferromagnet in Fig. 9.1

$T > |s|^{z\nu}$, i.e. at *higher* temperatures in the vicinity of the quantum critical point. To the left of the quantum critical region, we have a description of the dynamics and transport in terms of an effectively classical model of spin waves: this is the 'renormalized classical' regime of [23]. To the right of the quantum critical region, we again have a regime of classical dynamics, but now in terms of a Boltzmann equation for the triplon particles. A key property of quantum critical region is that there is no description in terms of either classical particles or classical waves at the times of order the typical relaxation time, τ_r, of thermal excitations. Instead, quantum and thermal effects are equally important, and involve the non-trivial dynamics of the fixed-point theory describing the quantum critical point. Note that while the fixed-point theory applies only at a single point ($\lambda = \lambda_c$) at $T = 0$, its influence broadens into the quantum critical region at $T > 0$. Because there is no characteristing energy scale associated with the fixed-point theory, $k_B T$ is the only energy scale available to determine τ_r at non-zero temperatures. Thus, in the quantum critical region [24, 25]

$$\tau_r = \mathcal{C} \frac{\hbar}{k_B T} \tag{9.8}$$

where $\mathcal{C}$ is a universal constant dependent only upon the universality class of the fixed point theory *i.e.* it is universal number just like the critical exponents. This value of τ_r determines the 'friction coefficients' associated with the dissipative relaxation of spin fluctuations in the quantum critical region. It is also important for the transport co-efficients associated with conserved quantities, and this will be discussed in Sect. 9.4

Let us now consider the similar $T > 0$ crossovers for the other models of Sect. 9.2.

The Néel-VBS transition of Sect. 9.2.2 has crossovers very similar to those in Fig. 9.3, with one important difference. The VBS state breaks a discrete lattice symmetry, and this symmetry remains broken for a finite range of non-zero temperatures. Thus, within the right-hand 'triplon gas' regime of Fig. 9.3, there is a phase transition line at a critical temperature T_{VBS}. The value of T_{VBS} vanishes very rapidly as $s \searrow 0$, and is controlled by the non-perturbative monopole effects which were briefly noted in Sect. 9.2.2.

For graphene, the discussion above applied to Fig. 9.2 leads to the crossover diagram shown in Fig. 9.4, as noted by Sheehy and Schmalian [26]. We have the Fermi liquid regimes of the electron- and hole-like Fermi surfaces on either side of the critical point, along with an intermediate quantum critical Dirac liquid. A new feature here is related to the logarithmic flow of the dimensionless 'fine structure constant' α controlling the Coulomb interactions, which was noted in Sect. 9.2.3. In the quantum critical region, this constant takes the typical value $\alpha \sim 1/\ln(1/T)$. Consequently for the relaxation time in Eq. 9.8 we have $\mathcal{C} \sim \ln^2(1/T)$. This time determines both the width of the electron spectral functions, and also the transport co-efficients, as we will see in Sect. 9.4.

9.4 Quantum Critical Transport

We now turn to the 'transport' properties in the quantum critical region: we consider the response functions associated with any globally conserved quantity. For the antiferromagnetic systems in Sect. 9.2.1 and 9.2.2, this requires consideration of the transport of total spin, and the associated spin conductivities and diffusivities. For graphene, we can consider charge and momentum transport. Our discussion below will also apply to the superfluid-insulator transition: for bosons in a periodic potential, this transition is described [27] by a field theory closely related to that in Eq. 9.3. However, we will primarily use a language appropriate to charge transport in graphene below. We will describe the properties of a generic strongly-coupled quantum critical point and mention, where appropriate, the changes due to the logarithmic flow of the coupling in graphene.

In traditional condensed matter physics, transport is described by identifying the low-lying excitations of the quantum ground state, and writing down 'transport equations' for the conserved charges carried by them. Often, these excitations have a particle-like nature, such as the 'triplon' particles of Fig. 9.3 or the electron or hole quasiparticles of the Fermi liquids in Fig. 9.4. In other cases, the low-lying excitations are waves, such as the spin-waves in Fig. 9.3, and their transport is described by a non-linear wave equation (such as the Gross-Pitaevski equation). However, as we have discussed in Sect. 9.3 neither description is possible in the quantum critical region, because the excitations do not have a particle-like or wave-like character.

Despite the absence of an intuitive description of the quantum critical dynamics, we can expect that the transport properties should have a universal character determined by the quantum field theory of the quantum critical point. In addition to describing single excitations, this field theory also determines the S-matrix of these excitations by the renormalization group fixed-point value of the couplings, and these should be sufficient to determine transport properties [28].

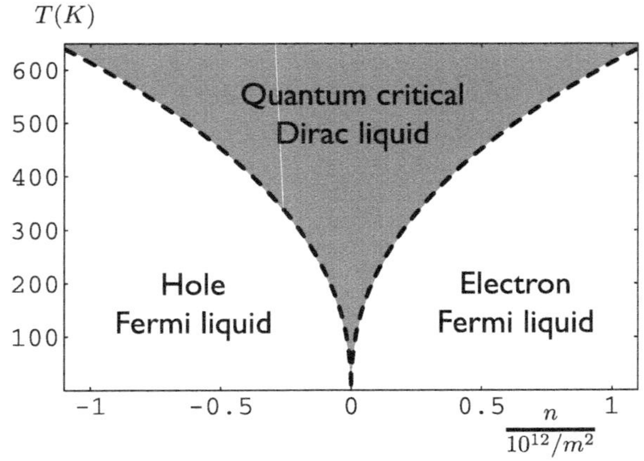

Fig. 9.4 Finite temperature crossovers of graphene as a function of electron density n (which is tuned by μ in Eq. 9.7) and temperature, T. Adapted from [26]

The transport co-efficients, and the relaxation time to local equilibrium, are not proportional to a mean free scattering time between the excitations, as is the case in the Boltzmann theory of quasiparticles. Such a time would typically depend upon the interaction strength between the particles. Rather, the system behaves like a "perfect fluid" in which the relaxation time is as short as possible, and is determined universally by the absolute temperature, as indicated in Eq. 9.8. Indeed, it was conjectured in [29] that the relaxation time in Eq. 9.8 is a generic lower bound for interacting quantum systems. Thus the non-quantum-critical regimes of all the phase diagrams in Sect. 9.3 have relaxation times which are all longer than Eq. 9.8.

The transport co-efficients of this quantum-critical perfect fluid also do not depend upon the interaction strength, and can be connected to the fundamental constants of nature. In particular, the electrical conductivity, σ, is given by (in two spatial dimensions) [28]

$$\sigma_Q = \frac{e^{*2}}{h}\Phi_\sigma, \tag{9.9}$$

where Φ_σ is a universal dimensionless constant of order unity, and we have added the subscript Q to emphasize that this is the conductivity for the case of graphene with the Fermi level at the Dirac point (for the superfluid-insulator transition, this would correspond to bosons at integer filling) with no impurity scattering, and at zero magnetic field. Here e^* is the charge of the carriers: for a superfluid-insulator transition of Cooper pairs, we have $e^* = 2e$, while for graphene we have $e^* = e$. The renormalization group flow of the 'fine structure constant' α of graphene to zero at asymptotically low T, allows an exact computation in this case [30–32]: $\Phi_\sigma \approx 0.05\ln^2(1/T)$. For the superfluid-insulator transition, Φ_σ is T-independent (this is the generic situation with non-zero fixed point values of the interaction [33]) but it has only been computed [28, 29] to leading order in expansions in $1/N$ (where N is the number of order parameter components) and in $3-d$ (where d is the spatial dimensionality). However, both expansions are neither straightforward nor rigorous, and require a physically motivated resummation of the bare perturbative expansion to all orders. It would therefore be valuable to have exact solutions of quantum critical transport where the above results can be tested, and we turn to such solutions in the next section.

In addition to charge transport, we can also consider momentum transport. This was considered in the context of applications to the quark-gluon plasma [34]; application of the analysis of [28] shows that the viscosity, η, is given by

$$\frac{\eta}{s} = \frac{\hbar}{k_B}\Phi_\eta, \tag{9.10}$$

where s is the entropy density, and again Φ_η is a universal constant of order unity. The value of Φ_η has recently been computed [35] for graphene, and again has a logarithmic T dependence because of the marginally irrelevant interaction: $\Phi_\eta \approx 0.008\ln^2(1/T)$.

We conclude this section by discussing some subtle aspects of the physics behind the seemingly simple result quantum-critical in Eq. 9.9. For simplicity, we will consider the case of a "relativistically" invariant quantum critical point in 2 + 1 dimensions (such as the field theories of Sects. 9.2.1 and 9.2.2, but marginally violated by graphene, a subtlety we ignore below). Consider the retarded correlation function of the charge density, $\chi(k, \omega)$, where $k = |\mathbf{k}|$ is the wavevector, and ω is frequency; the dynamic conductivity, $\sigma(\omega)$, is related to χ by the Kubo formula,

$$\sigma(\omega) = \lim_{k \to 0} \frac{-i\omega}{k^2} \chi(k, \omega). \tag{9.11}$$

It was argued in [28] that despite the absence of particle-like excitations of the critical ground state, the central characteristic of the transport is a crossover from collisionless to collision-dominated transport. At high frequencies or low temperatures, the limiting form for χ reduces to that at $T = 0$, which is completely determined by relativistic and scale invariance and current conversion upto an overall constant

$$\chi(k, \omega) = \frac{e^{*2}}{h} K \frac{k^2}{\sqrt{v^2 k^2 - (\omega + i\eta)^2}}, \qquad \sigma(\omega) = \frac{e^{*2}}{h} K; \;\; \hbar\omega \gg k_B T, \tag{9.12}$$

where K is a universal number [36, 37]. However, phase-randomizing collisions are intrinsically present in any strongly interacting critical point (above one spatial dimension) and these lead to relaxation of perturbations to local equilibrium and the consequent emergence of hydrodynamic behavior. So at low frequencies, we have instead an Einstein relation which determines the conductivity with

$$\chi(k, \omega) = e^{*2} \chi_c \frac{Dk^2}{Dk^2 - i\omega}, \qquad \sigma(\omega) = e^{*2} \chi_c D = \frac{e^{*2}}{h} \Theta_1 \Theta_2; \;\; \hbar\omega \ll k_{BT}, \tag{9.13}$$

where χ_c is the compressibility and D is the charge diffusion constant. Quantum critical scaling arguments show that the latter quantities obey

$$\chi_c = \Theta_1 \frac{k_B T}{h^2 v^2}, \qquad D = \Theta_2 \frac{h v^2}{k_B T}, \tag{9.14}$$

where $\Theta_{1,2}$ are universal numbers. A large number of papers in the literature, particularly those on critical points in quantum Hall systems, have used the collisionless method of Eq. 9.12 to compute the conductivity. However, the correct d.c. limit is given by Eq. 9.13, and the universal constant in Eq. 9.9 is given by $\Phi_\sigma = \Theta_1 \Theta_2$. Given the distinct physical interpretation of the collisionless and collision-dominated regimes, we expect that $K \neq \Theta_1 \Theta_2$. This has been shown in a resummed perturbation expansion for a number of quantum critical points [29].

9.5 Exact Results for Quantum Critical Transport

The results of Sect. 9.4 were obtained by using physical arguments to motivate resummations of perturbative expansions. Here we shall support the ad hoc assumptions behind these results by examining an exactly solvable model of quantum critical transport.

The solvable model may be viewed as a generalization of the gauge theory in Eq. 9.6 to the maximal possible supersymmetry. In 2 + 1 dimensions, this is known as $\mathcal{N} = 8$ supersymmetry. Such a theory with the U(1) gauge group is free, and so we consider the non-Abelian Yang-Millis theory with a SU(N) gauge group. The resulting supersymmetric Yang-Mills (SYM) theory has only one coupling constant, which is the analog of the electric charge g in Eq. 9.6. The matter content is naturally more complicated than the complex scalar z_α in Eq. 9.6, and also involves relativistic Dirac fermions as in Eq. 9.7. However all the terms in the action for the matter fields are also uniquely fixed by the single coupling constant g. Under the renormalization group, it is believed that g flows to an attractive fixed point at a non-zero coupling $g = g^*$; the fixed point then defines a supersymmetric conformal field theory in 2 + 1 dimensions (a SCFT3), and we are interested here in the transport properties of this SCFT3.

A remarkable recent advance has been the exact solution of this SCFT3 in the $N \to \infty$ limit using the AdS/CFT correspondence [38]. The solution proceeds by a dual formulation as a four-dimensional supergravity theory on a spacetime with uniform negative curvature: anti-de Sitter space, or AdS_4. Remarkably, the solution is also easily extended to non-zero temperatures, and allows direct computation of the correlators of conserved charges in real time. At $T > 0$ a black hole appears in the gravity, resulting in an AdS-Schwarzschild spacetime, and T is also the Hawking temperature of the black hole; the real time solutions also extend to $T > 0$.

The results of a full computation [39] of the density correlation function, $\chi(k, \omega)$ are shown in Figs. 9.5 and 9.6. The most important feature of these results is that the expected limiting forms in the collisionless (Eq. 9.12) and collision-dominated (Eq. 9.13) are obeyed. Thus the results do display the collisionless to collision-dominated crossover at a frequency of order $k_B T/\hbar$, as was postulated in Sect. 9.4.

An additional important feature of the solution is apparent upon describing the full structure of both the density and current correlations. Using spacetime indices ($\mu, \nu = t, x, y$) we can represent these as the tensor $\chi_{\mu\nu}(\mathbf{k}, \omega)$, where the previously considered $\chi \equiv \chi_{tt}$. At $T > 0$, we do not expect $\chi_{\mu\nu}$ to be relativistically covariant, and so can only constrain it by spatial isotropy and density conservation. Introducing a spacetime momentum $p_\mu = (\omega, \mathbf{k})$, and setting the velocity $v = 1$, these two constraints lead to the most general form

$$\chi_{\mu\nu}(\mathbf{k}, \omega) = \frac{e^{*2}}{h} \sqrt{p^2} \Big(P^T_{\mu\nu} K^T(k, \omega) + P^L_{\mu\nu} K^L(k, \omega) \Big) \tag{9.15}$$

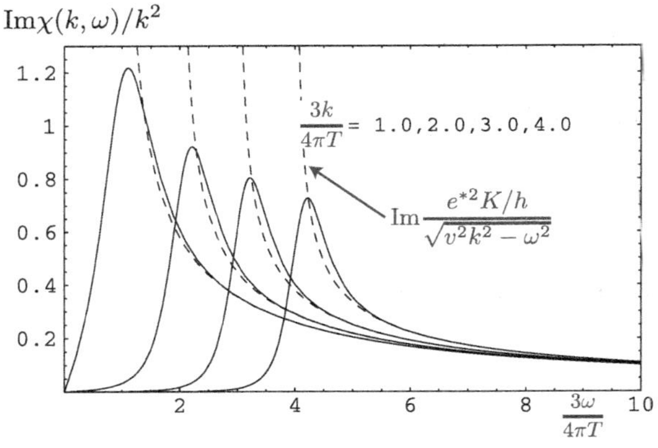

Fig. 9.5 Spectral weight of the density correlation function of the SCFT3 with $\mathcal{N} = 8$ supersymmetry in the collisionless regime

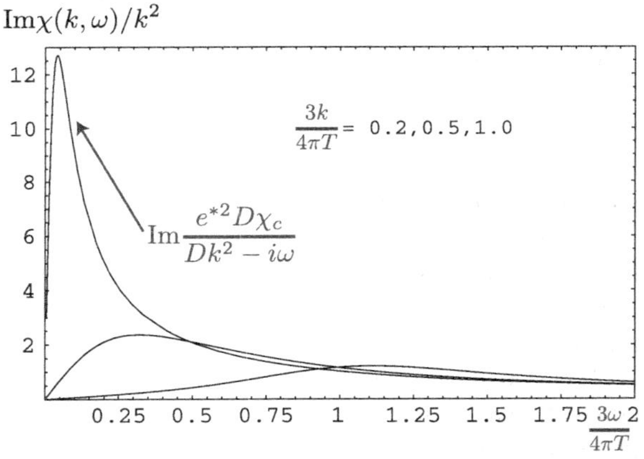

Fig. 9.6 As in Fig. 9.5, but for the collision-dominated regime

where $p^2 = \eta^{\mu\nu} p_\mu p_\nu$ with $\eta_{\mu\nu} = \text{diag}(-1, 1, 1)$, and $P^T_{\mu\nu}$ and $P^L_{\mu\nu}$ are orthogonal projectors defined by

$$P^T_{00} = P^T_{0i} = P^T_{i0} = 0, \; P^T_{ij} = \delta_{ij} - \frac{k_i k_j}{k^2}, \; P^L_{\mu\nu} = \left(\eta_{\mu\nu} - \frac{p_\mu p_\nu}{p^2}\right) - P^T_{\mu\nu}, \tag{9.16}$$

with the indices i, j running over the two spatial components. The two functions $K^{T,L}(k, \omega)$ define all the correlators of the density and the current, and the results in Eqs. 9.13 and 9.12 are obtained by taking suitable limits of these functions. We will also need below the general identity

$$K^T(0, \omega) = K^L(0, \omega), \tag{9.17}$$

which follows from the analyticity of the $T > 0$ current correlations at $\mathbf{k} = 0$.

The relations of the previous paragraph are completely general and apply to any theory. Specializing to the AdS-Schwarzschild solution of SYM3, the results were found to obey a simple and remarkable identity [39]:

$$K^L(k,\omega)K^T(k,\omega) = \mathcal{K}^2 \tag{9.18}$$

where $\mathcal{K}$ is a known pure number, independent of ω and k. It was also shown that such a relation applies to any theory which is equated to classical gravity on AdS_4, and is a consequence of the electromagnetic self-duality of its four-dimensional Maxwell sector. The combination of Eqs. 9.17 and 9.18 fully determines the $\chi_{\mu\nu}$ correlators at $\mathbf{k} = 0$: we find $K^L(0,\omega) = K^T(0,\omega) = \mathcal{K}$, from which it follows that the $\mathbf{k} = 0$ conductivity is frequency independent and that $\Phi_\sigma = \Theta_1\Theta_2 = K = \mathcal{K}$. These last features are believed to be special to theories which are equivalent to classical gravity, and not hold more generally.

We can obtain further insight into the interpretation of Eq. 9.18 by considering the field theory of the superfluid-insulator transition of lattice bosons at integer filling. As we noted earlier, this is given by the field theory in Eq. 9.3 with the field φ^a having two components. It is known that this two-component theory of relativistic bosons is equivalent to a dual relativistic theory, $\widetilde{\mathcal{S}}$ of vortices, under the well-known 'particle-vortex' duality [39, 40] considered the action of this particle-vortex duality on the correlation functions in Eq. 9.15, and found the following interesting relations:

$$K^L(k,\omega)\widetilde{K}^T(k,\omega) = 1, \quad K^T(k,\omega)\widetilde{K}^L(k,\omega) = 1 \tag{9.19}$$

where $\widetilde{K}^{L,T}$ determine the vortex current correlations in $\widetilde{\mathcal{S}}$ as in Eq. 9.15. Unlike Eq. 9.18, Eq. 9.19 does *not* fully determine the correlation functions at $\mathbf{k} = 0$: it only serves to reduce the four unknown functions $K^{L,T}$, $\widetilde{K}^{L,T}$ to two unknown functions. The key property here is that while the theories $\mathcal{S}_{LG}$ and $\widetilde{\mathcal{S}}$ are dual to each other, they are not equivalent, and the theory $\mathcal{S}_{LG}$ is not self-dual.

We now see that Eq. 9.18 implies that the classical gravity theory of SYM3 is self-dual under an analog of particle-vortex duality [39]. It is not expected that this self-duality will hold when quantum gravity corrections are included; equivalently, the SYM3 at finite N is expected to have a frequency dependence in its conductivity at $\mathbf{k} = 0$. If we apply the AdS/CFT correspondence to the superfluid-insulator transition, and approximate the latter theory by classical gravity on AdS_4, we immediately obtain the self-dual prediction for the conductivity, $\Phi_\sigma = 1$. This value is not far from that observed in numerous experiments, and we propose here that the AdS/CFT correspondence offers a rationale for understanding such observations.

9.6 Hydrodynamic Theory

The successful comparison between the general considerations of Sect. 9.4, and the exact solution using the AdS/CFT correspondence in Sect. 9.5, emboldens us to seek a more general theory of low frequency ($\hbar\omega \ll k_B T$) transport in the quantum critical regime. We will again present our results for the special case of a relativistic quantum critical point in 2 + 1 dimensions (a CFT3), but it is clear that similar considerations apply to a wider class of systems. Thus we can envisage applications to the superfluid-insulator transition, and have presented scenarios under which such a framework can be used to interpret measurements of the Nernst effect in the cuprates [41]. We have also described a separate set of applications to graphene [30–32]: while graphene is strictly not a CFT3, the Dirac spectrum of electrons leads to many similar results, especially in the inelastic collision-dominated regime associated with the quantum critical region. These results on graphene are reviewed in a separate paper [42], where explicit microscopic computations are also discussed.

Our idea is to relax the restricted set of conditions under which the results of Sect. 9.4 were obtained. We will work within the quantum critical regimes of the phase diagrams of Sect. 9.3 but now allow a variety of additional perturbations. First, we will move away from the particle-hole symmetric case, allow a finite density of carriers. For graphene, this means that μ is no longer pinned at zero; for the antiferromagnets, we can apply an external magnetic field; for the superfluid-insulator transition, the number density need not be commensurate with the underlying lattice. For charged systems, such as the superfluid-insulator transition or graphene, we allow application of an external magnetic field. Finally, we also allow a small density of impurities which can act as a sink of the conserved total momentum of the CFT3. In all cases, the energy scale associated with these perturbations is assumed to be smaller than the dominant energy scale of the quantum critical region, which is $k_B T$. The results presented below were obtained in two separate computations, associated with the methods described in Sects. 9.4 and 9.5, and are described in the two subsections below.

9.6.1 Relativistic Magnetohydrodynamics

With the picture of relaxation to local equilibrium at frequencies $\hbar\omega \ll k_B T$ developed in [28], we postulate that the equations of relativistic magnetohydrodynamics should describe the low frequency transport. The basic principles involved in such a hydrodynamic computation go back to the nineteenth century: conservation of energy, momentum, and charge, and the constraint of the positivity of entropy production. Nevertheless, the required results were not obtained until our recent work [41]: the general case of a CFT3 in the presence of a chemical potential, magnetic field, and small density of impurities is very intricate, and the guidance provided by the dual gravity formulation was very helpful to us. In this approach, we do not have quantitative knowledge of a few transport co-efficients, and this is complementary to our ignorance of the effective couplings in the dual gravity theory to be discussed in Sect. 9.6.2.

The complete hydrodynamic analysis can be found in [41]. The analysis is intricate, but is mainly a straightforward adaption of the classic procedure outlined by Kadanoff and Martin [43] to the relativistic field theories which describe quantum critical points. We list the steps:

1. Identify the conserved quantities, which are the energy-momentum tensor, $T^{\mu\nu}$, and the particle number current, J^{μ}.
2. Obtain the real time equations of motion, which express the conservation laws:
$$\partial_\nu T^{\mu\nu} = F^{\mu\nu} J_\nu, \quad \partial_\mu J^\mu = 0; \tag{9.20}$$
here $F^{\mu\nu}$ represents the externally applied electric and magnetic fields which can change the net momentum or energy of the system, and we have not written a term describing momentum relaxation by impurities.
3. Identify the state variables which characterize the local thermodynamic state—we choose these to be the density, ρ, the temperature T, and an average velocity u^{μ}.
4. Express $T^{\mu\nu}$ and J^{μ} in terms of the state variables and their spatial and temporal gradients; here we use the properties of the observables under a boost by the velocity u^{μ}, and thermodynamic quantities like the energy density, ε, and the pressure, P, which are determined from T and ρ by the equation of state of the CFT. We also introduce transport co-efficients associated with the gradient terms.
5. Express the equations of motion in terms of the state variables, and ensure that the entropy production rate is positive [44]. This is a key step which ensures relaxation to local equilibrium, and leads to important constraints on the transport co-efficients. In $d = 2$, it was found that situations with the velocity u^{μ} spacetime independent are characterized by only a *single* independent transport co-efficient [41]. This we choose to be the longitudinal conductivity at $B = 0$.
6. Solve the initial value problem for the state variables using the linearized equations of motion.
7. Finally, translate this solution to the linear response functions, as described in [43].

9.6.2 Dyonic Black Hole

Given the success of the AdS/CFT correspondence for the specific supersymmetric model in Sect. 9.5, we boldly assume a similar correspondence for a generic CFT3. We assume that each CFT3 is dual to a strongly-coupled theory of gravity on AdS_4. Furthermore, given the operators associated with the perturbations away from the pure CFT3 we want to study, we can also deduce the corresponding perturbations away from the dual gravity theory. So far, this correspondence is purely formal and not of much practical use to us. However, we now restrict our attention to the hydrodynamic, collision dominated regime, $\hbar\omega \ll k_B T$ of the CFT3. We would like to know the corresponding low energy effective theory describing the quantum gravity theory on AdS_4. Here, we make the simplest possible assumption: the effective theory is just the Einstein-Maxwell theory of general relativity and electromagnetism on AdS_4. As in Sect. 9.5, the temperature T of CFT3 corresponds to introducing a black hole on AdS_4 whose Hawking temperature is T. The chemical potential, μ, of the CFT3 corresponds to an electric charge on the black hole, and the applied magnetic field maps to a magnetic charge on the black hole. Such a dynoic black hole solution of the Einstein-Maxwell equations is, in fact, known: it is the Reissner-Nordstrom black hole.

We solved the classical Einstein-Maxwell equations for linearized fluctuations about the metric of a dyonic black hole in a space which is asymptotically AdS_4. The results were used to obtain correlators of a CFT3 using the prescriptions of the AdS/CFT mapping. As we have noted, we have no detailed knowledge of the strongly-coupled quantum gravity theory which is dual to the CFT3 describing the superfluid-insulator transition in condensed matter systems, or of graphene. Nevertheless, given our postulate that its low energy effective field theory essentially captured by the Einstein-Maxwell theory, we can then obtain a powerful set of results for CFT3s.

9.6.3 Results

In the end, we obtained complete agreement between the two independent computations in Sects. 9.6.1 and 9.6.2, after allowing for their distinct equations of state. This agreement demonstrates that the assumption of a low energy Einstein-Maxwell effective field theory for a strongly coupled theory of quantum gravity is equivalent to the assumption of hydrodynamic transport for $\hbar\omega \ll k_B T$ in a strongly coupled CFT3.

Finally, we turn to our explicit results for quantum critical transport with $\hbar\omega \ll k_B T$.

First, consider adding a chemical potential, μ, to the CFT3. This will induce a non-zero number density of carriers ρ. The value of ρ is defined so that the total charge density associated with ρ is $e^*\rho$. Then the electrical conductivity at a frequency ω is

$$\sigma(\omega) = \frac{e^{*2}}{h}\Phi_\sigma + \frac{e^{*2}\rho^2 v^2}{(\varepsilon + P)} \frac{1}{(-i\omega + 1/\tau_{\text{imp}})}. \tag{9.21}$$

In this section, we are again using the symbol v to denote the characteristic velocity of the CFT3 because we will need c for the physical velocity of light below. Here ε is the energy density and P is the pressure of the CFT3. We have assumed a small density of impurities which lead to a momentum relaxation time τ_{imp} [41, 45]. In general, Φ_σ, ρ, ε, P, and $1/\tau_{\text{imp}}$ will be functions of $\mu/k_B T$ which cannot be computed by hydrodynamic considerations alone. However, apart from Φ_σ, these quantities are usually amenable to direct perturbative computations in the CFT3, or by quantum Monte Carlo studies. The physical interpretation of Eq. 9.21 should be evident: adding a charge density ρ leads to an additional Drude-like contribution to the conductivity. This extra current cannot be relaxed by collisions between the unequal density of particle and hole excitations, and so requires an impurity relaxation mechanism to yield a finite conductivity in the d.c. limit.

Now consider thermal transport in a CFT3 with a non-zero μ. The d.c. thermal conductivity, κ, is given by

$$\kappa = \Phi_\sigma \left(\frac{k_B^2 T}{h}\right)\left(\frac{\varepsilon + P}{k_B T \rho}\right)^2, \tag{9.22}$$

in the absence of impurity scattering, $1/\tau_{\text{imp}} \to 0$. This is a Wiedemann-Franz-like relation, connecting the thermal conductivity to the electrical conductivity in the $\mu = 0$ CFT. Note that κ diverges as $\rho \to 0$, and so the thermal conductivity of the $\mu = 0$ CFT is infinite.

Next, turn on a small magnetic field B; we assume that B is small enough that the spacing between the Landau levels is not as large as $k_B T$. The case of large Landau level spacing is also experimentally important, but cannot be addressed by the present analysis. Initially, consider the case $\mu = 0$. In this case, the result Eq. 9.22 for the thermal conductivity is replaced by

$$\kappa = \frac{1}{\Phi_\sigma}\left(\frac{k_B^2 T}{h}\right)\left(\frac{\varepsilon + P}{k_B T B/(hc/e^*)}\right)^2 \tag{9.23}$$

also in the absence of impurity scattering, $1/\tau_{\text{imp}} \to 0$. This result relates κ to the electrical *resistance* at criticality, and so can be viewed as Wiedemann-Franz-like relation for the vortices. A similar $1/B^2$ dependence of κ appeared in the Boltzmann equation analysis of [46, 47], but our more general analysis applies in a wider and distinct regime [30–32], and relates the co-efficient to other observables.

We have obtained a full set of results for the frequency-dependent thermoelectric response functions at non-zero B and μ. The results are lengthy and we refer the reader to [41] for explicit expressions. Here we only note that the characteristic feature [41, 48] of these results is a new *hydrodynamic cyclotron resonance*. The usual cyclotron resonance occurs at the classical cyclotron frequency, which is independent of the particle density and temperature; further, in a Galilean-invariant system this resonance is not broadened by electron-electron interactions alone, and requires impurities for non-zero damping. The situation for our hydrodynamic resonance is very different. It occurs in a collision-dominated regime, and its frequency depends on the density and temperature: the explicit expression for the resonance frequency is

$$\omega_c = \frac{e^* B \rho v^2}{c(\varepsilon + P)}. \tag{9.24}$$

Further, the cyclotron resonance involves particle and hole excitations moving in opposite directions, and collisions between them can damp the resonance even in the absence of impurities. Our expression for this intrinsic damping frequency is [41, 48]

$$\gamma = \frac{e^{*2}}{h} \Phi_\sigma \frac{B^2 v^2}{c^2(\varepsilon + P)}, \tag{9.25}$$

relating it to the quantum-critical conductivity as a measure of collisions between counter-propagating particles and holes. We refer the reader to a separate discussion [30–32] of the experimental conditions under which this hydrodynamic cyclotron resonance may be observed.

9.7 *d*-Wave Superconductors

We now turn to the second class of strong-coupling problems outlined in Sect. 9.1: those involving quantum critical points with fermionic excitations. This section will consider the simpler class of problems in which the fermions have a Dirac spectrum, and the field-theoretic $1/N$ expansion does allow for substantial progress.

We will begin in Sect. 9.7.1 by an elementary discussion of the origin of these Dirac fermions. Then we will consider two quantum phase transitions, both involving a simple Ising order parameter. The first in Sect. 9.7.2, with time-reversal symmetry breaking, leads to a relativistic quantum field theory closely related to the Gross-Neveu model. The second model of Sect. 9.7.3 involves breaking of a lattice rotation symmetry, leading to "Ising-nematic" order. The theory for this model is not relativistically invariant: it is strongly coupled, but can be controlled by a traditional $1/N$ expansion.

We note that symmetry breaking transitions in graphene are also described by field theories similar to those discussed in this section [49, 50].

9.7.1 Dirac Fermions

We begin with a review of the standard BCS mean-field theory for a *d*-wave superconductor on the square lattice, with an eye towards identifying the fermionic Bogoliubov quasiparticle excitations. For now, we assume we are far from any QPT associated with SDW, Ising-nematic, or other broken symmetries. We consider the Hamiltonian

$$H_{tJ} = \sum_k \varepsilon_k c^\dagger_{k\alpha} c_{k\alpha} + J_1 \sum_{\langle ij \rangle} \mathbf{S}_i \cdot \mathbf{S}_j \tag{9.26}$$

where $c_{j\alpha}$ is the annihilation operator for an electron on site j with spin $\alpha = \uparrow, \downarrow$, $c_{k\alpha}$ is its Fourier transform to momentum space, ε_k is the dispersion of the electrons (it is conventional to choose $\varepsilon_k = -2t_1(\cos(k_x) + \cos(k_y)) - 2t_2(\cos(k_x + k_y) + \cos(k_x - k_y)) - \mu$, with $t_{1,2}$ the first/second neighbor hopping and μ the chemical potential), and the J_1 term is similar to that in Eq. 9.1 with

$$S_{ja} = \frac{1}{2} c^\dagger_{j\alpha} \sigma^a_{\alpha\beta} c_{j\beta} \tag{9.27}$$

and σ^a the Pauli matrices. We will consider the consequences of the further neighbor exchange interactions for the superconductor in Sect. 9.7.2 below. Applying the BCS mean-field decoupling to H_{tJ} we obtain the Bogoliubov Hamiltonian

$$H_{BCS} = \sum_k \varepsilon_k c^\dagger_{k\alpha} c_{k\alpha} - \frac{J_1}{2} \sum_{j\mu} \Delta_\mu \left(c^\dagger_{j\uparrow} c^\dagger_{j+\hat{\mu},\downarrow} - c^\dagger_{j\downarrow} c^\dagger_{j+\hat{\mu},\uparrow} \right) + \text{h.c.}. \tag{9.28}$$

For a wide range of parameters, the ground state energy optimized by a $d_{x^2-y^2}$ wavefunction for the Cooper pairs: this corresponds to the choice $\Delta_x = -\Delta_y = \Delta_{x^2-y^2}$. The value of $\Delta_{x^2-y^2}$ is determined by minimizing the energy of the BCS state

$$E_{BCS} = J_1 |\Delta_{x^2-y^2}|^2 - \int \frac{d^2k}{4\pi^2} [E_k - \varepsilon_k] \tag{9.29}$$

where the fermionic quasiparticle dispersion is

$$E_k = \left[\varepsilon_k^2 + \left| J_1 \Delta_{x^2-y^2} (\cos k_x - \cos k_y) \right|^2 \right]^{1/2}. \tag{9.30}$$

The energy of the quasiparticles, E_k, vanishes at the four points $(\pm Q, \pm Q)$ at which $\varepsilon_k = 0$. We are especially interested in the low energy quasiparticles in the vicinity of these points, and so we perform a gradient expansion of H_{BCS} near each of them. We label the points $\boldsymbol{Q}_1 = (Q, Q)$, $\boldsymbol{Q}_2 = (-Q, Q)$, $\boldsymbol{Q}_3 = (-Q, -Q)$, $\boldsymbol{Q}_4 = (Q, -Q)$ and write

$$c_{j\alpha} = f_{1\alpha}(\boldsymbol{r}_j) e^{i\boldsymbol{Q}_1 \cdot \boldsymbol{r}_j} + f_{2\alpha}(\boldsymbol{r}_j) e^{i\boldsymbol{Q}_2 \cdot \boldsymbol{r}_j} + f_{3\alpha}(\boldsymbol{r}_j) e^{i\boldsymbol{Q}_3 \cdot \boldsymbol{r}_j} + f_{4\alpha}(\boldsymbol{r}_j) e^{i\boldsymbol{Q}_4 \cdot \boldsymbol{r}_j}, \tag{9.31}$$

while assuming the $f_{1-4,\alpha}(\boldsymbol{r})$ are slowly varying functions of x. We also introduce the bispinors $\Psi_1 = (f_{1\uparrow}, f^\dagger_{3\downarrow}, f_{1\downarrow}, -f^\dagger_{3\uparrow})$, and $\Psi_2 = (f_{2\uparrow}, f^\dagger_{4\downarrow}, f_{2\downarrow}, -f^\dagger_{4\uparrow})$, and then express H_{BCS} in terms of $\Psi_{1,2}$ while performing a spatial gradient expansion. This yields the following effective action for the fermionic quasiparticles:

$$\mathcal{S}_\Psi = \int d\tau d^2r \left[\Psi_1^\dagger \left(\partial_\tau - i\frac{v_F}{\sqrt{2}}(\partial_x + \partial_y)\tau^z - i\frac{v_\Delta}{\sqrt{2}}(-\partial_x + \partial_y)\tau^x \right)\Psi_1 \right.$$
$$\left. + \Psi_2^\dagger \left(\partial_\tau - i\frac{v_F}{\sqrt{2}}(-\partial_x + \partial_y)\tau^z - i\frac{v_\Delta}{\sqrt{2}}(\partial_x + \partial_y)\tau^x \right)\Psi_2 \right]. \quad (9.32)$$

where the $\tau^{x,z}$ are 4×4 matrices which are block diagonal, the blocks consisting of 2×2 Pauli matrices. The velocities $v_{F,\Delta}$ are given by the conical structure of E_k near the Q_{1-4}: we have $v_F = |\nabla_k \varepsilon_k|_{k=Q_a}|$ and $v_\Delta = |J_1 \Delta_{x^2-y^2}\sqrt{2}\sin(Q)|$. In this limit, the energy of the Ψ_1 fermionic excitations is $E_k = (v_F^2(k_x+k_y)^2/2 + v_\Delta^2(k_x-k_y)^2/2)^{1/2}$ (and similarly for Ψ_2), which is the spectrum of massless Dirac fermions.

9.7.2 Time-Reversal Symmetry Breaking

We will consider a simple model in which the pairing symmetry of the superconductor changes from $d_{x^2-y^2}$ to $d_{x^2-y^2} \pm id_{xy}$. The choice of the phase between the two pairing components leads to a breaking of time-reversal symmetry. Studies of this transition were originally motivated by the cuprate phenomenology, but we will not explore this experimental connection here because the evidence has remained sparse.

The mean field theory of this transition can be explored entirely within the context of BCS theory, as we will review below. However, fluctuations about the BCS theory are strong, and lead to non-trivial critical behavior involving both the collective order parameter and the Bogoliubov fermions: this is probably the earliest known example [23, 51, 52] of the failure of BCS theory in two (or higher) dimensions in a superconducting ground state. At $T > 0$, this failure broadens into the "quantum critical" region.

We extend H_{tJ} in Eq. 9.26 so that BCS mean-field theory permits a region with d_{xy} superconductivity. With a J_2 second neighbor interaction, Eq. 9.26 is modified to:

$$\widetilde{H}_{tJ} = \sum_k \varepsilon_k c_{k\sigma}^\dagger c_{k\sigma} + J_1 \sum_{\langle ij \rangle} \mathbf{S}_i \cdot \mathbf{S}_j + J_2 \sum_{\text{nnn } ij} \mathbf{S}_i \cdot \mathbf{S}_j. \quad (9.33)$$

We will follow the evolution of the ground state of $\widetilde{H}_{tJ}$ as a function of J_2/J_1.

The mean-field Hamiltonian is now modified from Eq. 9.28 to

$$\widetilde{H}_{BCS} = \sum_k \varepsilon_k c_{k\sigma}^\dagger c_{k\sigma} - \frac{J_1}{2}\sum_{j,\mu} \Delta_\mu (c_{j\uparrow}^\dagger c_{j+\hat{\mu},\downarrow}^\dagger - c_{j\downarrow}^\dagger c_{j+\hat{\mu},\uparrow}^\dagger) + \text{h.c.}$$
$$- \frac{J_2}{2}\sum_{j,\nu}{}' \Delta_\nu (c_{j\uparrow}^\dagger c_{j+\hat{\nu},\downarrow}^\dagger - c_{j\downarrow}^\dagger c_{j+\hat{\nu},\uparrow}^\dagger) + \text{h.c.}, \quad (9.34)$$

where the second summation over v is along the diagonal neighbors $\hat{x}+\hat{y}$ and $-\hat{x}+\hat{y}$. To obtain d_{xy} pairing along the diagonals, we choose $\Delta_{x+y} = -\Delta_{-x+y} = \Delta_{xy}$. We summarize our choices for the spatial structure of the pairing amplitudes (which determine the Cooper pair wavefunction) in Fig. 9.7. The values of $\Delta_{x^2-y^2}$ and Δ_{xy} are to be determined by minimizing the ground state energy (generalizing Eq. 9.29)

$$E_{BCS} = J_1|\Delta_{x^2-y^2}|^2 + J_2|\Delta_{xy}|^2 - \int \frac{d^2k}{4\pi^2}[E_k - \varepsilon_k] \tag{9.35}$$

where the quasiparticle dispersion is now (generalizing Eq. 9.30)

$$E_k = \left[\varepsilon_k^2 + \left|J_1\Delta_{x^2-y^2}(\cos k_x - \cos k_y) + 2J_2\Delta_{xy}\sin k_x \sin k_y\right|^2\right]^{1/2}. \tag{9.36}$$

Notice that the energy depends upon the relative phase of $\Delta_{x^2-y^2}$ and Δ_{xy}: this phase is therefore an observable property of the ground state.

It is a simple matter to numerically carry out the minimization of Eq. 9.36, and the results for a typical choice of parameters are shown in Fig. 9.8 as a function J_2/J_1. One of the two amplitudes $\Delta_{x^2-y^2}$ or Δ_{xy} is always non-zero and so the ground state is always superconducting. The transition from pure $d_{x^2-y^2}$ superconductivity to pure d_{xy} superconductivity occurs via an intermediate phase in which *both* order parameters are non-zero. Furthermore, in this regime, their relative phase is found to be pinned to $\pm\pi/2$ i.e.

$$\arg(\Delta_{xy}) = \arg(\Delta_{x^2-y^2}) \pm \pi/2. \tag{9.37}$$

The reason for this pinning can be intuitively seen from Eq. 9.36: only for these values of the relative phase does the equation $E_k = 0$ never have a solution. In other words, the gapless nodal quasiparticles of the $d_{x^2-y^2}$ superconductor acquire a finite energy gap when a secondary pairing with relative phase $\pm\pi/2$ develops. By

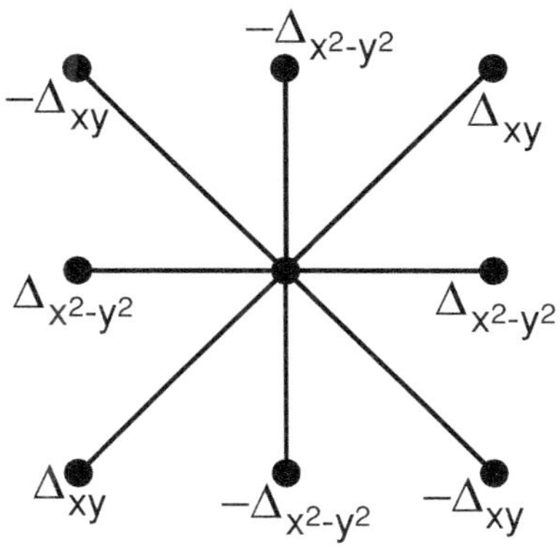

Fig. 9.7 Values of the pairing amplitudes, $-\langle c_{i\uparrow}c_{j\downarrow} - c_{i\downarrow}c_{j\uparrow}\rangle$ with i the central site, and j is one of its eight near neighbors

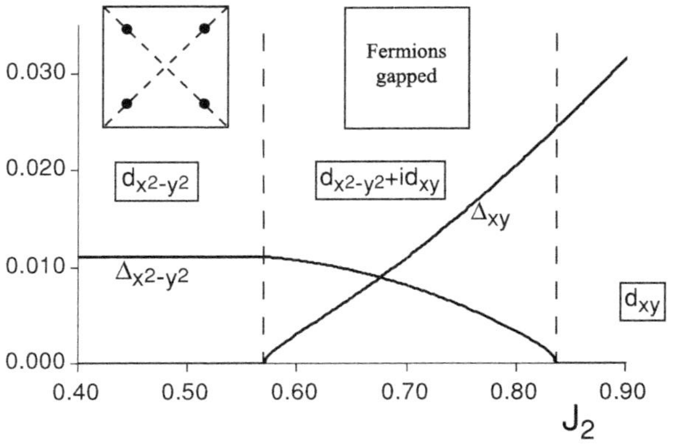

Fig. 9.8 BCS solution of the phenomenological Hamiltonian $\widetilde{H}_{tJ}$ in Eq. 9.33. Shown are the optimum values of the pairing amplitudes $|\Delta_{x^2-y^2}|$ and $|\Delta_{xy}|$ as a function of J_2 for $t_1 = 1$, $t_2 = -0.25$, $\mu = -1.25$, and J_1 fixed at $J_1 = 0.4$. The relative phase of the pairing amplitudes was always found to obey Eq. 9.37. The dashed lines denote locations of phase transitions between $d_{x^2-y^2}$, $d_{x^2-y^2} + id_{xy}$, and d_{xy} superconductors. The pairing amplitudes vanishes linearly at the first transition corresponding to the exponent $\beta_{BCS} = 1$ in Eq. 9.40. The Brillouin zone location of the gapless Dirac points in the $d_{x^2-y^2}$ superconductor is indicated by filled circles. For the dispersion ε_k appropriate to the cuprates, the d_{xy} superconductor is fully gapped, and so the second transition is ordinary Ising

a level repulsion picture, we can expect that gapping out the low energy excitations should help lower the energy of the ground state. The intermediate phase obeying Eq. 9.37 is called a $d_{x^2-y^2} + id_{xy}$ superconductor.

The choice of the sign in Eq. 9.37 leads to an overall two-fold degeneracy in the choice of the wavefunction for the $d_{x^2-y^2} + id_{xy}$ superconductor. This choice is related to the breaking of time-reversal symmetry, and implies that the $d_{x^2-y^2} + id_{xy}$ phase is characterized by the non-zero expectation value of a Z_2 Ising order parameter; the expectation value of this order vanishes in the two phases (the $d_{x^2-y^2}$ and d_{xy} superconductors) on either side of the $d_{x^2-y^2} + id_{xy}$ superconductor. As is conventional, we will represent the Ising order by a real scalar field ϕ. Fluctuations of ϕ become critical near both of the phase boundaries in Fig. 9.8. As we will explain below, the critical theory of the $d_{x^2-y^2}$ to $d_{x^2-y^2} + id_{xy}$ transition is *not* the usual ϕ^4 field theory which describes the ordinary Ising transition in three spacetime dimensions. (For the dispersion ε_k appropriate to the cuprates, the d_{xy} superconductor is fully gapped, and so the $d_{x^2-y^2} + id_{xy}$ to d_{xy} transition in Fig. 9.8 will be ordinary Ising.)

Near the phase boundary from $d_{x^2-y^2}$ to $d_{x^2-y^2} + id_{xy}$ superconductivity it is clear that we can identify

$$\phi = i\Delta_{xy}, \tag{9.38}$$

(in the gauge where $\Delta_{x^2-y^2}$ is real). We can now expand E_{BCS} in Eq. 9.35 for small ϕ (with $\Delta_{x^2-y^2}$ finite) and find a series with the structure [53, 54]

$$E_{BCS} = E_0 + s\phi^2 + v|\phi|^3 + \ldots, \tag{9.39}$$

where s, v are coefficients and the ellipses represent regular higher order terms in even powers of ϕ; s can have either sign, whereas v is always positive. Notice the non-analytic $|\phi|^3$ term that appears in the BCS theory—this arises from an infrared singularity in the integral in Eq. 9.35 over E_k at the four nodal points of the $d_{x^2-y^2}$ superconductor, and is a preliminary indication that the transition differs from that in the ordinary Ising model, and that the Dirac fermions play a central role. We can optimize ϕ by minimizing E_{BCS} in Eq. 9.39— this shows that $\langle\phi\rangle = 0$ for $s > 0$, and $\langle\phi\rangle \neq 0$ for $s<0$. So $s \sim (J_2/J_1)_c - J_2/J_1$ where $(J_2/J_1)_c$ is the first critical value in Fig. 9.8. Near this critical point, we find

$$\langle\phi\rangle \sim (s_c - s)^{\beta}, \tag{9.40}$$

where we have allowed for the fact that fluctuation corrections will shift the critical point from $s = 0$ to $s = s_c$. The present BCS theory yields the exponent $\beta_{BCS} = 1$; this differs from the usual mean-field exponent $\beta_{MF} = 1/2$, and this is of course due to the non-analytic $|\phi|^3$ term in Eq. 9.39.

We can now write down the required field theory of the onset of d_{xy} order. In addition to the order parameter ϕ, the field theory should also involve the low energy nodal fermions of the $d_{x^2-y^2}$ superconductor, as described by $\mathcal{S}_\Psi$ in Eq. 9.32. For the ϕ fluctuations, we write down the usual terms permitted near a phase transition with Ising symmetry:

$$\mathcal{S}_\phi = \int d^2r d\tau \left[\frac{1}{2}\left((\partial_\tau\phi)^2 + c^2(\partial_x\phi)^2 + c^2(\partial_y\phi)^2 + s\phi^2\right) + \frac{u}{24}\phi^4\right]. \tag{9.41}$$

Note that, unlike Eq. 9.39, we do not have any non-analytic $|\phi|^3$ terms in the action: this is because we have not integrated out the low energy Dirac fermions, and the terms in Eq. 9.41 are viewed as arising from high energy fermions away from the nodal points. Finally, we need to couple the ϕ and $\Psi_{1,2}$ excitations. Their coupling is already contained in the last term in Eq. 9.34: expressing this in terms of the $\Psi_{1,2}$ fermions using Eq. 9.31 we obtain

$$\mathcal{S}_{\Psi\phi} = \vartheta_{xy} \int d^2r d\tau \left[\phi\left(\Psi_1^\dagger \tau^y \Psi_1 - \Psi_2^\dagger \tau^y \Psi_2\right)\right], \tag{9.42}$$

where ϑ_{xy} is a coupling constant. The partition function of the full theory is now

$$\mathcal{Z}_{did} = \int \mathcal{D}\phi\mathcal{D}\Psi_1\mathcal{D}\Psi_2 \exp\left(-\mathcal{S}_\Psi - \mathcal{S}_\phi - \mathcal{S}_{\Psi\phi}\right), \tag{9.43}$$

where $\mathcal{S}_\Psi$ was in Eq. 9.32. It can now be checked that if we integrate out the $\Psi_{1,2}$ fermions for a spacetime independent ϕ, we do indeed obtain a $|\phi|^3$ term in the effective potential for ϕ.

We begin our analysis of Z_{did} by assuming that the transition is described by a fixed point with $\vartheta_{xy} = 0$: then the theory for the transition would be the ordinary ϕ^4 field theory $\mathcal{S}_\phi$, and the nodal fermions would be innocent spectators. The scaling dimension of ϕ at such a fixed point is $(1 + \eta_I)/2$ (where η_I is the anomalous order parameter exponent at the critical point of the ordinary three dimensional Ising model), while that of $\Psi_{1,2}$ is 1. Consequently, the scaling dimension of ϑ_{xy} is $(1 - \eta_I)/2 > 0$. This positive scaling dimension implies that ϑ_{xy} is relevant and the $\vartheta_{xy} = 0$ fixed point is unstable: the Dirac fermions are fully involved in the critical theory.

Determining the correct critical behavior now requires a full renormalization group analysis of Z_{did}. This has been described in some detail in [55], and we will not reproduce the details here. The main result we need for our purposes is that couplings ϑ_{xy}, u, v_F/c and v_Δ/c all reach *non-zero* fixed point values which define a critical point in a new universality class. These fixed point values, and the corresponding critical exponents, can be determined in expansions in either $(3 - d)$ [51, 52, 55] (where d is the spatial dimensionality) or $1/N$ [56] (where N is the number of fermion species). An important simplifying feature here is that the fixed point is actually relativistically invariant. Indeed the fixed point has the structure of the so-called Higgs-Yukawa (or Gross-Neveu) model which has been studied extensively in the particle physics literature [57] in a different physical context: quantum Monte Carlo simulation of this model also exist [58], and provide probably the most accurate estimate of the exponents.

The non-trivial fixed point has strong implications for the correlations of the Bogoliubov fermions. The fermion correlation function $G_1 = \langle \Psi_1 \Psi_1^\dagger \rangle$ obeys

$$G_1(k, \omega) = \frac{\omega + v_F k_x \tau^z + v_\Delta \tau^x}{(v_F^2 k_x^2 + v_\Delta^2 k_y^2 - \omega^2)^{(1-\eta_f)/2}} \tag{9.44}$$

at low frequencies for $s \geq s_c$. Away from the critical point in the $d_{x^2-y^2}$ superconductor with $s > s_c$, Eq. 9.44 holds with $\eta_f = 0$, and this is the BCS result, with sharp quasi-particle poles in the Green's function. At the critical point $s = s_c$ Eq. 9.44 holds with the fixed point values for the velocities (which satisfy $v_F = v_\Delta = c$) and with the anomalous dimension $\eta_f \neq 0$—the $(3 - d)$ expansion [51, 52] estimate is $\eta_f \approx (3 - d)/14$, and the $1/N$ expansion estimate [56] is $\eta_f \approx 1/(3\pi^2 N)$, with $N = 2$. This is clearly non-BCS behavior, and the fermionic quasiparticle pole in the spectral function has been replaced by a branch-cut representing the continuum of critical excitations. The corrections to BCS extend also to correlations of the Ising order ϕ: its expectation value vanishes as Eq. 9.40 with the Monte Carlo estimate $\beta \approx 0.877$ [58]. The critical point correlators of ϕ have the anomalous dimension $\eta \approx 0.754$ [58], which is clearly different from the very small value of the exponent η_I at the unstable $\vartheta_{xy} = 0$ fixed point. The value of β is related to η by the usual scaling law $\beta = (1 + \eta)\nu/2$, with $\nu \approx 1.00$ the correlation length exponent (which also differs from the exponent ν_I of the Ising model).

9.7.3 Nematic Ordering

We now consider an Ising transition associated with "Ising-nematic" ordering in the d-wave superconductor. This is associated with a spontaneous reduction of the lattice symmetry of the Hamiltonian from "square" to "rectangular". Our study is motivated by experimental observations of such a symmetry breaking in the cuprate superconductors [59–61].

The ingredients of such an ordering are actually already present in our simple review of BCS theory in Sect. 9.7.1. In Eq. 9.28, we introduce 2 variational pairing amplitudes Δ_x and Δ_y. Subsequently, we assumed that the minimization of the energy led to a solution with $d_{x^2-y^2}$ pairing symmetry with $\Delta_x = -\Delta_y = \Delta_{x^2-y^2}$. However, it is possible that upon including the full details of the microscopic interactions we are led to a minimum where the optimal solution also has a small amount of s-wave pairing. Then $|\Delta_x| \neq |\Delta_y|$, and we can expect all physical properties to have distinct dependencies on the x and y co-ordinates. Thus, one measure of the the Ising nematic order parameter is $|\Delta_x|^2 - |\Delta_y|^2$.

The derivation of the field theory for this transition follows closely our presentation in Sect. 9.7.2. We allow for small Ising-nematic ordering by introducing a scalar field ϕ and writing

$$\Delta_x = \Delta_{x^2-y^2} + \phi; \; \Delta_y = -\Delta_{x^2-y^2} + \phi. \tag{9.45}$$

The evolutions of the Dirac fermion spectrum under such a change is indicated in Fig. 9.9. We now develop an effective action for ϕ and the Dirac fermions $\Psi_{1,2}$. The result is essentially identical to that in Sect. 9.7.2, apart from a change in the structure of the Yukawa coupling. Thus we obtain a theory $\mathcal{S}_\Psi + \mathcal{S}_\phi + \overline{\mathcal{S}}_{\Psi\phi}$, defined by Eqs. 9.32 and 9.41, and where Eq. 9.42 is now replaced by

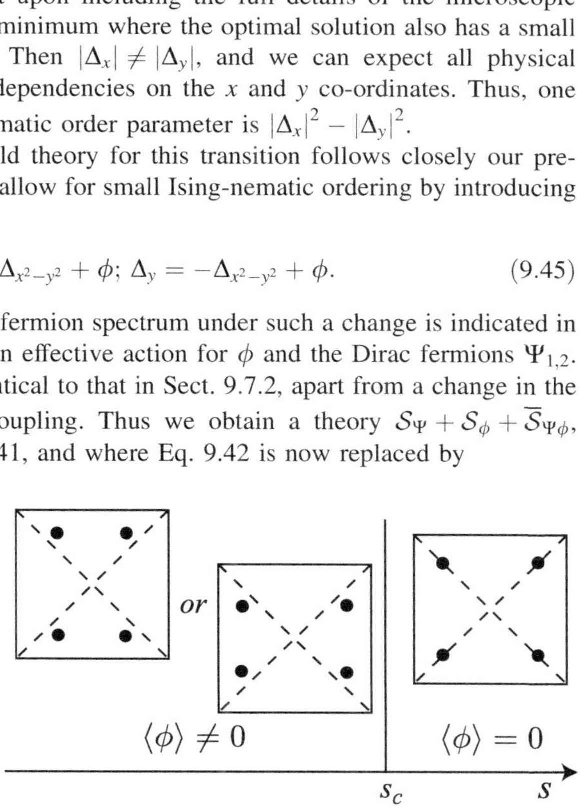

Fig. 9.9 Phase diagram of Ising nematic ordering in a d-wave superconductor as a function of the coupling s in $\mathcal{S}_\phi$. The filled circles indicate the location of the gapless fermionic excitations in the Brillouin zone. The two choices for $s < s_c$ are selected by the sign of $\langle\phi\rangle$.

$$\overline{\mathcal{S}}_{\Psi\phi} = \vartheta_I \int d^2r d\tau \Big[\phi\Big(\Psi_1^\dagger \tau^x \Psi_1 + \Psi_2^\dagger \tau^x \Psi_2\Big)\Big]. \tag{9.46}$$

The seemingly innocuous change between Eqs. 9.42 and 9.46 however has strong consequences. This is partly linked to the fact with $\overline{\mathcal{S}}_{\Psi\phi}$ cannot be relativistically invariant even after all velocities are adjusted to equal. A weak-coupling renormalization group analysis in powers of the coupling ϑ_I was performed in $(3-d)$ dimensions in [51, 52, 55], and led to flows to strong coupling with no accessible fixed point: thus no firm conclusions on the nature of the critical theory were drawn.

This problem remained unsolved until the recent works of [62, 63]. It is essential that there not be any expansion in powers of the coupling ϑ_I. This is because it leads to strongly non-analytic changes in the structure of the ϕ propagator, which have to be included at all stages. In a model with N fermion flavors, the $1/N$ expansion does avoid any expansion in ϑ_I. The renormalization group analysis has to be carried out within the context of the $1/N$ expansion, and this involves some rather technical analysis which is explained in [63]. In the end, an asymptotically exact description of the vicinity of the critical point was obtained. It was found that the velocity ratio v_F/v_Δ diverged logarithmically with energy scale, leading to strongly anisotropic 'arc-like' spectra for the Dirac fermions. Associated singularities in the thermal conductivity have also been computed [64].

9.8 Metals

This section considers symmetry breaking transitions in two-dimensional metals. Away from the quantum critical point, the phases will be ordinary Fermi liquids. We will be interested in the manner in which the Fermi liquid behavior breaks down at the quantum critical point. Our focus will be exclusively on two spatial dimensions: quantum phase transitions of metals in three dimensions are usually simpler, and the traditional perturbative theory appears under control.

In Sect. 9.7 the fermionic excitations had vanishing energy only at isolated nodal points in the Brillouin zone: see Figs. 9.8 and 9.9. Metals have fermionic excitations with vanishing energy along a *line* in the Brillouin zone. Thus we can expect them to have an even stronger effect on the critical theory. This will indeed be the case, and we will be led to problems with a far more complex structure. Unlike the situation in insulators and d-wave superconductor, many basic issues associated with ordering transition in two dimensional metals have not been fully resolved. The problem remains one of active research and is being addressed by many different approaches. In recent papers [65, 66], Metlitski and the author have argued that the problem is strongly coupled, and proposed field theories and scaling structures for the vicinity of the critical point. We will review the main ingredients for the transition involving Ising-nematic ordering in a metal. Thus the symmetry breaking will be just as in Sect. 9.7.3, but the fermionic spectrum will be quite different. Our study here is also motivated by experimental observations in the cuprate superconductors [59–61].

As in Sect. 9.7, let us begin by a description of the non-critical fermionic sector, before its coupling to the order parameter fluctuations. We use the band structure describing the cuprates in the over-doped region, well away from the Mott insulator. Here the electrons $c_{k\alpha}$ are described by the kinetic energy in Eq. 9.26, which we write in the following action

$$\mathcal{S}_c = \int d\tau \sum_{k} c_{k\alpha}^{\dagger}\left(\frac{\partial}{\partial\tau} + \varepsilon_k\right) c_{k\alpha}, \tag{9.47}$$

As in Sect. 9.7.3, we will have an Ising order parameter represented by the real scalar field ϕ, which is described as before by $\mathcal{S}_\phi$ in Eq. 9.41. Its coupling to the electrons can be deduced by symmetry considerations, and the most natural coupling (the analog of Eqs. 9.46) is

$$\mathcal{S}_{c\phi} = \frac{1}{V}\int d\tau \sum_{k,q} (\cos k_x - \cos k_y)\phi(q) c_{k+q/2,\alpha}^{\dagger} c_{k-q/2,\alpha}. \tag{9.48}$$

where V is the volume. The momentum dependent form factor is the simplest choice which changes sign under $x \leftrightarrow y$, as is required by the symmetry properties of ϕ. The sum over q is over small momenta, while that over k extends over the entire Brillouin zone. The theory for the nematic ordering transition is now described by $\mathcal{S}_c + \mathcal{S}_\phi + \mathcal{S}_{c\phi}$. The phase diagram as a function of the coupling s in $\mathcal{S}_\phi$ and temperature T is shown in Fig. 9.10. Note that there is a line of Ising phase transitions at $T = T_c$: this transition is in the same universality class as the classical two-dimensional Ising model. However, quantum effects and fermionic excitations are crucial at $T = 0$ critical point at $s = s_c$ and its associated quantum critical region.

A key property of Eq. 9.48 is that small momentum critical ϕ fluctuations can efficiently scatter fermions at every point on the Fermi surface. Thus the non-Fermi singularities in the fermion Green's function will extend to all points on the Fermi surface. This behavior is dramatically different from all the field theories we have met so far, all of which had singularities only at isolated points in momentum space. We evidently have to write down a long-wavelength theory which has singularities along a line in momentum space.

We describe the construction of [65] of a field theory with this unusual property. Pick a fluctuation of the order parameter ϕ at a momentum q. As shown in Fig. 9.11 this fluctuation will couple most efficiently to fermions near two points on the Fermi surface, where the tangent to the Fermi surface is parallel to q. A fermion absorbing momentum q at these points, changes its energy only by $\sim q^2$; at all other points on the Fermi surface the change is $\sim q$. Thus we are led to focus on different points on the Fermi surface for each direction of q. In the continuum limit, we will therefore need a separate field theory for each pair of points $\pm k_0$ on

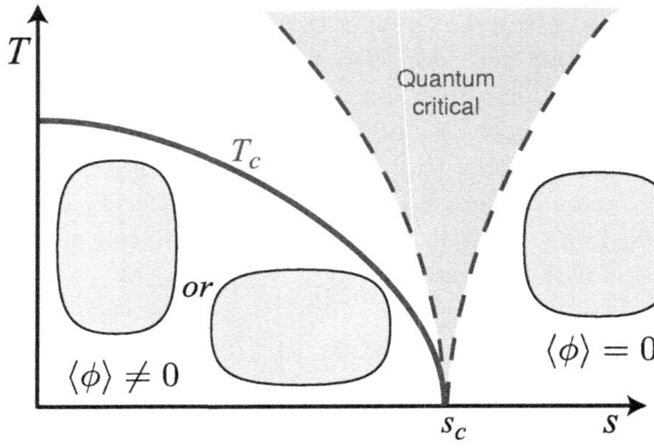

Fig. 9.10 Phase diagram of Ising nematic ordering in a metal as a function of the coupling s in $\mathcal{S}_\phi$ and temperature T. The Fermi surface for $s > 0$ is as in the overdoped region of the cuprates, with the shaded region indicating the occupied hole (or empty electron) states. The choice between the two quadrapolar distortions of the Fermi surface is determined by the sign of $\langle\phi\rangle$. The line of $T > 0$ phase transitions at T_c is described by Onsager's solution of the classical two-dimensional Ising model. We are interested here in the quantum critical point at $s = s_c$, which controls the quantum-critical region

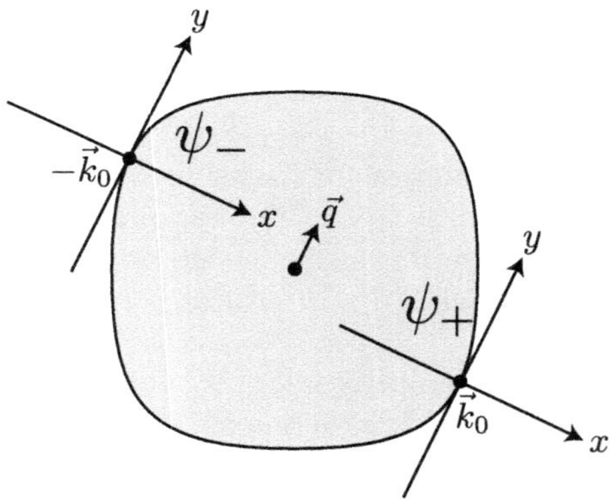

Fig. 9.11 A ϕ fluctuation at wavevector $\boldsymbol{q}$ couples most efficiently to the fermions $\psi_\pm$ near the Fermi surface points $\pm\boldsymbol{k}_0$

the Fermi surface. We conclude that the quantum critical point is described by an infinite number of field theories.

There have been earlier descriptions of Fermi surfaces by an infinite number of field theories [67–73]. However many of these earlier works differ in a crucial respect from the the theory to be presented here. They focus on the motion of fermions transverse to Fermi surface, and so represent each Fermi surface point by a 1 + 1 dimensional chiral fermion. Thus they have an infinite number of 1 + 1 dimensional field theories, labelled by points on the one dimensional Fermi surface. The original problem was 2 + 1 dimensional, and so this conserves the total dimensionality and the number of degrees of freedom. However, we have already argued above that the dominant fluctuations of the fermions are in a direction transverse to the Fermi surface, and so we believe this earlier approach is not suited for the vicinity of the nematic quantum critical point. As we will see in Sect. 9.8.1, we will need an infinite number of 2 + 1 dimensional field theories labelled by points on the Fermi surface. (Some of the earlier works included fluctuations transverse to the Fermi surface [72, 73], but did not account for the curvature of the Fermi surface; it is important to take the scaling limit at fixed curvature, as we will see.) Thus we have an emergent dimension and a redundant description of the degrees of freedom. We will see in Sect. 9.8.2 how compatibility conditions ensure that the redundancy does not lead to any inconsistencies. The emergent dimensionality suggests a connection to the AdS/CFT correspondence, as will be discussed in Sect. 9.8.5.

9.8.1 Field Theories

Let us now focus on the vicinity of the points $\pm\boldsymbol{k}_0$, by introducing fermionic field $\psi_\pm$ by

$$\psi_+(\boldsymbol{k}) = c_{\boldsymbol{k}_0+\boldsymbol{k}}, \quad \psi_-(\boldsymbol{k}) = c_{-\boldsymbol{k}_0+\boldsymbol{k}}. \tag{9.49}$$

Then we expand all terms in $\mathcal{S}_c + \mathcal{S}_\phi + \mathcal{S}_{c\phi}$ in spatial and temporal gradients. Using the co-ordinate system illustrated in Fig. 9.11, performing appropriate rescaling of co-ordinates, and dropping terms which can later be easily shown to be irrelevant, we obtain the 2 + 1 dimensional Lagrangian

$$\begin{aligned}\mathcal{L} = {} & \psi^\dagger_{+\alpha}\left(\zeta\partial_\tau - i\partial_x - \partial_y^2\right)\psi_{+\alpha} + \psi^\dagger_{-\alpha}\left(\zeta\partial_\tau + i\partial_x - \partial_y^2\right)\psi_{-\alpha} \\ & - \lambda\phi\left(\psi^\dagger_{+\alpha}\psi_{+\alpha} + \psi^\dagger_{-\alpha}\psi_{-\alpha}\right) + \frac{N}{2}(\partial_y\phi)^2 + \frac{Ns}{2}\phi^2.\end{aligned} \tag{9.50}$$

Here ζ, λ and s are coupling constants, with s the tuning parameter across the transition; we will see that all couplings apart from s can be scaled away or set equal to unity. We now allow the spin index $\alpha = 1, \ldots, N$, as we will be interested in the structure of the large N expansion. Note that Eq. 9.50 has the same basic structure as the models considered in Sect. 9.7, apart from differences in the spatial gradients and the matrix structure. We will see that these seemingly minor differences will completely change the physical properties and the nature of the large N expansion.

9.8.2 Symmetries

A first crucial property of $\mathcal{L}$ is that the fermion Green's functions do indeed have singularities along a line in momentum space, as was required by our discussion above. This singularity is a consequence of the invariance of $\mathcal{L}$ under the following transformation

$$\phi(x,y) \to \phi(x, y+\theta x), \quad \psi_s(x,y) \to e^{-is(\frac{\theta}{2}y+\frac{\theta^2}{4}x)}\psi_s(x, y+\theta x), \tag{9.51}$$

where θ is a constant. Here we have dropped time and α indices because they play no role, and $s = \pm$. We can view this transformation as one which performs a 'rotation' of spatial co-ordinates, moving the point $\boldsymbol{k}_0$ to neighboring points on the Fermi surface. We are not assuming the Fermi surface is circular, and so the underlying model is not rotationally invariant. However, we are considering a limiting case of a rotation, precisely analogous to the manner in which Galilean transformations emerge as a limiting case of a relativistic transformation (x behaves like time, and y as space, in this analogy). This 'Galilean' symmetry is an emergent symmetry of $\mathcal{L}$ for arbitrary shapes of the Fermi surface. It is not difficult to now show from (9.51) that the ϕ Green's function D, and the fermion Green's functions G_s obey the exact identities

$$D(q_x, q_y) = D(q_y) \tag{9.52}$$

$$G_s(q_x, q_y) = G(sq_x + q_y^2). \tag{9.53}$$

So we see that the Ψ_+ Green's function depends only on $q_x + q_y^2$. The singularities of this function appear when $q_x + q_y^2 = 0$, and this is nothing but the equation of the Fermi surface passing through the point $\boldsymbol{k}_0$ in Figs. 9.11 and 9.12. Thus we have established the existence of a line of singularities in momentum space.

The identities in Eq. 9.51 also help establish the consistency of our description in terms of an infinite number of 2 + 1 dimensional field theories. Consider the fermion Green's function at the point P in Fig. 9.12. This can be computed in terms of the 2 + 1 dimensional field theory defined at the point $\boldsymbol{k}_0$, or from that at a

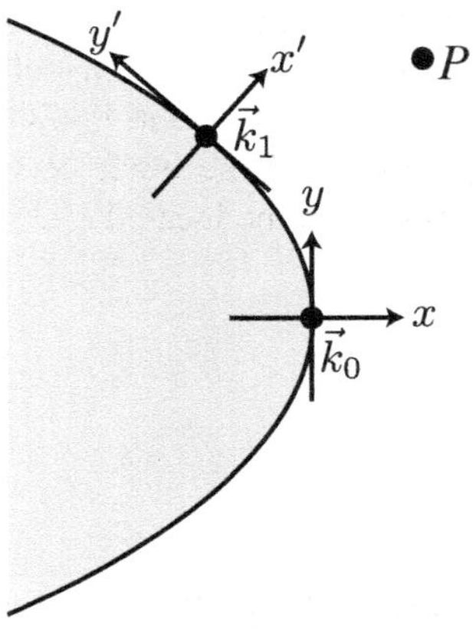

Fig. 9.12 The fermion correlator at the point P can be described either in terms of the 2 + 1 dimensional field theory at $\boldsymbol{k}_0$, or that at $\boldsymbol{k}_1$

neighboring point $\boldsymbol{k}_1$. Equation 9.53 ensures that both methods yield the same result. A little geometry [65] shows that $q_x + q_y^2$ is an invariant that measures the distance between P and the closest point on the Fermi surface: thus it takes the same value in the co-ordinates systems at $\boldsymbol{k}_0$ and $\boldsymbol{k}_1$, with $q_x + q_y^2 = q'_x + q'^2_y$, establishing the identity of the two computations.

9.8.3 Scaling Theory

Let us now discuss the behavior of $\mathcal{L}$ under renormalization group scaling transformations. The structure of the spatial gradient terms in the Lagrangian indicates that the rescaling of spatial co-ordinates should be defined by

$$x' = x/b^2, \quad y' = y/b. \tag{9.54}$$

The invariance in Eq. 9.53 implies that these scalings are exact, and the spatial anisotropy acquires no fluctuation corrections. Or, in other words

$$\text{dim}[y] = -1, \quad \text{dim}[x] = -2. \tag{9.55}$$

For now, let us keep the rescaling of the temporal co-ordinate general:

$$\text{dim}[\tau] = -z. \tag{9.56}$$

Note that the dynamic critical exponent z is defined relative to the spatial co-ordinate y tangent to the Fermi surface (other investigators sometimes define it relative to the co-ordinate x normal to the Fermi surface, leading to a difference in a factor of 2). We define the engineering dimensions of the fields so that coefficients of the y derivatives remain constant. Allowing for anomalous dimensions η_ϕ and η_ψ from loop effects we have

$$\text{dim}[\phi] = (1 + z + \eta_\phi)/2, \quad \text{dim}[\psi] = (1 + z + \eta_\psi)/2. \tag{9.57}$$

Using these transformations, we can examine the scaling dimensions of the couplings in $\mathcal{L}$ at tree level

$$\text{dim}[\zeta] = 2 - z - \eta_\psi, \quad \text{dim}[\lambda] = (3 - z - \eta_\phi - 2\eta_\psi)/2. \tag{9.58}$$

We will see in Sect. 9.8.4 that low order loop computations suggest that the anomalous dimensions η_ϕ and η_ψ are small, and that $z \approx 3$. Assuming these estimates are approximately correct, we see that the coupling ζ is strongly irrelevant. Thus we can send $\zeta \to 0$ in all our computations. However, we do not set $\zeta = 0$ at the outset, because the temporal derivative term is needed to define the proper analytic structure of the frequency loop integrals [74]. As we will discuss later, the limit $\zeta \to 0$, also dramatically changes the counting of powers of $1/N$ in the loop expansion [74].

Also note that these estimates of the scaling dimensions imply $\dim[\lambda] \approx 0$. Thus the fermion and order parameter fluctuations remain strongly coupled at all scales. Conversely, we can also say that the requirement of working in a theory with fixed λ implies that $z \approx 3$; this circumvents the appeal to loop computations for taking the $\zeta \to 0$ limit. With a near zero scaling dimension for λ, we cannot expand perturbatively in powers of λ. This features was also found in Sect. 9.7.3, but there we were able to use the $1/N$ expansion to circumvent this problem.

Moving beyond tree level considerations, we note that another Ward identity obeyed by the theory $\mathcal{L}$ allows us to fix the scaling dimension of ϕ exactly. This Ward identity is linked to the fact that ϕ appears in the Yukawa coupling like the x component of a gauge field coupled to the fermions [65]. The usual arguments associated with gauge invariance then imply that $\dim[\phi] = 2$ (the same as the scaling dimension of ∂_x), and that we can work in theory in which the "gauge coupling" λ set equal to unity at all scales. Note that with this scaling dimension, we have the exact relation

$$\eta_\phi = 3 - z. \tag{9.59}$$

Note also that Eq. 9.57 now implies that $\dim[\lambda] = \eta_\psi$ at tree level, which is the same as the tree level transformation of the spatial derivative terms. The latter terms have been set equal to unity by rescaling the fermion field, and so it is also consistent to set $\lambda = 1$ from now on.

We reach the remarkable conclusion that at the critical point $s = s_c$, $\mathcal{L}$ is independent of all coupling constants. The only parameter left is N, and we have no choice but to expand correlators in powers of $1/N$. The characterization of the critical behavior only requires computations of the exponents z and η_ψ, and associated scaling functions.

We can combine all the above results into scaling forms for the ϕ and Ψ Green's functions at the quantum critical point at $T = 0$. These are, respectively

$$D^{-1}(q_x, q_y, \omega) = q_y^{z-1}\mathcal{F}_D\left(\frac{\omega}{q_y^z}\right) \tag{9.60}$$

$$G^{-1}(q_x, q_y, \omega) = (q_x + q_y^2)^{1-\eta_\psi/2}\mathcal{F}_G\left(\frac{\omega}{(q_x + q_y^2)^{z/2}}\right), \tag{9.61}$$

where $\mathcal{F}_D$ and $\mathcal{F}_G$ are scaling functions.

9.8.4 Large N Expansion

We have come as far as possible by symmetry and scaling analyses alone on $\mathcal{L}$. Further results require specific computations of loop corrections, and these can only be carried out within the context of the $1/N$ expansion. At $N = \infty$, the ϕ propagator at criticality is

$$\frac{D^{-1}}{N} = q_y^2 + \frac{|\omega|}{4\pi|q_y|} \tag{9.62}$$

for imaginary frequencies ω. This is clearly compatible with Eq. 9.60 with $z = 3$. The leading correction to the fermion propagator comes from the self energy associated with one ϕ exchange, and this leads to

$$G_+^{-1} = -i\zeta\omega + q_x + q_y^2 - i\mathrm{sgn}(\omega)\frac{2}{\sqrt{3}(4\pi)^{2/3}N}|\omega|^{2/3}, \tag{9.63}$$

which is also compatible with Eq. 9.61 with $z = 3$ and $\eta_\psi = 0$. Notice also that as $\omega \to 0$, the $\zeta\omega$ term in Eq. 9.63 is smaller than the $|\omega|^{2/3}$ term arising from the self energy at order $1/N$; this relationship is equivalent to our earlier claim that ζ is irrelevant at long scales, and so we should take the limit $\zeta \to 0$ to obtain our leading critical scaling functions.

The structure of Eq. 9.63 also illustrates a key difficulty associated with the $\zeta \to 0$ limit. At $\zeta = 0$, the leading ω dependence of G^{-1} is $\sim |\omega|^{2/3}/N$. Feynman graphs which are sensitive to this ω dependence will therefore acquire additional factors of N, leading to a breakdown of the conventional counting of powers of $1/N$ in the higher loop graphs.

This breakdown of the $1/N$ expansion was investigated by Lee [74] for a 'single patch' theory with fermions only at $\boldsymbol{k}_0$ (and not at $-\boldsymbol{k}_0$). We see from Eq. 9.63 that the $1/N$ term in G^{-1} becomes important when $q_x = q_y = 0$ i.e. the fermion is precisely on the Fermi surface. Thus the power of N is maximized when fermions in all internal lines are on the Fermi surface. Such a Fermi surface restriction is satisfied only in a subspace of reduced dimension in the momentum space integral of any Feynman graph. Lee presented an algorithm for computing the dimensionality of this restricted subspace: he demonstrated that the power of $1/N$ was determined by the genus of the surface obtained after drawing the graph in a double-line representation. So determining the leading $1/N$ terms in Eqs. 9.62 and 9.63 requires summation of the infinite set of planar graphs. This problem remains unsolved, but the unexpected appearance of planar graphs does suggest an important role for the AdS/CFT correspondence.

The structure of the loop expansion for the 'two patch' theory with fermions at $\pm\boldsymbol{k}_0$, as written in Eq. 9.50, was studied in [65]. It was found that $z = 3$ was preserved upto three loops, but a small non-zero value for η_ψ did appear at three loop order. Also, the genus counting of powers of $1/N$ was found to break down, with larger powers of N appearing in some three loop graphs.

9.8.5 AdS/CFT Correspondence

There has been a great deal of recent work [75–93] investigating the structure of Fermi surfaces using the AdS/CFT correspondence. The results obtained so far do have features that resemble our results above for the Ising-nematic transition in two dimensional metals. However, the precise connection remains obscure, and is an important topic for future research. In particular, a microscopic understanding of the field content of the CFT dual of the AdS theory is lacking, although there has been interesting progress very recently [88, 92].

One of the main results of the analysis of [77, 78] is the general structure of the fermion Green's function obtained in a theory dual to a Reissner-Nordstrom black hole in AdS_4. This had the form

$$G^{-1}(\boldsymbol{k}, \omega) = -i\omega + v_F(|\boldsymbol{k}| - k_F) - c_1\omega^\theta, \tag{9.64}$$

where the momentum $\boldsymbol{k}$ is now measured from the origin of momentum space (and not from a Fermi surface), and the complex number c_1 and exponent θ are computable functions of the ultraviolet scaling dimension of the fermion field. The AdS theory only considers a circular Fermi surface, and for this geometry (after appropriate rescaling)

$$v_F(|\boldsymbol{k}| - k_F) = v_F(|\boldsymbol{q} + \boldsymbol{k}_0| - k_F) \approx q_x + q_y^2; \tag{9.65}$$

now Eq. 9.64 is seen to be strikingly similar to Eq. 9.63. Liu et al. [77], and Faulkner et al. [78] also argued that Eq. 9.64 was a generic property of the near horizon geometry of the Reissner-Nordstrom black hole: the geometry changes from AdS_4 near the boundary to $AdS_2 \times R^2$ near the black hole horizon.

It is interesting to compare the structure of the critical theory in the AdS/CFT framework to that found in the subsections above for the Ising-nematic transition in a metal. The latter was described by an infinite set of 2 + 1 dimensional field theories labelled by pairs of momenta on a one-dimensional Fermi surface *i.e.* a S^1/Z_2 set of 2 + 1 dimensional field theories. In the low-energy limit, the AdS/CFT approach yields [77, 78] a $AdS_2 \times R^2$ geometry: this can be interpreted as an infinite set of chiral 1 + 1 dimensional theories labelled by a R^2 set of two-dimensional momenta $\boldsymbol{k}$. It is notable, and perhaps significant, that both approaches have an emergent dimension not found in the underlying degrees of freedom. The Ising nematic theory began with a 2 + 1 dimensional Hamiltonian $\mathcal{S}_c + \mathcal{S}_\phi + \mathcal{S}_{c\phi}$ in Eqs. (9.41, 9.47, 9.48), and ended up with a S^1/Z_2 set of 2+1 dimensional field theories. In AdS/CFT, there is the emergent radial direction representing energy scale. These emergent dimensions imply redundant descriptions, and require associated consistency conditions: we explored such consistency conditions in Sect. 9.8.2, while in AdS/CFT the consistency conditions are Einstein's equations representing the renormalization group flow under changes of energy scale. It would be interesting to see if fluctuations about the classical gravity theory yield corrections to the $AdS_2 \times R^2$ geometry which clarify the connection to our Ising-nematic theory.

Acknowledgements Many of the ideas reviewed in Sect. 9.8 are due to Max Metlistki [65]; I also thank him for valuable comments on the manuscript. I thank the participants of the school for their interest, and for stimulating discussions. This research was supported by the National Science Foundation under grant DMR-0757145, by the FQXi foundation, and by a MURI grant from AFOSR.

References

1. Sachdev, S.: 24th Solvay Conference on Physics, Quantum Theory of Condensed Matter, Brussels, 11–13 October 2008, arXiv:0901.4103
2. Sachdev, S.: Lectures at the International Summer School on Fundamental Problems in Statistical Physics X, August–September 2001, Altenberg, Germany, Physica A **313**, 252 cond-mat/0109419 (2002)
3. Sachdev, S.: In: Carr, L. (ed.) Developments in Quantum Phase Transitions, arXiv:0910.1139
4. Troyer, M., Imada, M., Ueda, K.: J. Phys. Soc. Japan **66**, 2957 (1997)
5. Matsumoto, M., Yasuda, C., Todo, S., Takayama, H.: Phys. Rev. B **65**, 014407 (2002)
6. Wenzel, S., Janke, W.: Phys. Rev. B **79**, 014410 (2009)
7. Campostrini, M., Hasenbusch, M., Pelissetto, A., Rossi, P., Vicari, E.: Phys. Rev. B **65**, 144520 (2002)
8. Sandvik, A.W., Scalapino, D.J.: Phys. Rev. Lett. **72**, 2777 (1994)
9. Matsushita, Y., Gelfand, M.P., Ishii, C.: J. Phys. Soc. Jpn **66**, 3648 (1997)
10. Albuquerque, A.F., Troyer, M., Oitmaa, J.: Phys. Rev. B **78**, 132402 (2008)
11. Ruegg, Ch., Normand, B., Matsumoto, M., Furrer, A., McMorrow, D. F., Krämer, K.W., Güdel, H.-U., Gvasaliya, S.N., Mutka, H., Boehm, M.: Phys. Rev. Lett. **100**, 205701 (2008)
12. Read, N., Sachdev, S.: Phys. Rev. Lett. **62**, 1694 (1989)
13. Tamura, M., Nakao, A., Kato, R.: J. Phys. Soc. Japan **75**, 093701 (2006)
14. Shimizu, Y., Akimoto, H., Tsujii, H., Tajima, A., Kato, R.: Phys. Rev. Lett. **99**, 256403 (2007)
15. Senthil, T., Vishwanath, A., Balents, L., Sachdev, S., Fisher, M.P.A.: Science **303**, 190 (2004)
16. Sandvik, A.W.: Phys. Rev. Lett. **98**, 227202 (2007)
17. Lou, J., Sandvik, A.W., Kawashima, N.: arXiv:0908.0740
18. Melko, R.G., Kaul, R.K.: Phys. Rev. Lett. **100**, 017203 (2008)
19. Melko, R.G., Kaul, R.K.: Phys. Rev. B **78**, 014417 (2008)
20. Jiang, F.-J., Nyfeler, M., Chandrasekharan, S., Wiese, U.-J.: Stat. Mech. P02009 (2008)
21. Kuklov, A.B., Matsumoto, M., Prokof'ev, N.V., Svistunov, B.V., Troyer, M.: Phys. Rev. Lett. **101**, 050405 (2008)
22. Castro Neto, A.H., Guinea, F., Peres, N.M., Novoselov, K.S., Geim, A.K.: Rev. Mod. Phys. **81**, 109 (2009)
23. Chakravarty, S., Halperin, B.I., Nelson, D.R.: Phys. Rev. B **39**, 2344 (1989)
24. Sachdev, S., Ye, J.: Phys. Rev. Lett. **69**, 2411 (1992)
25. Chubukov, A.V., Sachdev, S., Ye J.: Phys. Rev. B **49**, 11919 (1994)
26. Sheehy, D.E., Schmalian, J.: Phys. Rev. Lett. **99**, 226803 (2007)
27. Fisher, M.P.A., Weichman, P.B., Grinstein, G., Fisher, D.S.: Phys. Rev. B **40**, 546 (1989)
28. Damle, K., Sachdev, S.: Phys. Rev. B **56**, 8714 (1997)
29. Sachdev, S.: Quantum Phase Transitions. Cambridge University Press, Cambridge, U.K. (1999)

30. Fritz, L., Schmalian, J., Müller, M., Sachdev, S.: Phys. Rev. B **78**, 085416 (2008)
31. Müller, M., Sachdev, S.: Phys. Rev. B **78**, 115419 (2008)
32. Müller, M., Fritz, L., Sachdev, S.: Phys. Rev. B **78**, 115406 (2008)
33. Kovtun, P., Yaffe, L.G.: For the case of neutral boson superfluids (but not charged systems like graphene), hydrodynamic 'long-time tails' cause the constants D and Θ_2 to acquire a weak logarithmic dependence on $\hbar\omega/k_BT$ at small ω in a sample with perfect momentum conservation. Phys. Rev. D. **68**, 025007 (2003)
34. Kovtun, P.K., Son, D.T., Starinets, A.: Phys. Rev. Lett. **94**, 11601 (2005)
35. Müller, M., Schmalian, J., Fritz, L.: Phys. Rev. Lett. **103**, 025301 (2009)
36. Fisher, M.P.A., Grinstein, G., Girvin, S.M.: Phys. Rev. Lett. **64**, 587 (1990)
37. Fisher, M.P.A.: Phys. Rev. Lett. **65**, 923 (1990)
38. Itzhaki, N., Maldacena, J.M., Sonnenschein, J., Yankielowicz, S.: Phys. Rev. D **58**, 046004 (1998)
39. Herzog, C.P., Kovtun, P., Sachdev, S., Son, D.T.: Phys. Rev. D **75**, 085020 (2007)
40. Dasgupta, C., Halperin, B.I.: Phys. Rev. Lett. **47**, 1556 (1981)
41. Hartnoll, S.A., Kovtun, P.K., Müller, M., Sachdev, S.: Phys. Rev. B **76**, 144502 (2007)
42. Müller, M., Fritz, L., Sachdev, S., Schmalian, J.: Advances in theoretical physics. In: Landau Memorial Conference, AIP Conference Proceedings, Chernogolovka 2008, vol. 1134, p. 170 (2009)
43. Kadanoff, L.P., Martin, P.C.: Annals of Physics **24**, 419–469 (1963)
44. Landau, L.D., Lifshitz, E.M.: Fluid Mechanics, Section 127. Butterworth-Heinemann, Oxford (1987)
45. Hartnoll, S.A., Herzog, C.P.: Phys. Rev. D **77**, 106009 (2008)
46. Bhaseen, M.J., Green, A.G., Sondhi, S.L.: Phys. Rev. Lett. **98**, 166801 (2007)
47. Bhaseen, M.J., Green, A.G., Sondhi, S.L.: Phys. Rev. B **79**, 094502 (2009)
48. Hartnoll, S.A., Herzog, C.P.: Phys. Rev. D **76**, 106012 (2007)
49. Herbut, I.F., Juričić, V., Roy, B.: Phys. Rev. B **79**, 085116 (2009)
50. Juričić, V., Herbut, I.F., Semenoff, G.W.: Phys. Rev. B **80**, 081405 (R) (2009)
51. Vojta, M., Zhang, Y., Sachdev, S.: Phys. Rev. Lett. **85**, 4940 (2000)
52. Vojta, M., Zhang, Y., Sachdev, S.: Phys. Rev. Lett. **100**, 089904 (2008)
53. Laughlin, R.B.: Phys. Rev. Lett. **80**, 5188 (1998)
54. Li, M.-R., Hirschfeld, P.J., Woelfle, P.: Phys. Rev. B **63**, 054504 (2001)
55. Vojta, M., Zhang, Y., Sachdev, S.: Int. J. Mod. Phys. B **14**, 3719 (2000)
56. Khveshchenko, D.V., Paaske, J.: Phys. Rev. Lett. **86**, 4672 (2001)
57. Rosenstein, B., Warr, B.J., Park, S.H.: Phys. Rep. **205**, 59 (1991)
58. Kärkkäinen, L., Lacaze, R., Lacock, P., Petersson, B.: Nucl. Phys. B **415**, 781 (1994)
59. Daou, R., Chang, J., LeBoeuf, D., Cyr-Choiniere, O., Laliberte, F., Doiron-Leyraud, N., Ramshaw, B.J., Liang, Ruixing, Bonn, D.A., Hardy, W.N., Taillefer, L.: Nature **463**, 519 (2010)
60. Ando, Y., Segawa, K., Komiya, S., Lavrov, A.N.: Phys. Rev. Lett. **88**, 137005 (2002)
61. Hinkov, V., Haug, D., Fauqué, B., Bourges, P., Sidis, Y., Ivanov, A., Bernhard, C., Lin, C.T., Keimer, B.: Science **319**, 597 (2008)
62. Kim, E.-A., Lawler, M.J., Oreto, P., Sachdev, S., Fradkin, E., Kivelson, S.A.: Phys. Rev. B **77**, 184514 (2008)
63. Huh, Y., Sachdev, S.: Phys. Rev. B **78**, 064512 (2008)
64. Fritz, L., Sachdev, S.: Phys. Rev. B **80**, 144503 (2009)
65. Metlitksi, M.A., Sachdev, S.: arXiv:1001.1153
66. Metlitski, M.A., Sachdev, S.: to appear
67. Luther, A.: Phys. Rev. B **19**, 320 (1979)
68. Houghton, A., Marston, J.B.: Phys. Rev. B **48**, 7790 (1993)
69. Kwon, H.-J., Houghton, A., Marston, J.B.: Phys. Rev. Lett. **73**, 284 (1994)
70. Haldane, F.D.M.: Course CXXI "Perspectives in Many-Particle Physics". In: Broglia, R.A., Schrieffer, J.R. (eds.) Proceedings of the International School of Physics "Enrico Fermi", North-Holland, Amsterdam (1994); arXiv:cond-mat/0505529

71. Castro Neto, A.H., Fradkin, E.H.: Phys. Rev. B **51**, 4084 (1995)
72. Lawler, M.J., Fernandez, V., Barci, D.G., Fradkin, E., Oxman, L. Phys. Rev. B **73**, 085101 (2006)
73. Lawler, M.J., Fradkin, E.: Phys. Rev. B **75**, 033304 (2007)
74. Lee, S.-S.: Phys. Rev. B **80**, 165102 (2009)
75. Lee, S.-S.: Phys. Rev. D **79**, 086006 (2009)
76. Cubrovic, M., Zaanen, J., Schalm, K.: Science **325**, 439 (2009)
77. Liu, H., McGreevy, J., Vegh, D.: arXiv:0903.2477
78. Faulkner, T., Liu, H., McGreevy, J., Vegh, D.: arXiv:0907.2694
79. Denef, F., Hartnoll, S.A., Sachdev, S.: Phys. Rev. D **80**, 126016 (2009)
80. Hartnoll, S.A., Hofman, D.M.: arXiv:0912.0008
81. Maity, D., Sarkar, S., Sircar, N., Sathiapalan, B., Shankar, R.: arXiv:0909.4051
82. Hung, L.-Y., Sinha, A.: arXiv:0909.3526
83. D'Hoker, E., Kraus, P.: arXiv:0911.4518
84. Goldstein, K., Kachru, S., Prakash, S., Trivedi, S.P.: arXiv:0911.3586
85. Faulkner, T., Horowitz, G.T., McGreevy, J., Roberts, M.M., Vegh, D.: arXiv:0911.3402
86. Nishioka, T., Ryu, S., Takayanagi, T.: arXiv:0911.0962
87. Gubser, S.S., Rocha, F.D., Talavera, P.: arXiv:0911.3632
88. Herzog, C.P., Klebanov, I.R., Pufu, S.S., Tesileanu, T.: arXiv:0911.0400
89. Hartman, T., Song, W., Strominger, A.: arXiv:0912.4265
90. Hartnoll, S.A., Polchinski, J., Silverstein, E., Tong, D.: arXiv:0912.1061
91. Albash, T., Johnson, C.V.: arXiv:1001.3700
92. Faulkner, T., Polchinski, J.: arXiv:1001.5049
93. Jensen, K., Karch, A., Thompson, E.G.: arXiv:1002.2447

Chapter 10
Introduction to Holographic Superconductors

Gary T. Horowitz

Abstract These lectures give an introduction to the theory of holographic superconductors. These are superconductors that have a dual gravitational description using gauge/gravity duality. After introducing a suitable gravitational theory, we discuss its properties in various regimes: the probe limit, the effects of backreaction, the zero temperature limit, and the addition of magnetic fields. Using the gauge/gravity dictionary, these properties reproduce many of the standard features of superconductors. Some familiarity with gauge/gravity duality is assumed. A list of open problems is included at the end.

10.1 Introduction

The name "holographic superconductor" suggests that one can look at a two (spatial) dimensional superconductor and see a three dimensional image. We will see that there is a class of superconductors for which this is true, but the image one "sees" is quite striking. It involves a charged black hole with nontrivial "hair". This remarkable connection between condensed matter and gravitational physics was discovered just a few years ago. It grew out of the gauge/gravity duality which has emerged from string theory [54, 26]. Although this duality was first formulated as a equivalence between a certain gauge theory and a theory of quantum gravity (and provided new insights into each of these theories), over the past decade it has been applied with notable success to other areas of physics as well. Many of these new applications are discussed in other lectures in this school. I will focus

G. T. Horowitz (✉)
Physics Department, UCSB, Santa Barbara, CA 93106, USA
e-mail: gary@physics.ucsb.edu

E. Papantonopoulos (ed.), *From Gravity to Thermal Gauge Theories: The Ads/CFT Correspondence*, Lecture Notes in Physics, 828,
DOI: 10.1007/978-3-642-04864-7_10,

on the application to superconductivity. These lectures will be heavily based on [36, 37, 44, 38]. For a more general discussion of applying gauge/gravity duality to condensed matter, see the excellent reviews by Hartnoll [33], Herzog [39] and McGreevy [55].

I will start with a brief introduction to superconductivity. In Sect. 10.2, I will introduce a simple model for a holographic superconductor. Most of our discussion will be devoted to exploring the consequences of this model, beginning in Sect. 10.3 with the probe limit—a simplification of the model which preserves most of the physics. In Sect. 10.4 we will discuss the full solution with backreaction. We next examine the ground state of the superconductor Sect. 10.5, and study its behavior when magnetic fields are added (Sect. 10.6). We conclude with a brief discussion of recent developments (Sect. 10.7) and a list of open problems (Sect. 10.8).

10.1.1 Superconductivity

It was noticed in the early part of the 20th century that the electrical resistivity of most metals drops suddenly to zero as the temperature is lowered below a critical temperature T_c. These materials were called superconductors.[1] A second independent property of these materials was the Meissner effect: A magnetic field is expelled when $T<T_c$. This is perfect diamagnetism and does not follow from the perfect conductivity (which alone would imply that a pre-existing magnetic field is trapped inside the sample). A phenomenological description of both of these properties was first given by the London brothers in 1935 with the simple equation $J_i \propto A_i$ [51]. Taking a time derivative yields $E_i \propto \partial J_i/\partial t$, showing that electric fields accelerate superconducting electrons rather than keeping their velocity constant as in Ohm's law with finite conductivity. Taking the curl of both sides and combining with Maxwell's equations yields $\nabla^2 B_i \propto B_i$ showing the decay of magnetic fields inside a superconductor.

In 1950, Landau and Ginzburg described superconductivity in terms of a second order phase transition whose order parameter is a complex scalar field φ [22]. The density of superconducting electrons is given by $n_s = |\varphi(x)|^2$. The contribution of φ to the free energy is assumed to take the form

$$F = \alpha(T - T_c)|\varphi|^2 + \frac{\beta}{2}|\varphi|^4 + \cdots \tag{10.1}$$

where α and β are positive constants and the dots denote gradient terms and higher powers of φ. Clearly for $T > T_c$ the minimum of the free energy is at $\varphi = 0$ while for $T<T_c$ the minimum is at a nonzero value of φ. This is just like the Higgs

---

[1] A good general reference is [63].

mechanism in particle physics, and is associated with breaking a $U(1)$ symmetry. The London equation follows from this spontaneous symmetry breaking [66].

A more complete theory of superconductivity was given by Bardeen, Cooper and Schrieffer in 1957 and is known as BCS theory [7]. They showed that interactions with phonons can cause pairs of elections with opposite spin to bind and form a charged boson called a Cooper pair. Below a critical temperature T_c, there is a second order phase transition and these bosons condense. The DC conductivity becomes infinite producing a superconductor. The pairs are not bound very tightly and typically have a size which is much larger than the lattice spacing. In the superconducting ground state, there is an energy gap Δ for charged excitations. This gap is typically related to the critical temperature by $\Delta \approx 1.7T_c$. The charged excitations are "dressed electrons" called quasiparticles. The gap in the spectrum results in a gap in the (frequency dependent) optical conductivity. If a photon of frequency ω hits the superconductor, it must produce two quasiparticles. The binding energy of the Cooper pair is very small, but the energy of each quasiparticle is Δ, so the gap in the optical conductivity is $\omega_g = 2\Delta \approx 3.5T_c$.

It was once thought that the highest T_c for a BCS superconductor was around 30 K. But in 2001, MgB_2 was found to be superconducting at 40 K and is believed to be described by BCS. Some people now speculate that BCS could describe a superconductor with $T_c = 200$ K^2.[2]

A new class of high T_c superconductors were discovered in 1986 [9]. They are cuprates and the superconductivity is along the CuO_2 planes. The highest T_c known today (at atmospheric pressure) is $T_c = 134$ K for a mercury, barium, copper oxide compound. If you apply pressure, T_c climbs to about 160 K. Another class of superconductors were discovered in 2008 based on iron and not copper [47]. The highest T_c so far is 56 K. These materials are also layered and the superconductivity is again associated with the two dimensional planes. They are called iron pnictides since they involve other elements like arsenic in the nitrogen group of the periodic table.

There is evidence that electron pairs still form in these high T_c materials, but the pairing mechanism is not well understood. Unlike BCS theory, it involves strong coupling. Gauge/gravity duality is an new tool to study strongly coupled field theories. In particular, it allows one to compute dynamical transport properties of strongly coupled systems at nonzero temperature. Condensed matter theorists have very few tools to do this. We will describe below the first steps toward applying this new tool to better understand high T_c superconductivity. I must stress at the beginning that we are still at the early stages of this endeavor. We will construct simple gravity models and show that they reproduce basic properties of superconductors. But our models are too crude to make detailed comparisons with any real-world material.

[2] Discussion at the KITP program on Quantum Criticality and the AdS/CFT Correspondence, July 2009.

10.2 A Gravitational Dual

How do we go about constructing a holographic dual for a superconductor? The minimal ingredients are the following. In the superconductor we need a notion of temperature. On the gravity side, that role is played by a black hole. Recall that in the 1970s Hawking (following work by Bekenstein and others) showed that stationary black holes are thermodynamic objects with a temperature T related to the surface gravity κ via $T = \kappa/2\pi$. In gauge/gravity duality, the Hawking temperature of the black hole is identified with the temperature of the dual field theory.[3] Since gauge/gravity duality traditionally requires that spacetime asymptotically approach anti de Sitter (AdS) space at infinity, we will be studying black holes in AdS. Unlike asymptotically flat black holes, these black holes have the property that at large radius, their temperature increases with their mass, i.e., they have positive specific heat, just like familiar nongravitational systems. There are also planar AdS black holes, which will be of most interest. These black holes always have positive specific heat.

In the superconductor, we also need a condensate. In the bulk, this is described by some field coupled to gravity. A nonzero condensate corresponds to a static nonzero field outside a black hole. This is usually called black hole "hair". So to describe a superconductor, we need to find a black hole that has hair at low temperatures, but no hair at high temperatures. More precisely, we need the usual Schwarzschild or Reissner–Nordstrom AdS black hole (which exists for all temperatures) to be unstable to forming hair at low temperature. At first sight, this is not an easy task. There are "no-hair theorems" which say that certain matter fields must be trivial outside a black hole (see, e.g., [10, 42]). The idea behind these theorems is simply that matter outside a black hole wants to fall into the horizon (or radiate out to infinity in the asymptotically flat case). However, there is no general "no-hair theorem". Each matter field must be considered separately. The result is a set of black hole uniqueness theorems showing that when gravity is coupled to certain matter fields, stationary black holes are uniquely characterized by their conserved charges: mass, angular momentum and electromagnetic charge. These theorems usually require linear matter fields or scalars with certain potentials $V(\phi)$. Counterexamples to a general no-hair theorem have been known since the early 1990s. (So our task is not impossible.) For example, it was shown that static Yang–Mills fields can exist outside the horizon [65].

String theory has many dilatonic black holes with scalar hair, and one might be tempted to try to use one of these to model a superconductor. But if the action includes a term like $e^{2a\phi}F_{\mu\nu}F^{\mu\nu}$, this is doomed to failure. In this case, F^2 is a source for ϕ, so all charged black holes have nonzero ϕ. This "secondary hair" is not what we want, since we want the hair to go away at high temperatures. (Theories with more general coupling $f(\phi)F_{\mu\nu}F^{\mu\nu}$ are possible candidates provided

[3] I will assume that the reader is familiar with the basics of gauge/gravity duality. If not, see other contributions in this book.

f does not have a linear term in ϕ. However, the example we will study uses a standard Maxwell action.) A general argument against AdS black holes developing scalar hair was given by Hertog [38]. He considered a real scalar field with arbitrary potential $V(\phi)$ (with negative extremum so AdS is a solution), and showed that neutral AdS black holes can have scalar hair if and only if AdS itself is unstable. This is clearly unacceptable.

A surprisingly simple solution to this problem was found by Gubser [25]. He argued that a charged scalar field around a *charged* black hole in AdS would have the desired property. Consider

$$S = \int d^4x\sqrt{-g}\left(R + \frac{6}{L^2} - \frac{1}{4}F_{\mu\nu}F^{\mu\nu} - |\nabla\Psi - iqA\Psi|^2 - m^2|\Psi^2|\right). \tag{10.2}$$

This is just general relativity with a negative cosmological constant $\Lambda = -3/L^2$, coupled to a Maxwell field and charged scalar with mass m and charge q. It is easy to see why black holes in this theory might be unstable to forming scalar hair: For an electrically charged black hole, the effective mass of Ψ is $m^2_{\text{eff}} = m^2 + q^2 g^{tt} A_t^2$. But the last term is negative, so there is a chance that m^2_{eff} becomes sufficiently negative near the horizon to destabilize the scalar field. Furthermore, as one lowers the temperature of a charged black hole, it becomes closer to extremality which means that g_{tt} is closer to developing a double zero at the horizon. This means that $|g^{tt}|$ becomes larger and the potential instability becomes stronger at low temperature.

We will see that black holes in this theory indeed develop scalar hair at low temperature. One might wonder why such a simple type of hair was not noticed earlier. One reason is that this does not work for asymptotically flat black holes. The AdS boundary conditions are crucial. One way to understand the difference is by the following quantum argument. Let Q_i be the initial charge on the black hole. If qQ_i is large enough, even maximally charged black holes with zero Hawking temperature create pairs of charged particles. This is simply due to the fact that the electric field near the horizon is strong enough to pull pairs of oppositely charged particles out of the vacuum via the Schwinger mechanism of ordinary field theory. The particle with opposite charge to the black hole falls into the horizon, reducing Q_i while the particle with the same sign charge as the black hole is repelled away. In asymptotically flat spacetime, these particles escape to infinity, so the final result is a standard Reissner–Nordstrom black hole with final charge $Q_f < Q_i$. In AdS, the charged particles cannot escape since the negative cosmological constant acts like a confining box, and they settle outside the horizon. This gas of charged particles is the quantum description of the hair. This quantum process has an entirely classical analog in terms of superradiance of the charged scalar field.

The four dimensional bulk theory (10.2) is dual to a 2 + 1 dimensional boundary theory. This is the right context to try to understand the superconductivity associated with two dimensional planes in the high T_c cuprates or iron pnictides. One can also study this theory in five dimensions to describe 3 + 1 dimensional superconductors. I should emphasize that at the moment we are not trying to derive the gravitational theory from string theory. The idea is to find a simple gravity theory with the properties we want, and analyze it using standard entries in the gauge/gravity duality dictionary. However, we will see later that this simple model can, in fact, be realized as a consistent truncation of string theory.

Before we proceed, I should comment on the following confusing point. In gauge/gravity duality, gauge symmetries in the bulk correspond to global symmetries in the dual field theory. So although the scalar hair breaks a local $U(1)$ symmetry in the bulk, the dual description consists of a condensate breaking a global $U(1)$ symmetry. Thus strictly speaking, the dual theory is a superfluid rather than a superconductor. The superfluid properties of this model have indeed been investigated [4, 8, 41]. However, one can still view the dual theory as a superconductor in the limit that the $U(1)$ symmetry is "weakly gauged".[4] In fact, most of condensed matter physics does not include dynamical photons, since their effects are usually small. In particular, BCS theory only includes the electrons and phonons. Electromagnetic fields are usually introduced as external fields, as we will do here. (Unfortunately, our electromagnetic fields will not be dynamical on the boundary.)

10.3 Probe Limit

If one rescales $A_\mu = \tilde{A}_\mu/q$ and $\Psi = \tilde{\Psi}/q$, then the matter action in (10.2) has a $1/q^2$ in front, so the backreaction of the matter fields on the metric is suppressed when q is large. The limit $q \to \infty$ with $\tilde{A}_\mu$ and $\tilde{\Psi}$ fixed is called the probe limit. It simplifies the problem but retains most of the interesting physics since the non-linear interactions between the scalar and Maxwell field are retained. In this section, we will explore this probe limit. We first discuss the formation of the condensate and then compute the conductivity. To simplify the presentation, we will drop the tildes.

10.3.1 Condensate

We start with the planar Schwarzschild anti-de Sitter black hole in four dimensions

$$ds^2 = -f(r)dt^2 + \frac{dr^2}{f(r)} + r^2(dx^2 + dy^2), \tag{10.3}$$

where

[4] This means that one imagines that the dual action includes terms like $|\nabla_i - ieA_i)\varphi|^2$ with very small charge e.

$$f = \frac{r^2}{L^2}\left(1 - \frac{r_0^3}{r^3}\right). \tag{10.4}$$

L is the AdS radius, and the Schwarzschild radius r_0 determines the Hawking temperature of the black hole:

$$T = \frac{3r_0}{4\pi L^2}. \tag{10.5}$$

In the probe approximation this metric is a fixed background in which we solve the Maxwell-scalar equations.

We assume a plane symmetric ansatz,

$$\Psi = \psi(r), \qquad A_t = \phi(r). \tag{10.6}$$

With $A_r = A_x = A_y = 0$, the Maxwell equations imply that the phase of ψ must be constant. Without loss of generality we therefore take ψ to be real. The Maxwell-scalar field equations reduce to the following coupled, nonlinear, ordinary differential equations:

$$\psi'' + \left(\frac{f'}{f} + \frac{2}{r}\right)\psi' + \frac{\phi^2}{f^2}\psi - \frac{m^2}{f}\psi = 0, \tag{10.7}$$

$$\phi'' + \frac{2}{r}\phi' - \frac{2\psi^2}{f}\phi = 0. \tag{10.8}$$

The key term in the first equation is $(\phi^2/f^2)\psi$. This comes in with the opposite sign of the mass term and will cause the scalar hair to form at low temperature.

We first consider the case $m^2 = -2/L^2$. It might seem strange to make the scalar field tachyonic, but this mass is perfectly allowed in gauge/gravity duality. First note that a tachyonic mass in field theory does not describe particles moving faster than the speed of light. Instead, it usually describes an instability. The value $\psi = 0$ is unstable and the field rolls off the potential. However, Breitenlohner and Freedman [9] showed that AdS_{d+1} spacetime is stable even with scalar fields with $m^2 < 0$ provided $m^2 \geq m_{BF}^2$ with

$$m_{BF}^2 = -\frac{d^2}{4L^2}. \tag{10.9}$$

This is because there is so much volume at large radius that the positive gradient energy can compensate for a negative m^2. Our choice of mass satisfies this bound and in fact corresponds to a conformally coupled scalar in AdS_4. The standard compactification of 11D supergravity on S^7 produces many fields with this mass.

We now consider the boundary conditions. At the horizon, one often argues that $\phi = A_t$ must vanish in order for $g^{\mu\nu}A_\mu A_\nu$ to remain finite. This is correct, but A_μ is gauge dependent, so it is not obvious that a diverging vector potential is a problem if the Maxwell field remains finite. A better argument for setting $A_t = 0$ at the horizon is the following.[5] The source for Maxwell's equations in the bulk is, of course, gauge invariant. But in a gauge in which ψ is real, the current is just $\psi^2 A_\mu$. Since the current must remain finite at the horizon, we need A_μ to remain finite, and hence $\phi = A_t = 0$. Even in Einstein–Maxwell theory (without charged sources) A_t must vanish at a static black hole horizon by the following argument: For describing thermal properties of the black hole, one should use the Euclidean solution. The Wilson loop of A_μ around the Euclidean time circle is finite and gauge invariant. If A_t is nonzero at the horizon, the Wilson loop is nonzero around a vanishing circle which implies that the Maxwell field is singular.

There is another constraint that must be satisfied in order for the solution to be smooth at the horizon. Mutliplying (10.7) by f and evaluating at $r = r_0$ one finds $f'\psi' = m^2\psi$, so $\psi(r_0)$ and $\psi'(r_0)$ are not independent. As a result, even though we start with two second order equations which have a four parameter family of solutions, there is only a two parameter subfamily which is regular at the horizon. They can be labelled by $\psi(r_0)$ and $\phi'(r_0)$. (Note that $\phi'(r_0)$ is essentially the electric field at the horizon: $(\phi')^2 = F_{\mu\nu}F^{\mu\nu}$.)

We now turn to the boundary condition at infinity. Asymptotically:

$$\psi = \frac{\psi^{(1)}}{r} + \frac{\psi^{(2)}}{r^2} + \cdots . \tag{10.10}$$

and

$$\phi = \mu - \frac{\rho}{r} + \cdots . \tag{10.11}$$

Usually, normalizability requires that the leading coefficient in ψ must vanish. However, since we have chosen a mass close the the BF bound, even the leading term in ψ is normalizable. In this case, one has a choice: One can consider solutions with $\psi^{(1)} = 0$ or $\psi^{(2)} = 0$. For definiteness, we will mostly consider standard boundary conditions, $\psi^{(1)} = 0$.

After imposing this asymptotic boundary condition, we have a one parameter family of solutions, which can be found numerically [36]. (See [24] for an accurate analytic approximation and [50] for an analytic solution of a related model.) We will not present any plots of $\psi(r)$ and $\phi(r)$ since they look rather boring. It is easy to see from (10.8) that ϕ is a monotonic function. It starts at zero at the horizon and at any local extremum $\phi'' \propto \phi$. So it cannot have a positive maximum or a negative minimum. If it starts increasing away from the horizon, it continues to increase and asymptotically approaches the constant μ. $\psi(r)$ does not have to be monotonic. There is a discrete infinite family of solutions for $\psi(r)$ that satisfy our asymptotic boundary conditions. They can be labelled by the number of times ψ vanishes. It is believed that only the lowest solution which monotically decreases from $\psi(r_0)$ to

---

[5] I thank Karl Landsteiner for suggesting this.

zero is stable, although I do not think a stability analysis has been performed yet. Despite the fact that the solutions of interest are simple monotonic functions, they have important consequences for the dual field theory as we now discuss.

Since we have not started with a consistent truncation of string theory, we do not have a detailed microscopic description of the dual field theory. Nevertheless, basic elements of the gauge/gravity duality dictionary allow us to say the following. The dual theory is a 2 + 1 dimensional conformal field theory (CFT) at temperature T given by (10.5). The local gauge symmetry in the bulk corresponds to a global $U(1)$ symmetry in the CFT. The asymptotic behavior of the bulk solution determines certain properties of the dual field theory. For example, from (10.11), μ is the chemical potential and ρ is the charge density. It may seem strange that our superconductor has a nonzero charge density. After all, ordinary superconductors are electrically neutral. Perhaps the best analogy is to say that our holographic superconductor is modeling the electrons, but does not include the (positively charged) atomic lattice. Indeed, our model has translational symmetry, and there is no sign of any lattice. From a practical standpoint, we need the charge since neutral black holes in the bulk are not unstable to forming scalar hair. In addition, without the charge (or chemical potential), the dual theory is scale invariant and cannot have a phase transition.

The dual theory also has an operator charged under the $U(1)$ which is dual to ψ. Since we have chosen m^2 close to the BF bound, there are two possible operators depending on how one quantizes ψ in the bulk.[6] [49] If the modes are defined with the standard boundary condition (faster falloff) for ψ in the bulk, the dual operator has dimension two. In this case, a nonzero $\psi^{(1)}$ corresponds to adding a source for this operator in the CFT, and a nonzero $\psi^{(2)}$ corresponds to a nonzero expectation value[7]

$$O_2 = \psi^{(2)}. \tag{10.12}$$

Since we want the condensate to turn on without being sourced, we have set $\psi^{(1)} = 0$. There is an alternative quantization of ψ in the bulk in which the roles of $\psi^{(1)}$ and $\psi^{(2)}$ are reversed. ψ is now dual to a dimension one operator. If one wishes to study this case, one should impose the boundary condition $\psi^{(2)} = 0$.

We want to know how the condensate O_2 behaves as a function of temperature. Before presenting the results we have to discuss an important scaling symmetry of our problem. In any conformal field theory on R^n, one can change the temperature by a simple rescaling. In the bulk, this is reflected in the statement that the rescaling

$$r \to ar, \quad (t,x,y) \to (t,x,y)/a, \quad r_0 \to ar_0 \tag{10.13}$$

---

[6] We will work in the classical limit, but the correspondence applies to the full quantum theory.

[7] This normalization of O differs from that of [36] by a factor of $\sqrt{2}$.

leaves the form of the black hole (10.3) invariant with $f \to a^2 f$. It is easy to check that the Maxwell-scalar field equations (10.7, 10.8) are invariant under this rescaling if $\phi \to a\phi$ (so $A = \phi dt$ is invariant) and ψ is unchanged $\psi \to \psi$. Rather than discuss this trivial dependence on temperature which simply reflects the scaling dimension, we are interested in the dependence of a dimensionless measure of the condensate as a function of a dimensionless measure of the temperature. It is convenient to use the chemical potential to fix a scale and consider $\sqrt{O_2}/\mu$ as a function of T/μ. When one does this, one finds that the condensate is nonzero only when T/μ is small enough. Setting T_c equal to the critical temperature when the condensate first turns on, we get Fig. 10.1.

This curve is qualitatively similar to that obtained in BCS theory, and observed in many materials, where the condensate rises quickly as the system is cooled below the critical temperature and goes to a constant as $T \to 0$. Near T_c, there is a square root behavior $O_2 = 100T_c^2(1 - T/T_c)^{1/2}$. This is the standard behavior predicted by Landau–Ginzburg theory.

A nonzero condensate means that the black hole in the bulk has developed scalar hair. One can compute the free energy (Euclidean action) of these hairy configurations and compare with the solution $\psi = 0, \phi = \rho(1/r_0 - 1/r)$ with describes a black hole with the same charge or chemical potential, but no scalar hair. It turns out that the free energy is always lower for the hairy configurations and becomes equal as $T \to T_c$ [36]. The difference of free energies scales like $(T_c - T)^2$ near the transition, showing that this is a second order phase transition. Actually, you can compare the free energy at fixed charge or fixed chemical potential. These correspond to two different ensembles. In both cases, the free energy is lower for the hairy configuration.

We now generalize to other masses. Recall that the asymptotic behavior of a scalar field of mass m in AdS_4 is

$$\psi = \frac{\psi_-}{r^{\lambda_-}} + \frac{\psi_+}{r^{\lambda_+}} + \cdots \tag{10.14}$$

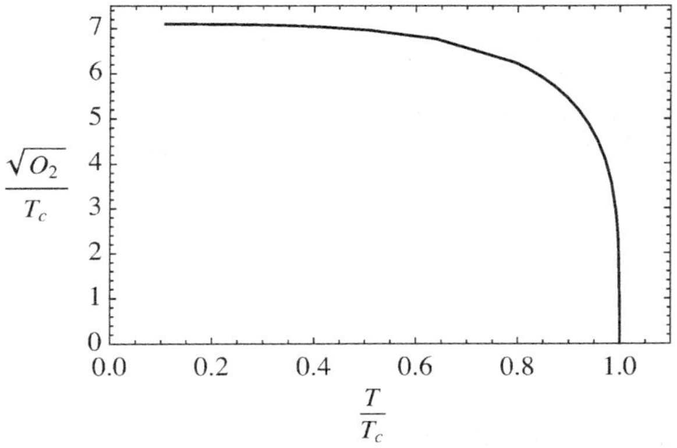

Fig. 10.1 The condensate as a function of temperature. The critical temperature is proportional to the chemical potential

where

$$\lambda_{\pm} = \frac{1}{2}\left(3 \pm \sqrt{9 + 4(mL)^2}\right). \tag{10.15}$$

For $m^2 \geq m_{BF}^2 + L^{-2}$, only the mode with λ_+ falloff is normalizable, and so we interperet $\psi_+ = O$, where O is the expectation value of the operator dual to the scalar field, and λ_+ is the dimension of the operator. ψ_- is dual to a source for this operator, so we only consider solutions with the standard boundary condition $\psi_- = 0$. For $m_{BF}^2 < m^2 < m_{BF}^2 + L^{-2}$ one can also consider the alternative boundary condition $\psi_+ = 0$, and the dual operator has dimension λ_-.

Figure 10.2 shows the results for the condensate for various masses. It is convenient to label the curves in terms of the dimension of the condensate λ, rather than the mass to distinguish the two possible cases when $m^2 = -2/L^2$. The masses range from the BF bound to $m^2 = 0$. Similar behavior is found for $m^2 > 0$ [48].

The qualitative behavior is the same as before. In all cases, there is a critical temperature T_c (proportional to the chemical potential) above which the condensate is zero. Near the critical temperature, $O \propto (T_c - T)^{1/2}$. In all but one case, the condensate saturates as $T \to 0$. The exceptional case is the dimension one curve which starts to grow at low temperature. When the condensate becomes very large, the backreaction on the bulk metric can no longer be neglected. We will see later that in the full solution with backreaction, the condensate approaches a finite limit at zero temperature.

It turns out that the ratio T_c/μ increases as the dimension of the operator decreases. For $\lambda > 3/2$, one can understand this since a smaller λ corresponds to a smaller m^2 making is easier for the scalar hair to form. However, this continues for $\lambda < 3/2$ when one must increase m^2 (and use the alternative boundary conditions) to decrease the dimension. In fact, as observed in [14] , T_c/μ appears to diverge as λ approaches the unitarity bound, 1/2. The reason for this is not clear. But the

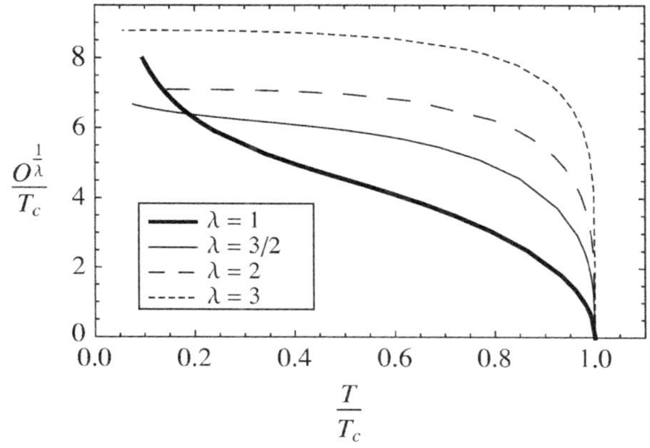

Fig. 10.2 Condensates with different dimension, λ, as a function of temperature. The condensate tends to increase with λ. Figure is taken from [44]

lesson is that, at least in this simple model, to have a higher temperature superconductor, one should lower the dimension of the condensate.

10.3.2 Conductivity

We want to compute the optical conductivity, i.e. the conductivity as a function of frequency. By symmetry, it suffices to consider just the conductivity in the x direction. According to the gauge/gravity dictionary, this is obtained by solving for fluctuations in the Maxwell field in the bulk. Maxwell's equation with zero spatial momentum and time dependence $e^{-i\omega t}$ gives the following equation for $A_x(r)$:

$$A_x'' + \frac{f'}{f}A_x' + \left(\frac{\omega^2}{f^2} - \frac{2\psi^2}{f}\right)A_x = 0. \tag{10.16}$$

We want to solve this with ingoing wave boundary conditions at the horizon, since this corresponds to causal propagation on the boundary, i.e., yields the retarded Green's function [61]. Asymptotically,

$$A_x = A_x^{(0)} + \frac{A_x^{(1)}}{r} + \cdots \tag{10.17}$$

The gauge/gravity dictionary says the limit of the electric field in the bulk is the electric field on the boundary: $E_x = -\dot{A}_x^{(0)}$, and the expectation value of the induced current is the first subleading term: $J_x = A_x^{(1)}$. From Ohm's law we get:

$$\sigma(\omega) = \frac{J_x}{E_x} = -\frac{J_x}{\dot{A}_x^{(0)}} = -\frac{iJ_x}{\omega A_x^{(0)}} = -\frac{iA_x^{(1)}}{\omega A_x^{(0)}}. \tag{10.18}$$

The real part of the conductivity is given in Fig. 10.3 for the case $\lambda = 2$. Above the critical temperature, the conductivity is constant [40]. As you start to lower the

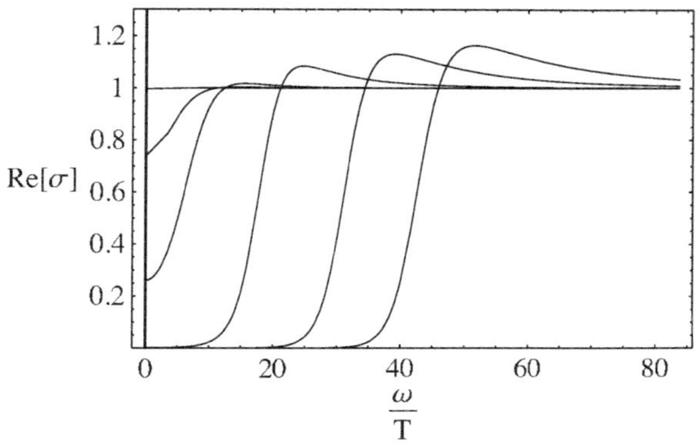

Fig. 10.3 The formation of a gap in the real part of the conductivity as the temperature is lowered below the critical temperature. The curves describe successively lower temperatures. There is also a delta function at $\omega = 0$. Figure is for the dimension two condensate and is taken from [36]

temperature below T_c a gap opens up at low frequency. When these curves were first obtained, it was thought that $\mathrm{Re}[\sigma]$ was given by $e^{-\Delta/T}$ at small ω, which is what one expects from a BCS type description with an energy gap Δ. This would imply that at $T = 0, \mathrm{Re}[\sigma]$ should vanish inside the gap. We will see that this is not the case.

There is also a delta function at $\omega = 0$ for all $T < T_c$. This cannot be seen from a numerical solution of the real part, but it can be seen by looking for a pole in $\mathrm{Im}[\sigma]$. A simple argument for this comes from the Drude model of a conductor. Suppose we have charge carriers with mass m, charge e, and number density n in a normal conductor. They satisfy

$$m\frac{dv}{dt} = eE - m\frac{v}{\tau} \tag{10.19}$$

where the last term is a damping term and τ is the relaxation time due to scattering. The current is $J = env$, so if $E(t) = Ee^{-i\omega t}$, the conductivity is

$$\sigma(\omega) = \frac{k\tau}{1 - i\omega\tau} \tag{10.20}$$

where $k = ne^2/m$. So

$$\mathrm{Re}[\sigma] = \frac{k\tau}{1 + \omega^2\tau^2}, \qquad \mathrm{Im}[\sigma] = \frac{k\omega\tau^2}{1 + \omega^2\tau^2}. \tag{10.21}$$

For superconductors, $\tau \to \infty$, so $\mathrm{Re}[\sigma] \propto \delta(\omega)$ and $\mathrm{Im}[\sigma] \propto 1/\omega$. A more general derivation comes from the Kramers–Kronig relations. These relate the real and imaginary parts of any causal quantity, such as the conductivity, when expressed in frequency space. One of the relations is

$$\mathrm{Im}[\sigma(\omega)] = -\frac{1}{\pi}\mathcal{P}\int_{-\infty}^{\infty} \frac{\mathrm{Re}[\sigma(\omega')]d\omega'}{\omega' - \omega}. \tag{10.22}$$

From this formula we can see that the real part of the conductivity contains a delta function, if and only if the imaginary part has a pole. One finds that there is indeed a pole in $\mathrm{Im}[\sigma]$ at $\omega = 0$ for all $T < T_c$.

Figure 10.4 shows the low temperature limit of the optical conductivity. The solid line denotes the real part and the dashed line denotes the imaginary part. The pole at $\omega = 0$ is clearly visible.

A finite conductivity implies dissipation. In the bulk, this is reflected by the ingoing wave boundary conditions at the horizon. A standard normalizable perturbation of the Maxwell field would decay and get swallowed by the black hole. We maintain a constant amplitude and purely harmonic time dependence by driving the mode with an applied electric field at the boundary. Dissipation normally causes a system to heat up, and indeed if we had included the backreaction of the Maxwell perturbation on the metric, the flow of energy into the horizon would cause the black hole to grow and increase its temperature.

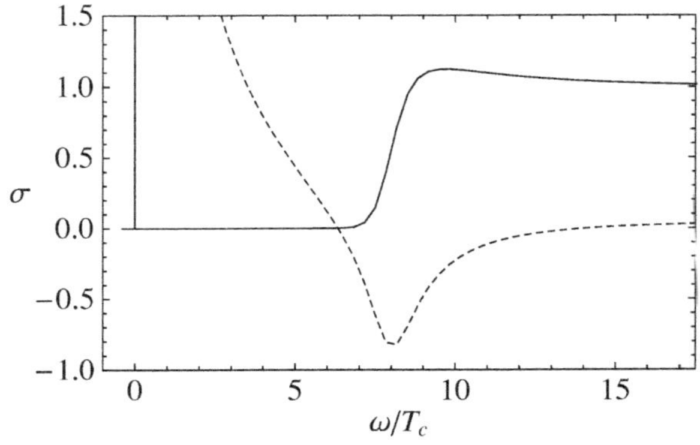

Fig. 10.4 The low temperature limit of the optical conductivity for the dimension two condensate. The solid line denotes the real part and the dashed line denotes the imaginary part. Figure is taken from [44]

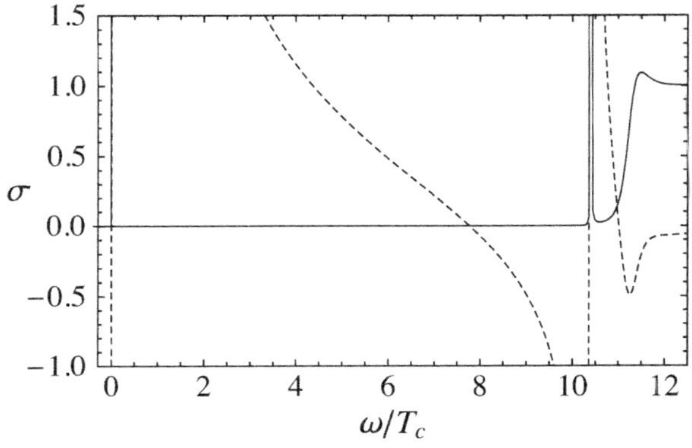

Fig. 10.5 The low temperature limit of the conductivity for the dimension 3/2 condensate. Note the extra spike that appears inside the gap. Figure is taken from [44]

When you approach the BF bound something interesting happens. As shown in Fig. 10.5, a new spike appears inside the gap. This looks like a bound state of quasiparticles with the binding energy given by distance between the pole and the edge of the gap. However, one must keep in mind that quasiparticles are a weak coupling concept and may not be well defined at strong coupling. We will have more to say about this spike in the next section.

Although we have focussed on the case of $2+1$ dimensional superconductors with a $3+1$ dimensional bulk dual, everything we have done is easily generalized to higher dimensions [44]. The results are qualitatively the same.

A robust feature that holds in both $2+1$ and $3+1$ superconductors and all $\lambda > \lambda_{BF}$ is that

$$\frac{\omega_g}{T_c} \approx 8 \tag{10.23}$$

with deviations of less than 10%.[8] This is more than twice the weakly coupled BCS value of 3.5. Remarkably, measurements of this ratio in the high T_c superconductors give roughly this value [23]!

10.4 Full Solution with Backreaction

We now discuss the full solution to action (10.2) including the backreaction on the spacetime geometry. We start with the following ansatz for the metric

$$ds^2 = -g(r)e^{-\chi(r)}dt^2 + \frac{dr^2}{g(r)} + r^2(dx^2 + dy^2) \tag{10.24}$$

and use the same ansatz for the matter fields as before:

$$A = \phi(r)dt, \quad \Psi = \psi(r). \tag{10.25}$$

The equations of motion are[9]:

$$\psi'' + \left(\frac{g'}{g} - \frac{\chi'}{2} + \frac{2}{r}\right)\psi' + \frac{q^2\phi^2 e^{\chi}}{g^2}\psi - \frac{m^2}{g}\psi = 0, \tag{10.26}$$

$$\phi'' + \left(\frac{\chi'}{2} + \frac{2}{r}\right)\phi' - \frac{2q^2\psi^2}{g}\phi = 0, \tag{10.27}$$

$$\chi' + r\psi'2 + \frac{rq^2\phi^2\psi^2 e^{\chi}}{g^2} = 0, \tag{10.28}$$

$$g' + \left(\frac{1}{r} - \frac{\chi'}{2}\right)g + \frac{r\phi'2e^{\chi}}{4} - 3r + \frac{rm^2\psi^2}{2} = 0. \tag{10.29}$$

The first two are the scalar and Maxwell equations as before. The last two are the two independent components of Einstein's equation. (There are three nonzero components of Einstein's equation but only two are independent.) Note that the equations for g and χ are first order, and χ is monotonically decreasing. These equations are invariant under a scaling symmetry analogous to (10.13):

$$r \to ar\,, \quad (t,x,y) \to (t,x,y)/a\,, \quad g \to a^2 g\,, \quad \phi \to a\phi\,. \tag{10.30}$$

---

[8] A few caveats should be noted: If one adds scalar interactions in the bulk by introducing a more general potential $V(\Psi)$, the gap in the low temperature optical conductivity can become much less pronounced, so that this ratio becomes ill defined [28, 29]. We will see in the next section that it also becomes ill defined at small q. Even when it is well defined, it is modified by higher order curvature corrections in bulk [24].

[9] We will set $L = 1$ for the rest of our discussion.

When the horizon is at nonzero r, this can be used to set $r_0 = 1$. These equations are also invariant under

$$e^{\chi} \to a^2 e^{\chi}, \quad t \to at, \quad \phi \to \phi/a. \tag{10.31}$$

This symmetry can be used to set $\chi = 0$ at the boundary at infinity, so the metric takes the standard AdS form asymptotically.

The qualitative behavior of the solutions is similar to the probe limit, i.e., the probe limit indeed captures most of the physics. However, there are a few important differences [37]. First, the apparent divergence in the dimension one condensate at low T is gone. Figure 10.6 shows a plot of qO_1 as a function of temperature. We have multiplied the condensate by the charge q since this is the quantity which is represented by the probe limit at large q. For q of order one, the curves are similar to the other condensates in the probe limit. But for large q, they grow at low temperature. One can show that for all q, the condensate still approaches a finite limit at $T = 0$. It is not clear why the dimension one condensate behaves differently in the probe limit from other masses.

The second difference is more surprising. As explained earlier, the origin of the instability responsible for the scalar hair is the coupling of the charged scalar to the charge of the black hole. It was therefore expected that as $q \to 0$ the instability would turn off. This is not what happens. A nearly extremal Reissner–Nordstrom AdS black hole remains unstable to forming neutral scalar hair, provided that m^2 is close to the Breitenlohner–Freedman (BF) bound. This means that there is a new source of instability which can be understood as follows: An extremal Reissner–Nordstrom AdS black hole has a near horizon geometry $AdS_2 \times R^2$. The BF bound for AdS_{d+1} is $m_{BF}^2 = -d^2/4$. So scalars which are slightly above the BF bound for AdS_4, can be below the bound for AdS_2. This instability to forming neutral scalar hair is not associated with superconductivity (or superfluidity) since it does not break a $U(1)$ symmetry. At most it breaks a Z_2 symmetry corresponding to $\psi \to -\psi$. Its interpretation in the dual field theory is not clear.

One can make the following general argument for when an extremal Reissner–Nordstrom AdS black hole will be unstable to forming scalar hair [14]. Consider a

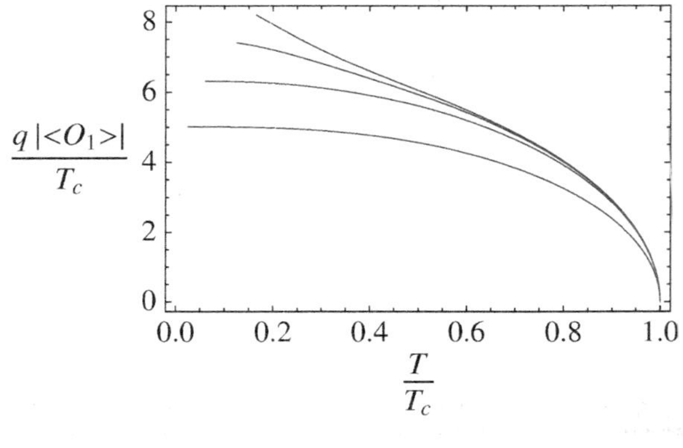

Fig. 10.6 From *bottom* to *top*, the various curves correspond to $q = 1, 3, 6$, and 12. Figure is taken from [37]

scalar field with mass m and charge q in the near horizon geometry of an extremal Reissner–Nordstrom AdS black hole. Its field equation reduces to a wave equation in AdS_2 with effective mass $m_{eff}^2 = (m^2 - 2q^2)/6$. The $-2q^2$ is the usual coupling to the electric charge and the factor of six comes from the difference between the radius of curvature in AdS_4 and AdS_2. The instability to form scalar hair at low temperature is then just the instability of scalar fields below the BF bound for AdS_2: $m_{BF}^2 = -1/4$. Thus the condition for instability is

$$m^2 - 2q^2 < -3/2. \qquad (10.32)$$

Of course, the mass must be above the four-dimensional BF bound, $m^2 > -9/4$, so that the asymptotic AdS_4 geometry remains stable. Equation 10.32 is a sufficient condition to guarantee instability, but it is not necessary.

To compute the conductivity, we again perturb the Maxwell field in the bulk. Assuming zero momentum and harmonic time dependence, the perturbed Maxwell field A_x now couples to the perturbed metric component g_{tx}. Physically, this is what one expects from the standpoint of the dual field theory. We are applying an electric field and inducing a current. The current carries momentum, so T_{tx} should be nonzero. This requires that the metric perturbation g_{tx} must also be nonzero.[10] However, this also means that one has to solve for the thermal conductivity at the same time as the electrical conductivity. Fortunately, for homogeneous perturbations with harmonic time dependence, one can solve for g_{tx} in terms of A_x and get

$$A_x'' + \left[\frac{g'}{g} - \frac{\chi'}{2}\right]A_x' + \left[\left(\frac{\omega^2}{g^2} - \frac{\phi'2}{g}\right)e^{\chi} - \frac{2q^2\psi^2}{g}\right]A_x = 0\,. \qquad (10.33)$$

This is similar to (10.16) except for the form of the metric and the extra term $\phi'2/g$ coming from the metric perturbation g_{tx}. The conductivity is still

$$\sigma(\omega) = -\frac{i}{\omega}\frac{A_x^{(1)}}{A_x^{(0)}}\,. \qquad (10.34)$$

The results for the optical conductivity are qualitatively similar to the probe limit, and shown in Fig. 10.7. There are a few differences. One is that the conductivity in the normal phase is no longer constant. Another difference is that in the low temperature limit, the gap in Re$\sigma(\omega)$ for small ω becomes less pronounced at small q. The robust feature $\omega_g/T_c \approx 8$ seen in the probe limit continues to hold for $q > 3$, but is less robust for $q<3$ mainly because the gap becomes less well defined. Perhaps the most important difference is that Reσ now has a delta function contribution at $\omega = 0$ even in the normal phase when $T > T_c$. This infinite DC conductivity is not superconductivity, but just a consequence of translational invariance. A translationally invariant, charged system does not have a finite DC conductivity because application of an electric field will cause uniform acceleration. One does not see this

[10] I thank Hong Liu for this comment.

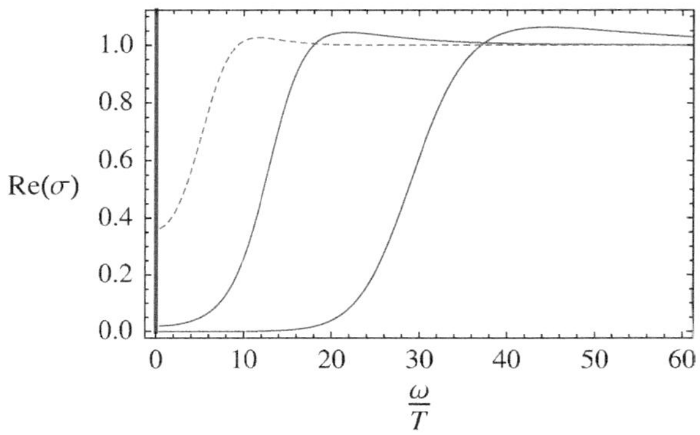

Fig. 10.7 Conductivity for the dimension two condensate with $q = 3$. The *dashed line* is the real part of the conductivity at $T = T_c$ and the solid lines are the same conductivities at successively lower temperature $T/T_c = 0.651, 0.304$. There is a delta function at the origin in all cases. Figure is taken from [37]

in the probe limit, since we fixed the gravitational background which implicitly breaks translation invariance. One indication of this is that we get an electric current without momentum flow as we mentioned earlier. Although there is a delta function at $\omega = 0$ for all temperature, its coefficient grows as the temperature is lowered below T_c. This indicates the presence of a new contribution coming from the onset of superconductivity.

10.4.1 Reformulation of the Conductivity

Equation 10.33 can be simplified by introducing a new radial variable

$$dz = \frac{e^{\chi/2}}{g} dr. \tag{10.35}$$

At large $r, dz = dr/r^2$, and we can choose the additive constant so that $z = -1/r + O(1/r^2)$. Since g vanishes at least linearly at a horizon and χ is monotonically decreasing, the horizon corresponds to $z = -\infty$. In terms of z, (10.33) takes the form of a standard Schrödinger equation with energy ω^2:

$$-A_{x,zz} + V(z)A_x = \omega^2 A_x \tag{10.36}$$

where

$$V(z) = g[\phi_{,r}^2 + 2q^2\psi^2 e^{-\chi}]. \tag{10.37}$$

From the known asymptotic behavior of the solution near infinity, one can show that $V(0) = 0$ if the dimension of the condensate, λ, is greater than one, $V(0)$ is a nonzero constant if $\lambda = 1$, and $V(z)$ diverges as $z \to 0$ if $1/2 < \lambda < 1$. One can also show that the potential always vanishes at the horizon [45].

We want to solve (10.36) with boundary conditions at $z = -\infty$ corresponding to waves propagating to the left. This corresponds to ingoing boundary conditions at the horizon which is needed to extract causal results. The easiest way to do this is to first extend the definition of the potential to all z by setting $V = 0$ for $z > 0$. An incoming wave from the right will be partly transmitted and partly reflected by the potential barrier. Since the transmitted wave is purely ingoing at the horizon, this satisfies our desired boundary conditions. Writing the solution for $z > 0$ as $A_x = e^{-i\omega z} + \mathcal{R}e^{i\omega z}$, we clearly have $A_x(0) = 1 + \mathcal{R}$ and $A_{x,z}(0) = -i\omega(1 - \mathcal{R})$. In terms of $z, A_x^{(1)} = -A_{x,z}(0)$, so from (10.34)

$$\sigma(\omega) = \frac{1 - \mathcal{R}}{1 + \mathcal{R}}. \tag{10.38}$$

The conductivity is directly related to the reflection coefficient, with the frequency simply giving the incident energy [45]! The qualitative behavior of $\sigma(\omega)$ is now clear. Let us first assume that the dimension of the condensate is $\lambda \geq 1$ so that V is bounded. At frequencies below the height of the barrier, the probability of transmission will be small, $\mathcal{R}$ will be close to one, and $\sigma(\omega)$ will be small. At frequencies above the height of the barrier, $\mathcal{R}$ will be very small and $\sigma(\omega) \sim 1$ (the normal state value). Clearly the size of the gap in $\sigma(\omega)$ is set by the height of the barrier: $\omega_g \sim \sqrt{V_{\max}}$. The case $1/2 < \lambda < 1$ is qualitatively similar. Even though the potential is not bounded, $\sqrt{V}$ is integrable, so there is still tunneling through the barrier.

As one lowers the temperature, the condensate ψ increases and the potential becomes both higher and wider (see Fig. 10.8). This causes an increasing exponential suppression which was erroneously interpreted as evidence for $\mathrm{Re}\,\sigma \sim e^{-\Delta/T}$ in the original papers on holographic superconductors. However, as $T \to 0$, the potential approaches a finite, limiting form which still vanishes at $z = -\infty$. This means that there will always be nonzero tunnelling probability and hence a nonzero conductivity even at small frequency. In other words, there is no hard gap in the optical conductivity at $T = 0$. This conclusion continues to hold if the $m^2|\Psi|^2$ term in our bulk lagrangian is replaced by a general potential $V(\Psi)$. The fact that there is not a hard gap is consistent with calculations of the specific heat which show that it vanishes as a power law at low temperature and not exponentially.

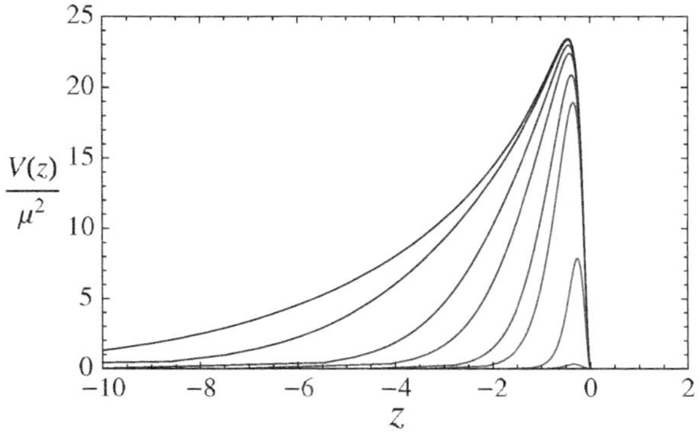

Fig. 10.8 Schrödinger potential for $\lambda = 2, q = 10$. The *top* curve as $T = 0$ and the *bottom* curve has $T = T_c$. Figure is taken from [45]

As discussed in Sect. 3.2, in order for Reσ to have a delta function at $\omega = 0$ representing the infinite DC conductivity, one needs Im σ to have a pole at $\omega = 0$. It is easy to see that this is indeed the case, for any positive Schrödinger potential $V(z)$ that vanishes at $z = -\infty$. Imagine solving (10.36) with $\omega = 0$, and $A_x = 1$ at $z = -\infty$. (This represents the normalizable solution and is the zero frequency limit of the ingoing wave boundary condition.) Since $A_{x,zz} > 0$, the solution will be monotonically increasing. At $z = 0, A_x^{(0)} = A_x(0)$ and $A_x^{(1)} = -A_{x,z}(0)$. These are both real and nonzero. From (10.34) it then follows that Im σ has a pole at $\omega = 0$. Even in the normal phase when $\psi = 0$, the potential is nonzero due to the contribution from the electric field. This is precisely the term which arises due to the backreaction on the metric and is absent in the probe approximation. This is another way to see why the probe limit does not have a delta function at $\omega = 0$ in the normal phase while the full backreacted solution does. Note that the Schrödinger potential remains finite in the probe limit $q \to \infty$, since $q\psi$ is held fixed.

This approach also explains the spike in the conductivity that was seen in Fig. 10.5. At low frequency, the incoming wave from the right is almost entirely reflected. If the potential is high enough, one can raise the frequency so that about one wavelength fits between the potential and $z = 0$. In this case, the reflected wave can interfere destructively with the incident wave and cause its amplitude at $z = 0$ to be exponentially small. This produces a spike in the conductivity. If one can raise the frequency so that two wavelengths fit between the potential and $z = 0$ one gets a second spike, etc. More precisely, using standard WKB matching formula, spikes will occur when there exists ω satisfying

$$\int_{-z_0}^{0} \sqrt{\omega^2 - V(z)}dz + \frac{\pi}{4} = n\pi \tag{10.39}$$

for some integer n, where $V(-z_0) = \omega^2$. It is now clear that the spike which we saw in the probe approximation with m^2 saturating the BF bound, also appears in the full backreacted solution, and for some m^2 slightly above the BF bound. When the spikes were first seen, it was speculated that they corresponded to vector normal modes of the hairy black hole. It is now clear that they are not true normal modes even at $T = 0$, since A_x does not actually vanish at infinity. The actual modes all have complex frequency and correspond to familiar quasinormal modes. In other words, there are no bound states in this potential (with boundary condition $A_x = 0$ at $z = 0$) since the potential vanishes at $z = -\infty$.

10.5 Zero Temperature Limit

The extremal Reissner–Nordstrom AdS black hole has large entropy at $T = 0$. If this was dual to a condensed matter system, it would mean the ground state was highly degenerate. The extremal limit of the hairy black holes dual to the superconductor is not like Reissner–Nordstrom. It has zero horizon area, consistent with a nondegenerate ground state [45]. One might wonder how this could be the limit of the $T > 0$ black holes since we have seen that for all $T \neq 0$, one can scale r so that $r_0 = 1$. The point is simply that the dimensionless ratio $r_0/\mu \to 0$ as $T \to 0$.

The extremal limit also has zero charge inside the horizon (except in the case $q = 0$ when the scalar hair cannot carry the charge). This is an expected consequence of the horizon having zero area. If the black hole tried to keep a nonzero charge as the temperature was lowered and its horizon shrunk to zero, the electric field on the surface would grow arbitrarily large. Eventually this would pair create charged particles via the Schwinger mechanism and neutralize the black hole. This is a quantum argument involving pair creation, but it has a classical analog in terms of a superradiant instability for charged scalar fields.

The near horizon behavior of the zero temperature solution depends on the mass and charge of the bulk scalar field. We will discuss two classes of solutions below. In both cases, one can solve for the solutions analytically near $r = 0$. These leading order solutions depend on a free parameter which can be adjusted so that the solution asymptotically satisfies the desired boundary condition. In the cases we study, the solution is typically not smooth at $r = 0$, but since they are the limit of smooth black holes, they are physically allowed. If one modifies the potential for Ψ so that it has more than one extremum, then smooth, zero temperature solutions exist in which Ψ rolls from one extremum to another [28, 29].

10.5.1 $m^2 = 0$

This corresponds to a marginal, dimension three operator developing a nonzero expectation value in the dual superconductor. To determine the leading order behavior near $r = 0$, we try an ansatz

$$\phi = r^{2+\alpha}, \quad \psi = \psi_0 - \psi_1 r^{2(1+\alpha)}, \quad \chi = \chi_0 - \chi_1 r^{2(1+\alpha)}, \quad g = r^2(1 - g_1 r^{2(1+\alpha)}) \tag{10.40}$$

We have used the scaling symmetries (10.30) and (10.31) to set the coefficients in ϕ and g to one. Substituting this into the field equations and equating the dominant terms for small r (assuming $\alpha > -1$), one finds:

$$q\psi_0 = \left(\frac{\alpha^2 + 5\alpha + 6}{2}\right)^{1/2}, \quad \chi_1 = \frac{\alpha^2 + 5\alpha + 6}{4(\alpha + 1)} e^{\chi_o} \tag{10.41}$$

$$g_1 = \frac{\alpha + 2}{4} e^{\chi_o}, \quad \psi_1 = \frac{q e^{\chi_o}}{2(2\alpha^2 + 7\alpha + 5)} \left(\frac{\alpha^2 + 5\alpha + 6}{2}\right)^{1/2}. \tag{10.42}$$

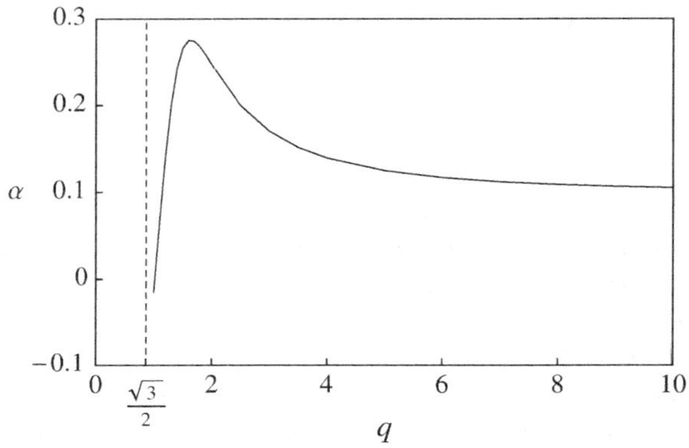

Fig. 10.9 Values of α for various charges. Note that it approaches a constant as $q \to \infty$. Figure taken from [45]

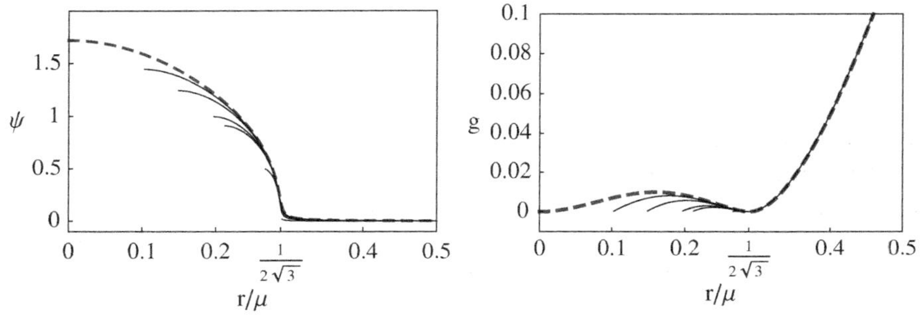

Fig. 10.10 Zero temperature, $\lambda = 3$ and $q = 1$ solution (*dashed line*), compared to successively lower temperature hairy black holes (*solid lines*). Note that g almost has a double zero at $r/\mu = 1/2\sqrt{3}$ which is the extremal horizon for Reissner–Nordstrom AdS. In the limit $q \to \sqrt{3}/2$ the solution becomes extremal Reissner–Nordstrom AdS with all hair behind the horizon. Figure is taken from [45]

One can now numerically integrate this solution to large radius and adjust α so that the solution for ψ is normalizable. One finds that this is possible provided $q^2 > 3/4$. This is consistent with the condition for instability (10.32). The value of α depends weakly on q (see Fig. 10.9). In all cases, $|\alpha| < .3$. Figure 10.10 shows the results for ψ and g for the zero temperature solution, and shows how the $T > 0$ solutions approach it as $T \to 0$.

Near $r = 0$, χ approaches a constant and $g = r^2$. Thus the metric approaches AdS_4 with the same value of the cosmological constant as the asymptotic region. The extremal horizon is just the Poincare horizon of AdS_4. The scalar field approaches a constant and the Maxwell field vanishes. In terms of the dual field theory, this means that the full conformal symmetry is restored in the infrared.

These solutions are not singular at $r = 0$ since all curvature invariants remain finite. They can be viewed as static, charged domain walls. Even though they are not singular, when $\alpha \neq 0$ the solutions are not C^∞ across $r = 0$. Some derivatives of the curvature will blow up. However, there is a special value of the charge, $q = 1.018$, where $\alpha = 0$. This solution is completely smooth across the horizon. It is not clear what the significance of this value is for the dual field theory.

10.5.2 $q^2 > |m^2|/6$ $(m^2 < 0)$

We try the following ansatz near $r = 0$:

$$\psi = A(-\log r)^{1/2}, \quad g = g_0 r^2(-\log r), \quad \phi = \phi_0 r^\beta(-\log r)^{1/2} \tag{10.43}$$

where we have used the radial scaling symmetry (10.30) to set an arbitrary length scale in the logarithm to one. The behavior of χ is determined by (10.28) and whether we expect $r\psi'2$ or $rq^2\phi^2\psi^2e^\chi/g^2$ to dominate. We assume $r\psi'2$ dominates, so

$$e^\chi = K(-\log r)^{A^2/4} \tag{10.44}$$

with K a constant of integration. The second term in (10.28) is indeed negligible provided $\beta > 1$. Equating the dominant terms in the equations of motion leads to

$$A = 2, \quad g_0 = -\frac{2}{3}m^2, \quad \beta = -\frac{1}{2} \pm \frac{1}{2}\left(1 - \frac{48q^2}{m^2}\right)^{1/2}. \tag{10.45}$$

It is clear that for appropriate metric signature we need positive g_0 which tells us this ansatz is only appropriate for negative m^2. Since we require $\beta > 1$, only the plus sign in (10.45) is allowed and we require

$$q^2 > -m^2/6. \tag{10.46}$$

With these restrictions, our near horizon solution is

$$\psi = 2(-\log r)^{1/2}, \quad g = (2m^2/3)r^2 \log r, \quad e^\chi = -K \log r \tag{10.47}$$

$$\phi = \phi_0 r^\beta(-\log r)^{1/2}, \tag{10.48}$$

The one remaining free parameter is ϕ_0. This parameter can be tuned so that the asymptotic boundary condition on ψ is satisfied.

The horizon at $r = 0$ has a mild singularity. The scalar field diverges logarithmically and the metric takes the form (after rescaling t)

$$ds^2 = r^2(-dt^2 + dx_i dx^i) + \frac{dr^2}{g_0 r^2(-\log r)}. \tag{10.49}$$

Notice that Poincare invariance is restored near the horizon, but not the full conformal invariance. (Some early indications of emergent Poincare symmetry were found in [28, 29].) Introducing a new radial coordinate, $\tilde{r} = -2(-\log r)^{1/2}/g_0^{1/2}$ the metric becomes

$$ds^2 = e^{-g_0 \tilde{r}^2/2}(-dt^2 + dx_i dx^i) + d\tilde{r}^2 \tag{10.50}$$

near the horizon which is now at $\tilde{r} = -\infty$.

10.6 Adding Magnetic Fields

As we mentioned in the introduction, one of the characteristic properties of superconductors is that they expel magnetic fields. However, a superconductor can expel a magnetic field only up to a point. A sufficiently strong field will destroy the superconductivity. One defines the thermodynamic critical field B_c by setting the energy it takes to expel the magnetic field equal to the difference in free energy between the normal and superconducting states:

$$\frac{B_c^2(T)\, V}{8\pi} = F_n(T) - F_s(T) \tag{10.51}$$

where V is the volume. Superconductors are divided into two classes depending on how they make the transition from a superconducting to a normal state as the magnetic field is increased. In type I superconductors, there is a first order phase transition at $B = B_c$, above which magnetic field lines penetrate uniformly and the material no longer superconducts. In type II superconductors, there is a more gradual second order phase transition. The magnetic field starts to penetrate the superconductor in the form of vortices with quantized flux when $B = B_{c1} < B_c$. The vortices become more dense as the magnetic field is increased, and at an upper critical field strength, $B = B_{c2} > B_c$, the material ceases to superconduct.

We now show that our 2 + 1 dimensional holographic superconductors are type II. (Interestingly enough, the high T_c superconductors are also type II.) We will view this superconductor as a thin film superconductor in 3 + 1 dimensions with a perpendicular magnetic field. Suppose the 2 + 1 dimensional sample is a disk of radius R. In order for the disk to expel the magnetic field, it must produce a current circulating around the perimeter. Solving Maxwell's equations, this current will expel a field not only in the area πR^2 of the disk but in a larger volume of size $V_3 \sim R^3$. Since the difference in free energies between the normal and superconducting state scales like R^2, in the large R (thermodynamic) limit, the superconductor does not have enough free energy available to expel a magnetic field. Thus magnetic fields of any non-vanishing strength will penetrate a thin superconducting film and $B_c = 0$.

To show that the holographic superconductor is type II, we start in the normal phase with a large applied magnetic field, and slowly lower B. We will see that parts of the system start to become superconducting at a value $B_{c2} > 0$ [1, 44]. To model the normal phase with an applied magnetic field, we use a black hole in AdS with both electric and magnetic charges (but no scalar hair). The magnetic field in the bulk corresponds to applying a background magnetic field on the boundary. The metric takes the form (10.24) with $\chi = 0$ and

$$g(r) = r^2 - \frac{1}{4rr_0}\left(4r_0^4 + \rho^2 + B^2\right) + \frac{1}{4r^2}\left(\rho^2 + B^2\right). \tag{10.52}$$

The black hole temperature is

$$T = \frac{12r_0^4 - \rho^2 - B^2}{16\pi r_0^3}. \tag{10.53}$$

The vector potential is

$$A = \rho\left(\frac{1}{r_0} - \frac{1}{r}\right) dt + Bxdy. \tag{10.54}$$

Since we are interested in the onset of superconductivity which is a second order phase transition, the condensate will be small and we can ignore the backreaction of ψ on the metric. (Note that this is different from the probe approximation since the backreaction of the Maxwell field is included here.) Since the gauge potential now depends on x, we separate variables as follows

$$\psi(r, x, y) = R(r)X(x)e^{iky}. \tag{10.55}$$

The equation for $X(x)$ turns out to be a simple harmonic oscillator centered at k/qB with width $(qB)^{-1/2}$. The eigenvalues are $2qB(n + \frac{1}{2})$ but only the lowest mode is expected to be stable. Note that the momentum k shifts the origin of the harmonic oscillator, but does not change its energy. This is just the degeneracy of Landau levels. The radial equation becomes:

$$(r^2 gR')' + \left[\frac{q^2\rho^2}{g}\left(\frac{r}{r_0} - 1\right) + qB - m^2r^2\right]R = 0. \tag{10.56}$$

One can now choose B so that there is a normalizable solution. This defines the upper critical field B_{c2}. One finds that B_{c2} is a decreasing function of temperature and vanishes at $T = T_c$. This is what one expects: At lower temperatures, a superconductor can support a larger magnetic field. Due to the degeneracy in k, the actual condensate involves a superposition of these modes for all k and forms a lattice of vortices. Maeda et al. [53] have recently studied the free energy as a function of the two parameters which govern the shape of the vortex array and shown that the minimum of the free energy at long wavelength corresponds to a triangular array which is what is predicted by the Landau–Ginzburg model.

10.6.1 London Equation

As mentioned in Sect. 10.2, the boundary theory does not have dynamical Maxwell fields, so one cannot actually see the Meisner effect. However, one can show that in the superconducting state, static magnetic fields induce currents whose backreaction would cancel the applied magnetic field [52, 37]. In fact, one can show that these currents are proportional to A_i and thus reproduce London's equation.

To make life easier, in this section we shall work in the probe limit ($q \to \infty$) in which the metric is kept fixed to be simply the Schwarzschild AdS black hole (10.3). Assume that we have solved for A_t and ψ in this background as described in Sect. 3.1. We then wish to add a static perturbation A_x that has nonzero momentum. We will take the momentum to lie in a direction orthogonal to A_μ, so we can consistently perturb the gauge field without sourcing any other fields. This perturbation corresponds to adding a static magnetic field in the bulk, and the asymptotic value of this magnetic field corresponds to the magnetic field added to the boundary theory. As before, the first subleading term in $A_x(r)$ corresponds to the induced current.

Rather than doing this calculation explicitly, we can obtain the answer by relating it to the calculation in Sect. 3.2. Let A_x have both momentum and frequency dependence of the form $A_x \sim e^{-i\omega t + iky}$. The radial equation for $A_x(r)$ reduces to

$$(fA_x')' + \left(\frac{\omega^2}{f} - \frac{k^2}{r^2}\right)A_x = 2\psi^2 A_x\,, \tag{10.57}$$

where $'$ denotes differentiation with respect to r. We studied this equation with $k = 0$ in Sect. 3.2, and found a pole in $\mathrm{Im}\sigma[\omega]$ at $\omega = 0$. Let the residue of this pole be n_s, the superfluid density. From (10.18) this means that $A_x^{(1)} = -n_s A_x^{(0)}$. But $J_x = A_x^{(1)}$ and $A_x^{(0)}$ is just the vector potential in the boundary theory, so

$$J_x(\omega, k = 0) = -n_s A_x(\omega, k = 0). \tag{10.58}$$

It is clear from the form of Eq. 10.57 that the limits $\omega \to 0$ and $k \to 0$ must commute. To compute n_s, we simply set both ω and k to zero and solve (10.57). Thus for small k, we directly obtain the London equation

$$J_i(\omega, k) = -n_s A_i(\omega, k) \tag{10.59}$$

where we have used the isotropy of space. This equation is clearly not gauge invariant. It is expected to hold in a gauge where $\nabla_i A^i = 0$.

10.6.2 Correlation Length

The gauge/gravity dictionary says that the retarded Green's function (for J_x) in the dual field theory is

$$G^R(\omega, k) = \frac{A_x^{(1)}(\omega, k)}{A_x^{(0)}(\omega, k)}. \tag{10.60}$$

We can define a correlation length ξ by expanding

$$\operatorname{Re} G^R(0, k) = -n_s(1 + \xi^2 k^2 + \cdots) \tag{10.61}$$

By solving (10.57) numerically, including the k dependence, one can compute this correlation length. One finds that it diverges near the critical temperature like $\xi T_c \approx 0.1(1 - T/T_c)^{-1/2}$ [44]. A similar divergence is found in a correlation length obtained from fluctuations in the condensate [52].

10.6.3 Vortices

As we discussed above, at the onset of superconductivity when $B = B_{c2}$, there is a lattice of vortices. However, at this point, the condensate is small everywhere and the equations could be linearized. It is of interest to find the bulk solution describing a vortex lattice away from $B = B_{c2}$. The first step is to find the single vortex solution. This corresponds to a something like a cosmic string stretching from the black hole horizon to infinity. However, there is a crucial difference between a cosmic string and the dual of a vortex. A standard cosmic string has a fixed proper radius. This means that its size in the x, y coordinates used in (10.24) goes to zero at large r. In terms of the boundary theory, this is a point-like object. We want a solution in which the cosmic string is a fixed size in the x, y coordinates, so it has a finite radius in the superconductor. This looks more like a funnel in the bulk with a proper radius that grows as one increases r.

Such a solution has recently been found in the probe limit [56].[11] Let us write the background metric using polar coordinates for the flat transverse space:

$$ds^2 = -f(r)dt^2 + \frac{dr^2}{f(r)} + r^2\left(d\tilde{r}^2 + \tilde{r}^2 d\varphi^2\right). \tag{10.62}$$

Assume an ansatz

$$\Psi = \psi(r, \tilde{r})e^{in\varphi}, \quad A_t = A_t(r, \tilde{r}), \quad A_\varphi = A_\varphi(r, \tilde{r}). \tag{10.63}$$

[11] See [2, 3] for another approach.

Substituting this into the field equations, one obtains a set of nonlinear PDE's. The condensate is now a function of radius $\tilde{r}$. It vanishes at $\tilde{r} = 0$ which represents the center of the vortex, and approaches a constant at large $\tilde{r}$. This shows that there is no superconductivity inside the vortex. Since the Maxwell field is not dynamical on the boundary, the magnetic field is not localized inside the vortex as one would expect. Instead, the authors of [56] work in finite volume and assume a uniform magnetic field on the boundary with total flux equal to $2\pi n$ as expected for the vortex.

10.7 Recent Developments

So far, we have discussed planar black holes in the bulk. Spherical black holes have the same instability and describe spherical superconductors. One can add rotation to a spherical black hole, and the effect of this rotation on the dual superconductor has recently been studied [62]. It was found that for $T < T_c$ (the transition temperature at zero rotation), there is a critical value of the rotation which destroys the superconductivity in analogy to the critical magnetic field.

The bulk gravitational theory that we have studied was not derived from string theory. It was just the simplest model that could describe a dual superconductor, and we have seen that it works rather well. It predicts that a charged condensate forms at low temperature (in a second order phase transition), that the DC conductivity is infinite, and that there is a gap in the optical conductivity at low frequency—all basic properties of superconductors. However, to go beyond this and have a more detailed microscopic understanding of the dual superconductor, one needs to embed this model in string theory. A linearized instability of the type needed to describe a holographic superconductor was found in a string compactification in [14]. This has now been extended to a full description by two different groups. Gauntlett et al. [20, 21] realized the $m^2 = -2, q = 2$ model in M theory. In other words, they found a consistent truncation of eleven dimensional supergravity in which the four dimensional fields were just a metric, Maxwell field and charged scalar with this mass and charge.[12] Gubser et al. [32] realized the same model in one higher dimension (a five dimensional bulk which is dual to a $3+1$ dimensional superconductor) with $m^2 = -3, q = 2$ in type IIB string theory. Both groups used Sasaki–Einstein compactifications with $U(1)$ symmetry, where the charged scalar is related to the size of the U(1) fibration.

There are two main differences between the models obtained from string theory and our simple model. The first is that the kinetic term for the charged scalar involves nonminimal coupling. (This was investigated earlier by Franco et al. [18, 19].) The second difference is that the scalar potentials $V(\psi)$ coming from

---

[12] In general, there is an additional neutral scalar, but for purely electrically charged black holes, this field can be set to zero.

supergravity are more complicated than the simple mass terms we have considered so far. This does not affect the onset of superconductivity: since the scalar field is small near T_c, the critical temperature is just determined by the mass term in the potential. However, the low temperature limit is different. The potentials coming from supergravity have more than one extremum. In this case, the zero temperature limit of the hairy black hole is a smooth domain wall in which the scalar rolls from one extremum in the UV to another extremum in the IR [49]. The IR limit is another copy of AdS with, in general, a different cosmological constant. These examples all have emergent conformal symmetry in the IR.

The optical conductivity in these theories never show a pronounced gap at low temperature. Typically, $\text{Re}\sigma(\omega) = k\omega^\delta$ for small ω with a coefficient k of order one. From our Schrödinger equation reformulation this is a bit surprising. Why is not there an exponential suppression due to tunnelling through the barrier? The answer is simply that the effective potential in the Schrödinger equation never gets very large in these solutions since all fields are order one.

Even though embedding in string theory makes it possible to study the dual field theory in more detail, this is still difficult in $2+1$ dimensions. It is somewhat easier in $3+1$ dimensions since the dual theories are related to ordinary gauge theory. One particularly interesting aspect of the $3+1$ superconductor is that the condensate includes a term bilinear in the fermions, like a Cooper pair [26].

It was shown in [27] that one can replace the Maxwell field and charged scalar in the bulk with an $SU(2)$ gauge field and still have a dual description of a superconductor. The main difference is the following: The symmetry of a superconductor refers to the energy gap above the Fermi surface. The holographic superconductors we have discussed here all have S-wave symmetry since they do not prefer a direction. With an $SU(2)$ gauge field in the bulk, one obtains a P-wave superconductor.[13] This leads to another way to realize holographic superconductors (in the probe limit) in string theory: one can use the $SU(2)$ gauge field realized on a pair of branes [5, 6, 58].

This is a rapidly evolving field and there have been several developments since the Milos summer school in September 2009, including:

- A key property of condensed matter systems is, of course, the atomic lattice. This gives rise to the phonons in the BCS theory which cause the electrons to pair. Our simple model has translational symmetry and no sign of a lattice. However, some interesting ideas on how to include a lattice have been discussed by Kachru et al. [46].
- Some superconductors exhibit striped phases associated with charged density waves. It was shown in [57] that the Chern–Simons term in five dimensional

[13] One cannot compute the energy gap directly. Indeed, as we have seen, there probably is not a strict gap. The interpretation as a P-wave superconductor comes from the fact that there is a vector order parameter and the conductivity is strongly anisotropic in a manor consistent with P-wave nodes on the Fermi surface.

supergravity can lead to an instability at nonzero momentum in which a spatially modulated condensate forms.
- The response of fermions to the superconducting phase has been studied [13, 17, 31]. It was found that with a suitable coupling between the bulk fermions and scalar, there are stable quasiparticles with a gap [17].

10.8 Conclusions and Open Problems

For 90 years after its discovery in 1915, general relativity was viewed as a theory of gravity. It has proved very successful in describing a wide range of gravitational phenomena from the bending of light to gravitational waves and black holes. But in the last few years we have seen that this same theory can describe other areas of physics as well, including superconductivity. This is all due to the magic of gauge/gravity duality. Although the full power of gauge/gravity duality relates a quantum theory of gravity (indeed string theory) in the bulk to a nongravitational theory on the boundary, we have worked in a large N limit in which the bulk theory is just classical general relativity. This large N limit also explains how we can have spontaneous symmetry breaking in a $2+1$ dimensional field theory, in apparent contradiction to the Coleman–Mermin–Wagner theorem. The large N limit suppresses fluctuations in the fields.[14]

It is natural to ask how surprised one should be that general relativity can reproduce the basic properties of superconductors. After all, Weinberg [66] has shown that much of the phenomenology of superconductivity follows just from the spontaneous breaking of the $U(1)$ symmetry. Once we have found the instability that leads to charged scalar hair, does not everything else follow? There are indications that something deeper is going on. For example, order one dimensionless ratios can be computed and compared with experiment. In particular, the ratio $\omega_g/T_c \approx 8$ discussed in Sect. 3.2, is close to the observed value. This does not follow from symmetry arguments alone.

Since our bulk dual of a superconductor just involves gravity interacting with a Maxwell field and a charged scalar, there is a superficial similarity to a Landau–Ginzburg description. However, it is important to keep in mind two key differences. First, the low temperature instability must be put in by hand in the Landau–Ginzburg model, whereas it arises naturally in our gravitational description. Indeed, we have seen that there are two physically distinct instabilities which can trigger the phase transition. Second, the Landau–Ginzburg model is only valid near the transition temperature, since it involves a power series in the order parameter φ. To go beyond $T \approx T_c$, one would need to specify an entire potential $V(\varphi)$. Initially, our bulk theory also had the freedom to add an arbitrary potential $V(\Psi)$. We chose just a mass term for simplicity. However,

[14] This is not an issue for the $3+1$ dimensional superconductor, which can be holographically described by the bulk gravitational theory (10.2) in one higher dimension.

once one embeds the bulk theory into string theory, the potential is fixed and is no longer arbitrary.

10.8.1 Open Problems

We close with a list of open problems.[15] They are roughly ordered in difficulty with the easier problems listed first. (Of course, this is my subjective impression. With the right approach, an apparently difficult problem may become easy!)

1. In the probe limit below the critical temperature, there is an infinite discrete set of solutions for ψ which are all regular on the horizon and satisfy the required boundary condition at infinity. (When backreaction is included, there is only a finite number of such solutions.) They can be labelled by the number of nodes (zeros) they contain. It is widely believed that only the lowest solution with no nodes is stable. Is this true in the theory we have discussed where the scalar potential $V(\psi)$ just contains a mass term? Is it true if the theory contains scalar self interactions, like the potentials derived from string theory?
2. We have seen that an extreme Reissner–Nordstrom AdS black hole is unstable to forming neutral scalar hair if the scalar has mass close to the BF bound (9). Under evolution, as the black hole is developing scalar hair, the total charge and mass at infinity are conserved. Since the scalar field can carry energy, but not charge, the final black hole (which need not be extremal) has the same charge, but less mass than the initial extremal limit. How is this consistent with the original black hole being extremal? What is the extremal limit in this Einstein–Maxwell-scalar theory? A related question is: what is the interpretation of this neutral scalar hair in the dual field theory?
3. We studied magnetic fields in 2 + 1 dimensional superconductors in Sect. 10.6. What happens if you add a magnetic field to 3 + 1 superconductors? One immediate difference is that B_c is no longer zero. Are the holographic superconductors still type II? The new magnetic brane solutions in [15, 16] may be useful.
4. As discussed in Sect. 6.2, a vortex solution has recently been found in the probe approximation. Find the vortex solution with backreaction. More generally, find the solution describing a lattice of vortices in the bulk. This would correspond to a magnetically charged black hole in which the magnetic flux is confined to "cosmic strings" stretching from the horizon to infinity. To solve this problem, one must solve nonlinear partial differential equations.
5. Recall that the symmetry of a superconductor refers to the energy gap on the Fermi surface. The holographic superconductors we have discussed here all have S-wave symmetry since they do not prefer a direction. P-wave holographic superconductors have been found [27, 60]. The high T_c cuprates are

[15] I thank Sean Hartnoll for suggesting some of these problems.

known to have D-wave symmetry. Can one find a holographic superconductor with D-wave symmetry? It is natural to try to condense a charged spin two field. Interestingly enough, the newly discovered iron pnictides appear to have S-wave symmetry, but it is complicated by multiple Fermi surfaces [12].

6. It was discovered recently that there are materials in which the (DC) conductivity goes to zero for $T < T_c$ [64]. This is the opposite of a superconductor and is called a "superinsulator". Can one find a holographic description of a superinsulator? It is tempting to use the fact that under electromagnetic duality in AdS_4, the conductivity transforms as $\sigma \rightarrow -1/\sigma$ [35]. However, one must start with a bulk action which is invariant under electromagnetic duality. The action (10.2) is not, since it explicitly involves A_μ.
7. As we discussed in Sect. 10.7, it has recently been shown that the holographic superconductors can be derived from string theory, so in some cases the dual microscopic theory is known. Can one understand the pairing mechanism in these cases (if indeed "pairing mechanism" is the right concept at strong coupling)? This remains one of the main open questions in understanding the high T_c materials.
8. As mentioned above, the large N limit is responsible for allowing spontaneous symmetry breaking in our 2 + 1 dimensional field theory. Can one show that away from this limit, massless fluctuations lead to infrared diverges which destroy the long range order?
9. In gauge/gravity duality, one takes a large N limit to justify using classical general relativity in the bulk. What is the analog of this large N limit in condensed matter systems? In other words, what types of materials are likely to have a (tractable) dual gravitational gravitational description? (See [59] for a discussion of some of the issues.)
10. The high temperature cuprate superconductors satisfy a simple scaling law relating the superfluid density, the normal state (DC) conductivity and the critical temperature [43]. Can this be given a dual gravitational interpretation?

Acknowledgement It is a pleasure to thank my collaborators, Sean Hartnoll, Chris Herzog, and Matt Roberts for teaching me many of the results described here. I also thank Hartnoll and Roberts for comments on these lecture notes. Finally, I thank the organizers and participants of the 5th Aegean Summer School, "From Gravity to Thermal Gauge Theories: the AdS/CFT Correspondence" for stimulating discussions. This work is supported in part by NSF grant number PHY-0855415.

References

1. Albash, T., Johnson, C.V.: A holographic superconductor in an external magnetic field. JHEP **0809**, 121 (2008)
2. Albash, T., Johnson, C.V.: Phases of holographic superconductors in an external magnetic field. Phys. Rev. D arXiv:0906.0519 [hep-th]

3. Albash, T., Johnson, C.V.: Vortex and droplet engineering in holographic superconductors. Phys. Rev. D **80**, 126009 (2009)
4. Amado, I., Kaminski, M., Landsteiner, K.: Hydrodynamics of holographic superconductors. JHEP **0905**, 021 (2009)
5. Ammon, M., Erdmenger, J., Kaminski, M., Kerner, P.: Superconductivity from gauge/gravity duality with flavor. Phys. Lett. B **680**, 516 (2009)
6. Ammon, M., Erdmenger, J., Kaminski, M., Kerner, P.: Flavor Super- conductivity from gauge/gravity duality. JHEP **0910**, 067 (2009)
7. Bardeen, J., Cooper, L.N., Schrieffer, J.R.: Theory of superconductivity. Phys. Rev. **108**, 1175 (1957)
8. Basu, P., Mukherjee, A., Shieh, H.H.: Supercurrent: vector hair for an AdS black hole. Phys. Rev. D **79**, 045010 (2009)
9. Bednorz, J.G., Muller, K.A.: Possible high T_c superconductivity in the Ba-La-Cu-O system. Z. Phys. B **64**, 189 (1986)
10. Bekenstein, J.D.: Black hole hair: twenty-five years after. (1996) [arXiv:gr-qc/9605059]
11. Breitenlohner, P., Freedman D.Z.: Stability in gauged extended supergravity. Ann. Phys. **144**, 249 (1982)
12. Chen, T.Y., Tesanovic, Z., Liu, R.H., Chen, X.H., Chien, C.L.: A BCS-like gap in the superconductor SmFeAsOF. Nature **453**, 1224 (2008)
13. Chen, J.W., Kao, Y.J., Wen, W.Y.: Peak-dip-hump from holographic superconductivity. arXiv:0911.2821 [hep-th]
14. Denef, F., Hartnoll, S.A.: Landscape of superconducting membranes. Phys. Rev. D **79**, 126008 (2009)
15. D'Hoker, E., Kraus, P.: Magnetic brane solutions in AdS. JHEP **0910**, 088 (2009)
16. D'Hoker, E., Kraus, P.: Charged magnetic brane solutions in AdS_5 and the fate of the third law of thermodynamics. (2009) arXiv:0911.4518 [hep-th]
17. Faulkner, T., Horowitz, G.T., McGreevy, J., Roberts, M.M., Vegh, D.: Photoemission 'experiments' on holographic superconductors. arXiv:0911.3402 [hep-th]
18. Franco, S., Garcia-Garcia, A., Rodriguez-Gomez, D.: A general class of holographic superconductors. arXiv:0906.1214 [hep-th]
19. Franco, S., Garcia-Garcia, A., Rodriguez-Gomez, D.: A holographic approach to phase transitions. arXiv:0911.1354 [hep-th]
20. Gauntlett, J.P., Sonner, J., Wiseman, T.: Holographic superconductivity in M-Theory. Phys. Rev. Lett. **103**, 151601 (2009)
21. Gauntlett, J.P., Sonner, J., Wiseman, T.: Quantum criticality and holographic superconductors in M-theory. arXiv:0912.0512 [hep-th]
22. Ginzburg, V.L., Landau, L.D.: On the theory of superconductivity. Zh. Eksp. Teor. Fiz. **20**, 1064 (1950)
23. Gomes, K.K., Pasupathy, A.N., Pushp, A., Ono, S., Ando, Y., Yazdani, A..: Visualizing pair formation on the atomic scale in the high-Tc superconductor $Bi_2Sr_2CaCu_2O_{8+\delta}$. Nature **447**, 569 (2007)
24. Gregory, R., Kanno, S., Soda, J.: Holographic superconductors with higher curvature corrections. JHEP **0910**, 010 (2009)
25. Gubser, S.S.: Breaking an Abelian gauge symmetry near a black hole horizon. Phys. Rev. D **78**, 065034 (2008)
26. Gubser, S.S., Klebanov, I.R., Polyakov, A.M.: Gauge theory correlators from non-critical string theory. Phys. Lett. B **428**, 105 (1998)
27. Gubser, S.S., Pufu, S.S.: The gravity dual of a p-wave superconductor. JHEP **0811**, 033 (2008)
28. Gubser, S.S., Nellore, A.: Low-temperature behavior of the Abelian Higgs model in anti-de Sitter space. JHEP **0904**, 008 (2009)
29. Gubser, S.S., Nellore, A.: Ground states of holographic superconductors. Phys. Rev. D **80**, 105007 (2009)

30. Gubser, S.S., Pufu, S.S., Rocha, F.D.: Quantum critical superconductors in string theory and M-theory. Phys. Lett. B **683**, 201 (2010)
31. Gubser, S.S., Rocha, F.D., Talavera, P.: Normalizable fermion modes in a holographic superconductor. arXiv:0911.3632 [hep-th]
32. Gubser, S.S., Herzog, C.P., Pufu, S.S., Tesileanu, T.: Superconductors from superstrings. Phys. Rev. Lett. **103**, 141601 (2009)
33. Hartnoll, S.A.: Lectures on holographic methods for condensed matter physics. Class. Quant. Grav. **26**, 224002 (2009)
34. Hartnoll, S.A.: Quantum critical dynamics from black holes. arXiv:0909.3553 [cond-mat.str-el]
35. Hartnoll, S.A., Herzog, C.P.: Ohm's Law at strong coupling: S duality and the cyclotron resonance. Phys. Rev. D **76**, 106012 (2007)
36. Hartnoll, S.A., Herzog, C.P., Horowitz, G.T.: Building a holographic superconductor. Phys. Rev. Lett. **101**, 031601 (2008)
37. Hartnoll, S.A., Herzog, C.P., Horowitz, G.T.: Holographic superconductors. JHEP **0812**, 015 (2008)
38. Hertog, T.: Towards a novel no-hair theorem for black holes. Phys. Rev. D **74**, 084008 (2006)
39. Herzog C.P.: Lectures on holographic superfluidity and superconductivity. J. Phys. A **42**, 343001 (2009)
40. Herzog, C.P., Kovtun, P., Sachdev, S., Son, D.T.: Quantum critical transport, duality, and M-theory. Phys. Rev. D **75**, 085020 (2007)
41. Herzog, C.P., Kovtun, P.K., Son, D.T.: Holographic model of superfluidity. Phys. Rev. D **79**, 066002 (2009)
42. Heusler, M.: No-hair theorems and black holes with hair. Helv. Phys. Acta **69**, 501 (1996)
43. Homes, C.C. et al.: Universal scaling relation in high-temperature superconductors. Nature **430** (2004) 539
44. Horowitz, G.T., Roberts, M.M.: Holographic superconductors with various condensates. Phys. Rev. D **78**, 126008 (2008)
45. Horowitz, G.T., Roberts, M.M.: Zero temperature limit of holographic superconductors. JHEP **0911**, 015 (2009)
46. Kachru, S., Karch, A., Yaida, S.: Holographic lattices, dimers, and glasses. Phys. Rev. D **81**, 026007 (2010)
47. Kamihara, Y., Watanabe, T., Hirano, M., Hosono, H. : J. Am. Chem. Soc. **130**, 3296 (2008)
48. Kim,Y., Ko, Y., Sin, S.J.: Density driven symmetry breaking and butterfly effect in holographic superconductors. Phys. Rev. D **80**, 126017 (2009)
49. Klebanov, I.R., Witten, E.: AdS/CFT correspondence and symmetry breaking. Nucl. Phys. B **556**, 89 (1999)
50. Koutsoumbas, G., Papantonopoulos, E., Siopsis, G.: Exact gravity dual of a gapless superconductor. JHEP **0907**, 026 (2009)
51. London, F., London, H.: Proc. Roy. Soc. (London) **A149**, 71 (1935)
52. Maeda, K., Okamura, T.: Characteristic length of an AdS/CFT superconductor. Phys. Rev. D **78**, 106006 (2008)
53. Maeda, K., Natsuume, M., Okamura, T.: Vortex lattice for a holographic superconductor. Phys. Rev. D **81**, 026002 (2010)
54. Maldacena, J.M.: The large N limit of superconformal field theories and supergravity. Adv. Theor. Math. Phys. **2**, 231 (1998) [Int. J. Theor. Phys. **38**, 1113 (1999)]
55. McGreevy, J.: Holographic duality with a view toward many-body physics. arXiv:0909.0518 [hep-th]
56. Montull, M., Pomarol, A., Silva, P.J.: The holographic superconductor vortex. Phys. Rev. Lett. **103**, 091601 (2009)
57. Nakamura, S., Ooguri, H., Park, C.S.: Gravity dual of spatially modulated phase. arXiv:0911.0679 [hep-th]
58. Peeters, K., Powell, J., Zamaklar, M.: Exploring colourful holographic superconductors. JHEP **0909**, 101 (2009)

59. Polchinski, J.: KITP talk during the workshop on quantum criticality and AdS/CFT correspondence, July 21, 2009
60. Roberts, M.M., Hartnoll, S.A.: Pseudogap and time reversal breaking in a holographic superconductor. JHEP **0808**, 035 (2008)
61. Son, D.T., Starinets, A.O.: Minkowski-space correlators in AdS/CFT correspondence: recipe and applications. JHEP **0209**, 042 (2002)
62. Sonner, J.: A Rotating holographic superconductor. Phys. Rev. D **80**, 084031 (2009)
63. Tinkham, M.: Introduction to Superconductivity, 2nd (edn.) Dover, New York (1996)
64. Vinokur, V., Baturina, T., Fistul, M., Mironov, A., Baklanov, M., Strunk, C.: Superinsulator and quantum synchronization. Nature **452**, 613 (2008)
65. Volkov, M.S., Galtsov, D.V.: NonAbelian Einstein Yang-Mills black holes. JETP Lett. **50**, 346 (1989) [Pisma Zh. Eksp.Teor. Fiz. **50**, 312 (1989)]
66. Weinberg, S.: Superconductivity for particular theorists. Prog. Theor. Phys. Suppl. **86**, 43 (1986)
67. Witten, E.: Anti-de Sitter space and holography. Adv. Theor. Math. Phys. **2**, 253 (1998)

Chapter 11
Flavor Superconductivity and Superfluidity

Matthias Kaminski

Abstract In these lecture notes we derive a generic holographic string theory realization of a p-wave superconductor/superfluid. For this purpose we also review basic D-brane physics, gauge/gravity methods at finite temperature, key concepts of superconductivity and recent progress in distinct realizations of holographic superconductors and superfluids. The gravity dual is a D3/D7-brane construction yielding a superconducting or superfluid vector-condensate. The corresponding gauge theory is 3 + 1-dimensional $\mathcal{N} = 2$ supersymmetric Yang-Mills theory with $SU(N_c)$ color and $SU(2)$ flavor symmetry. It shows a second order phase transition to a phase in which a $U(1)$ subgroup of the $SU(2)$ symmetry is spontaneously broken and typical superconductivity signatures emerge. For example an infinite dc conductivity, a conductivity (pseudo-)gap and the Meissner-Ochsenfeld effect are found. Condensates of this nature are comparable to those recently found experimentally in p-wave superconductors such as a *ruthenate compound*. A string picture of the pairing mechanism and condensation is given using the exact knowledge of the corresponding field theory degrees of freedom.

11.1 Introduction

In these lecture notes[1] we generically construct a holographic p-wave superconductor. This introductory section serves to explain why this particular setup is of

[1] These lecture notes are based on joint research work with Martin Ammon, Johanna Erdmenger, Patrick Kerner and Felix Rust.

M. Kaminski (✉)
Department of Physics, Princeton University, Princeton, NJ 08544, USA
e-mail: mkaminsk@princeton.edu

E. Papantonopoulos (ed.), *From Gravity to Thermal Gauge Theories: The Ads/CFT Correspondence,* Lecture Notes in Physics, 828,
DOI: 10.1007/978-3-642-04864-7_11,

interest both from a string-theoretical as well as from a condensed matter physics point of view. In order to guide the unexperienced reader in Sect. 11.2 we will describe the basic concepts in words and motivate the project. Detailing this overview we build the full holographic setup from scratch in a self-contained manner in Sect. 11.3. Section 11.4 briefly introduces common holographic methods and then summarizes the major results known for the thermodynamics and the spectrum of D-brane systems. With these outcomes in the back of our minds we will understand how our system develops a new ground state and why it can be interpreted as a ρ meson superfluid or superconductor. We compute and discuss thermodynamic observables, conductivities and the spectrum of our holographic superconductor in Sect. 11.5. Finally, we put together all of our findings in order to draw a string theory picture for the *pairing mechanism* in our p-wave superconductor. These notes are self-contained and widely complementary to the notes focusing on holographic s-wave superconductors [1, 2].

11.1.1 String Motivation

The setup of intersecting D-branes and especially the particular D3-D7 construction presented here is interesting because of its diverse applications. It has been successfully used to model strongly coupled particle physics phenomena such as a *deconfinement*, or rather *meson melting* phase transition for fundamental matter and transport coefficients in the quark gluon plasma experimentally created at the RHIC collider in Brookhaven (see [3] for an introductory review and further references). On the other hand it has recently been used to model strongly correlated electron systems as they appear in superfluids and superconductors in the realm of condensed matter physics [4, 5]. All these applications are also crucial checks of the basic principles and methods coming from the conjectured gauge/gravity correspondence. Receiving physically meaningful outcomes when applying these holographic methods to various systems accessible by experiment, strengthens our confidence in the gauge/gravity conjecture.

From the *string-theoretic point of view* this particular setup is interesting because of its simplicity, uniqueness (explained below) and the naturalness with which the symmetry is broken spontaneously here. The flavor[2] superfluid/superconductor[3] examined here was the first generic *top-down* string theory realization of a superfluid/superconducting phase. In contrast to this the pioneering papers on

[2] The term "flavor" superconductor stems from earlier applications of this setup to model strongly correlated high energy systems such as the quark gluon plasma. In the present case the name is not important and possibly misguiding, since it is really only essential that the system has a non-Abelian $SU(2)$ symmetry.

[3] We are using the terms superfluid and superconductor interchangably here. For the considered phenomena this distinction does not make any difference. Some details to this distinction are given in Sect. 11.2.1.

holographic superconductors [6, 7] had exclusively treated gravity toy models which were not directly obtained from string theory, i.e. *bottom-up models*. Therefore the gauge/gravity correspondence could not be used to identify the exact gauge theory dual. This fact obstructed the interpretation of the outcomes. Furthermore these toy models had few restrictions on their parameters such that in principle a large parameter space needed to be scanned, see e.g. [8]. Our string-derived flavor superfluid/superconductor overcomes those two problems: First the field theory degrees of freedom are exactly known since it is simply $\mathcal{N} = 2$ supersymmetric Yang-Mills theory with an $SU(2)$ flavor-symmetry. Second, the values of parameters, such as for example the dimension of the condensing operator, in our setup are severely restricted by their string theoretic derivation. In this sense this setup is "unique" compared to the big parameter space to be scanned in bottom-up toy models. Later, other top-down string realizations have been suggested and for example involve a consistent truncation of type IIB supergravity with a chemical potential for the R-charge [9], and domain-wall solutions interpolating between AdS solutions with distinct radii which may be lifted to IIB supergravity or eleven-dimensional supergravity [10].

11.1.2 Condensed Matter Motivation

From a *condensed-matter physics point of view* our flavor superconductor is highly interesting because it reproduces features which have been measured in experiments with unconventional superconductors, such as p-wave superconductivity, a system of strongly-correlated particles, a pseudo-gap in the frequency-dependent conductivity (found in high temperature d-wave superconductors). Most of these phenomena lack a widely-accepted microscopic explanation by conventional approaches, so there is the hope that gauge/gravity can shed some light on the nature of these systems. These systems usually contain strongly correlated electrons, so the dual weak gravity description is in principle accessible. Our system has a vector operator which condenses upon breaking a *residual Abelian* flavor symmetry spontaneously. This gives the vector order parameter as described below. So there is a preferred spatial direction in the superfluid/superconducting condensate. This is exactly the situation recently found experimentally in the *p-wave* (explained below in Sect. 11.2.1) superconductor Sr_2RuO_4 [11]. These materials are investigated with great excitement in the condensed-matter community because they are hoped to be usable for *quantum computing* [12]. The reason is that the p-wave structure implies the presence of non-Abelian quasi-particles in the *ruthenate compound*. These non-Abelian quasi-particles can be used as the states to be manipulated in a *topological quantum computer*. The biggest practical obstacle for quantum computation are the errors which may occur during a calculation due to materials being not ideal. Topological quantum computers minimize this source of error because they carry out operations by *braiding* the non-Abelian quasi-particles in a Hilbert-subspace containing degenerate ground states. Due to an energy gap between this subspace and the rest of the Hilbert space it virtually decouples from all local perturbations [13]. Another confirmed and well-studied condensed-matter example for an emerging p-wave structure is superfluid ^{3}He-A [14].

11.2 Superconductivity and Holography

This section is a primer on the subject of spontaneous symmetry breaking, superconductivity and superfluidity in the holographic context of the gauge/gravity correspondence. Little formalism is used, while we introduce all the necessary concepts.

11.2.1 Basics of Superconductivity and Our Field Theory Idea

Let us review the essential concepts of superconductivity and understand how to build a p-wave superconductor. We are going to need this knowledge in order to appreciate the fact that our holographic setup reproduces this behavior in great detail.

11.2.1.1 Superconductor Basics and the P-Wave

Superconductivity is the phenomenon associated with infinite dc conductivity in materials at low temperatures. It is caused by the formation of a *charged condensate* in which directed currents do not experience resistivity. A defining criterion for superconductivity is the *Meissner-Ochsenfeld effect* described below. In *conventional superconductors* the superconducting condensate consists of electron pairs called *Cooper pairs*. So there are two simultaneous steps: the fermionic electrons have to pair up to form bosonic Cooper pairs, and these pairs do condense. This condensation happens in a second order phase transition (at vanishing magnetic field) when the temperature is lowered through its critical value T_c. Due to the requirement for the fermionic state to be antisymmetric there are only certain symmetry combinations allowed for the two electron state describing a Cooper pair. As seen from Table 11.1 the name "*p-wave*" superconductor refers to those pairs in which the relative orbital angular momentum between the two electrons is $L = 1$, the spatial part of the parity is odd and the spin state is a triplet.

The mechanism pairing electrons in conventional superconductors is well described by a mean field theory approach and the microscopic *BCS-theory*. Recall that condensed matter systems are conveniently described in terms of lattices with many (about 10^{23}) sites. Then conventional BCS-theory tells us that the lattice is slightly deformed by the presence of an electron. This deformation can be

Table 11.1 Nomenclature of superconducting states

Orbital angular momentum	Name	Parity of spatial part	Spin state
0	s-wave	even	singlet
1	p-wave	odd	triplet
2	d-wave	even	singlet

described by a quasi-particle excitation, a *phonon*. We can imagine the phonon to create a small potential well near the electron in which another electron can be caught. So lattice vibrations (phonons) mediate a weakly attractive interaction between the electrons which then form pairs. Conventional BCS-theory with phonons is only valid at low temperatures because around 30 K the phononic lattice vibrations caused by the temperature already destroy the weakly attractive interaction between the electrons. Note that BCS-theory itself does not depend on the origin of the attractive interaction.

A superconductor is simply a charged superfluid

The crucial defining property for any kind of superconductivity or superfluidity is that a symmetry is spontaneously broken. In superfluids this symmetry is global while in superconductors it is local, i.e. a gauge symmetry. Therefore in superconductors there are a few additional effects related to the gauge symmetry and the corresponding gauge field. But besides that those two phenomena are very similar. Especially in both cases there is a *Goldstone boson* created for each spontaneously broken symmetry of the describing field theory. For a broken global symmetry this Goldstone boson survives and is visible in the spectrum as a hydrodynamic mode [15]. For a broken local symmetry however the Goldstone boson is eaten by the gauge field which couples to the charge belonging to the broken symmetry. In superconductors this causes the gauge boson, i.e. the photon for the broken electromagnetic $U(1)$ to become massive. Since these heavy photons can travel only an exponentially small distance, the electromagnetic interaction becomes short-ranged. Therefore magnetic fields, which can be thought of as consisting of photons, can only penetrate the system up to a certain distance, the *penetration depth*. This is called *Meissner-Ochsenfeld effect* and it is a defining criterion for superconductivity. In the Anderson-Higgs mechanism particles acquire a mass by the same mechanism. Thus it is sometimes described as the superfluidity of the vacuum. See [16] for a more precise review.

There is a class of experimentally well-studied but theoretically less understood *unconventional superconductors*, such as copper or ruthenate compounds. Some of these materials show superconducting phases at high temperatures[4] up to 138 K.

[4] Sr_2RuO_4 is superconducting at low temperatures around 2 K but the pairing mechansim is not microscopically understood. The simplified argument is that the phonon-interaction of conventional Cooper pairing is isotropic, thus not providing an anisotropic p-wave structure.

The conventional BCS-theory does not apply to such high temperatures as mentioned above. So the biggest mystery remains to understand the *pairing mechanism* of electrons in these high temperature superconductors. Pairing and condensation need not occur at the same temperature here. Most important for the application of gauge/gravity duality: In these *unconventional superconductors* the coupling strength of the electrons to each other is strong. Thus these are experimentally accessible systems governed by strongly coupled field theory.

Another clear signature for superconductivity is of course an infinite dc conductivity. Together with that there is a conductivity gap in the frequency-dependent conductivity. This is caused by the fact that the conductivity at small frequencies or energies vanishes until there is enough energy to break up one Cooper pair. From that energy on the material is a normal conductor with individual electrons being the charge carriers. In unconventional superconductors there is a surprising phenomenon called *pseudo-gap*. This means that the experiments carried out in unconventional superconductors show a gap in the conductivity even at and above the transition temperature where the superconducting condensate forms. Inside this pseudo-gap the conductivity does not drop to zero but to a small finite value.

11.2.1.2 How to Build a Feld Theory with P-wave Superconductivity

As stressed above the crucial thing to do in order to get a superconductor or superfluid is to spontaneously break a symmetry. We are going to accomplish this by allowing our system to develop a charged condensate.

Let us assume a particle physics point of view for a while. In these notes we will focus on a 3 + 1-dimensional $\mathcal{N} = 2$ supersymmetric $SU(N_c)$ Yang-Mills theory at temperature T, consisting of a $\mathcal{N} = 4$ gauge multiplet as well as N_f massive $\mathcal{N} = 2$ supersymmetric hypermultiplets (ψ_i, ϕ_i). The hypermultiplets give rise to the flavor degrees transforming in the fundamental representation of the gauge group. The action is written down explicitly for instance in [17]. In particular, we work in the large N_c limit with $N_f \ll N_c$ at strong coupling, i.e. with $\lambda \gg 1$, where $\lambda = g^2_{\mathrm{YM}} N_c$ is the 't Hooft coupling constant. In the following we will consider only two flavors, i.e. $N_f = 2$. The flavor degrees of freedom are called u and d. If the masses of the two flavor degrees are degenerate, the theory has a global $U(2)$ flavor symmetry, whose overall $U(1)_B$ subgroup can be identified with the baryon number.

In the following we will consider the theory at finite isospin chemical potential μ, which is introduced as the source of the operator

$$J^3_0 \propto \bar{\psi}\sigma^3\gamma_0\psi + \phi\sigma^3\partial_0\phi = n_u - n_d \,, \tag{11.1}$$

where $n_{u/d}$ is the charge density of the isospin fields, $(\phi_u, \phi_d) = \phi$ and $(\psi_u, \psi_d) = \psi$. σ^i are the Pauli matrices. A non-zero vev $\langle J^3_0 \rangle$ introduces an isospin density as discussed in [18]. The isospin chemical potential μ explicitly breaks the $U(2) \simeq U(1)_B \times SU(2)_I$ flavor symmetry down to $U(1)_B \times U(1)_3$, where $U(1)_3$ is generated by the unbroken generator σ^3 of the $SU(2)_I$. Under the $U(1)_3$ symmetry the fields with index u and d have positive and negative charge, respectively.

However, the theory is unstable at large isospin chemical potential [18]. The new phase is sketched in Fig. 11.1. We show in this lecture (see also [4]), that the new phase is stabilized by a non–vanishing vacuum expectation value of the current component

$$J_3^1 \propto \bar{\psi}\sigma^1\gamma_3\psi + \phi\sigma^1\partial_3\phi = \bar{\psi}_u\gamma_3\psi_d + \bar{\psi}_d\gamma_3\psi_u + \text{bosons}\,. \tag{11.2}$$

This current component breaks both the $SO(3)$ rotational symmetry as well as the remaining Abelian $U(1)_3$ flavor symmetry spontaneously. The rotational $SO(3)$ is broken down to $SO(2)_3$, which is generated by rotations around the x^3 axis. Due to the non-vanishing vev for J_3^1, flavor charged vector mesons condense and form a superfluid. Let us emphasize that we do not describe a color superconductor on the field theory side, since the condensate is a gauge singlet. Figure 11.1 shows a "not accessible" parameter region in which we get divergent quantities. This is due to the fact that at such large charge densities we would need to take in account the backreaction of the D7-branes on the AdS geometry.

In a condensed matter context our model can be considered as a holographic p-wave superconductor in the following way. The global $U(1)_3$ in our model is the analog of the local $U(1)_{\text{em}}$ symmetry of electromagnetic interactions. So far in all holographic models of superconductors the breaking of a global symmetry on the field theory side is considered. In our model, the current J^3 corresponds to the electric current J_{em}. The condensate $\langle J_3^1\rangle$ breaks the $U(1)_3$ spontaneously. Therefore it can be viewed as the superconducting condensate, which is analogous

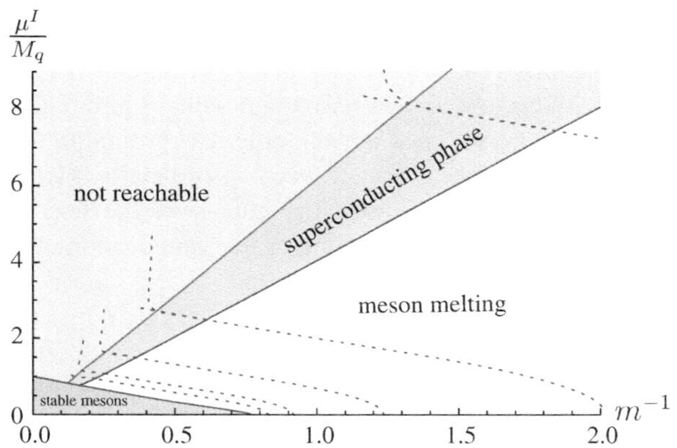

Fig. 11.1 Phase diagram of our field theory depending on the temperature parametrized by $m^{-1} \propto T$ and on the isospin chemical potential μ^I scaled by the quark mass M_q. At low temperatures and chemical potentials a phase of stable mesons forms which melts at higher μ^I and m^{-1}. Above a critical density we find a superconducting or superfluid phase with a vector order parameter

to the Cooper pairs. Since the condensate $\langle J_3^1 \rangle$ transforms as a vector under spatial rotations, it is a p-wave superconductor. Strictly speaking, for a superconductor interpretation it would be necessary to charge the superfluid, i.e. gauge the global $U(1)_3$ symmetry which is broken spontaneously in our model. A spontaneously broken global symmetry corresponds to a superfluid. However, as we mentioned before many features of superconductivity do not depend on whether the $U(1)_3$ is gauged. One exception to this is the Meissner–Ochsenfeld effect. To generate the currents expelling the magnetic field, the $U(1)_3$ symmetry has to be gauged. This matter is discussed further below in Sect. 11.5.4 where we will be able to see the onset of the Meissner–Ochsenfeld effect within our holographic setup.

11.2.2 Holographic Realization

Holographic superconductors have first been studied in [6, 7, 19, 20]. The initial idea presented in [19] was that the Abelian Higgs model coupled to gravity with a negative cosmological constant provides a charged scalar condensate near but outside a charged black hole horizon. The charged condensate spontaneously breaks the Abelian gauge symmetry of the theory. Later studies revealed that this breaking also occurs in setups with neutral black holes but with a negative mass for the charged scalar [6]. The basic idea for a holographic p-wave superconductor has been outlined in [7]. Let us review the basic ideas.

11.2.2.1 Holographic Superconductor Basics

In order to spontaneously break a symmetry we need a gravity theory with a gauge symmetry. Note that this is not the usual gauge symmetry $SU(N)$ of the correspondence where $N \to \infty$, but an additional one with a finite rank, for example an Abelian $U(1)$. Furthermore we need a black hole background in order to introduce finite temperature in the dual field theory, see [3] for details. Most importantly we need a charged condensate hovering in the bulk over the horizon. To be more precise, we need a bulk field ϕ charged under the gauge symmetry which is to be broken. ϕ has to have the typical expansion $\phi = \phi_{\text{normalizable}} + \phi_{\text{non-normalizable}} + \dots$ near the AdS boundary, but with $\phi_{\text{non-normalizable}} \equiv 0$. Why do we require this particular structure? Recall that in the gauge/gravity correspondence the normalizable mode is identified with the field theory *expectation value* $\langle \mathcal{O}_\phi \rangle$ of the operator $\mathcal{O}_\phi$ dual to the gravity field ϕ. This is our condensate in the field theory, which we want to be non-zero. The normalizable mode on the other hand is dual to a *source* in the field theory. Therefore we want it to vanish since it would break the symmetry *explicitly*. The role of the field ϕ providing the condensate could be played for example by a charged scalar or by one component of a gauge field. These concepts are illustrated in the following example.

Example Consider the bottom-up toy model p-wave superconductor introduced in [7]. There we have an Einstein Yang-Mills theory with a negative cosmological constant

$$S \sim \int \mathcal{D}^4 x \left[R - \frac{1}{4}(F^a_{\mu\nu})^2 + \frac{6}{L^2}, \right] \tag{11.3}$$

with F being the field strength of an $SU(2)$ gauge field A. This $SU(2)$ in this case is the gauge group that we want to break. Actually we break only a $U(1)$ subgroup of it. But let us not worry about the details now. Only note that this theory is placed in an AdS_4 charged black hole background. Our gauge field A now plays the role of the field ϕ providing the condensate. The operator dual to the gauge field A is the electromagnetic current $\mathcal{O}_\phi = J^\mu$. The boundary behavior of the gravity field's components according to the equations of motion derived from Eq. 11.3 is given by

$$A_t^3 = \mu + \frac{d}{r}, \quad A_x^1 = 0 + \frac{\langle J_x^1 \rangle}{r}, \tag{11.4}$$

with the radial AdS coordinate r. In principle we could allow more non-vanishing components but this combination turns out to be both sufficient and consistent. As usual in thermal AdS/CFT the temporal component A_t introduces a chemical potential μ in the dual field theory. This chemical potential sources the corresponding operator $J^t = d$ which is simply the charge density of the $SU(2)$ charge in the field theory at the boundary. This explicitly breaks $SU(2) \to U(1)_3$. However the spatial component A_x^1 spontaneously breaks the remaining $U(1)_3$ with the condensate $\langle J_x^1 \rangle$.

11.2.2.2 How to Build a Gravity Dual to P-Wave Superconductivity

As mentioned several times before, we need a vector condensate for our p-wave superconductor. Conveniently the previous example already showed us what the structure for the dual gravity theory has to be in order to realize a vector condensate. Unfortunately that example has not been derived from string theory. Probably the most difficult task in building a holographic superconductor is finding a gravity setup with all the necessary features, which is actually stable and thermodynamically favored. We are going to see that the system of intersecting D3 and D7 branes provides exactly such a stable configuration. Since we are going to review Dp/Dq-brane systems below, let us here spot only those features which are of importance to superconductivity. The $SU(2)$ "flavor" gauge group which we are going to break (partly spontaneously) is created by using two coincident D7 branes. By introducing a gauge field A_μ^a living on these D7-branes and giving its temporal component a non-trivial profile in one of the flavor directions, we introduce a chemical potential in the dual field theory. This is completely analogous to the introducing the non-trivial A_t^3 in the previous example. All this is going to take place in the background of a stack of N_c black D3-branes. They introduce the temperature into the dual field theory. Finally, the gravity field which is going to break the residual flavor symmetry is going to be a spatial component of the gauge field just as A_x^1 in the previous example. Our analysis of the thermodynamic potentials is going to show that this phase is thermodynamically preferred over the phase without the symmetry-breaking condensate. Furthermore it is stable against all obvious gauge field fluctuations.

11.3 Holographic Setup

In this section we carry out exactly the D3-D7 brane construction outlined in Sect. 11.2.2. The result will be a gravity setup being holographically dual to a p-wave superfluid/superconductor. But let us first review how to add flavor to the gauge/gravity correspondence.

11.3.1 Flavor from Intersecting Branes

Let us imagine for this subsection that we want to use gauge/gravity in order to model QCD or the quark gluon plasma state of matter produced at the RHIC Brookhaven heavy-ion collider. The original AdS/CFT conjecture does not include matter in the fundamental representation of the gauge group but only adjoint matter. In order to come closer to a QCD-like behavior one can therefore investigate how to incorporate quarks and their bound states in this section. We focus on the main results of [21] and [22], however for a concise review the reader is referred to [17].

Since AdS/CFT has been discovered a lot of modifications of the original conjecture have been proposed and analyzed. This is always achieved by modifying the gravity theory in an appropriate way. For example the metric on which the gravity theory is defined may be changed to produce chiral symmetry breaking in the dual gauge theory [23, 24]. Other modifications put the gauge theory at finite temperature and produce confinement [25]. Besides the introduction of finite temperature the inclusion of fundamental matter, i.e. quarks, is the most relevant extension for us since we are aiming at a qualitative description of strongly coupled QCD effects at finite temperature. These effects are similar to the ones observed at the RHIC heavy ion collider.

11.3.1.1 Adding Flavor to AdS/CFT

The change we have to make on the gravity side in order to produce fundamental matter on the gauge theory side is the introduction of a small number N_f of

D7-branes. These are also called *probe branes* since their backreaction on the geometry originally produced by the stack of N D3-branes is neglected. Strings within this D3/D7-setup now have the choice of starting (ending) on the D3- or alternatively on the D7-brane. Note that the two types of branes share the four Minkowski directions $0, 1, 2, 3$ in which also the dual gauge theory will extend on the boundary of AdS as visualized in Fig. 11.2.

The configuration of one string ending on N coincident D3-branes produces an $SU(N)$ gauge symmetry of rotations in color space. Similarly the N_f D7-branes generate a $U(N_f)$ flavor gauge symmetry. We will call the strings starting on the stack of Dp-branes and ending on the stack of Dq-branes $p-q$ strings. The original 3–3 strings are unchanged while the 3–7- or equivalently 7–3 strings are interpreted as quarks on the gauge theory side of the correspondence. This can be understood by looking at the 3–3 strings again. They come in the adjoint representation of the gauge group which can be interpreted as the decomposition of a bifundamental representation $(N^2 - 1) \oplus 1 = N \otimes \bar{N}$. So the two string ends on the D3-brane are interpreted as one giving the fundamental, the other giving the anti-fundamental representation in the gauge theory. In contrast to this the 3–7 string has only one end on the D3-brane stack corresponding to a single fundamental representation which we interpret as a single quark in the gauge theory.

We can also give mass to these quarks by separating the stack of D3-branes from the D7-branes in a direction orthogonal to both branes. Now 3–7 strings are forced to have a finite length L which is the minimum distance between the two brane stacks. On the other hand a string is an object with tension and if it assumes a minimum length, it needs to have a minimum energy being the product of its length and tension. The dual gauge theory object is the quark and it now also has a minimum energy which we interpret as its mass $M_q = L/(2\pi\alpha')$.

The 7–7 strings decouple from the rest of the theory since their effective coupling is suppressed by N_f/N. In the dual gauge theory this limit corresponds to neglecting quark loops which is often called *the quenched approximation*. Nevertheless, they are important for the description of mesons as we will see below.

Let us be a bit more precise about the fundamental matter introduced by 3–7 strings. The gauge theory introduced by these strings (in addition to the original setup) gives a $\mathcal{N} = 2$ supersymmetric $U(N)$ gauge theory containing N_f fundamental hypermultiplets.

	0	1	2	3	4	5	6	7	8	9
D3	x	x	x	x						
D7	x	x	x	x	x	x	x	x		

Fig. 11.2 Coordinate directions in which the Dp-branes extend are marked by 'x'. D3- and D7-branes always share the four Minkowski directions and may be separated in the $8, 9$-directions which are orthogonal to both brane types

Fig. 11.3 This figure summarizes our coordinates and indices

	AdS_5					S^3
coord. names	x^0	x^1	x^2	x^3	ϱ	$-$
	$\mu, \nu, \ldots$					
indices		$i, j, \ldots$			ϱ	
	0	1	2	3	4	

11.3.1.2 D7 Embeddings and Meson Excitations

Mesons correspond to fluctuations of the D7-branes[5] embedded in the $AdS_5 \times S^5$-background generated by the D3-branes. From the string-point of view these fluctuations are fluctuations of the hypersurface on which the $7-7$ strings can end, hence these are small oscillations of the $7-7$ string ends. The $7-7$ strings again lie in the adjoint representation of the flavor gauge group for the same reason which we employed above to argue that $3-3$ strings are in the adjoint of the (color) gauge group. Mesons are the natural objects in the adjoint flavor representation. Vector mesons correspond to fluctuations of the gauge field on the D7-branes.

Before we can examine mesons as D7-fluctuations we need to find out how the D7-branes are embedded into the 10-dimensional geometry without any fluctuations. Such a stable configuration needs to minimize the effective action. The effective action to consider is the world volume action of the D7-branes which is composed of a Dirac-Born-Infeld and a topological Chern-Simons part

$$\begin{aligned} S_{\mathrm{D7}} = & - T_{\mathrm{D7}} \int \mathcal{D}^8 \sigma e^{-\Phi} \sqrt{-\det\left\{ P[g+B]_{\alpha\beta} + (2\pi\alpha')F_{\alpha\beta} \right\}} \\ & + \frac{(2\pi\alpha')^2}{2} T_{\mathrm{D7}} \int P[C_4] \wedge F \wedge F. \end{aligned} \tag{11.5}$$

The preferred coordinates to examine the fluctuations of the D7 are obtained from the standard AdS coordinates

$$\mathcal{D}s^2 = \frac{R^2}{\varrho^2}\mathcal{D}\varrho^2 + \frac{\varrho^2}{R^2}(-\mathcal{D}t^2 + \mathcal{D}\mathbf{x}^2), \tag{11.6}$$

with the AdS radius R and the dimensionful radial AdS coordinate ϱ.

Exercise Show that the standard AdS metric (11.6) transforms to

$$\mathcal{D}s^2 = \frac{r^2}{R^2}\mathcal{D}\mathbf{x}^2 + \frac{R^2}{r^2}(\mathcal{D}\varrho^2 + \varrho^2\mathcal{D}\Omega_3{}^2 + \mathcal{D}w_5{}^2 + \mathcal{D}w_6^2), \tag{11.7}$$

[5] To be precise the fluctuations correspond to the mesons with spins 0, 1/2 and 1 [22, 26].

under the transformation$\varrho^2 = {w_1}^2 + \cdots + {w_4}^2$, $r^2 = \varrho^2 + {w_5}^2 + {w_6}^2$, *where* $\mathbf{x}$ is a four vector in Minkowski directions$0, 1, 2, 3$ *and* R is the AdS radius. The coordinate$r$is the radial AdS coordinate whileϱ is the radial coordinate on the coincident D7-branes.

Let us follow [22]: For a static D7 embedding with vanishing field strength F on the D7 world volume the equations of motion are

$$0 = \frac{\mathcal{D}}{\mathcal{D}\varrho}\left(\frac{\varrho^3}{\sqrt{1 + {w'_5}^2 + {w'_6}^2}}\frac{\mathcal{D}w_{5,6}}{\mathcal{D}\varrho}\right), \tag{11.8}$$

where $w_{5,6}$ denotes that these are two equations for the two possible directions of fluctuation. Since (11.8) is the equation of motion of a supergravity field in the bulk, the solution near the AdS boundary takes the standard form with a non-normalizable and a normalizable mode, or source and expectation value respectively

$$w_{5,6} = m + \frac{c}{\varrho^2} + \cdots, \tag{11.9}$$

with m being the quark mass acting as a source and c being the expectation value of the operator which is dual to the field $w_{5,6}$. While c can be related to the scaled quark condensate $c \propto \langle\bar{q}q\rangle(2\pi\alpha')^3$.

If we now separate the D7-branes from the stack of D3-branes the quarks become massive and the radius of the S^3 on which the D7 is wrapped becomes a function of the radial AdS coordinate r. The separation of stacks by a distance m modifies the metric induced on the D7 $P[g]$ such that it contains the term $R^2\varrho^2/(\varrho^2 + m^2)\mathcal{D}{\Omega_3}^2$. This expression vanishes at a radius $\varrho^2 = r^2 - m^2 = 0$ such that the S^3 shrinks to zero size at a finite AdS radius.

Fluctuations about these w_5 and w_6 embeddings give scalar and pseudoscalar mesons. We take

$$w_5 = 0 + 2\pi\alpha'\chi, \quad w_6 = m + 2\pi\alpha'\varphi \tag{11.10}$$

After plugging these into the effective action (11.5) and expanding to quadratic order in fluctuations we can derive the equations of motion for φ and χ. As an example we consider scalar fluctuations using an Ansatz

$$\varphi = \phi(\varrho)e^{i\mathbf{k}\cdot\mathbf{x}}\mathcal{Y}_l(S^3), \tag{11.11}$$

where $\mathcal{Y}_l(S^3)$ are the scalar spherical harmonics on the S^3, ϕ solves the radial part of the equation and the exponential represents propagating waves with real momentum $\mathbf{k}$. We additionally have to assume that the mass-shell condition

$$M^2 = -\mathbf{k}^2 \tag{11.12}$$

is valid. Solving the radial part of the equation we get the hypergeometric function $\phi \propto F(-\alpha, -\alpha + l + 1; (l + 2); \frac{-\varrho^2}{m^2})$ and the parameter

$$\alpha = -\frac{1 - \sqrt{1 - \mathbf{k}^2 R^4/m^2}}{2} \tag{11.13}$$

summarizes a factor appearing in the equation of motion. In general this hypergeometric function may diverge if we take $\varrho \to \infty$. But since this is not compatible with our linearization of the equation of motion in small fluctuations, we further demand normalizability of the solution. This restricts the sum of parameters appearing in the hypergeometric function to take the integer values

$$n = \alpha - l - 1\,, \quad n = 0, 1, 2, \ldots . \tag{11.14}$$

With this quantization condition we determine the scalar meson mass spectrum to be

$$M_{\text{(pseudo)scalar}} = \frac{2m}{R^2}\sqrt{(n + l + 1)(n + l + 2)}\,, \tag{11.15}$$

where n is the radial excitation number found for the hypergeometric function. Similarly we can determine pseudoscalar masses and find the same formula (11.15). For vector meson masses we need to consider fluctuations of the gauge field A appearing in the field strength F in Eq. 11.5. The formula for vector mesons (corresponding to e.g. the ϱ-meson of QCD) is

$$M_{\text{v}} = \frac{2m}{R^2}\sqrt{(n + l + 1)(n + l + 2)}\,. \tag{11.16}$$

Note that the scalar, pseudoscalar and vector mesons computed within this framework show identical mass spectra. Further fluctuations corresponding to other mesonic excitations can be found in [22, 26].

11.3.1.3 Brane Embeddings at Finite Temperature

In order to get a finite temperature in the dual field theory, the gravity theory needs to be put into a black hole or black brane background geometry. It was found in [24, 27, 28] that at finite temperature our N_f probe flavor branes can be embedded in two distinct ways. There are high temperature configurations called *black hole embeddings* in which part of the brane falls into the black hole horizon. On the other hand there are low-temperature configurations called in which the brane stays outside the black hole horizon. These two configurations are separated by a geometric transition, i.e. the configuration in which the brane just barely touches the black hole horizon. This geometric transition corresponds to the meson melting transition for the fundamental matter of the dual field theory. Note that the adjoint matter of this field theory is always deconfined in this setup.

At finite charge density however there is only one kind of embedding and that is the black hole embedding. The heuristic argument is that introducing a finite charge on the brane there have to be field lines for the associated field strength. These lines have to end somewhere. Our setup has rotational symmetry in the spatial directions. Imagine the radial AdS coordinate and field lines running along it. If they are supposed to end somewhere, there has to be a horizon. Otherwise they would all meet in the origin at $\rho = 0$. Note that field lines ending at the horizon means, that we can interpret the horizon as being charged. In these lecture notes we will exclusively deal with non-vanishing density and thus only encounter black hole embeddings.

11.3.2 Background and Brane Configuration

We aim to have a field theory dual at finite temperature. This is holographically accomplished by placing the gravity theory in a black hole or black brane background. Here the black hole's Hawking temperature can be identified with the field theory temperature. We consider asymptotically $AdS_5 \times S^5$ space-time. The $AdS_5 \times S^5$ geometry is holographically dual to the $\mathcal{N} = 4$ Super Yang-Mills theory with gauge group $SU(N_c)$. The dual description of a finite temperature field theory is an AdS black hole. We use the coordinates of [29] to write the AdS black hole background in Minkowski signature as

$$\mathcal{D}s^2 = \frac{\varrho^2}{2R^2}\left(-\frac{f^2}{\tilde{f}}\mathrm{d}t^2 + \tilde{f}\mathrm{d}\mathbf{x}^2\right) + \left(\frac{R}{\varrho}\right)^2(\mathrm{d}\varrho^2 + \varrho^2\mathrm{d}\Omega_5^2), \tag{11.17}$$

with $\mathrm{d}\Omega_5^2$ the metric of the unit 5-sphere and

$$f(\varrho) = 1 - \frac{\varrho_H^4}{\varrho^4}, \quad \tilde{f}(\varrho) = 1 + \frac{\varrho_H^4}{\varrho^4}, \tag{11.18}$$

where R is the AdS radius, with $R^4 = 4\pi g_s N_c\, \alpha'^2 = 2\lambda\, \alpha'^2$.

Exercise Show that the metric (11.17) can be obtained from the standard finite temperature AdS metric

$$\mathcal{D}s^2 = \frac{(\pi TR)^2}{u}[-f(u)\mathcal{D}t^2 + \mathcal{D}\mathbf{x}^2] + \frac{R^2}{4u^2f(u)}\mathcal{D}u^2 + R^2\mathcal{D}\Omega_5{}^2; \quad f(u) = 1 - u^2, \tag{11.19}$$

by the transformation$(r_0\rho)^2 = 2r^2 + \sqrt{r^4 - r_0{}^4}$,*with* r being the original radial AdS coordinate and$r_0$the location of the black hole horizon.

Exercise The temperature of the black hole given by (11.17) may be determined by demanding regularity of the Euclidean section. Show that it is given by

$$T = \frac{\varrho_H}{\pi R^2}. \tag{11.20}$$

In the following we may use the dimensionless coordinate $\rho = \varrho/\varrho_H$, which covers the range from the event horizon at $\rho = 1$ to the boundary of the AdS space at $\rho \to \infty$. To include fundamental matter, we embed N_f coinciding D7-branes into the ten-dimensional space-time as illustrated in Fig. 11.2. These D7-branes host flavor gauge fields A_μ with gauge group $U(N_f)$. To write down the DBI action for the D7-branes, we introduce spherical coordinates $\{r, \Omega_3\}$ in the 4567-directions and polar coordinates $\{L, \phi\}$ in the 89-directions [29]. The angle between these two spaces is denoted by θ $(0<\theta<\pi/2)$. The six-dimensional space in the 456789-directions is given by

$$\begin{aligned} \mathrm{d}\varrho^2 + \varrho^2 \mathrm{d}\Omega_5^2 &= \mathrm{dr}^2 + \mathrm{r}^2\mathrm{d}\Omega_3^2 + \mathrm{dL}^2 + \mathrm{L}^2\mathrm{d}\phi^2 \\ &= \mathrm{d}\varrho^2 + \varrho^2(\mathrm{d}\theta^2 + \cos^2\theta \mathrm{d}\phi^2 + \sin^2\theta \mathrm{d}\Omega_3^2), \end{aligned} \tag{11.21}$$

where $r = \varrho \sin\theta, \varrho^2 = r^2 + L^2$ and $L = \varrho\cos\theta$. D3- and D7-branes always share the four Minkowski directions and may be separated in the 8, 9-directions which are orthogonal to both brane types. That separation is dual to the mass of the fundamental flavor fields in the dual gauge theory, i.e. the *quarks*.

Due to the $SO(4)$ rotational symmetry in the 4567 directions, the embedding of the D7-branes only depends on the radial coordinate ρ. Defining $\chi = \cos\theta$, we parametrize the embedding by $\chi = \chi(\rho)$ and choose $\phi = 0$ using the $SO(2)$ symmetry in the 89-direction. The induced metric G on the D7-brane probes is then

$$ds^2(G) = \frac{\varrho^2}{2R^2}\left(-\frac{f^2}{\tilde{f}}\mathrm{dt}^2 + \tilde{\mathrm{f}}\mathrm{d}\mathbf{x}^2\right) + \frac{R^2}{\varrho^2}\frac{1-\chi^2+\varrho^2(\partial_\varrho\chi)^2}{1-\chi^2}\mathrm{d}\varrho^2 + \mathrm{R}^2(1-\chi^2)\mathrm{d}\Omega_3^2. \tag{11.22}$$

The square root of the determinant of G is given by

$$\sqrt{-G} = \frac{\sqrt{h_3}}{4}\varrho^3 f\tilde{f}(1-\chi^2)\sqrt{1-\chi^2+\varrho^2(\partial_\varrho\chi)^2}, \tag{11.23}$$

where h_3 is the determinant of the 3-sphere metric.

As in [18] we introduce a $SU(2)$ isospin chemical potential μ by a non-vanishing time component of the non-Abelian background field on the D7-brane. The generators of the $SU(2)$ gauge group are given by the Pauli matrices σ^i. Due to the gauge symmetry, we may rotate the flavor coordinates until the chemical potential lies in the third flavor direction,

$$\mu = \lim_{\rho\to\infty} A_0^3(\rho). \tag{11.24}$$

This non-zero gauge field breaks the $SU(2)$ gauge symmetry down to $U(1)_3$ generated by the third Pauli matrix σ^3. The spacetime symmetry on the boundary is still $SO(3)$. Notice that the Lorentz group $SO(3,1)$ is already broken down to $SO(3)$ by the finite temperature. In addition, we consider a further non-vanishing background gauge field which stabilizes the system for large chemical potentials. Due to the symmetry of our setup we may choose $A_3^1 \mathrm{d}x^3\sigma^1$ to be non-zero. To obtain an isotropic configuration in the field theory, this new gauge field A_3^1 only depends on ρ. Due to this two non-vanishing gauge fields, the field strength tensor on the branes has the following non-zero components,

$$F^1_{\varrho 3} = -F^1_{3\varrho} = \partial_\varrho A^1_3 \qquad (7-7\text{strings}), \tag{11.25}$$

$$F^3_{\varrho 0} = -F^3_{0\varrho} = \partial_\varrho A^3_0 \qquad (3-7\text{strings}), \tag{11.26}$$

$$F^2_{03} = -F^2_{30} = \frac{\gamma}{\sqrt{\lambda}} A^3_0 A^1_3 \qquad (\text{interaction}). \tag{11.27}$$

The labels behind those equations refer to the sort of strings which generate the corresponding gauge fields. The field strength F^2_{03} can be understood as an interaction term between 7–7 and 3–7 strings. We derive this interpretation in Sect. 11.6.1.

11.3.3 DBI Action and Equations of Motion

In this section we calculate the equations of motion which determine the profile of the D7-brane probes and of the gauge fields on these branes. A discussion of the gauge field profiles, which we use to give a geometrical interpretation of the stabilization of the system and the pairing mechanism, may be found in Sect. 11.6.1.

The DBI action determines the shape of the brane embeddings, i.e. the scalar fields ϕ, as well as the configuration of the gauge fields A on these branes. We consider the case of $N_f = 2$ coincident D7-branes for which the non-Abelian DBI action reads [30]

$$S_{\mathrm{DBI}} = -T_{D7}\mathrm{Str}\int \mathrm{d}^8\xi\sqrt{\det \mathrm{Q}}\Big[\det\Big(\mathrm{P_{ab}}\Big[\mathrm{E}_{\mu\nu} + \mathrm{E}_{\mu\mathrm{i}}(\mathrm{Q}^{-1}-\delta)^{\mathrm{ij}}\mathrm{E}_{\mathrm{j}\nu}\Big] + 2\pi\alpha'\mathrm{F_{ab}}\Big)\Big]^{\frac{1}{2}} \tag{11.28}$$

with

$$Q^i_j = \delta^i_j + i2\pi\alpha'[\Phi^i, \Phi^k]E_{kj} \tag{11.29}$$

and P_{ab} the pullback to the Dp-brane, where for a Dp-brane in d dimensions we have $\mu, \nu = 0, \ldots, (d-1), a, b = 0, \ldots, p, i, j = (p+1), \ldots, (d-1), E_{\mu\nu} = g_{\mu\nu} + B_{\mu\nu}$. In our case we set $p = 7, d = 10, B \equiv 0$. As in [18] we can simplify this action significantly by using the spatial and gauge symmetries present in our setup. The action becomes

$$S_{\mathrm{DBI}} = -T_{D7}\int \mathrm{d}^8\xi \mathrm{Str}\sqrt{|\det(\mathrm{G}+2\pi\alpha'\mathrm{F})|} \tag{11.30}$$

$$\begin{aligned} =& -T_{D7}\int \mathrm{d}^8\xi\sqrt{-\mathrm{G}}\mathrm{Str}\Big[1+\mathrm{G}^{00}\mathrm{G}^{44}\left(\mathrm{F}^3_{\varrho 0}\right)^2\left(\sigma^3\right)^2+\mathrm{G}^{33}\mathrm{G}^{44}\left(\mathrm{F}^1_{\varrho 3}\right)^2\left(\sigma^1\right)^2 \\ &+G^{00}G^{33}\left(F^2_{03}\right)^2\left(\sigma^2\right)^2\Big]^{\frac{1}{2}} \end{aligned} \tag{11.31}$$

where in the second line the determinant is calculated. Due to the symmetric trace, all commutators between the matrices σ^i vanish. It is known that the symmetrized trace prescription in the DBI action is only valid up to fourth order in α' [31, 32]. However the corrections to the higher order terms are suppressed by N_f^{-1} [33] (see also [34]). Here we use two different approaches to evaluate the non-Abelian DBI action (11.30). First, we modify the symmetrized trace prescription by omitting the commutators of the generators σ^i and then setting $(\sigma^i)^2 = 1$ (see Sect. 11.3.3.1 below). This prescription makes the calculation of the full DBI action feasible. This prescription is not verified in general but we obtain physically reasonable results as discussed in Sects. 11.5.1 and 11.5.2. Second, we expand the non-Abelian DBI action to fourth order in the field strength F (see Sect. 11.3.3.2). Here it should be noted that in general the higher terms of this expansion need not be smaller than the leading ones. However, we again get physical results in our specific case which confirm this approach. We further motivate the validity of our two approaches below.

11.3.3.1 Adapted Symmetrized Trace Prescription

Using the adapted symmetrized trace prescription defined above, the action becomes

$$\begin{aligned} S_{\mathrm{DBI}} =& -T_{D7}N_f\int \mathrm{d}^8\xi\sqrt{-\mathrm{G}}\Big[1+\mathrm{G}^{00}\mathrm{G}^{44}\left(\mathrm{F}^3_{\varrho 0}\right)^2+\mathrm{G}^{33}\mathrm{G}^{44}\left(\mathrm{F}^1_{\varrho 3}\right)^2 \\ &+G^{00}G^{33}\left(F^2_{03}\right)^2\Big]^{\frac{1}{2}} \\ =& -\frac{T_{D7}N_f}{4}\int \mathrm{d}^8\xi\; \varrho^3 f\tilde{f}(1-\chi^2)\Upsilon(\rho,\chi,\tilde{A}), \end{aligned} \tag{11.32}$$

with

$$\Upsilon(\rho,\chi,\tilde{A}) = \left[1-\chi^2+\rho^2(\partial_\rho\chi)^2 - \frac{2\tilde{f}}{f^2}(1-\chi^2)\left(\partial_\rho\tilde{A}_0^3\right)^2 + \frac{2}{\tilde{f}}(1-\chi^2)\left(\partial_\rho\tilde{A}_3^1\right)^2 \right.$$
$$\left. - \frac{2\gamma^2}{\pi^2\rho^4 f^2}(1-\chi^2+\rho^2(\partial_\rho\chi)^2)\left(\tilde{A}_0^3\tilde{A}_3^1\right)^2\right]^{\frac{1}{2}}, \tag{11.33}$$

where the dimensionless quantities $\rho = \varrho/\varrho_H$ and $\tilde{A} = (2\pi\alpha')A/\varrho_H$ are used. To obtain first order equations of motion for the gauge fields which are easier to solve numerically, we perform a Legendre transformation. Similarly to [18, 29] we calculate the electric displacement p_0^3 and the magnetizing field p_3^1 which are given by the conjugate momenta of the gauge fields A_0^3 and A_3^1,

$$p_0^3 = \frac{\delta S_{\mathrm{DBI}}}{\delta(\partial_\varrho A_0^3)}, \quad p_3^1 = \frac{\delta S_{\mathrm{DBI}}}{\delta(\partial_\varrho A_3^1)},. \tag{11.34}$$

In contrast to [18, 29, 35, 36], the conjugate momenta are not constant any more but depend on the radial coordinate ϱ due to the non-Abelian term $A_0^3 A_3^1$ in the DBI action. For the dimensionless momenta $\tilde{p}_0^3$ and $\tilde{p}_3^1$ defined as

$$\tilde{p} = \frac{p}{2\pi\alpha' N_f T_{D7}\varrho_H^3}, \tag{11.35}$$

we get

$$\tilde{p}_0^3 = \frac{\rho^3\tilde{f}^2(1-\chi^2)^2\partial_\rho\tilde{A}_0^3}{2f\Upsilon(\rho,\chi,\tilde{A})}, \quad \tilde{p}_3^1 = -\frac{\rho^3 f(1-\chi^2)^2\partial_\rho\tilde{A}_3^1}{2\Upsilon(\rho,\chi,\tilde{A}))}. \tag{11.36}$$

Finally, the Legendre-transformed action is given by

$$\tilde{S}_{\mathrm{DBI}} = S_{\mathrm{DBI}} - \int \mathrm{d}^8\xi\left[\left(\partial_\varrho \mathrm{A}_0^3\right)\frac{\delta S_{\mathrm{DBI}}}{\delta\left(\partial_\varrho \mathrm{A}_0^3\right)} + \left(\partial_\varrho \mathrm{A}_3^1\right)\frac{\delta S_{\mathrm{DBI}}}{\delta\left(\partial_\varrho \mathrm{A}_3^1\right)}\right] \tag{11.37}$$

$$= -\frac{T_{D7}N_f}{4}\int \mathrm{d}^8\xi\varrho^3 f\tilde{f}(1-\chi^2)\sqrt{1-\chi^2+\rho^2(\partial_\rho\chi)^2}\; V(\rho,\chi,\tilde{A},\tilde{p}), \tag{11.38}$$

with

$$V(\rho,\chi,\tilde{A},\tilde{p}) = \left[\left(1-\frac{2\gamma^2}{\pi^2\rho^4 f^2}\left(\tilde{A}_0^3\tilde{A}_3^1\right)^2\right)\right.$$
$$\left.\times\left(1+\frac{8\left(\tilde{p}_0^3\right)^2}{\rho^6\tilde{f}^3(1-\chi^2)^3} - \frac{8\left(\tilde{p}_3^1\right)^2}{\rho^6\tilde{f}f^2(1-\chi^2)^3}\right)\right]^{\frac{1}{2}}. \tag{11.39}$$

Then the first order equations of motion for the gauge fields and their conjugate momenta are

$$\partial_\rho \tilde{A}_0^3 = \frac{2f\sqrt{1-\chi^2+\rho^2(\partial_\rho\chi)^2}}{\rho^3\tilde{f}^2(1-\chi^2)^2}\tilde{p}_0^3 W(\rho,\chi,\tilde{A},\tilde{p}), \tag{11.40}$$

$$\partial_\rho \tilde{A}_3^1 = -\frac{2\sqrt{1-\chi^2+\rho^2(\partial_\rho\chi)^2}}{\rho^3 f(1-\chi^2)^2}\tilde{p}_3^1 W(\rho,\chi,\tilde{A},\tilde{p}), \tag{11.41}$$

$$\partial_\rho \tilde{p}_0^3 = \frac{\tilde{f}(1-\chi^2)\sqrt{1-\chi^2+\rho^2(\partial_\rho\chi)^2}c^2}{2\pi^2\rho f W(\rho,\chi,\tilde{A},\tilde{p})}\left(\tilde{A}_3^1\right)^2\tilde{A}_0^3, \tag{11.42}$$

$$\partial_\rho \tilde{p}_3^1 = \frac{\tilde{f}(1-\chi^2)\sqrt{1-\chi^2+\rho^2(\partial_\rho\chi)^2}c^2}{2\pi^2\rho f W(\rho,\chi,\tilde{A},\tilde{p})}\left(\tilde{A}_0^3\right)^2\tilde{A}_3^1, \tag{11.43}$$

with

$$W(\rho,\chi,\tilde{A},\tilde{p}) = \sqrt{\frac{1-\frac{2\gamma^2}{\pi^2\rho^4 f^2}\left(\tilde{A}_0^3\tilde{A}_3^1\right)^2}{1+\frac{8\left(\tilde{p}_0^3\right)^2}{\rho^6\tilde{f}^3(1-\chi^2)^3}-\frac{8\left(\tilde{p}_3^1\right)^2}{\rho^6\tilde{f}f^2(1-\chi^2)^3}}}. \tag{11.44}$$

For the embedding function χ we get the second order equation of motion

$$\begin{aligned}
\partial_\rho\left[\frac{\rho^5 f\tilde{f}(1-\chi^2)(\partial_\rho\chi)}{\sqrt{1-\chi^2+\rho^2(\partial_\rho\chi)^2}}V\right] = &-\frac{\rho^3 f\tilde{f}\chi}{\sqrt{1-\chi^2+\rho^2(\partial_\rho\chi)^2}}\left[\left[3(1-\chi^2)+2\rho^2(\partial_\rho\chi)^2\right]V\right. \\
&\left.-\frac{24\left(1-\chi^2+\rho^2(\partial_\rho\chi)^2\right)}{\tilde{f}^3\rho^6(1-\chi^2)^3}W\left(\left(\tilde{p}_0^3\right)^2-\frac{\tilde{f}^2}{f^2}\left(\tilde{p}_3^1\right)\right)\right].
\end{aligned} \tag{11.45}$$

We solve the equations numerically and determine the solution by integrating the equations of motion from the horizon at $\rho = 1$ to the boundary at $\rho = \infty$. The initial conditions may be determined by the asymptotic expansion of the gravity fields near the horizon

$$\tilde{A}_0^3 = \frac{c_0}{\sqrt{\left(1-\chi_0^2\right)^3+c_0^2}}(\rho-1)^2+\mathcal{O}\left((\rho-1)^3\right), \tag{11.46}$$

$$\tilde{A}_3^1 = b_0+\mathcal{O}\left((\rho-1)^3\right), \tag{11.47}$$

$$\tilde{p}_0^3 = c_0+\frac{\gamma^2 b_0^2 c_0}{8\pi^2}(\rho-1)^2+\mathcal{O}\left((\rho-1)^3\right), \tag{11.48}$$

$$\tilde{p}_3^1 = +\mathcal{O}\Big(\left(\rho - 1\right)^3\Big), \tag{11.49}$$

$$\chi = \chi_0 - \frac{3\chi_0(1-\chi_0^2)^3}{4[(1-\chi_0^2)^3 + c_0^2]}(\rho - 1)^2 + \mathcal{O}\Big(\left(\rho - 1\right)^3\Big), \tag{11.50}$$

where the terms in the expansions are arranged according to their order in $\rho - 1$. For the numerical calculation we consider the terms up to sixth order in $\rho - 1$. The three independent parameter b_0, c_0 and χ_0 may be determined by field theory quantities defined via the asymptotic expansion of the gravity fields near the boundary,

$$\tilde{A}_0^3 = \tilde{\mu} - \frac{\tilde{d}_0^3}{\rho^2} + \mathcal{O}(\rho^{-4}), \tag{11.51}$$

$$\tilde{A}_3^1 = -\frac{\tilde{d}_3^1}{\rho^2} + \mathcal{O}(\rho^{-4}), \tag{11.52}$$

$$\tilde{p}_0^3 = \tilde{d}_0^3 + \mathcal{O}(\rho^{-4}), \tag{11.53}$$

$$\tilde{p}_3^1 = -\tilde{d}_3^1 + \frac{\gamma^2\tilde{\mu}^2\tilde{d}_3^1}{4\pi^2\rho^2} + \mathcal{O}(\rho^{-4}), \tag{11.54}$$

$$\chi = \frac{m}{\rho} + \frac{c}{\rho^3} + \mathcal{O}(\rho^{-4}). \tag{11.55}$$

According to the AdS/CFT dictionary, μ is the isospin chemical potential. The parameters $\tilde{d}$ are related to the vev of the flavor currents J by

$$\tilde{d}_0^3 = \frac{2^{\frac{5}{2}}\langle J_0^3\rangle}{N_f N_c\sqrt{\lambda}T^3}, \quad \tilde{d}_3^1 = \frac{2^{\frac{5}{2}}\langle J_3^1\rangle}{N_f N_c\sqrt{\lambda}T^3} \tag{11.56}$$

and m and c to the bare quark mass M_q and the quark condensate $\langle\bar{\psi}\psi\rangle$,

$$m = \frac{2M_q}{\sqrt{\lambda}T}, \quad c = -\frac{8\langle\bar{\psi}\psi\rangle}{\sqrt{\lambda}N_f N_c T^3}, \tag{11.57}$$

respectively. There are two independent physical parameters, e. g. m and μ, in the grand canonical ensemble. From the boundary asymptotics (11.51), we also obtain that there is no source term for the current J_3^1. Therefore as a non-trivial result we find that the $U(1)_3$ symmetry is always broken spontaneously. In contrast, in the related works on p-wave superconductors in $2+1$ dimensions [7, 37], the spontaneous breaking of the $U(1)_3$ symmetry has to be put in by hand by setting the source term for the corresponding operator to zero. With the constraint $\tilde{A}_3^1|_{\rho\to\infty} = 0$ and the two independent physical parameters, we may fix the three independent parameters of the near-horizon asymptotics and obtain a solution to the equations of motion.

11.3.3.2 Expansion of the DBI Action

We now outline the second approach which we use. Expanding the action (11.30) to fourth order in the field strength F yields

$$S_{\rm DBI} = -T_{D7}N_f \int {\rm d}^8\xi\sqrt{-{\rm G}}\left[1 + \frac{\mathcal{T}_2}{2} - \frac{\mathcal{T}_4}{8} + \cdots\right], \qquad (11.58)$$

where $\mathcal{T}_i$ consists of the terms with order i in F. To calculate the $\mathcal{T}_i$, we use the following results for the symmetrized traces

$$2\sigma: \quad {\rm Str}\left[\left(\sigma^{\rm i}\right)^2\right] = {\rm N_f}, \qquad (11.59)$$

$$4\sigma: {\rm Str}\left[\left(\sigma^{\rm i}\right)^4\right] = {\rm N_f}\,, \quad {\rm Str}\left[\left(\sigma^{\rm i}\right)^2\left(\sigma^{\rm j}\right)^2\right] = \frac{\rm N_f}{3}, \qquad (11.60)$$

where the indices i,j are distinct. Notice that the symmetric trace of terms with unpaired σ matrices vanish, e.g. ${\rm Str}[\sigma^{\rm i}\sigma^{\rm j}] = {\rm N_f}\delta^{\rm ij}$. The $\mathcal{T}_i$ are given in the appendix of [5].

To perform the Legendre transformation of the above action, we determine the conjugate momenta as in (11.34). However, we cannot easily solve these equations for the derivative of the gauge fields since we obtain two coupled equations of third degree. Thus we directly calculate the equations of motion for the gauge fields on the D7-branes. The equations are given in the appendix of [5].

To solve these equations, we use the same strategy as in the adapted symmetrized trace prescription discussed above. We integrate the equations of motion from the horizon at $\rho = 1$ to the boundary at $\rho = \infty$ numerically. The initial conditions may be determined by the asymptotic behavior of the gravity fields near the horizon

$$\tilde{A}_0^3 = a_2(\rho-1)^2 + \mathcal{O}\left((\rho-1)^3\right), \qquad (11.61)$$

$$\tilde{A}_3^1 = b_0 + \mathcal{O}\left((\rho-1)^3\right), \qquad (11.62)$$

$$\chi = \chi_0 + \frac{3(a_2^4 + 4a_2^2 - 8)\chi_0}{4(3a_2^4 + 4a_2^2 + 8)}(\rho-1)^2 + \mathcal{O}\left((\rho-1)^3\right). \qquad (11.63)$$

For the numerical calculation we use the asymptotic expansion up to sixth order. As in the adapted symmetrized trace prescription, there are again three independent parameters a_2, b_0, χ_0. Since we have not performed a Legendre transformation, we trade the independent parameter c_0 in the asymptotics of the conjugate momenta $\tilde{p}_0^3$ in the symmetrized trace prescription with the independent parameter a_2 (cf. asymptotics in Eq. 11.46). However, the three independent parameters may again be determined in field theory quantities which are defined by the asymptotics of the gravity fields near the boundary

$$\tilde{A}_0^3 = \mu - \frac{\tilde{d}_0^3}{\rho^2} + \mathcal{O}(\rho^4), \tag{11.64}$$

$$\tilde{A}_3^1 = -\frac{\tilde{d}_3^1}{\rho^2} + \mathcal{O}(\rho^4), \tag{11.65}$$

$$\chi = \frac{m}{\rho} + \frac{c}{\rho^3} + \mathcal{O}(\rho^4). \tag{11.66}$$

The independent parameters $\mu, \tilde{d}_0^3, \tilde{d}_3^1, m, c$ are given by field theory quantities as presented in (11.56) and (11.57). Again we find that there is no source term for the current J_3^1, which implies spontaneous symmetry breaking. Therefore the independent parameters in both prescriptions are the same and we can use the same strategy to solve the equations of motion as described below (11.57).

Exercise In [38] only the leading order of the action (11.58) quadratic in field fluctuations was considered. For a specific chemical potential an analytic solution with non-zero A_0^3 and A_3^1 can be found. Show that the analytic solution found there (adapted to our AdS_5 case)

$$A_0^3 = \mu(1 - \rho^4), \qquad A_3^1 = \epsilon \frac{\rho^4}{(1+\rho^4)^2}, \tag{11.67}$$

indeed solves the equations of motion derived from the action expanded to quadratic order for the chemical potential $\mu = 4$. ϵ is a constant formed from the coupling constant and the vacuum expectation value of the dual current $\langle J_3^1 \rangle$. Note: This exercise requires some work.

11.4 D-Brane Thermodynamics and Spectrum

In this section we briefly review the results obtained for the thermodynamics and the spectrum of our setup with a baryonic, and later an isospin chemical potential.

11.4.1 Baryon Chemical Potential

Figure 11.4 shows the phase diagram of the field theory dual to the gravity setup we have constructed in the previous Sect. 11.3. In this field theory we have introduced a

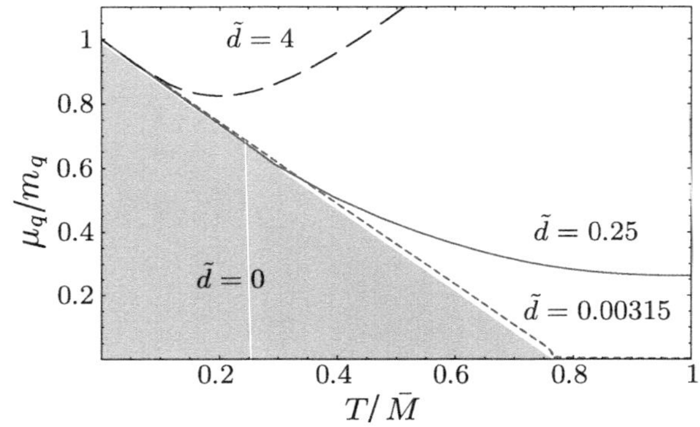

Fig. 11.4 The phase diagram in the canonical ensemble plotted against the variables of the grand canonical ensemble. On the axes the scaled chemical potential μ_q/M_q, with the quark mass M_q is shown versus the scaled temperature $T/\bar{M} = m^{-1}$ from [39]

baryonic chemical potential μ_q which is shown on the vertical axis scaled by the quark mass M_q defined by Eq. 11.57. It is introduced by a non trivial background gauge field solving the equation of motion derived from the DBI-action and asymptoting to the chemical potential near the boundary as $\lim_{\rho \to \rho_B} A_t(\rho) = \mu_q$. The horizontal axis shows the scaled temperature $T/\bar{M} = m^{-1}$. Recall that m is the asymptotic value of the D7-brane embedding near the AdS boundary. In other words it is the source term for a quark condensate, or the non-normalizable mode of the brane embedding function. Here however we use it merely as a temperature scale. The lines in the diagram in Fig. 11.4 are lines of equal baryon density $\tilde{d}$. At low temperature and chemical potential there is a triangle-shaped phase of zero density. Its diagonal borderline to the white region with finite baryon density is the location of the so-called *meson-melting transition*. This is the transition where the fundamental matter melts, i.e. quark bound states, the mesons, acquire an increasing decay width becoming quasi-particles. In the grey phase we therefore have stable mesons with zero decay width. The deeper we go into the white phase away from the transition line, the mesons melt. We will also see this in the spectral functions computed below. As explained at the end of Sect. 11.3.1 we are going to stay entirely in the white phase at finite temperature where the brane embeddings are of black hole type, which means that part of them falls into the black hole horizon (see Sect. 11.3.1).

11.4.1.1 Correlator Recipe

We need to examine fluctuations of the fields in our setup in order to determine its spectrum and stability. There might be a fluctuation which is tachyonic and therefore could destabilize the whole system. For this purpose we quickly review how to compute real-time correlation functions in gauge/gravity.

Let us work along the example of a gauge field fluctuation a_μ. This appears in the action, in our case the Dirac-Born-Infeld action, in the field strength $F = \mathcal{D}a + a \wedge a$. Note that a background gauge field A would appear in this field strength too, as we will see in later sections. For simplicity we consider Abelian gauge field fluctuations without background A, i.e. $F_{\mu\nu} = \partial_\mu a_\nu - \partial_\nu a_\mu$. The action then reads something like

$$S \sim \int \mathcal{D}^8 \xi \sqrt{P[g_{\mu\nu}] + F_{\mu\nu}}, \tag{11.68}$$

with the P being the pullback of the metric g to the flavor brane. Expanding this action to quadratic order in fluctuations a, we get the linearized equation of motion for example for the spatial fluctuation a_y which in Fourier space looks like this

$$0 = \partial_\rho{}^2 a_y + \frac{\partial_\rho\left[\sqrt{-g}g^{yy}g^{\rho\rho}\right]}{\sqrt{-g}g^{yy}g^{\rho\rho}}\partial_\rho a_y + \frac{g^{tt}}{g^{\rho\rho}} w^2 a_y, \tag{11.69}$$

where w is the dimensionless frequency of the fluctuation. Also for other fields we end up with a second order differential equation that we need to solve. Usually these equations have singular (at the horizon ρ_H) coefficients which need to be regularized by an appropriate ansatz. In order to find the most singular behavior solving this equation, we plug the ansatz $(\rho - \rho_H)^\beta$ into Eq. 11.69 and expand. The leading order is a quadratic equation which can be solved for β giving β_{in} or β_{out}. Only one of those two solutions, β_{in} describes a fluctuation falling into the horizon, the other one is outgoing. We discard the outgoing one because nothing is supposed to leave a classical black hole. β is sometimes called the *indicial exponent*. Now we plug $a = (\rho - \rho_H)^{\beta_{\text{in}}} F(\rho)$, with $F(\rho) = \sum_{n=0}^{\infty} f_n (\rho - \rho_H)^n$ being a regular function of ρ, into the equation of motion (11.69). Note that there might be logarithmic terms present as explained in general for example in [40] and discussed in detail in [3]. Picking the fixes one of the two boundary conditions. The other boundary condition may be fixed by choosing the normalization of $F(\rho)$, i.e. the value f_0. All the higher f_n depend recursively on f_0 and the indicial exponent β. Now we solve the equation for F analytically or numerically. The correlator is then obtained from the quadratic part of the on-shell action, which in our case has the structure $S_{\text{on-shell}} \sim \int \mathcal{D}\rho B(\rho) a \partial_\rho a$, where B is a function of ρ depending on metric coefficients. The recipe developed in [41, 42] tells us to strip the boundary values a^{bdy} off from the fields $a = a^{\text{bdy}} \mathcal{A}$ and identify what remains of the integrand with the Green function at the boundary $\rho = \rho_B$

$$G^R_{\mu\nu} = \lim_{\rho \to \rho_B} B(\rho) \mathcal{A}_\mu \partial_\rho \mathcal{A}_\nu. \tag{11.70}$$

Now we only need to plug in our solutions $a = a^{\text{bdy}} \mathcal{A}$. More details on the analytic and numerical procedures are explained in [3].

11.4.1.2 Spectral Function and Quasi-Normal Modes

The spectral function $\mathcal{R}$ is obtained from the imaginary part of the Green function $\mathcal{R} = -2\mathrm{ImG}^{\mathrm{R}}$. It encodes the spectrum of our thermal field theory. In particular in Fig. 11.5 we are able to identify pretty stable quasi-particle excitations at a low finite temperature parametrize by χ_0 and at finite baryon density $\tilde{d} = 0.25$. Lowering the temperature these quasiparticles approach the line spectrum (11.16) indicated here as dashed vertical lines. The reason is that at lower temperature our theory restores its original supersymmetry. This also tells us that the vector quasi-particles we see in the thermal spectral function are vector mesons. Our mesons melt in the finite temperature, finite density phase as mentioned above in the discussion of the phase diagram (11.4).

Quasi Normal Modes (QNM)

The difference between the zero temperature line spectrum and the finite temperature spectrum of quasi-particle excitations lies in the nature of the corresponding eigenmodes. In the zero temperature case there is no black hole, thus no dissipation on the gravity side. The system has well-defined normal modes at real frequencies $w \in \mathbf{R}$. At finite temperature however the (quasi) eigenfrequencies are complex $w \in \mathbf{C}$ owing to the dissipation into the field theory plasma or in the dual gravity picture: dissipation into the black hole. The modes travelling with these (quasi) eigenfrequencies are called *quasi-normal modes (QNM)*. We can roughly think of quasi-normal modes as being those solutions to the gravity fluctuation equations which vanish at the AdS boundary. These QNMs have been found to be identical to the poles in the field theory Green function. Therefore the location of the QNMs is closely related to the location of the peaks in the spectral function. At least some of the quasi-normal modes create quasi-particle peaks in the spectral function. In the zero temperature limit we can think of those QNMs as approaching the real frequency axis and reaching it in the limit, becoming real-valued.

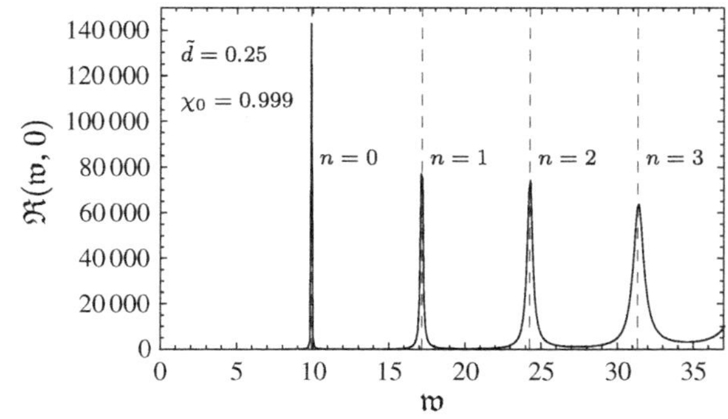

Fig. 11.5 The spectral function $\mathcal{R}$ (in units of $N_f N_c T^2/4$) at finite baryon density $\tilde{d}$ versus dimensionless frequency w. At large χ_0 here the peaks approach the dashed drawn line spectrum given by (11.16)

The corresponding quasiparticles become stable which means line-shaped in the spectrum. A more complete picture of QNMs is given in [43].

11.4.2 Isospin Chemical Potential

We can equally well introduce a chemical potential with $SU(2)$—let's call it isospin—structure, represented by the Pauli matrices σ^i. Choosing the chemical potential to point along the third flavor direction this background boils down to having two copies of the Abelian background gauge field $A_t(\rho)$ explored above in the following way

$$\mu^3 = A_t(\rho)\sigma^3 = \begin{pmatrix} A_t(\rho) & 0 \\ 0 & -A_t(\rho) \end{pmatrix}. \tag{11.71}$$

However we get some interesting new signatures through the $SU(2)$ structure. For example the spectrum shown in Fig. 11.6 shows a triplet splitting for our mesons. In particular we observe a splitting of the line expected at the lowest meson mass at $w = 4.5360$ ($n = 0$). The resonance is shifted to lower frequencies for $\mathcal{R}_{XY}$ and to higher ones for $\mathcal{R}_{YX}$, while it remains in place for $\mathcal{R}_{E^3E^3}$. The second meson resonance peak ($n = 1$) shows a similar behavior. So the different flavor combinations propagate differently and have distinct quasi-particle resonances. This behavior is analogous to that of QCD's ρ meson which is vector meson and a triplet under the isospin $SU(2)$ of QCD. Thus we have modelled the melting process of vector mesons in a quark gluon plasma at finite isospin density.

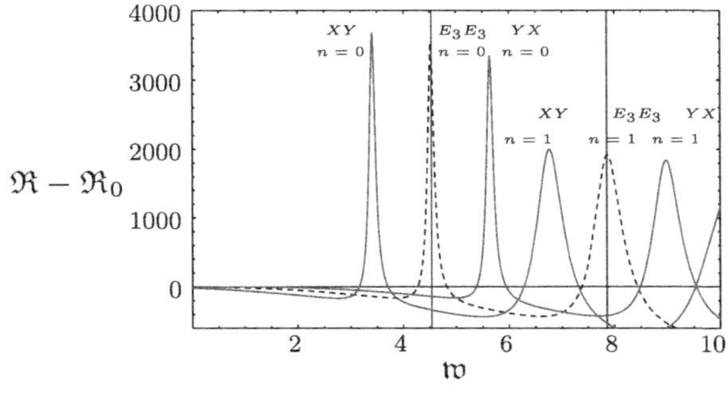

Fig. 11.6 A comparison between the finite temperature part of the spectral functions $\mathcal{R}_{XY}$ and $\mathcal{R}_{YX}$ (solid lines) in the two flavor directions X and Y transversal to the chemical potential is shown in units of $N_c T^2 T_r/4$ for large quark mass to temperature ratio $\chi_0 = 0.99$ and $\tilde{d} = 0.25$. The spectral function $\mathcal{R}_{E^3E^3}$ along the $a = 3$-flavor direction is shown as a dashed line

11.4.3 Instabilities and the New Phase

What does all this have to do with our condensed matter motivation? The crucial thing is that the isospin setup described above develops an instability at large enough isospin density. This means that the modes X and Y develop quasi-normal modes (QNM) which have a positive imaginary part, i.e. which are enhanced instead of being damped. This suggests that exactly the mesons corresponding to the X, Y fluctuations condense. This instability comes about naturally if we think about the fact, that we are trying to push more and more charge density into a confined volume. In particular the 3–7 strings charging the D7-brane are located at the black hole horizon as motivated earlier in Sect. 11.3.1. Our current setup does not allow the strings to move into the bulk because we are forcing all the background fields they would create there to be zero. Up to now we have required $A_0^3 \neq 0$ and $A_\mu^a \equiv 0$ for all other field components. But we are going to relax that restriction by allowing a non-trivial A_3^1. We will see below that this is sufficient to stabilize the theory in a new phase which we will prove to be superconducting/ superfluid. This setup naturally produces a p-wave structure since we have seen above that the condensing vector mesons have a triplet structure. According to Table 11.1 this implies a p-wave (at least to leading order).

11.5 Signatures of Super-Something

Finally we put together everything we have learned about D-branes, superconductors and holographic methods. We discuss the holographic results and interpret them in the condensed matter context. Section 11.5.1 starts with the thermodynamics and Sect. 11.5.2 continues with the details of the fluctuation computation. The spectrum and conductivity is Examined in Sect. 11.5.3 and all the signatures will be pointing to the fact that *we have indeed created a holographic p-wave superconductor/superfluid.*

11.5.1 Thermodynamics of the Broken Phase

Figure 11.7 shows the background field configuration. The different curves correspond to the temperatures $T = T_c$ and $T \approx 0.9T_c$. The plots are obtained at zero quark mass $m = 0$ and by using the adapted symmetrized trace prescription. Similar plots may also be obtained at finite mass $m \neq 0$ and by using the DBI action expanded to fourth order in F. These plots show the same features: (top left) The gauge field $\tilde{A}_0^3$ increases monotonically towards the boundary. At the boundary, its value is given by the dimensionless chemical potential $\tilde{\mu}$. The gauge field $\tilde{A}_3^1$ is zero for $T \geq T_c$. For $T < T_c$, its value is non-zero at the horizon and

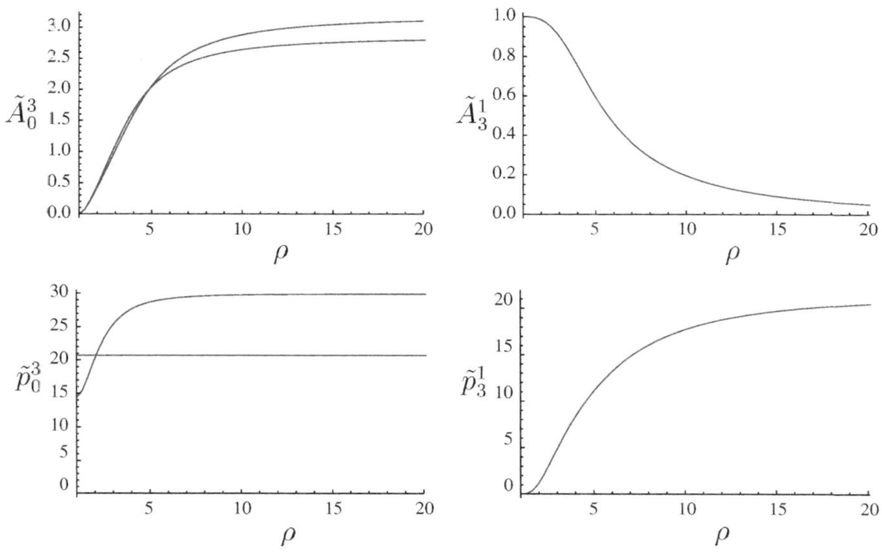

Fig. 11.7 Profiles of the relevant dimensionless gauge fields $\tilde{A}$ on the D7-branes and their dimensionless conjugate momenta $\tilde{p}$ versus the dimensionless AdS radial coordinate ρ near the horizon at $\rho = 1$. (*top left*) the lower curve is $T = 0.9T_c$, (*bottom left*) the constant is at $T = T_c$ since there are no charges in the bulk, the other curve shows that the charge proportional to $\tilde{p}_0^3$ increases towards the boundary

decreases monotonically towards the boundary where its value has to be zero. (top right) The conjugate momentum $\tilde{p}_0^3$ of the gauge field $\tilde{A}_0^3$ is constant for $T \geq T_c$. For $T < T_c$, its value increases monotonically towards the boundary. Its boundary value is given by the dimensionless density $\tilde{d}_0^3$. (bottom left) The conjugate momentum $\tilde{p}_3^1$ of the gauge field $\tilde{A}_3^1$ is zero for $T \geq T_c$. For $T < T_c$, its value increases monotonically towards the boundary. Its boundary value is given by the dimensionless density $-\tilde{d}_3^1$. (bottom right)

All thermodynamic quantities are determined in terms of the relevant thermodynamic potential. We use the grand potential W_7 in the grand canonical and the free energy F_7 in the canonical ensemble. Both stem from the D7-brane action. They are related through a Legendre transformation in the background gauge field A_0^3 on the gravity side. We obtain both potentials due to the gauge/gravity dictionary from the Euclideanized gravity on-shell action according to $Z = \mathcal{E}^{-I_{\text{on-shell}}}$, with the partition function Z of the boundary field theory. So the for example $W_7 = TS^{\text{euclideanized}}_{\text{D7,on-shell}}$.

Comparison of the grand potentials in Fig. 11.8 then shows that the new phase with a finite value for A_3^1 is thermodynamically preferred below a temperature $T = T_c$ and does not exist above. The transition seems numerically smooth, i.e. it is a *continuous phase transition*. In these respects both computation schemes, the

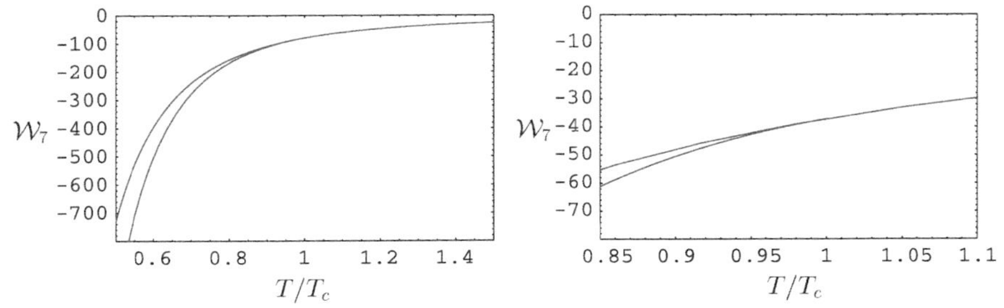

Fig. 11.8 Grand canonical potential computed **a** from the symmetrized trace prescription and **b** from the expanded DBI action, both at vanishing quark mass $M_q = 0$. The qualitative behavior agrees for both prescriptions: At $T = T_c$ the energy curve splits into a lower and a higher energy branch. The branch with lower energy in both cases is the one with a finite new condensate $\tilde{d}_3^1$ below T_c

symmetrized trace prescription as well as the DBI expansion to fourth order give the same qualitative behavior. Figure 11.9 confirms this result and at least numerically determines the transition to be of second order. The order parameter $\tilde{d}_3^1$ vanishes with a critical exponent of $1/2$ to numerical accuracy. It has been explicitly verified that this condensation of vector particles occurs before, i.e. at higher temperature than the condensation of the scalars in our theory (see remarks in [5]). If this was QCD the scalar pions would condense before the vector mesons did. However, there might be other instabilities and also other symmetry breaking configurations may be possible.

Using some intuition to define the superconducting density $\tilde{d}_s = (\tilde{d}_0^3 - c_0)/\tilde{d}_0^3$, we find that it vanishes linearly at the critical temperature as shown in Fig. 11.10

All these signatures are those of a superconducting/superfluid phase transition. The order parameter $\tilde{d}_3^1$ has vector structure by construction, which implies the p-wave.

We can also compute the specific heat which the flavor branes contribute to the theory as $C_7 = -T\partial^2 F_7/\partial T^2$. In Fig. 11.11 the higher curve corresponds to the normal phase with $A_3^1 = 0$, while the lower one corresponds to the superconducting phase with $A_3^1 \neq 0$. Note that the total specific heat is always positive although the flavor brane contribution is negative. The divergences near $T = 0$ in both phases can be attributed to the missing backreaction in our setup.[6] We read off from the numerical result that near the critical temperature, the dimensionless specific heat $\mathcal{C}_7 = 32C_7/(\lambda N_f N_c T^3)$ is constant in the superconducting phase.

This implies that the dimensionful specific heat C_7 is proportional to T^3. This temperature dependence is characteristic for Bose liquids. There is also a finite jump in the specific heat at T_c.

[6] The backreaction of the gauge field on the geometry, i.e. on the Einstein equations for metric components has been considered in [44].

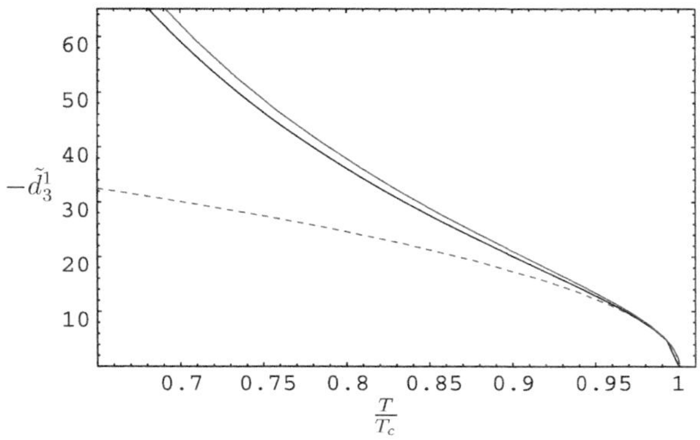

Fig. 11.9 The order parameter $\tilde{d}_3^1$ defined in (11.51) as obtained from the adapted symmetrized trace prescription versus temperature: The case of vanishing quark mass (*top solid curve*) shows the same behavior near T_c as that where $\mu/M_q = 3$ is fixed (*lower solid curve*). Both curves go to zero with a critical exponent of 1/2 near T_c, as visualized by the fit $55(1 - T/T_c)^{1/2}$ (*dashed curve*)

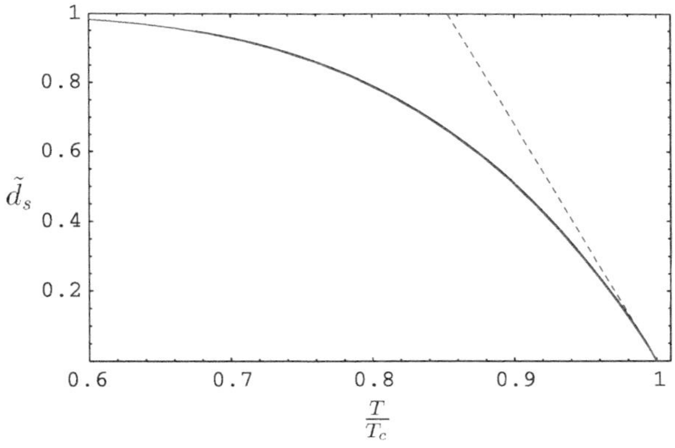

Fig. 11.10 Superconducting density $\tilde{d}_s = (\tilde{d}_0^3 - c_0)/\tilde{d}_0^3$ versus temperature T: In both, the massless and the massive case at $\mu/M_q = 3$ (both curves coincide on the plot), the superconducting density $\tilde{d}_s$ vanishes linearly at the critical temperature. This is visualized by the fit $6.8(1 - T/T_c)$ (*dashed blue curve*)

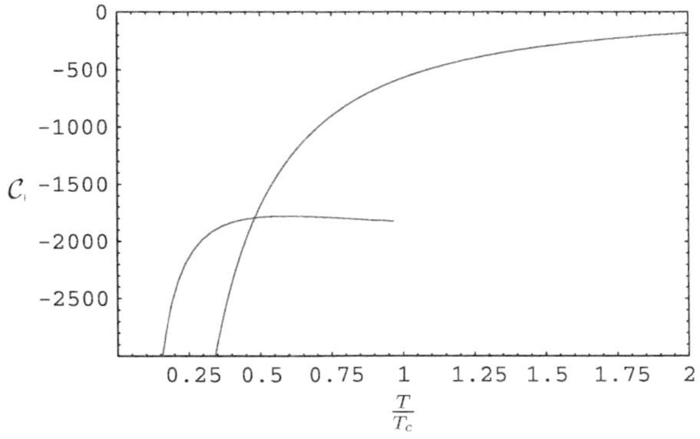

Fig. 11.11 The flavor brane contribution to the specific heat as computed from the adapted symmetrized trace prescription in the massless case. The lower curve starting at $T = T_c$ is the heat capacity in the superconducting phase. The other curve is computed in the normal phase

11.5.2 Fluctuations in the Broken Phase

Let us now investigate fluctuations of this system. We are going to see that the computation of the non-Abelian DBI-action is a very subtle issue. We argue for a novel prescription which actually gives reasonable results. The full gauge field $\hat{A}$ on the branes consists of the field A and fluctuations a,

$$\hat{A} = A_0^3 \tau^3 \mathcal{D}t + A_3^1 \tau^1 \mathcal{D}x_3 + a_\mu^a \tau^a \mathcal{D}x^\mu, \tag{11.72}$$

where τ^a are the $SU(2)$ generators. The linearized equations of motion for the fluctuations a are obtained by expanding the DBI action in a to second order. We will analyze the fluctuations a_2^3 and $X = a_2^1 + \mathrm{i}a_2^2$, $Y = a_2^1 - \mathrm{i}a_2^2$.

Including these fluctuations, the DBI action reads

$$S = -T_7 \int \mathcal{D}^8 \xi \operatorname{Str} \sqrt{\det\left[\mathrm{G} + (2\pi\alpha')\hat{\mathrm{F}}\right]}, \tag{11.73}$$

with the non-Abelian field strength tensor

$$\hat{F}_{\mu\nu}^a = F_{\mu\nu}^a + \check{F}_{\mu\nu}^a, \tag{11.74}$$

where the background is collected in

$$F_{\mu\nu}^a = 2\partial_{[\mu}A_{\nu]}^a + \frac{\gamma}{\sqrt{\lambda}} f^{abc} A_\mu^b A_\nu^c, \tag{11.75}$$

and all terms containing fluctuations in the gauge field are summed in

$$\check{F}_{\mu\nu}^a = 2\partial_{[\mu}a_{\nu]}^a + \frac{\gamma}{\sqrt{\lambda}} f^{abc} a_\mu^b a_\nu^c + \frac{\gamma}{\sqrt{\lambda}} f^{abc} (A_\mu^b a_\nu^c + a_\mu^b A_\nu^c). \tag{11.76}$$

Index anti-symmetrization is always defined with a factor of two in the following way $\partial_{[\mu}A_{\nu]} = (\partial_\mu A_\nu - \partial_\nu A_\mu)/2$.

11.5.2.1 Adapted Symmetrized Trace Prescription

In this section we use the adapted symmetrized trace prescription to determine the fluctuations about the background we discussed in Sect. 11.3.3.1. To obtain the linearized equations of motion for the fluctuations a, we expand the action (11.73) to second order in fluctuations,

$$\begin{aligned} S^{(2)} = -T_7 \int \mathcal{D}^8 \xi \operatorname{Str} \Bigg[& \sqrt{-\mathcal{G}} + \frac{(2\pi\alpha')}{2} \sqrt{-\mathcal{G}} \mathcal{G}^{\mu\nu} \mathrm{F}_{\nu\mu}. \\ & - \frac{(2\pi\alpha')^2}{4} \sqrt{-\mathcal{G}} \mathcal{G}^{\mu\mu'} \check{F}_{\mu'\nu} \mathcal{G}^{\nu\nu'} \check{F}_{\nu'\mu} + \frac{(2\pi\alpha')^2}{8} \sqrt{-\mathcal{G}} \left(\mathcal{G}^{\mu\nu} \check{F}_{\nu\mu}\right)^2 \Bigg]. \end{aligned} \tag{11.77}$$

As in [39], we collect the metric and gauge field background in the tensor $\mathcal{G} = G + (2\pi\alpha')F$. Using the Euler-Lagrange equation, we get the linearized equation of motion for fluctuations a^d_κ in the form

$$0 = \partial_\lambda \mathrm{Str}\left[\sqrt{-\mathcal{G}}\tau^{\mathrm{d}}\left\{\mathcal{G}^{[\kappa\lambda]} + (2\pi\alpha')\left(\mathcal{G}^{\mu[\kappa}\mathcal{G}^{\lambda]\nu} + \frac{1}{2}\mathcal{G}^{\mu\nu}\mathcal{G}^{[\kappa\lambda]}\right)\mathrm{F}_{\nu\mu}\right\}\right]$$
$$- \mathrm{Str}\left[\frac{\mathrm{c}}{\sqrt{\lambda}}\mathrm{f}^{\mathrm{abd}}\tau^{\mathrm{a}}\sqrt{-\mathcal{G}} \times \left\{\mathcal{G}^{[\kappa\lambda]}(\mathrm{a}+\mathrm{A})^{\mathrm{b}}_{\lambda} + (2\pi\alpha')\left(\mathcal{G}^{\mu[\kappa}\mathcal{G}^{\lambda]\nu}\right.\right.\right.$$
$$\left.\left.\left. + \frac{1}{2}\mathcal{G}^{\mu\nu}\mathcal{G}^{[\kappa\lambda]}\right)\check{F}_{\nu\mu}A^b_\lambda\right\}\right]. \tag{11.78}$$

Note that the linearized version of the fluctuation field strength used in Eq. 11.78 is given by

$$\check{F}^a_{\mu\nu} = 2\partial_{[\mu}a^a_{\nu]} + \frac{\gamma}{\sqrt{\lambda}}f^{abc}(A^b_\mu a^c_\nu + a^b_\mu A^c_\nu) + \mathcal{O}(a^2). \tag{11.79}$$

In our specific case the background tensor in its covariant form is given by

$$\mathcal{G}_{\mu\nu} = G_{\mu\nu}\tau^0 + (2\pi\alpha')\left(2\partial_\varrho A^3_0\delta_{4[\mu}\delta_{\nu]0}\tau^3 + 2\partial_\varrho A^1_3\delta_{4[\mu}\delta_{\nu]3}\tau^1 + 2\frac{\gamma}{\sqrt{\lambda}}A^3_0A^1_3\delta_{0[\mu}\delta_{\nu]3}\tau^2\right). \tag{11.80}$$

Inversion yields the contravariant form needed to compute the explicit equations of motion. The inverse of $\mathcal{G}$ is defined as $\mathcal{G}^{\mu\nu}\mathcal{G}_{\nu\mu'} = \delta^\mu_{\mu'}\tau^0$.[7] The non-zero components of $\mathcal{G}^{\mu\nu}$ may be found in the appendix of [5].

Fluctuations in a^3_2: For the fluctuation a^3_2 with zero spatial momentum, we obtain the equation of motion

$$0 = (a^3_2)'' + \frac{\partial_\rho H}{H}(a^3_2)' - \left[\frac{4\varrho^4_H}{R^4}\left(\frac{\mathcal{G}^{33}}{\mathcal{G}^{44}}(\mathcal{M}^1_3)^2 + \frac{\mathcal{G}^{00}}{\mathcal{G}^{44}}w^2\right)\right.$$
$$\left. -16\frac{\partial_\rho\left(\frac{H}{\rho^4f^2}\tilde{A}^3_0(\partial_\rho\tilde{A}^3_0)(\mathcal{M}^1_3)^2\right)}{H\left(1 - \frac{2c^2}{\pi^2\rho^4f^2}(\tilde{A}^1_3\tilde{A}^3_0)^2\right)}\right]a^3_2, \tag{11.81}$$

with $\mathcal{M}^1_3 = \gamma\tilde{A}^1_3/(2\sqrt{2}\pi)$ and $H = \sqrt{\mathcal{G}}\,G^{22}\mathcal{G}^{44}$.

Fluctuations in $X = a^1_2 + \mathrm{i}a^2_2$, $Y = a^1_2 - \mathrm{i}a^2_2$: For the fluctuations X and Y with zero spatial momentum, we obtain the coupled equations of motion

[7] We calculate the inverse of $\mathcal{G}$ by ignoring the commutation relation of the τ's because of the symmetrized trace. It is important that $\tau^a\tau^b$ must not be simplified to $\epsilon^{abc}\tau^c$ since the symmetrization is not the same for these two expressions.

$$
\begin{aligned}
0 = X'' &+ \frac{\partial_\rho H}{H} X' - \frac{4\varrho_H^4}{R^4}\left[\frac{\mathcal{G}^{00}}{\mathcal{G}^{44}}\left(w - \mathcal{M}_0^3\right)^2 + \frac{\mathcal{G}^{\{03\}}}{\mathcal{G}^{44}} \mathcal{M}_3^1 w\right] X + \frac{4\varrho_H^4}{R^4}\left[\frac{\mathcal{G}^{\{03\}}}{\mathcal{G}^{44}} \mathcal{M}_3^1 \mathcal{M}_0^3\right. \\
&\left. + \frac{R^2}{4\varrho_H^2} \frac{\partial_\rho\left[\sqrt{-\mathcal{G}} G^{22} \mathcal{G}^{\{34\}} \mathcal{M}_3^1\right]}{H} - \frac{\mathcal{G}^{33}}{2\mathcal{G}^{44}}\left(\mathcal{M}_3^1\right)^2\right](X - Y) + \frac{4\varrho_H^2}{R^2} \frac{\mathcal{G}^{\{04\}}}{\mathcal{G}^{44}} w Y' \\
&+ \frac{2\varrho_H^2}{R^2} \frac{\partial_\rho\left[\sqrt{-\mathcal{G}} G^{22} \mathcal{G}^{\{04\}}\left(w + \mathcal{M}_0^3\right)\right]}{H} Y,
\end{aligned}
\tag{11.82}
$$

where the component of the inverse background tensor may be found in the appendix of [5] (just like the corresponding formula for Y which is only different from (11.82) by a few signs), index symmetrization is defined $\mathcal{G}^{\{ij\}} = (\mathcal{G}^{ij} + \mathcal{G}^{ji})/2$ and $\mathcal{M}_0^3 = \gamma\tilde{A}_0^3/(2\sqrt{2}\pi)$.

11.5.2.2 Expansion of the DBI Action

In this section we determine the equation of motion for the fluctuation a_2^3 in the background determined by the DBI action expanded to fourth order in F (see Sect. 11.3.3.2). To obtain the quadratic action in the field a_2^3, we first have to expand the DBI action (11.73) to fourth order in the full gauge field strength $\hat{F}$, and expand the result to second order in a_2^3. Due to the symmetries of our setup, the equation of motion for the fluctuation a_2^3 at zero spatial momentum decouples from the other equations of motion, such that we can write down an effective Lagrangian for the fluctuation a_3^2. This effective Lagrangian is given in the appendix of [5]. The equation of motion for a_2^3 with zero spatial momentum determined by the Euler-Lagrange equation is given by

$$
\begin{aligned}
0 = (a_2^3)'' &+ \frac{\partial_\rho \mathcal{H}}{\mathcal{H}}(a_2^3)' - \frac{\varrho_H^4}{R^4}\left[4\left(\frac{\mathcal{H}^{00}}{\mathcal{H}^{44}} w^2 + \frac{\mathcal{H}^{33}}{\mathcal{H}^{44}}(\mathcal{M}_3^1)^2\right)\right. \\
&\left. + \frac{8}{3} \frac{\partial_\rho[\sqrt{-G} G^{00} G^{22} G^{33} G^{44} \tilde{A}_0^3 (\partial_\rho \tilde{A}_0^3)(\mathcal{M}_3^1)^2]}{\mathcal{H}}\right] a_2^3,
\end{aligned}
\tag{11.83}
$$

where $\mathcal{H} = \sqrt{-G} G^{22} \mathcal{H}^{44}$. We introduce the factors $\mathcal{H}^{ij}$ which may be found in the appendix of [5] to emphasize the similarity to the equation of motion obtained by the adapted symmetrized trace prescription (11.81).

11.5.3 Conductivity and Spectrum

We calculate the frequency-dependent conductivity $\sigma(\omega)$ using the Kubo formula,

$$\sigma(\omega) = \frac{i}{\omega} G^R(\omega, q = 0), \qquad (11.84)$$

where G^R is the retarded Green function of the current J_2^3 dual to the fluctuation a_2^3, which we calculate using the method obtained in [41]. The current J_2^3 is the analog to the electric current since it is charged under the $U(1)_3$ symmetry. In real space it is transverse to the condensate. Since this fluctuation is the only one which transforms as a vector under the $SO(2)$ rotational symmetry, it decouples from the other fluctuations of the system.

Exercise Prove Eq. 11.84 for the current J_2^3 assuming that the gauge gravity correspondence is correct. Recall that in regular electrodynamics $\sigma = J/E$ with the electric current J and the electric field E. Also recall that the two-point Green function for a current J dual to the gauge field A can holographically be written in the form $G^R \propto \partial_\rho A/A$.

11.5.3.1 (Pseudo) Gap

An analysis of the imaginary part of the conductivity using Kramers-Kronig relations shows a delta peak at vanishing frequency $w = 0$ in the real part. The frequency-dependent conductivity is shown for our two distinct computation schemes in Fig. 11.12. Independent from the scheme we see an energy gap develop and grow while the temperature is decreased. The temperature of the black hole horizon induced on the D7-branes is proportional to the inverse particle mass parameter $m^{-1} \propto T$. Therefore from the trivial flat brane embedding at $m = 0$ we get an infinite temperature in Fig. 11.12b. Both schemes show the development of peaks in the conductivities. The peaks coming from fluctuations around the

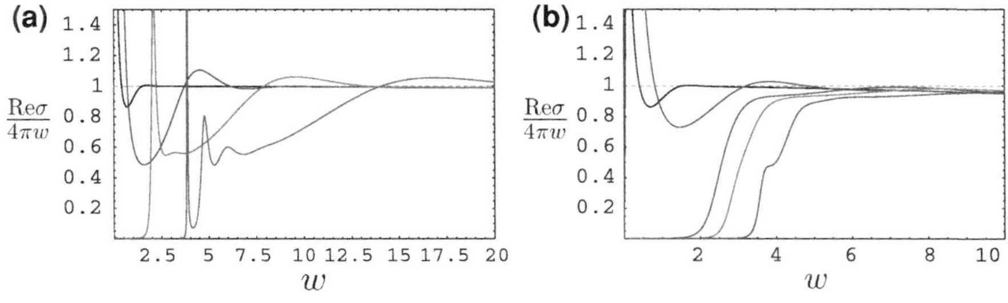

Fig. 11.12 Conductivity **a** from the symmetrized trace prescription at $T/T_c = 10, 1, 0.5, 0.28$ (*from left to right*), **b** from the DBI action expanded to fourth order at $T/T_c = \infty, 1, 0.6, 0.5, 0.39$

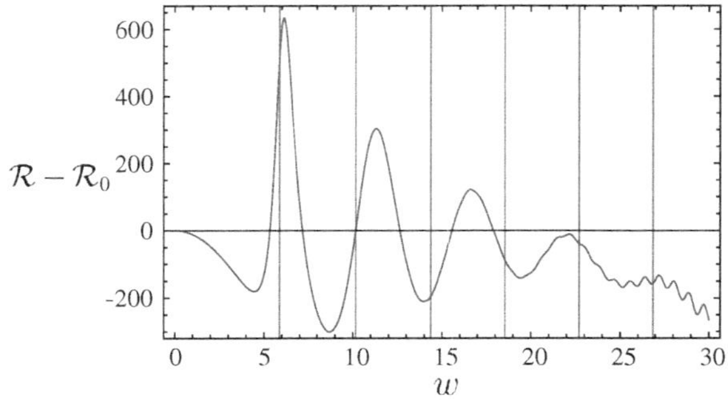

Fig. 11.13 Finite temperature part of the spectral function $\mathcal{R} - \mathcal{R}_0$ with $\mathcal{R}_0 = 4\pi w^2$in units of $N_f N_c T^2/8$ versus the dimensionless frequency $w = \omega/(2\pi T)$ at finite quark mass $m = 2.842$ and chemical potential $\tilde{\mu} = 3.483$. The grey lines correspond to the supersymmetric mass spectrum calculated in [22]

symmetrized trace prescription background are a lot more pronounced. Taking into account only second order terms in the expanded DBI action as in [38] would hide the peak structures completely. Therefore we conclude that these peaks are higher order effects. Since the conductivity is closely related to the spectral functions, we interpret the peaks as quasiparticles just as described in Sects. 11.4.1.2 and 11.4.2. In particular these are again vector mesons. This identification is confirmed by the spectral function in Fig. 11.13. The peaks in that figure are identical with those in the conductivity and they approach the supersymmetric line spectrum for vector mesons, as in Sect. 11.4.2.

11.5.3.2 Dynamical Mass Generation

Even if we choose the two D7 branes to coincide with the stack of D3-branes, i.e. if we choose the quark mass to vanish, we observe the quasi-particle peaks mentioned before. This is due to a Higgs-like mechanism which dynamically generates masses for the bulk field fluctuations, which in turn give massive quasi-particles in the boundary theory. In the bulk our fields A_0^3 and A_3^1 break the $SU(2)$ symmetry spontaneously since the bulk action is still $SU(2)$-invariant. Thus there are three Nambu-Goldstone bosons which are immediately eaten by the bulk gauge fields, giving them mass. This can be seen explicitly in the action where the following mass terms for the gauge field fluctuations appear: $(A_0^3)^2(a^{1,2})^2$ and $(A_3^1)^2(a^3)^2$.

11.5.4 Meissner-Ochsenfeld-Effect

The Meissner effect is a distinct signature of conventional and unconventional superconductors. It is the phenomenon of expulsion of external magnetic fields. An induced current in the superconductor generates a magnetic field counter-acting the external magnetic field H. In AdS/CFT we are not able to observe the generation of counter-fields since the symmetries on the boundary are always global. Nevertheless, we can study their cause, i.e. the current induced in the superconductor. As usual [45–48] the philosophy here is to weakly gauge the boundary theory afterwards. In order to investigate how an external magnetic field influences our p-wave superconductor, we have two choices. Either we introduce the field along the spatial z-direction $H_3^3\tau^3$ or equivalently one along the y-direction i.e. $H_2^3\tau^3$. Both are "aligned" with the spontaneously broken $U(1)_3$-flavor direction.

As an example here we choose a non-vanishing $H_3^3\tau^3$. This requires inclusion of some more non-vanishing field strength components in addition to those given in Eq. 11.25. In particular we choose $A_0^3(\varrho,x)$, $A_3^1(\varrho,x)$ and $A_2^3(x) = x^1\, H_3^3$ yielding the additional components[8]

$$H_3^3 = F_{12}^3 = -F_{21}^3 = \partial_1 A_2^3, \tag{11.85}$$

$$F_{13}^1 = -F_{31}^1 = \partial_1 A_3^1, \tag{11.86}$$

$$F_{23}^2 = -F_{32}^2 = \frac{\gamma}{\sqrt{\lambda}} A_2^3 A_3^1, \tag{11.87}$$

$$F_{10}^3 = -F_{01}^3 = \partial_1 A_0^3. \tag{11.88}$$

Recall that the radial AdS-direction is designated by the indices ϱ or 4 synonymously. Amending the DBI-action (11.30) with the additional components (11.85), we compute the determinant in analogy to Eq. 11.30. We then choose to expand the new action to second order in F, i.e. we only consider terms being at most quadratic in the fields. This procedure gives the truncated DBI action

$$\begin{aligned} S_{\mathrm{DBI}} = &- T_{D7}\, N_f \int \mathrm{d}^8\xi \sqrt{-\mathrm{G}} \Bigg[1 + \frac{(2\pi\alpha')^2}{2} \Big(\mathrm{G}^{00}\mathrm{G}^{33}(\mathrm{F}_{03}^2)^2 + \mathrm{G}^{33}\mathrm{G}^{44}(\mathrm{F}_{\varrho 3}^1)^2 \\ &+ G^{00}G^{44}(F_{\varrho 0}^3)^2 + (G^{33})^2(F_{13}^1)^2 + (G^{33})^2(F_{23}^2)^2 \\ &+ G^{00}G^{33}(F_{10}^3)^2 + (G^{33})^2(F_{12}^3)^2 \Big) + \cdots \Bigg]. \end{aligned} \tag{11.89}$$

Respecting the symmetries and variable dependencies in our specific system, this can be written as

[8] Close to the phase transition, it is consistent to drop the dependence of the field H_3^3 on ϱ. Away from the phase transition the ϱ dependence must be included. From the boundary asymptotics it will be possible to extract the magnetic field and the magnetization of the superconductor.

$$S_{\mathrm{DBI}} = -T_{D7}N_f \int \mathrm{d}^8\xi\sqrt{-\mathrm{G}}\Bigg[1 + \frac{1}{2}\Big(\mathrm{G}^{00}\mathrm{G}^{33}(\partial_{\bar{1}}\tilde{A}_0^3)^2 + \mathrm{G}^{33}\mathrm{G}^{44}(\partial_\varrho\tilde{A}_3^1)^2. \\ + G^{00}G^{44}(\partial_\varrho\tilde{A}_0^3)^2 + (G^{33})^2(\partial_{\bar{1}}\tilde{A}_3^1)^2 + (G^{33})^2(\bar{H}_3^3)^2 \\ + G^{00}G^{33}\frac{{\varrho_H}^4\gamma^2}{(2\pi\alpha')^2\lambda}(\tilde{A}_0^3\tilde{A}_3^1)^2 + (G^{33})^2\frac{{\varrho_H}^4\gamma^2}{(2\pi\alpha')^2\lambda}(\tilde{A}_3^1\bar{H}_3^3)^2\bar{x}^2\Big) + \cdots\Bigg], \tag{11.90}$$

with the convenient redefinitions

$$\tilde{A} = \frac{2\pi\alpha'}{\varrho_H}A\,, \quad x = \varrho_H\bar{x}\,, \quad \bar{H}_3^3 = 2\pi\alpha' H_3^3\,, \quad \varrho = \varrho_H\rho. \tag{11.91}$$

Rescaling the $\bar{x}$-coordinate once more

$$\tilde{x} = \sqrt{\frac{\bar{H}_3^3{\varrho_H}^2\gamma}{2\pi\alpha'\sqrt{\lambda}}}\,\bar{x}, \tag{11.92}$$

the equations of motion derived from the action (11.90) take a simple form

$$\begin{aligned} 0 &= \partial_\rho^2\tilde{A}_0^3 + \frac{\partial_\rho\left[\sqrt{-G}G^{00}G^{44}\right]}{\sqrt{-G}G^{00}G^{44}}\partial_\rho\tilde{A}_0^3 + \frac{\gamma\tilde{H}_3^3}{2\sqrt{2}\pi}\frac{G^{33}}{G^{44}}\partial_{\tilde{x}}\tilde{A}_0^3 - \frac{\gamma^2}{2\pi^2}\frac{G^{33}}{G^{44}}\tilde{A}_0^3(\tilde{A}_3^1)^2, \\ 0 &= \partial_\rho^2\tilde{A}_3^1 + \frac{\partial_\rho\left[\sqrt{-G}G^{33}G^{44}\right]}{\sqrt{-G}G^{33}G^{44}}\partial_\rho\tilde{A}_3^1 + \frac{\gamma\tilde{H}_3^3}{2\sqrt{2}\pi}\frac{G^{33}}{G^{44}}\left[\partial_{\tilde{x}}^2\tilde{A}_3^1 - \tilde{x}^2\tilde{A}_3^1\right] \\ &\quad - \frac{\gamma^2}{2\pi^2}\frac{G^{00}}{G^{44}}(\tilde{A}_0^3)^2\tilde{A}_3^1. \end{aligned} \tag{11.93}$$

Here all metric components are to be evaluated at $R = 1$ and $\varrho \to \rho$.

We aim at decoupling and solving the system of partial differential equations (11.93) by the product ansatz

$$\tilde{A}_3^1(\rho,\tilde{x}) = v(\rho)\,u(\tilde{x}). \tag{11.94}$$

For this ansatz to work, we need to make two assumptions: First we assume that $\tilde{A}_0^3$ is constant in $\tilde{x}$. Second we assume that A_3^1 is small, which clearly is the case near the transition $T \to T_c$. Our second assumption prevents A_0^3 from receiving a dependence on $\tilde{x}$ through its coupling to $A_3^1(\rho,\tilde{x})$. These assumptions allow to write the second equation in (11.93) as

$$0 = \partial_\rho^2 v(\rho) + \frac{\partial_\rho\left[\sqrt{-G}G^{33}G^{44}\right]}{\sqrt{-G}G^{33}G^{44}}\partial_\rho v(\rho) - \frac{\gamma^2}{2\pi^2}\frac{G^{00}}{G^{44}}(\tilde{A}_0^3)^2 v(\rho) \\ + \frac{\gamma\tilde{H}_3^3}{2\sqrt{2}\pi}\frac{G^{33}}{G^{44}}v(\rho)\,\frac{\partial_{\tilde{x}}^2 u(\tilde{x}) - \tilde{x}^2 u(\tilde{x})}{u(\tilde{x})}. \tag{11.95}$$

All terms but the last one are independent of $\tilde{x}$, so the product ansatz (11.94) is consistent only if

$$\frac{\partial_{\tilde{x}}^2 u(\tilde{x}) - \tilde{x}^2 u(\tilde{x})}{u(\tilde{x})} = C, \tag{11.96}$$

where C is a constant. The differential Eq. 11.96 has a particular solution if $C = -(2n+1)\,, n \in \mathrm{N}$. The solutions for $u(\tilde{x})$ are Hermite functions

$$u_n(\tilde{x}) = \frac{e^{-\frac{|\tilde{x}|^2}{2}}}{\sqrt{n!2^n\sqrt{\pi}}} H_n(\tilde{x})\,, \quad H_n(\tilde{x}) = (-1)^n e^{\frac{|\tilde{x}|^2}{2}} \frac{d^n}{dx^n} e^{-\frac{|\tilde{x}|^2}{2}}, \tag{11.97}$$

which have Gaussian decay at large $|\tilde{x}| \gg 1$. Choosing the lowest solution with $n = 0$ and $H_0 = 1$, which has no nodes, is most likely to give the configuration with lowest energy content. So the system we need to solve is finally given by

$$0 = \partial_\rho^2 \tilde{A}_0^3 + \frac{\partial_\rho\left[\sqrt{-G}G^{00}G^{44}\right]}{\sqrt{-G}G^{00}G^{44}} \partial_\rho \tilde{A}_0^3 \frac{\gamma^2}{2\pi^2} \frac{G^{33}}{G^{44}} \tilde{A}_0^3 (u_0(\tilde{x}) v(\rho))^2, \tag{11.98}$$

$$0 = \partial_\rho^2 v + \frac{\partial_\rho\left[\sqrt{-G}G^{33}G^{44}\right]}{\sqrt{-G}G^{33}G^{44}} \partial_\rho v - \frac{\gamma^2}{2\pi^2} \frac{G^{00}}{G^{44}} (\tilde{A}_0^3)^2 v - \frac{\gamma \tilde{H}_3^3}{2\sqrt{2}\pi} \frac{G^{33}}{G^{44}} v. \tag{11.99}$$

Asymptotically near the horizon the fields take the form

$$\tilde{A}_0^3 = +a_2(\rho - 1)^2 + \mathcal{O}((\rho - 1)^3), \tag{11.100}$$

$$v = b_0 + \frac{b_0 H_3^3}{4} (\rho - 1)^2 + \mathcal{O}((\rho - 1)^{-3}), \tag{11.101}$$

while at the boundary we obtain

$$\tilde{A}_0^3 = \tilde{\mu} + \frac{\tilde{d}_0^3}{\rho^2} + \mathcal{O}(\rho^{-4}), \tag{11.102}$$

$$v = +\frac{\tilde{d}_3^1}{\rho^2} + \mathcal{O}(\rho^{-4}). \tag{11.103}$$

We succeed in finding numerical solutions $v(\rho)$ and $A_0^3(\rho)$ to the set of equations (11.98) obeying the asymptotics given by Eqs. 11.100 and 11.102. These numerical solutions are used to approach the phase transition from the superconducting phase by increasing the magnetic field. We map out the line of critical temperature-magnetic field pairs in Fig. 11.14. In this way we obtain a phase diagram displaying the Meissner effect. The critical line in Figs. 11.14 and 11.15 separates the phase with and without superconducting condensate $\tilde{d}_3^1$.

We emphasize that this is a background calculation involving no fluctuations. Complementary to the procedure described above we also confirmed the phase

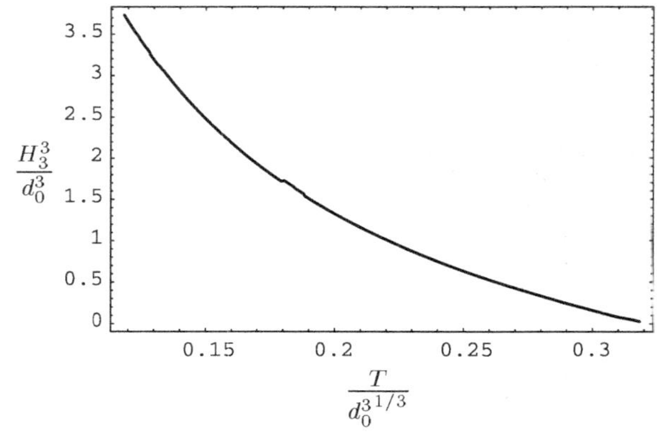

Fig. 11.14 The line of critical magnetic field versus critical temperature. Below this line the external magnetic field coexists with the superconducting condensate. Above the line the superconducting condensate vanishes. We set the spatial position arbitrarily to $\tilde{x} = 0.1$ since the critical line does not depend on $\tilde{x}$

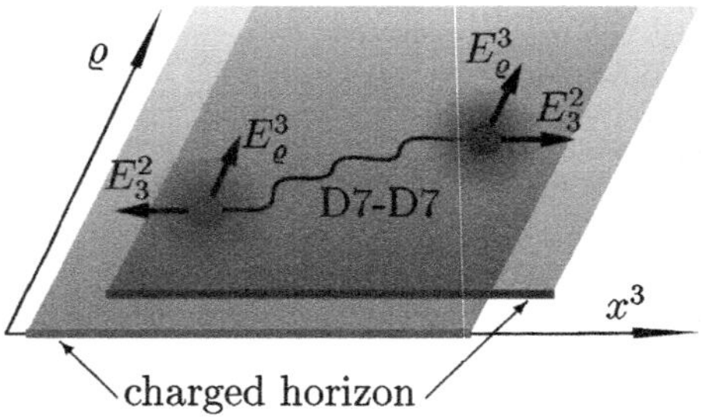

Fig. 11.15 Sketch of our string setup: The figure shows the two coincident D7 branes stretched from the black hole horizon to the boundary as a green and a blue plane, respectively. Strings spanned from the horizon of the AdS black hole to the D7-branes induce a charge at the horizon [18, 29, 35]. However, above a critical charge density, the strings charging the horizon recombine to D7–D7 strings. These D7–D7 strings are shown in the figure. Whereas the fundamental strings stretched between the horizon and the D7-brane are localized near the horizon, the D7–D7 strings propagate into the bulk balancing the flavorelectric and gravitational, i.e. tension forces (see text). Thus these D7–D7 strings distribute the isospin charges along the AdS radial coordinate, leading to a stable configuration of reduced energy. This configuration of D7–D7 strings corresponds to a superconducting condensate

diagram using the instability of the normal phase against fluctuations. Starting at large magnetic field and vanishing condensate $\tilde{d}_3^1$, we determine for a given magnetic field H_3^3 the temperature $T_c(H_3^3)$ at which the fluctuation a_1^3 becomes unstable. That instability signals the condensation process into the superconducting phase.

The presence of the coexistence phase below the critical line, where the system is still superconducting despite the presence of an external magnetic field, is the signal of the Meissner effect in the case of a global symmetry considered here. If we now weakly gauged the flavor symmetry at the boundary, the superconducting current J_3^1 would generate a magnetic field opposite to the external field. Thus the phase observed is a necessary condition in the case of a global symmetry for finding the standard Meissner effect when gauging the symmetry.

11.6 Interpretation and Conclusion

Our findings merge to a string theory picture of the *pairing mechanism* and the subsequent condensation process in Sect. 11.6.1. Finally we summarize what we have learned about holographic p-wave super-somethings and propose some future territories to be conquered.

11.6.1 String Theory Picture

We now develop a string theory interpretation, i.e. a geometrical picture, of the formation of the superconducting/superfluid phase, for which the field theory is discussed in Sect. 11.2.1. We show that the system is stabilized by dynamically generating a non-zero vev of the current component J_3^1 dual to the gauge field A_3^1 on the brane. Moreover, we find a geometrical picture of the pairing mechanism which forms the condensate $\langle J_3^1 \rangle$, the Cooper pairs. Let us first describe the unstable configuration in absence of the field A_3^1. As known from [18, 29, 35], the non-zero field A_0^3 is induced by fundamental strings which are stretched from the D7-brane to the horizon of the black hole. In the subsequent we call these strings 'horizon strings'. Since the tension of these strings would increase as they move to the boundary, they are localized at the horizon, i.e. the horizon is effectively charged under the isospin charge given by (11.1). By increasing the horizon string density, the isospin charge on the D7-brane at the horizon and therefore the energy of the system grows. In [18], the critical density was found beyond which this setup becomes unstable. In this case, the strings would prefer to move towards the boundary due to the repulsive force on their charged endpoints generated by the flavorelectric field $E_\varrho^3 = F_{0\varrho}^3 = -\partial_\varrho A_0^3$.

The setup is now stabilized by the new non-zero field A_3^1. This field is induced by D7-D7 strings moving in the x^3 direction. This movement of the strings may be interpreted as a current in x^3 direction which induces the magnetic field $B_{3\varrho}^1 = F_{3\varrho}^1 = -\partial_\varrho A_3^1$. Moreover, the non-Abelian interaction between the D7–D7 strings and the horizon strings induces a flavorelectric field $E_3^2 = F_{30}^2 = \gamma/\sqrt{\lambda} A_0^3 A_3^1$.

From the profile of the gauge fields and their conjugate momenta (see Fig. 11.7) we obtain the following: For $A_3^1 = 0$, i.e. in the normal phase ($T \geq T_c$), the isospin density $\tilde{d}_0^3$ is exclusively generated at the horizon by the horizon strings. This can also be understood by the profile of the conjugate momenta p_0^3 (see Fig. 11.7c). We interpret $p_0^3(\rho^*)$ as the isospin charge located between the horizon at $\rho = 1$ and a fictitious boundary at $\rho = \rho^*$. In the normal phase, the momentum p_0^3 is constant along the radial direction ρ (see Fig. 11.7c, blue curve), and therefore the isospin density is exclusively generated at the horizon. In the superconducting phase where $A_3^1 \neq 0$, i.e. $T < T_c$, the momentum p_0^3 is not constant any more. Its value increases monotonically towards the boundary and asymptotes to $\tilde{d}_0^3$ (see Fig. 11.7c, *red curve*). Thus the isospin charge is also generated in the bulk and not only at the horizon. This decreases the isospin charge at the horizon and stabilizes the system.

Now we describe the string dynamics which distributes the isospin charge into the bulk. Since the field A_3^1 induced by the D7–D7 strings is non-zero in the superconducting phase, these strings must be responsible for stabilizing this phase. In the normal phase, there are only horizon strings. In the superconducting phase, some of these strings recombine to form D7–D7 strings which correspond to the non-zero gauge field A_3^1 and carry isospin charge.[9] There are two forces acting on the D7–D7 strings, the flavorelectric force induced by the field E_ϱ^3 and the gravitational force between the strings and the black hole. The flavorelectric force points to the boundary while the gravitational force points to the horizon. The gravitational force is determined by the change in effective string tension, which contains the ϱ dependent warp factor. The position of the D7–D7 strings is determined by the equilibrium of these two forces. Therefore the D7–D7 strings propagate from the horizon into the bulk and distribute the isospin charge.

Since the D7–D7 strings induce the field A_3^1, they also generate the density $\tilde{d}_3^1$ dual to the condensate $\langle J_3^1 \rangle$, the Cooper pairs. This density $\tilde{d}_3^1$ is proportional to the D7–D7 strings located in the bulk, in the same way as the density $\tilde{d}_0^3$ counts the strings which carry isospin charge [29]. This suggest that we can also interpret $p_3^1(\rho^*)$ as the number of D7–D7 strings which are located between the horizon at $\rho = 1$ and the fictitious boundary at $\rho = \rho^*$. The momentum p_3^1 is always zero at the horizon and increases monotonically in the bulk (see Fig. 11.7d). Thus there are no D7–D7 strings at the horizon.

> The double importance of the D7-D7 strings is given by the fact that they are both responsible for stabilizing the superconducting phase by lowering the isospin charge density at the horizon, as well as being the dual of the Cooper pairs since they break the $U(1)_3$ symmetry. In QCD-language they correspond to quarks pairing up to form charged vector mesons which condense subsequently.

11.6.2 Summary

In conclusion we have derived from the top (string theory) down to the gravity theory a holographic p-wave superconductor. Thus we were able to directly

---

[9] Note that the D7–D7 strings are of the same order as the horizon strings, namely N_f/N_c, since they originate from the DBI action [35].

identify the degrees of freedom in the boundary field theory, allowing us to translate geometric features directly into field theory features. In particular we have found a string theory picture of the pairing mechanism. The Cooper pairs are modeled by strings spanned between the two flavor D7-branes corresponding to quasi-particles in the vector bi-fundamental representation, i.e. vector mesons. The dual thermal field theory is 3 + 1-dimensional $\mathcal{N} = 2$ supersymmetric Yang-Mills theory with $SU(N_c)$ color and $SU(2)$ flavor symmetry coupled to an $\mathcal{N} = 4$ gauge multiplet. It shows a conductivity gap at low temperatures. A pseudo-gap forms even above T_c. The onset of the Meissner-Ochsenfeld effect is visible and in the conductivity spectrum we find massive quasi-particles even at vanishing quark mass. Their masses are generated through a Higgs-like mechanism in the bulk.

Note that our results can also be interpreted using this very setup as a model for the quark gluon plasma as introduced in Sect. 11.3.1. In that case we have found a flavor superfluid phase. This should not be confused with the color-superconducting phase theoretically found in QCD at high *baryon density*.

11.6.3 Outlook

Some directions for promising future investigation are the study of critical exponents near the phase transition. Transport coefficients such as the speeds of second, fourth and other sounds can be determined. For some of these it will be necessary or convenient to use the backreacted solutions for this D3/D7 system given in [44]. Similar to earlier semi-classical drag computations it might be instructive to compute the drag on the various strings near the superconducting phase transition. It would be interesting to see what happens to the Fermi surface formed by the background fermions when the superconducting/superfluid phase is entered. Adventurous spirits may also take this setup as a serious model for p-wave superconductors such as the ruthenate compounds mentioned in the introduction. The geometric insights we gain from the dual gravity description can in principle be directly translated to precise field theory statements. Again this pleasant fact is due to knowing the field theory degrees of freedom exactly by using the gauge/gravity correspondence. A different setup in which vector mesons condense is the Sakai-Sugimoto model as shown in [49]. It could be illuminating to study all aforementioned effects in this model as well.

Acknowledgements I thank Johanna Erdmenger for discussions as well as kind support, and Martin Ammon, Patrick Kerner, Felix Rust for fruitful collaboration. I thank Ronny Thomale for instructive discussions, and especially Christopher Herzog and Andy O'Bannon for very helpful comments on these notes. This work is supported in part by the *"Deutsche Forschungsgemeinschaft" (DFG)*. Last but certainly not least, I thank the organizers and participants of the *Fifth Aegean Summer School "From Gravity to Thermal Gauge Theory: The AdS/CFT Correspondence"*.

References

1. Herzog, C.P.: J. Phys. A **42**, 343001 (2009). doi:10.1088/1751-8113/42/34/343001
2. Hartnoll, S.A.: Class. Quant. Grav. **26**, 224002 (2009). doi:10.1088/0264-9381/26/22/224002
3. Kaminski, M.: Fortsch. Phys. **57**, 3 (2009). doi:10.1002/prop.200810571
4. Ammon, M., Erdmenger, J., Kaminski, M., Kerner, P.: Phys. Lett. B**680**, 516 (2009). doi: 10.1016/j.physletb.2009.09.029
5. Ammon, M., Erdmenger, J., Kaminski, M., Kerner, P.: J. High Energy Phys. **10**, 067 (2009). doi:10.1088/1126-6708/2009/10/067
6. Hartnoll, S.A., Herzog, C.P., Horowitz, G.T.: Phys. Rev. Lett. **101**, 031601 (2008). doi: 10.1103/PhysRevLett.101.031601
7. Gubser, S.S., Pufu, S.S.: J. High Energy Phys. **11**, 033 (2008). doi:10.1088/1126-6708/2008/11/033
8. Horowitz, G.T., Roberts, M.M.: Phys. Rev. D **78**, 126008 (2008). doi:10.1103/PhysRevD.78.126008
9. Gubser, S.S., Herzog, C.P., Pufu, S.S., Tesileanu, T.: Phys. Rev. Lett. **103**, 141601 (2009). doi:10.1103/PhysRevLett.103.141601
10. Gubser, S.S., Pufu, S.S., Rocha, F.D.: Phys. Lett. B **683**, 201 (2010). doi: 10.1016/j.physletb.2009.12.017
11. Maeno, Y., et al.: Nature **372**, 532 (1994). doi:10.1038/372532a0
12. Tewari, S., Sarma, S.D., Nayak, C., Zhang, C., Zoller, P.: Phys. Rev. Lett. **98**, 010506 (2007). doi:10.1103/PhysRevLett.98.010506
13. Nayak, C., Simon, S.H., Stern, A., Freedman, M., Sarma, S.D.: Rev. Mod. Phys. **80**, 1083 (2008)
14. Vollhardt, D., Woelfle, P.: Superfluid Phases of Helium Three. CRC Press (1990)
15. Amado, I., Kaminski, M., Landsteiner, K.: J. High Energy Phys. **05**, 021 (2009). doi: 10.1088/1126-6708/2009/05/021
16. Greiter, M.: Ann. Phys. **319**, 217 (2005). doi:10.1016/j.aop.2005.03.008
17. Erdmenger, J., Evans, N., Kirsch, I., Threlfall, E.: Eur. Phys. J. A **35**, 81 (2008). doi: 10.1140/epja/i2007-10540-1
18. Erdmenger, J., Kaminski, M., Kerner, P., Rust, F.: J. High Energy Phys. **11**, 031 (2008). doi: 10.1088/1126-6708/2008/11/031
19. Gubser, S.S.: Phys. Rev. D **78**, 065034 (2008); [arXiv:0801.2977 [hep-th]]
20. Gubser, S.S.: Phys. Rev. Lett. **101**, 191601 (2008). doi:10.1103/PhysRevLett.101.191601
21. Karch, A., Katz, E.: J. High Energy Phys. **06**, 043 (2002)
22. Kruczenski, M., Mateos, D., Myers, R.C., Winters, D.J.: J. High Energy Phys. **07**, 049 (2003)
23. Constable, N.R., Myers, R.C.: J. High Energy Phys. **11**, 020 (1999)
24. Babington, J., Erdmenger, J., Evans, N.J., Guralnik, Z., Kirsch, I.: Phys. Rev. D **69**, 066007 (2004). doi:10.1103/PhysRevD.69.066007
25. Witten, E.: Adv. Theor. Math. Phys. **2**, 505 (1998); hep-th/9803131
26. Kirsch, I.: J. High Energy Phys. **09**, 052 (2006)
27. Mateos, D., Myers, R.C., Thomson, R.M.: Phys. Rev. Lett. **97**, 091601 (2006). doi: 10.1103/PhysRevLett.97.091601
28. Kirsch, I.: Fortsch. Phys. **52**, 727 (2004); doi:10.1002/prop.200410169
29. Kobayashi, S., Mateos, D., Matsuura, S., Myers, R.C., Thomson, R.M.: J. High Energy Phys. **02**, 016 (2007)
30. Myers, R.C.: J. High Energy Phys. **12**, 022 (1999)
31. Tseytlin, A.A.: Nucl. Phys. B **501**, 41 (1997). doi:10.1016/S0550-3213(97)00354-4
32. Hashimoto, A., Taylor, W.: Nucl. Phys. B **503**, 193 (1997). doi:10.1016/S0550-3213(97)00399-4
33. Constable, N.R., Myers, R.C., Tafjord, O.: Phys. Rev. D **61**:106009 (2000). doi: 10.1103/PhysRevD.61.106009

34. Myers, R.C., Wapler, M.C.: J. High Energy Phys. **12**, 115 (2008). doi:10.1088/1126-6708/2008/12/115
35. Karch, A., O'Bannon, A.: J. High Energy Phys. **11**:074 (2007). doi:10.1088/1126-6708/2007/11/074
36. Mateos, D., Matsuura, S., Myers, R.C., Thomson, R.M.: J. High Energy Phys. **11**, 085 (2007). doi:10.1088/1126-6708/2007/11/085
37. Roberts, M.M., Hartnoll, S.A.: J. High Energy Phys. **08**, 035 (2008). doi:10.1088/1126-6708/2008/08/035
38. Basu, P., He, J., Mukherjee, A., Shieh, H.H.: J. High Energy Phys. **11**, 070 (2009). doi:10.1088/1126-6708/2009/11/070
39. Erdmenger, J., Kaminski, M., Rust, F.: Phys. Rev. D **77**, 046005 (2008). doi:10.1103/PhysRevD.77.046005
40. Bender, C.M., Orszag, S.: Advanced Mathematical Methods for Scientists and Engineers. Springer (1978)
41. Son, D.T., Starinets, A.O.: J. High Energy Phys. **09**, 042 (2002); hep-th/0205051
42. Herzog, C.P., Son, D.T.: J. High Energy Phys. **03**, 046 (2003); hep-th/0212072
43. Berti, E., Cardoso, V., Starinets, A.O.: Class. Quant. Grav. **26**, 163001 (2009). doi:10.1088/0264-9381/26/16/163001
44. Ammon, M., Erdmenger, J., Grass, V., Kerner P., O'Bannon, A.: Phys. Lett. B **686**, 192 (2010); [arXiv:0912.3515 [hep-th]]
45. Nakano, E., Wen, W.Y.: Phys. Rev. D **78**, 046004 (2008). doi:10.1103/PhysRevD.78.046004
46. Albash, T., Johnson, C.V.: J. High Energy Phys. **09**, 121 (2008). doi:10.1088/1126-6708/2008/09/121
47. Maeda, K., Okamura, T.: Phys. Rev. D **78**, 106006 (2008). doi:10.1103/PhysRevD.78.106006
48. Hartnoll, S.A., Herzog, C.P., Horowitz, G.T.: J. High Energy Phys. **12**, 015 (2008). doi:10.1088/1126-6708/2008/12/015
49. Aharony, O., Peeters, K., Sonnenschein, J., Zamaklar, M.: J. High Energy Phys. **02**, 071 (2008). doi:10.1088/1126-6708/2008/02/071

Chapter 12
Holographic Torsion and the Prelude to Kalb–Ramond Superconductivity

Anastasios C. Petkou

Abstract We discuss the holographic implications of torsional degrees of freedom in the context of AdS_4/CFT_3, emphasizing in particular the physical interpretation of the latter as carriers of the non-trivial gravitational magnetic field, i.e. the part of the magnetic field not determined by the frame field. As a concrete example we present a new exact four-dimensional gravitational background with torsion and argue that it corresponds to the holographic dual of a 3D system undergoing parity symmetry breaking. Finally, we compare our new gravitational background with known wormhole solutions—with and without cosmological constant—and argue that they can all be unified under an intriguing "Kalb–Ramond superconductivity" framework.

12.1 Introduction and Summary of the Results

AdS_4/CFT_3 is currently emerging as a novel paradigm of holography that has qualitatively different properties from the more familiar AdS_5/CFT_4 correspondence. Particularly intriguing is the recent accumulation of evidence that AdS_4/CFT_3 can be used to describe a plethora of phenomena in $2+1$ dimensional systems, such as quantum criticality [1, 2], Quantum Hall transitions [3–6], superconductivity [7–11], supefluidity [12, 13] and spontaneous symmetry breaking [14–16]. This has given rise to a whole new research area that goes under the name of AdS/C(ondensed) M(atter) T(heory). Furthermore, AdS_4/CFT_3 is the appropriate setup to study the holographic consequences of generalized electric–magnetic duality of gravity and higher-spin gauge fields [17–22].

A. C. Petkou (✉)
Department of Physics, University of Crete, Heraklion 71003 Greece
e-mail: petkou@physics.uoc.gr

E. Papantonopoulos (ed.), *From Gravity to Thermal Gauge Theories: The Ads/CFT Correspondence,* Lecture Notes in Physics, 828,
DOI: 10.1007/978-3-642-04864-7_12,

In the absence of an explicit AdS_4/CFT_3 correspondence example[1] various toy models have been used to study its general qualitative aspects. This work presents yet another model of AdS_4/CFT_3 which possesses a novel feature. Namely, it can describe the gravity dual of parity symmetry breaking in a 3D system. However, this is not our only aim. We also wish to shed light into torsion from a holographic point of view. The study of torsion is an interesting subject in itself that poses formal and phenomenological challenges.[2] In the context of string theory, torsion is omnipresent through antisymmetric tensor fields, therefore AdS_4/CFT_3 provides the basic setup where it can be holographically investigated.

This review presents in a slightly expanded form the results of [34]. We consider a simple toy model where torsion is introduced via the topological Nieh–Yan class. In particular, we consider the modification of the Einstein–Hilbert action with a negative cosmological constant by the Nieh–Yan class, the latter having a spacetime-dependent coefficient. In the context of the $3 + 1$-split formalism for gravity [17] we emphasize that the torsional degrees of freedom carry the non-trivial 'gravitational magnetic field.' In pure gravity the gravitational magnetic field is fully determined by the frame field and hence torsion vanishes. In our model, the spacetime dependence of the Nieh–Yan coefficient makes some of the components of the magnetic field dynamical and as a consequence torsional degrees of freedom enter the theory. Our toy model is simple enough such that only one of the torsional degrees of freedom becomes dynamical. This degree of freedom can be either carried by a pseudoscalar, in which case our model is equivalent to a massless pseudoscalar coupled to gravity, or by a two-form gauge potential. In the latter case our model becomes equivalent to a Kalb–Ramond field coupled to gravity.

Next, we find an exact solution of the equations of motion in Euclidean signature. Our metric ansatz is that of a bulk domain wall (DW). The solution, the *torsion DW*, has two distinct asymptotically AdS_4 regimes along the "radial" coordinate. The pseudoscalar has a kink-like profile and it is finite at both of the asymptotic regimes. Our torsion DW can be viewed as a generalization of the axionic wormhole solution of [35] in the case of non-zero cosmological constant. See also [36] for recent work on AdS wormholes.

Having in mind the holographic interpretation of our model we focus mainly on the case where the torsional degree of freedom is carried by a pseudoscalar field. Following standard holographic recipes we find that the torsion DW is the gravity dual of a 3D system that possesses two distinct parity breaking vacua. The two vacua are distinguished by the relative sign of the pseudoscalar order parameter. Our bulk picture suggests that the transition from one vacuum to the other can be done by a marginal deformation of the boundary theory. In the Appendix we

---

[1] The recently suggested field theoretic models for M2 branes [23–28] are important steps towards the understanding of the boundary side of AdS_4/CFT_3.

[2] See [29–31] for recent reviews and [32, 33] for other recent works.

suggest that the above qualitative properties can be realized in the boundary by the 3D Gross–Neveu model coupled to U(1) gauge fields.

Further, we point out that the bulk physics of our DW solution bears some intriguing resemblance to the standard Abrikosov vortex in superconducting systems. There is a natural mapping of the parameters of the torsion DW to those of the Abrikosov vortex. We show that the gravitational parameter that is interpreted as an order parameter satisfies a ϕ^4-like equation and this motivates us to suggest that the cosmological constant is related to the "distance from the critical temperature" as $\Lambda \sim T_c - T$. However, there is an important difference in that the Abrikosov vortex is a one-dimensional defect while our DW is codimension one i.e. three-dimensional in AdS_4. We also discuss multi-DW configurations and DW condensation and show that H-flux supports bubbles of flat spacetime.

Quite intriguing is our result that DW condensation occurs at a critical value of the magnetic field. This motivates us to reconsider the known Euclidean solutions of an Einstein-axion system with [36] and without [35] cosmological constant. These are wormhole solutions whose salient properties include a quantized electric and (possibly) magnetic flux. Moreover, the $\Lambda \neq 0$ solutions possess a lower bound on their electric flux. Using our intuition that Λ plays the role of "temperature", we place the known wormhole and DW solutions on a ("Temperature","Magnetic Field") graph and observe that it resembles a standard superconductivity graph. We call such a system a "Kalb–Ramond superconductor" and we will present more details on its properties in a forthcoming work [43].

12.2 Torsion as the Non-trivial Magnetic Field of Gravity

In this section, we discuss the physical interpretation of torsion which is that it carries the non-trivial magnetic degrees of freedom of gravity, namely those that are not determined by the frame field (or, equivalently, by the metric in a second order formulation). To motivate things we recall the first order formalism of electromagnetism in the presence of an x-dependent θ-angle, in a non-trivial background here taken to be AdS_4. Then we present the $3+1$-split formalism for gravity introduced in [17]. This formalism is a refined form of the standard ADM formalism, which however unveils the physical importance of the gravitational torsional d.o.f. As we will see, such a point of view is crucial in order to understand the holographic interpretation of torsion.

12.2.1 *Electromagnetism with a $\theta(x)$-Angle in AdS_4*

The vierbeins and metric of AdS_4 are

$$e^0 = dt, \quad e^\alpha = e^{-t/L} dx^\alpha, \quad ds^2 = \sigma_\perp dt^2 + e^{-2t/L} \eta_{\alpha\beta} dx^\alpha dx^\beta, \tag{12.1}$$

with $\alpha, \beta = 1, 2, 3$. Throughout this work we are being flexible with the both the overall signature and also the nature (spacelilke or timelike) of the t-direction i.e. we set $\eta_{\alpha\beta} = \text{diag}(1, 1, \sigma_3), \sigma_\perp \sigma_3 = \sigma = \pm 1$. The gauge potential and field strength are one-forms

$$A = A_0 dt + \tilde{A} \quad F = -dt \wedge E + \tilde{F} = \frac{1}{2} F_{ab} e^a \wedge e^b, \tag{12.2}$$

where $a, b = 0, 1, 2, 3$ and $E = -F_{0\alpha} e^\alpha$ is the electric field. The tilde will always denote quantities along the three directions $1, 2, 3$. With the above definitions we find

$$*_4 F = dt \wedge *_3 \tilde{F} - \sigma_\perp *_3 E, \quad dA = dt \wedge (\dot{\tilde{A}} - \tilde{d} A_0) + \tilde{d}\tilde{A}. \tag{12.3}$$

Recall e.g. that $*_3 e^i = \frac{1}{2} \epsilon^i{}_{jk} e^j \wedge e^k$ and $*_3 e^i \wedge e^k = \epsilon^{ij}{}_k e^k$. Note also that $*_4^2 = \sigma_\perp, *_3^2 = \sigma_3$ and $\epsilon_{0123} = 1$. The first order action is

$$I = \int -dA \wedge *_4 F + \frac{1}{2} F \wedge *_4 F + \frac{\theta}{2} dA \wedge dA. \tag{12.4}$$

Notice that due to the x-dependance of θ the last term in (12.4) is not a total derivative and will give contributions to the e.o.m. After some work the action above takes the more familiar form

$$I = \sigma_\perp \int dt \wedge \Big[\dot{\tilde{A}} \wedge \big(*_3 E + \sigma_\perp \theta \tilde{d}\tilde{A}\big) + \frac{1}{2}\big(E \wedge *_3 E - \sigma_\perp \tilde{F} \wedge *_3 \tilde{F}\big) \\ + A_0 \big(\tilde{d} *_3 E + \sigma_\perp \tilde{d}\theta \wedge \tilde{d}\tilde{A}\big) \Big]. \tag{12.5}$$

This gives the Hamiltonian e.o.m.

$$E = -\dot{\tilde{A}}, \quad \tilde{d}B = -\sigma_\perp (\dot{*_3 E}) - \dot{\theta}\sigma_3 *_3 B - \tilde{d}\theta \wedge E, \tag{12.6}$$

where we have defined the magnetic field (also a one-form) as

$$\tilde{F} = \tilde{d}\tilde{A} \equiv \sigma_3 (*_3 B). \tag{12.7}$$

We also have the Gauss law and Bianchi identity, respectively

$$\tilde{d} *_3 E + \sigma_\perp \tilde{d}\theta \wedge \tilde{d}\tilde{A} = 0, \quad \tilde{d} *_3 B = 0. \tag{12.8}$$

It is straightforward to show that the above give the Maxwell equations in the more familiar form

$$\nabla \times \boldsymbol{E} = -\sigma_3 \frac{\partial \boldsymbol{B}}{\partial t}, \quad \nabla \cdot \boldsymbol{E} = -\sigma_\perp \sigma_3 \nabla\theta \cdot \boldsymbol{B}, \tag{12.9}$$

$$\nabla \times \boldsymbol{B} = -\sigma \frac{\partial \boldsymbol{E}}{\partial t} - \Big[\dot{\theta}\boldsymbol{B} + \nabla\theta \times \boldsymbol{E}\Big], \quad \nabla \cdot \boldsymbol{B} = 0. \tag{12.10}$$

We summarize the effects of an x-dependent θ-angle in electromagnetism:

- The modification of the canonical momentum as we see in (5)

$$*_3 E \mapsto *_3 E + \sigma\theta *_3 B. \tag{12.11}$$

- The presence of a source term for Gauss law.

In particular, there is no additional d.o.f. introduced by the θ-angle. We will compare this situation with gravity in the following.

12.3 Details on the 3 + 1-Split Formalism

In this section, we present a concise version of the 3 + 1-split formalism of [17] for gravity in the presence of non-zero cosmological constant. We consider a globally hyperbolic Lorentzian manifold $\mathcal{M}$ and take the Einstein–Hilbert action with cosmological constant in the first-order Palatini formalism as

$$S_{\mathrm{EH}} = -\frac{1}{32\pi G}\int_{\mathcal{M}} \epsilon_{abcd}\left(R^{ab} + \frac{\Lambda}{2} e^a \wedge e^b\right) \wedge e^c \wedge e^d. \tag{12.12}$$

This is thus equivalent to the standard second-order gravitational action

$$S_{\mathrm{second}} = -\frac{1}{16\pi G}\int d^4x\sqrt{-g}(R + 6\Lambda), \tag{12.13}$$

and hence the cosmological constants is related to the parameter Λ as $\Lambda_{\mathrm{cosm.}} = -3\Lambda$. The curvature and torsion two-forms are defined in terms of the vielbein e^a and spin-connection ω^{ab} as

$$R^{ab} = d\omega^{ab} + \omega^a{}_c \wedge \omega^{cb}, \quad T^a = de^a + \omega^a{}_b \wedge e^b. \tag{12.14}$$

We define as before $\eta_{ab} = \mathrm{diag}(\sigma_\perp, +, +, \sigma_3)$, where $\sigma_\perp\sigma_3 = \sigma = \pm 1$, $\sigma_\perp^2 = \sigma_3^2 = 1$ and set $\Lambda = \sigma_\perp/\ell^2$ such that $\Lambda < 0$ $(\Lambda > 0)$ yields the de Sitter (Anti-de Sitter) vacuum. Next, we split the vielbein and the spin connection as

$$e^0 = Ndt, \quad e^\alpha = N^\alpha dt + \tilde{e}^\alpha, \tag{12.15}$$

$$\omega^{0\alpha} = q^{0\alpha}dt + \sigma_\perp K^\alpha, \quad \omega^{\alpha\beta} = -\epsilon^{\alpha\beta\gamma}\left(Q_\gamma dt + B_\gamma\right). \tag{12.16}$$

The novelty of the formalism is the introduction of the gravitational electric K^α and magnetic fields B^α, which are both vector-valued one-forms on the slices. We then find for the torsion

$$T^\alpha = \tilde{T}^\alpha + dt \wedge \left\{\dot{\tilde{e}}^\alpha - \tilde{d}N^\alpha + NK^\alpha - \sigma\epsilon^\alpha{}_{\beta\gamma}Q^\beta e^\gamma - \sigma\epsilon^\alpha{}_{\beta\gamma}N^\beta B^\gamma\right\}, \tag{12.17}$$

$$T^0 = \sigma_\perp K_\alpha \wedge \tilde{e}^\alpha + dt \wedge \left\{ -\tilde{d}N - \sigma_\perp N_\alpha K^\alpha + q^0{}_\beta \tilde{e}^\beta \right\}, \tag{12.18}$$

and we write

$$R^a{}_b = \tilde{R}^a{}_b + dt\omega r^a{}_b, \tag{12.19}$$

$$\tilde{R}^0{}_\alpha = \sigma_\perp (\tilde{d}K_\alpha + K_\beta \wedge \tilde{\omega}^\beta{}_\alpha) \equiv \sigma_\perp (\tilde{D}K)_\alpha, \tag{12.20}$$

$$\tilde{R}^\alpha{}_\beta = {}^{(3)}R^\alpha{}_\beta - \sigma_\perp K^\alpha \wedge K_\beta, \tag{12.21}$$

with

$${}^{(3)}R^{\alpha\beta} = \sigma \left[\epsilon^{\alpha\beta\gamma} dB_\gamma - \sigma_\perp B^\alpha \omega B^\beta \right], \tag{12.22}$$

and

$$\begin{aligned} 2\epsilon_{\alpha\beta\gamma} r^{0\alpha} \wedge \tilde{e}^\beta \wedge \tilde{e}^\gamma = & 2\sigma_\perp \epsilon_{\alpha\beta\gamma} \dot{K}^\alpha \wedge \tilde{e}^\beta \wedge \tilde{e}^\gamma + 4Q_\alpha K_\beta \wedge \tilde{e}^\beta \wedge \tilde{e}^\alpha \\ & + 4q^{0\alpha} \left[\epsilon_{\alpha\beta\gamma} \tilde{T}^\beta \wedge \tilde{e}^\gamma \right]. \end{aligned} \tag{12.23}$$

After some tedious but straightforward calculations we find

$$\begin{aligned} S_{\mathrm{EH}} = & -\frac{\sigma_\perp}{8\pi G} \int dt \wedge \left\{ \dot{\tilde{e}}^\alpha \wedge (4\sigma_\perp \epsilon_{\alpha\beta\gamma} K^\gamma \wedge \tilde{e}^\beta) - 4\sigma_\perp N^\alpha \epsilon_{\alpha\beta\gamma} (\tilde{D}K)^\beta \wedge \tilde{e}^\gamma \right. \\ & + 2N\epsilon_{\alpha\beta\gamma} \left({}^{(3)}R^{\alpha\beta} - \sigma_\perp K^\alpha \wedge K^\beta - \frac{\Lambda}{3} \tilde{e}^\alpha \wedge \tilde{e}^\beta \right) \wedge \tilde{e}^\gamma \\ & \left. - 4q^{0\alpha} \epsilon_{\alpha\beta\gamma} \tilde{T}^\beta \omega \tilde{e}^\gamma + 4Q_\alpha \tilde{e}^\alpha \omega K_\beta \omega \tilde{e}^\beta \right\} - S_{\mathrm{GH}} \end{aligned} \tag{12.24}$$

where the last term is exactly the usual Gibbons–Hawking surface term

$$S_{\mathrm{GH}} = -\frac{1}{16\pi G} \int_{\partial\mathcal{M}} \left(q^{0\alpha} dt + \sigma_\perp K^\alpha \right) \wedge \epsilon_{\alpha\beta\gamma} \tilde{e}^\beta \wedge \tilde{e}^c. \tag{12.25}$$

Adding then the Gibbons–Hawking term in (12.24) we obtain

$$\begin{aligned} \hat{S}_{\mathrm{EH}} = & -\frac{\sigma_\perp}{8\pi G} \int_{\mathcal{M}} dt \wedge \left\{ -K_\alpha \wedge \dot{\Sigma}^\alpha + N\tilde{W}_\alpha \wedge \tilde{e}^\alpha + \sigma_\perp \hat{Q} \wedge K_\beta \wedge \tilde{e}^\beta \right. \\ & \left. + \sigma_\perp q^{0\alpha} \tilde{D}\Sigma_\alpha - N^\alpha \epsilon_{\alpha\beta\gamma} \tilde{D}K^\beta \wedge \tilde{e}^\gamma \right\}, \end{aligned} \tag{12.26}$$

where $\hat{Q} \equiv Q_\alpha \tilde{e}^\alpha$. We have introduced the two-form

$$\tilde{W}_\alpha \equiv \rho_\alpha - \frac{1}{2} \epsilon_{\alpha\beta\gamma} K^\beta \wedge K^\gamma + \frac{1}{\ell^2} \Sigma_\alpha. \tag{12.27}$$

and have defined the oriented surface element as

$$\Sigma^\alpha = *_3 \tilde{e}^\alpha = \frac{1}{2} \epsilon^\alpha{}_{\beta\gamma} \tilde{e}^\beta \wedge \tilde{e}^\gamma, \tag{12.28}$$

with $*_3$ the three-dimensional Hodge dual defined in terms of $\tilde{e}^\alpha$ only. The three-dimensional component of the curvature two-form

$$\rho_\alpha = \tilde{d}B_\alpha + \frac{1}{2}\epsilon_{\alpha\beta\gamma}B^\beta \wedge B^\gamma, \tag{12.29}$$

is made out of B^α only. Recall (i.e. 19) that $\tilde{D}$ is a covariant derivative with respect to the one-form field B^α as

$$\tilde{D}V^\alpha = \tilde{d}V^\alpha + \epsilon^\alpha_{\ \beta\gamma}B^\beta \wedge V^\gamma, \tag{12.30}$$

if V^α is a generic vector-valued one-form (with respect to either SO(3) or SO(2, 1) depending on whether $\sigma_\perp = \mp 1$, respectively) defined on Σ_t. Comparing the action (12.26) to the electromagnetic action (12.4) motivates calling the vector-valued one-forms K^α and B^α the "electric" and "magnetic" fields, respectively.

The action (12.26) is stationary on-shell when $\delta\tilde{e}^\alpha = 0$ in the boundary, i.e. it provides a good Dirichlet variational principle with respect to the vielbein. The form of the action (12.26) appears to indicate that the proper conjugate dynamical variables are Σ^α (or, equivalently, $\tilde{e}^\alpha$) and K^α. It has been shown in [19, 20] that the proper identification of the dynamical variables is slightly more involved than this. The remaining fields $\{N, N^\alpha, q^{0\alpha}, \hat{Q}, B^\alpha\}$ enter the action as Lagrange multipliers of the following constraints:

$$-8\pi G\sigma_\perp \frac{\delta S}{\delta N} = \tilde{W}_\alpha \wedge \tilde{e}^\alpha = 0, \tag{12.31}$$

$$-8\pi G\sigma_\perp \frac{\delta S}{\delta N^\alpha} = -\epsilon_{\alpha\beta\gamma}\tilde{D}K^\beta \wedge \tilde{e}^\gamma = 0, \tag{12.32}$$

$$-8\pi G\sigma_\perp \frac{\delta S}{\delta q^{0\alpha}} = \sigma_\perp \tilde{D}\Sigma_\alpha = \sigma_\perp \epsilon_{\alpha\beta\gamma}\tilde{T}^\beta \wedge \tilde{e}^\gamma = 0, \tag{12.33}$$

$$-8\pi G\sigma_\perp \frac{\delta S}{\delta \hat{Q}} = \sigma_\perp K_\alpha \wedge \tilde{e}^\alpha = 0, \tag{12.34}$$

$$-8\pi G\sigma_\perp \frac{\delta S}{\delta B^\alpha} = N\tilde{T}^\alpha + \left(\tilde{d}N + \sigma_\perp K_\beta N^\beta - \hat{q}\right) \wedge \tilde{e}^\alpha = 0, \tag{12.35}$$

where $\hat{q} \equiv q^0{}_\alpha \tilde{e}^\alpha$. The exterior multiplication of (12.35) by $\epsilon_{\alpha\beta\gamma}\tilde{e}^\gamma$ gives, by virtue of (12.33),

$$\tilde{d}N + \sigma_\perp K_\beta N^\beta - \hat{q} = 0, \tag{12.36}$$

and hence we obtain the zero torsion condition

$$\tilde{T}^{\alpha} = \tilde{D}\tilde{e}^{\alpha} = 0, \tag{12.37}$$

The last equation unveils the physical meaning of the gravitational magnetic field B^{α}: it is a Lagrange multiplier which is algebraically related to the vielbein via the vanishing of torsion (12.37). This is exactly analogous to electromagnetism and gives an important hint regarding the relevance of torsion to holography and gravitational duality [34].

12.3.1 The Analog of θ-Angle in Gravity

There is a number of topological terms built from the gravitational dynamical variables that one may consider in four dimension. These are all of potential interest to holography because being total derivatives they may induce interesting boundary effects. We may parameterize these terms as follows (writing all possible $SO(3,1)$-invariant four-forms constructed from $e^a, R^a{}_b, T^a$):

$$I_{\text{top}} = n\int_M C_{\text{NY}} + 2\gamma^{-1}\int_M C_{\text{Im}} + p\int_M P_4 + q\int_M E_4, \tag{12.38}$$

where

$$C_{\text{NY}} = T^a \wedge T_a - R_{ab} \wedge e^a \wedge e^b = d(T^a \wedge e_a), \tag{12.39}$$

is the Nieh–Yan form. The parameter γ is often referred to as the Immirzi parameter with

$$C_{\text{Im}} = R^a{}_b \wedge e^b \wedge e_a. \tag{12.40}$$

The remaining objects are the Pontryagin form

$$P_4 = -\frac{1}{8\pi^2} R^a{}_b \omega R^b{}_a = -\frac{1}{8\pi^2} d\left(\omega^a{}_b \wedge R^b{}_a - \frac{1}{3}\omega^a{}_b\omega\omega^b{}_c\omega\omega^c{}_a\right), \tag{12.41}$$

and the Euler form

$$E_4 = -\frac{1}{32\pi^2}\epsilon_{abcd} R^{ab}\omega R^{cd}. \tag{12.42}$$

We note that $P_4 + \frac{\sigma_\perp a^2}{4\pi^2} C_{\text{NY}}$ and $C_{\text{NY}} - C_{\text{Im}}$ are actually $SO(3,2)$ invariants [30]. These terms become very interesting even in gravity if we allow the coefficients to become fields (for some interesting recent literature on the matter see [32, 33]).

12.3.2 Torsion and the Magnetic Field of Gravity

In pure Eistein–Hilbert gravity the torsional d.o.f. are not dynamical and are carried by the magnetic field B^α. This is seen for example if we recall the definition of the non-trivial 'spatial' torsion as

$$\tilde{T}^\alpha = \tilde{d}\tilde{e}^\alpha - \sigma\epsilon^{\alpha\beta\gamma} B_\beta \wedge \tilde{e}_\gamma. \tag{12.43}$$

Moreover, it is seen from (12.18) that the radial component of torsion T^0 is determined by $\tilde{e}^\alpha$ and K^α. Notice that (12.43) implies that the tensor $B_{\alpha\beta}$ is odd under 'spatial' parity, hence its trace $B^\alpha_{\ \alpha}$ is a pseudoscalar. Hence, although a priori the torsional degrees of freedom are not connected with the pair of conjugate variables $\tilde{e}^\alpha$ and K^α, they are not dynamical as there is no kinetic term for B^α. Rather, they enter (12.26) algebraically and as such they give via (12.35) and (12.36) the algebraic zero torsion condition (12.37) by virtue of which the magnetic field is related to the frame field. This is the gravitational analogue of the electromagnetic case where the magnetic field is related to the gauge potential via the Bianchi identity.

Consider now adding to the Einstein–Hilbert action the Nieh–Yan class C_{NY} with a constant coefficient θ. Over a compact manifold, the NY class is a topological invariant and takes integer values[3] [30]. Having in mind holography, we are interested here in manifolds with boundary. In particular, the $3+1$ split has been set up so that the boundary is a constant-t slice. The NY term reduces to a boundary contribution. The explicit calculation yields

$$\mathcal{I}_{\mathrm{NY}} \equiv -2\sigma_\perp\theta \int C_{\mathrm{NY}} = 2\sigma_\perp\theta \int dt \wedge \left[2\epsilon_{\alpha\beta\gamma}\dot{\tilde{e}}^\alpha \wedge \tilde{e}^\beta \wedge B^\gamma + \epsilon_{\alpha\beta\gamma}\dot{B}^\alpha \wedge \tilde{e}^\beta \wedge \tilde{e}^\gamma\right] \tag{12.44}$$

Adding (12.44) to (12.26) we obtain

$$\hat{S}_{\mathrm{EH}} + \mathcal{I}_{\mathrm{NY}} \propto \int dt \wedge \big(\dot{\tilde{e}}^\alpha \wedge (-4\sigma_\perp\epsilon_{\alpha\beta\gamma}\tilde{e}^\beta \wedge [K^\gamma - \theta B^\gamma]) + 2\sigma_\perp\theta\epsilon_{\alpha\beta\gamma}\dot{B}^\alpha \wedge \tilde{e}^\beta \wedge \tilde{e}^\gamma + \text{constraint terms}\big). \tag{12.45}$$

Notice that the $\mathcal{I}_{\mathrm{NY}}$ term has two effects. One is to modify the canonical momentum variable $K^\alpha \mapsto K^\alpha - \theta B^\alpha$. This is analogous to the effect of the θ-angle in the canonical description of electromagnetism in Sect. 2.1. The other is to provide a kinetic term for the singlet component of the magnetic field [one easily verifies that only $B^\alpha_{\ \alpha}$ contributes in the second term in the first line of (12.45)]. This second effect has no analogue in electromagnetism. Taking the variation of (12.45) with respect to B^α, one finds that the zero torsion condition still holds.

[3] More precisely, $C_{\mathrm{NY}}/(2\pi L)^2$ is integral, as it is equal to the difference of two Pontryagin forms, one $SO(3,2)$ and one $SO(3,1)$.

This is expected of course since the $\mathcal{I}_{\rm NY}$ term is purely a boundary term. As a consequence, the true dynamical variables remain $\tilde{e}^{\alpha}$ and K^{α}. However, the holography is slightly modified. The variation of (12.45) gives on-shell

$$\delta\big(\hat{S}_{\rm EH}+\mathcal{I}_{\rm NY}\big)_{\rm onshell}\propto\int\limits_{\partial\mathcal{M}}\delta\tilde{e}^{\alpha}\wedge\big(-4\sigma_{\perp}\epsilon_{\alpha\beta\gamma}\tilde{e}^{\beta}\wedge[K^{\gamma}-\theta B^{\gamma}]\big)_{\rm onshell}. \qquad (12.46)$$

After the appropriate subtraction of divergences [19, 20], (12.46) yields a modified boundary energy momentum tensor. The modification is due to the term $4\sigma_{\perp}\theta\epsilon_{\alpha\beta\gamma}\tilde{e}^{\beta}\wedge B^{\gamma}$ which is parity odd and corresponds to the unique symmetric, conserved and traceless tensor of rank two and scaling dimension three that can be constructed from the three-dimensional metric [38]

$$T^{\rm bdry}_{\alpha\beta}\mapsto T^{\rm bdry}_{\alpha\beta}+\theta T^{\rm top}_{\alpha\beta},\quad T^{\rm top}_{\alpha\beta}\propto\epsilon_{\ell m(\alpha}\partial_{\ell}\partial^{2}g_{\beta)m} \qquad (12.47)$$

where $g_{\alpha\beta}$ being the boundary metric. It is the exact analogue of the topological spin-1 current constructed from the 3D gauge potential [38, 42].

The form of the action (12.45) unveils an intriguing possibility. The above holographic interpretation was based on the zero torsion condition that connects B^{α} to the frame field. However, to get the zero torsion condition from (12.45) we needed to integrate by parts the last term in the first line. Hence, if θ were t-dependent, the torsion would no longer be zero and the trace $B^{\alpha}{}_{\alpha}$ would become a proper dynamical degree of freedom independent of $\tilde{e}^{\alpha}$. In such a case the holographic interpretation of (12.45) would change. The new bulk degree of freedom would couple to a new pseudoscalar boundary operator. As a consequence, we have the possibility to probe additional aspects of the boundary physics and describe new 2 + 1 dimensional phenomena. That we do in the next section.

12.4 The Nieh–Yan Models

12.4.1 General Aspects

In the previous section we sketched a mechanism by which torsional degrees of freedom become dynamical. In particular, we have argued that the addition of the Nieh–Yan class with a space–time-dependent coefficient in the Einstein–Hilbert action makes dynamical one pseudoscalar degree of freedom which is connected to the trace of the gravitational magnetic field. Adding boundary terms to the bulk action corresponds to a canonical transformation. Consequently, by adding boundary terms we can change the canonical interpretation and the variational principle. Consider first the action

$$I'_{\rm NY}=\hat{S}_{\rm EH}[e,\omega]+I_{\rm GH}[e,\omega]+2\int\limits_{M}F(x)C_{\rm NY}, \qquad (12.48)$$

where F is a pseudoscalar 'axion' field with no kinetic term. If $F \equiv -\sigma_\perp\theta$ were a constant, this theory would be equivalent to that studied in the last section. With $F = F(x)$, we have additional terms in the action involving gradients of F. If we perform the 3 + 1 split on this action, we will find that $\tilde{e}^\alpha$ and B^α are canonical coordinates, and their conjugate momenta will depend on F.

The action as given may be supplemented by additional boundary terms. Such boundary terms are analogous to the Gibbons–Hawking term in gravity, but here involve the torsional degrees of freedom. In particular, we can replace $I'_{\rm NY}$ by

$$I_{\rm NY} = \hat{S}_{\rm EH}[e,\omega] + I_{\rm GH}[e,\omega] - 2\int_M dF \omega T_a \omega e^a. \tag{12.49}$$

This action is such that $\tilde{e}^\alpha$ and F are canonical coordinates with appropriate boundary conditions, while B^α appears in the momentum conjugate to F. To investigate this theory, we note that the variation of the action takes the form

$$\begin{aligned}\delta I_{\rm NY} = &2\int_M \delta e^d \omega \left[\epsilon_{abcd} e^b \omega \left(R^{cd} - \frac{1}{3}\Lambda e^c \omega e^d\right) + 2dF\omega T_d\right] \\ &+ 2\int_M \delta\omega^{ab}\omega\left[\epsilon_{abcd}T^c \omega e^d + dF\omega e_b \omega e_a\right] + 2\int_M \delta F\ C_{\rm NY} \\ &+ 2\int_M d[\delta e^a \wedge \left(\epsilon_{abcd}e^b \omega\omega^{cd} - dF\omega e_a\right) - T_a \omega e^a \delta F]. \end{aligned}\tag{12.50}$$

A non-trivial configuration of F sources a particular component of the torsion. Indeed the classical equations of motion can be manipulated to yield in the bulk

$$T^a \omega e_a = 3 *_4 dF, \tag{12.51}$$

where $*_4$ denotes the Hodge-$*$ operation. However, as d'Auria and Regge [37] showed, this classical system is equivalent to a pseudoscalar coupled to torsionless gravity.

$$I_{\rm PS} = \hat{S}_{\rm EH}[e,\mathring{\omega}] + I_{\rm GH}[e,\mathring{\omega}] - 3\int_M dF\omega *_4 dF. \tag{12.52}$$

This comes about as follows. We write the connection as $\omega = \mathring{\omega} + \Omega$, where $\mathring{\omega}$ is torsionless, and insert the equation of motion (12.51). The latter becomes an equation[4] for Ω, and we obtain (12.52).

A massless pseudoscalar field coupled to torsionless gravity is holographically dual to composite pseudoscalar operators of dimensions $\Delta = 3, 0$ in the boundary. The usual holographic dictionary then says that only the $\Delta = 3$ operator appears in

---

[4] Explicitly this is $\Omega^a{}_b = \frac{\sigma}{4}\epsilon^{acd}{}_b \partial_c F e_d$.

the boundary theory since only this is above the unitarity bound of the 3D conformal group $SO(3,2)$. A scalar operator with dimension $\Delta = 0$ would simply correspond to a constant in the boundary. Hence, the sensible holographic interpretation of the massless bulk pseudoscalar is that its leading behaviour determines the marginal coupling of a $\Delta = 3$ operator; the expectation value of the operator itself is determined by the subleading behaviour of the bulk pseudoscalar.

Another equivalent formulation of this bulk theory is obtained by writing

$$*_4 dF = \frac{1}{3} H. \tag{12.53}$$

with H a three-form field. This is the parameterization that would be most familiar from string theory, as the system simply corresponds to an antisymmetric two-form field. In this formulation, we write

$$\begin{aligned} I_{\text{KR}} &= \hat{S}_{\text{EH}}[e, \overset{\circ}{\omega}] + I_{\text{GH}}[e, \overset{\circ}{\omega}] + \frac{1}{3} \int_M H \omega *_4 H + \sqrt{\frac{2}{3}} \int_M C \omega d *_4 H \\ &= \hat{S}_{\text{EH}}[e, \overset{\circ}{\omega}] + I_{\text{GH}}[e, \overset{\circ}{\omega}] - \frac{1}{2} \int_M dC \wedge *_4 dC + \int_M d(C \omega *_4 dC). \end{aligned} \tag{12.54}$$

In the first equation, C appears as a Lagrange multiplier for the 'Gauss constraint' and in the second expression, we have solved for the H equation of motion in the bulk, which is just $H = \sqrt{\frac{3}{2}} dC$.

12.4.2 The 3 + 1-Split of the Pseudoscalar Nieh–Yan Model

To investigate the holographic aspects of our model it is most useful to use the 'radial quantization' in which we think of the radial coordinate as 'time' t. The Nieh–Yan deformation gives

$$\begin{aligned} -2 \int dF \wedge T^a \wedge e_a = 2 \int dt \omega \{ & -\dot{F} \tilde{T}_\alpha \omega \tilde{e}^\alpha - \dot{\tilde{e}}_\alpha \omega \tilde{d} F \omega \tilde{e}^\alpha \\ & + N[2 \tilde{d} F \omega K_\alpha \omega \tilde{e}^\alpha] + N^\alpha [2 \tilde{d} F \omega \tilde{T}_\alpha] + Q^\alpha [-\sigma \epsilon_{\alpha\beta\gamma} \tilde{d} F \omega \tilde{e}^\beta \omega \tilde{e}^\gamma] \}. \end{aligned} \tag{12.55}$$

We see that the F field makes a contribution to the constraints, and has a conjugate momentum proportional to the scalar part of the torsion (the part transverse to the radial direction). The full bulk action becomes

$$\begin{aligned} I = \int dt \wedge \Bigg(& \dot{\tilde{e}}^\alpha \wedge (4 \sigma_\perp \epsilon_{\alpha\beta\gamma} K^\gamma \wedge \tilde{e}^\beta - 2 \tilde{d} F \wedge \tilde{e}_\alpha) - 2 \dot{F} (\tilde{e}^\alpha \wedge \tilde{T}_\alpha). \\ & + N \left\{ 2 \epsilon_{\alpha\beta\gamma} \left({}^{(3)}R^{\alpha\beta} - \sigma_\perp K^\alpha \wedge K^\beta - \frac{\Lambda}{3} \tilde{e}^\alpha \wedge \tilde{e}^\beta \right) \wedge \tilde{e}^\gamma + 4 \tilde{d} F \wedge K_\alpha \wedge \tilde{e}^\alpha \right\} \end{aligned}$$

$$
\begin{aligned}
&+4N^{\alpha}\Big\{-\sigma_{\perp}\epsilon_{\alpha\beta\gamma}(\tilde{D}K)^{\beta}\wedge\tilde{e}^{\gamma}+\tilde{d}F\wedge\tilde{T}_{\alpha}\Big\}\\
&+4Q^{\alpha}\left\{(K_{\beta}\wedge\tilde{e}^{\beta})\wedge\tilde{e}_{\alpha}-\frac{1}{2}\sigma\epsilon_{\alpha\beta\gamma}\tilde{d}F\wedge\tilde{e}^{\beta}\wedge\tilde{e}^{\gamma}\right\}\\
&+4q^{0}{}_{\alpha}\{\epsilon^{\alpha}{}_{\beta\gamma}\tilde{T}^{\beta}\wedge\tilde{e}^{\gamma}\}\Bigg).
\end{aligned}
\tag{12.56}
$$

We notice that the Q-constraint term can be written in the form

$$
4Q_{\alpha}\tilde{e}^{\alpha}\omega\big(K_{\beta}\wedge\tilde{e}^{\beta}-\sigma *_{3}\tilde{d}F\big). \tag{12.57}
$$

Because of this constraint (which relates the antisymmetric part of the extrinsic curvature to the vorticity of F), the momentum conjugate to $\tilde{e}^{\alpha}$ is symmetric, i.e.

$$
\begin{aligned}
\Pi_{\alpha}&=4\sigma_{\perp}\epsilon_{\alpha\beta\gamma}K^{\gamma}\wedge\tilde{e}^{\beta}-2\tilde{d}F\wedge\tilde{e}_{\alpha}\\
&=4\sigma_{\perp}\left(\epsilon_{\alpha\beta\gamma}K^{\gamma}\wedge\tilde{e}^{\beta}-\frac{1}{2}\sigma_{3}*_{3}(K_{\beta}\omega\tilde{e}^{\beta})\omega\tilde{e}_{\alpha}\right).
\end{aligned}
\tag{12.58}
$$

When written out in components, one finds that the antisymmetric part $K_{[\alpha\beta]}$ cancels

$$
\Pi_{\alpha}=4\sigma_{\perp}(K_{(\beta\alpha)}-trK\ \eta_{\beta\alpha})\tilde{e}^{\beta}. \tag{12.59}
$$

This result is consistent with the fact noted above, that the system may be equivalently described as a pseudoscalar field coupled to torsionless gravity. Moreover, if we take the deDonder gauge $d^{\dagger}\tilde{e}^{\alpha}=0$, the torsion constraint implies that B is symmetric.

The $q^{0}{}_{\alpha}$ constraint yields $\tilde{T}^{\beta}{}_{\alpha\beta}=0$. Out of the nine components of $\tilde{T}$, which transform as **5** + **3** + **1** under $SO(3)$ (or $SO(2,1)$), this sets the triplet to zero (the **5** also vanishes on an equation of motion). The momentum conjugate to F is given by

$$
\Pi_{F}=-2\epsilon^{\alpha\beta\gamma}\tilde{T}_{\alpha\beta\gamma}. \tag{12.60}
$$

This is the singlet part of the torsion, which has become dynamical in this description of the theory, in the sense that it is canonically conjugate to F.

12.5 The Torsion Domain Wall

We will now simplify the analysis by taking a coordinate basis and looking for solutions of the form

$$
\tilde{e}^{\alpha}=e^{A(t)}dx^{\alpha},\quad N=1,\quad N^{\alpha}=0, \tag{12.61}
$$

and we will further suppose that $F = F(t)$. In this case K^α and B^α reduce to one degree of freedom each as a result of the constraints

$$K_\alpha = k\tilde{e}_\alpha, \qquad B_\alpha = b\tilde{e}_\alpha, \tag{12.62}$$

and one finds $\Pi_A = -4\sigma_\perp k$ and $\Pi_F = 2\sigma b$. The action then takes the following relatively simple Hamiltonian form

$$I_{\rm NY} \propto \int dt\, d^3x\, e^{3A(t)} \left[\dot{A}\Pi_A + \dot{F}\Pi_F - \left(\frac{1}{2}\sigma_3\Pi_F^2 + \frac{1}{8}\sigma_\perp\Pi_A^2 + \frac{2}{3}\Lambda\right)\right]. \tag{12.63}$$

and the equations of motion give

$$\begin{aligned} &\dot{\Pi}_A = 3\dot{F}\Pi_F, \quad \dot{\Pi}_F + 3\Pi_F\dot{A} = 0, \quad \Pi_A = 4\sigma_\perp\dot{A}, \quad \Pi_F = \sigma_3\dot{F}, \\ &\Pi_A^2 + 4\sigma\Pi_F^2 + \frac{16}{3}\sigma_\perp\Lambda = 0. \end{aligned} \tag{12.64}$$

These equations of motion could of course alternatively be obtained by considering the theory in the form (12.52). It is convenient to rescale $F(t) = \frac{1}{3}\Theta(t)$. Then the equations of motion can be put in the form

$$\dot{A} + 3\dot{A}^2 - 3a^2 = 0, \quad \dot{A} = \frac{1}{12}\sigma\dot{\Theta}^2, \quad \dot{\Theta} + 3\dot{\Theta}\dot{A} = 0. \tag{12.65}$$

where we have set $\Lambda = -3\sigma_\perp a^2$ with $a = 1/L$. These are of the standard form of domain wall equations that have appeared numerous times in the AdS/CFT literature. However, there is a crucial difference. Notice that the first two of (12.65) imply

$$\dot{A}^2 + \frac{1}{36}\sigma\dot{\Theta}^2 - a^2 = 0. \tag{12.66}$$

For Euclidean signature ($\sigma = \sigma_3 = 1$) the second term in (12.66) has *positive* sign in contrast to most of the other holographic studies. This is due to the fact that in passing from Lorentzian to Euclidean signature the pseudoscalar kinetic term acquires the 'wrong sign' [39]. This property allows for a remarkable exact solution to the above system of non-linear equations in Euclidean signature, which we refer to as the *torsion DW*. To obtain it we define

$$h(t) = \dot{A}(t), \tag{12.67}$$

at which point we have

$$\dot{h} = \frac{1}{12}\dot{\Theta}^2, \quad \dot{h} + 3(h^2 - a^2) = 0. \tag{12.68}$$

The general solution is of the form (Fig. 12.1)

$$h(t) = a \tanh 3a(t - t_0) \tag{12.69}$$

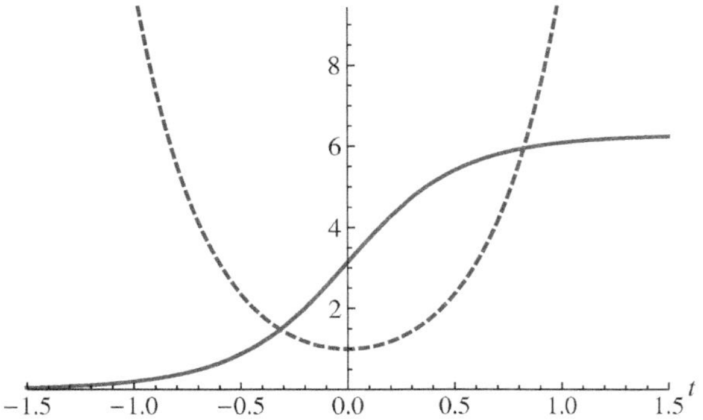

Fig. 12.1 Horizon versus parameter r_0 for $g = 3$ (*left panel*) and $g = 0.0005$ (*right panel*)

and we then have

$$\Pi_F = \dot{F} = \pm 2\sqrt{a^2 - h^2(t)} = \pm 2a \text{ sech } 3a(t - t_0) \tag{12.70}$$

which gives

$$\Theta(t) = \Theta_0 \pm 4 \arctan\left(e^{3a(t-t_0)}\right). \tag{12.71}$$

The $\pm$ sign corresponds to kink/antikink and we will without loss of generality choose the $+$ sign. We may also solve for

$$e^{A(t)} = \alpha(2\cosh 3a(t - t_0))^{1/3} \tag{12.72}$$

The parameter α is an arbitrary positive integration constant that sets the overall scale of the spatial part of the metric. t_0 may be interpreted as the position of the DW; when $t_0 = 0$ the torsion DW sits in the middle between the two asymptotically AdS_4 regimes. Below, we will discuss the interesting holographic interpretation of the torsion DW.

Note the curvature and torsion of this solution:

$$R^\alpha{}_\beta = -\dot{F}\dot{A}\epsilon^\alpha{}_{\beta\gamma} dt\omega e^\gamma - a^2 e^\alpha \omega e_\beta, \tag{12.73}$$

$$R^\alpha{}_0 = \left(\dot{h} + h^2\right) dt\omega e^\alpha - \frac{1}{2}\dot{F}\dot{A}\epsilon^\alpha{}_{\beta\gamma} e^\beta \omega e^\gamma, \tag{12.74}$$

$$T^\alpha = -\frac{1}{2}\dot{F}\epsilon^\alpha{}_{\beta\gamma} e^\beta \omega e^\gamma, \tag{12.75}$$

$$T^0 = 0. \tag{12.76}$$

These are non-singular for all $t \in (-\infty, \infty)$. The torsion DW solution has divergent action, but this divergence is cancelled by boundary counterterms, the same counterterms which render the action of AdS_4 finite. To see this, the energy of the torsion DW can be computed by evaluating the Euclidean action on the solution. Introducing a cutoff at $t = \pm L$, we find

$$I_{tv,\text{on-shell}} = 4a^2 \int \epsilon_{\alpha\beta\gamma} dx^\alpha \omega dx^\beta \omega dx^\gamma \int dt e^{3A(t)} \tag{12.77}$$

$$= \left(6 \int \widehat{Vol_3}\right) \cdot \left(\frac{4}{3} a\alpha^3 e^{3aL} + \cdots\right), \tag{12.78}$$

where the ellipsis contains terms that vanish when the cutoff is removed. As in pure AdS_4, an appropriate counterterm is of the form [40, 41]

$$I_{c.t.} = -\frac{4a}{3} \int_{\partial M} \epsilon_{\alpha\beta\gamma} \tilde{e}^\alpha \omega \tilde{e}^\beta \omega \tilde{e}^\gamma. \tag{12.79}$$

In the present case, we have such a counterterm on *each* asymptotic boundary, and thus we find

$$I_{c.t.} = -2\frac{2a}{3}\alpha^3 e^{3aL} \cdot \left(6 \int \widehat{Vol_3}\right), \tag{12.80}$$

which exactly cancels the divergent energy of the torsion DW.

Furthermore, we note that in the Kalb–Ramond representation, the solution has

$$H = \dot{\Theta} Vol_3 = \pm 6 a\alpha^3 \widehat{Vol_3} \equiv \hat{H} \widehat{Vol_3}, \tag{12.81}$$

where $\widehat{Vol_3} = \frac{1}{6} \epsilon_{\alpha\beta\gamma} dx^\alpha \omega dx^\beta \omega dx^\gamma$. This corresponds to a 'topological quantum number' of the kink

$$\int *_4 H = \pm \Delta\Theta = \pm 2\pi. \tag{12.82}$$

12.6 The Gravity Dual of Parity Symmetry Breaking

The holographic interpretation of the torsion DW is rather interesting. To study this, we set to zero without loss of generality the integration constant $\Theta_0 = 0$ and pick the plus sign in (12.70), (12.71). Next we need the asymptotic expansion of the vierbein which reads

$$\tilde{e}^\alpha = 2^{-1/3} \alpha e^{\pm a(t-t_0)} \left(1 + \frac{1}{3} e^{\mp 6a(t-t_0)} + \cdots\right) dx^\alpha \quad \text{for } t \to \pm\infty. \tag{12.83}$$

This shows that our solution is asymptotically anti-de Sitter for both $t \to \pm\infty$. The two asymptotic AdS spaces have the same cosmological constant. From this expansion we could read the expectation value of the renormalized boundary energy momentum tensor which would be given by the coefficient of the $e^{\pm 3at}$ term (see e.g. [19, 20]). Such a term is missing in (12.83), hence the expectation value of the boundary energy momentum tensor is zero.

It is not immediately apparent how to interpret these two asymptotic regimes. Are they truly distinct, or should they be identified in some way? We note that the pseudoscalar behaves in these asymptotic regimes as

$$\Theta(t) \to 4e^{-3a(t-t_0)} - \frac{4}{3}e^{-9a(t-t_0)} + \cdots \quad \text{for } t \to -\infty, \tag{12.84}$$

$$\Theta(t) \to 2\pi - 4e^{3a(t-t_0)} + \frac{4}{3}e^{9a(t-t_0)} + \cdots \quad \text{for } t \to +\infty. \tag{12.85}$$

From the above we confirm that $\Theta(t)$ is dual to a dimension $\Delta = 3$ boundary pseudoscalar that we denote $\mathcal{O}_3$. In each one of the asymptotically AdS regimes, the leading constant behavior of $\Theta(t)$ corresponds to the source (i.e., coupling constant) for $\mathcal{O}_3$ and the subleading term proportional to $e^{\mp 3a(t-t_0)}$ to the expectation value $\langle \mathcal{O}_3 \rangle$. The two asymptotic regimes are distinguished by the behavior of Θ. In fact, the essential difference is *parity*.

We can now describe the holography of our torsion DW. In the $t \to -\infty$ boundary sits a three-dimensional CFT at a parity breaking vacuum state. The order parameter is the expectation value of the pseudoscalar which is $\langle \mathcal{O}_3 \rangle = 4$ in units of the AdS radius. The expectation value breaks of course the conformal invariance of the boundary theory. Then, the theory is deformed by the same pseudoscalar operator $g\mathcal{O}_3$ where g is a marginal coupling. The torsion DW provides the holographic description of that deformation. Nevertheless, our solution should not be interpreted in terms of the usual holographic renormalization group flow. In our case, at $t \to +\infty$ the space becomes AdS with the *same* radius as at $t \to -\infty$. Hence, the two boundary theories have the same 'central charges'.[5]

We suggest that instead of interpreting the solution in terms of an RG flow, we should think of it as a domain wall transition between two inequivalent vacua of a single theory. This statement is supported by the behavior of $\Theta(t)$ in the two asymptotic regimes. For $t \to \infty$ the pseudoscalar asymptotes to the configuration (12.85). The interpretation is now that when the marginal coupling takes the fixed value $g_* = 2\pi$ we are back to the *same* CFT (i.e. having the same central charge) however in a distinct parity breaking vacuum such that $\langle \mathcal{O}_3 \rangle = -4$. In others words, the two asymptotic AdS regimes seem to describe two distinct parity breaking vacua of the same theory. The two vacua are distinguished by the expectation value of the parity breaking order parameter being $\langle \mathcal{O}_3 \rangle = \pm 4$. Quite remarkably, we also seem to find that starting in one of the two vacua, we can

[5] We use "central charge" in $d = 3$ for a quantity that counts the massless degrees of freedom at the fixed point. Such a quantity may be taken to be the coefficient in the two-point function of the energy momentum tensor or the coefficient of the free energy density. There is no conformal anomaly in $d = 3$.

reach the other by a marginal deformation with a *fixed* value of the deformation parameter.

Since the marginal operator is of dimension $\Delta = 3$ and parity odd, we tentatively identify it with a Chern–Simons operator of a boundary gauge field. In this case the torsion DW induces the T-transformation in the boundary CFT [38, 42]. In Appendix we will argue that the 3D Gross–Neveu model coupled to abelian gauge fields exhibits a large-N vacuum structure that matches our holographic findings. Although our bulk model is extremely simple to provide details for its possible holographic dual, we regard this remarkable similarity as strong qualitative evidence that our torsion DW is the gravity dual of the 'tunnelling' between different parity breaking vacua in three dimensions. However, in a three-dimensional quantum field theory, we do not expect that tunnelling can occur because of large volume effects, and distinct vacua remain orthogonal. Thus, referring to the torsion DW as a tunnelling event should be taken figuratively. We leave to future work a more careful study of the boundary interpretation of the torsion DW solution. An interpretation will depend on the precise topology of the boundary [43]. Moreover, embedding our model into M-theory could provide additional clues regarding its holography.

12.7 Physics in the Bulk: The Superconductor Analogy

The bulk interpretation of the exact solution is also interesting. Because the pseudoscalar field undergoes $\Theta(t) \to \Theta(t) + 2\pi$ under t goes from $-\infty$ to $+\infty$, the exact solution corresponds to a topological kink. It satisfies

$$\int dt\dot{\Theta} = 2\pi$$

In Fig. 12.2, we plot the solution.

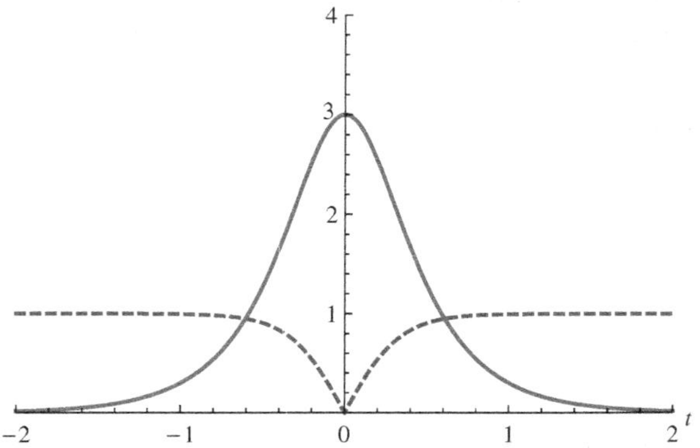

Fig. 12.2 The *blue dashed line* is $|h(t)|$, resembling the order parameter of a superconductor, while the *solid red line* is Π_F, analogous to the magnetic induction of an Abrikosov DW

12.7.1 Torsion Domain Wall Versus Abrikosov Vortex

The gravity DW solution (12.68–12.71) bears some resemblance to the Abrikosov vortex of superconducting systems. To avoid confusion we emphasize here that this is *not* a holographic interpretation i.e. the Abrikosov vortex (or more precisely, domain wall) is in the bulk. In this section, we will explore this and point out some possibly interesting features. The first thing to notice is that the plot in Fig. 12.2 is identical to the profile of an Abrikosov vortex (see for example Fig. 5.1 in Ref. [44].) The codimension differs and this is expected; the torsion DW supports a three-form field strength in contrast to a two-form field strength supported by the Abrikosov vortex. Nevertheless, there is a correspondence between our radial t-direction and the radial direction in the Abrikosov vortex, and $|h|$ and Π_F correspond to the condensate and magnetic induction of the superconductor, respectively. Table 12.1 summarizes the correspondence. In this correspondence, since the order parameter is $h = \dot{A}$, the superconducting phase (constant order parameter) corresponds to AdS_4, while the normal phase corresponds to flat space ($h = 0$). Far away from the core of the torsion DW, the geometry is asymptotically AdS, but at the core the spatial slice (at $t \to t_0$) becomes flat. To see this, note that if we think of the system as a pseudoscalar coupled to torsionless gravity, the torsion DW has $\overset{\circ}{\omega}{}^{\alpha}{}_{\beta} = 0$ and $\overset{\circ}{\omega}{}^{\alpha}{}_{0} = \dot{A}\tilde{e}^{\alpha}$, and so

$$\overset{\circ}{R}{}^{\alpha}{}_{\beta} = -h^2\tilde{e}^{\alpha}\omega\tilde{e}_{\beta}, \tag{12.86}$$

$$\overset{\circ}{R}{}^{\alpha}{}_{0} = (\dot{h} + h^2)dt\omega\tilde{e}^{\alpha}, \tag{12.87}$$

$$\overset{\circ}{T}{}^{\alpha} = 0. \tag{12.88}$$

Thus, at the core, we find that the Riemann tensor has components

$$R^{\alpha}{}_{0\alpha 0} \to -3a^2\alpha, \tag{12.89}$$

$$R^{\alpha}{}_{\beta\alpha\beta} \to 0. \tag{12.90}$$

This behavior is in line with an Abrikosov vortex in which there is normal phase at the core and superconducting phase away from the core.

Table 12.1 Abrikosov vortex versus torsion DW

Abrikosov vortex	Torsion DW
Order parameter Φ	Order parameter $\lvert h\rvert = \lvert\dot{A}\rvert$
$T - T_c$	Λ
Magnetic induction B	Π_F
Magnetic field H	$\hat{H}$
$\mathbb{Z}$-quantized magnetic flux	$\mathbb{Z}_2$-quantized electric flux

The analogue of the magnetic field is what we have called $\hat{H}$, proportional to the constant α^3. In the DW, the magnetic induction, analogous to Π_F, has a penetration length $\lambda \sim 1/3a$, and the coherence length of the order parameter is $\xi \sim 1/6a$. The penetration and coherence lengths are obtained by looking at the exponential fall-off of these quantities in the core of the DW, away from their values in the superconducting phase.

The torsion DW also has a quantized flux $\int *_4 H = \Delta\Theta = 2\pi$. This flux is independent of any parameters of the solution and of any rescaling of fields in the theory. Thus, this is an analogue of the quantized magnetic flux in superconductivity.

Finally, note the following interesting feature. If we take a derivative of the second equation in (12.68), we arrive at

$$\ddot{h} - 6\Lambda h - 18h^3 = 0. \tag{12.91}$$

This looks like a Landau–Ginzburg equation of motion of an effective ϕ^4 theory. This leads us to interpret

$$\Lambda = -3\sigma_\perp a^2 = -3\sigma_\perp \frac{1}{L^2} \sim T_c - T \tag{12.92}$$

Of course, there is no real temperature in the case of the torsion DW, but we note that this implies that the penetration and coherence lengths diverge as

$$(T - T_c)^{1/2} \tag{12.93}$$

i.e. with a mean field theory critical exponent $1/2$, as in superconductivity.

12.7.2 Domain Wall Condensation

In the last section, we noted that there is a strong analogue between the torsion DW solution and superconductivity. It is intriguing to carry the analogy further and consider multi-DW configurations. We have noted that at the core of the torsion DW, the spatial sections are flat. Thus, one might imagine that if it was favourable for torsion DWs to condense, as DWs do in Type I superconductors, then finite regions of normal phase (corresponding to $\Lambda = 0$) would be obtained. We will argue below that this can in fact occur, although the system appears not to be unstable.

To understand the physics involved, the first step is to consider a configuration of two DWs. In the superconductivity literature, this is a standard computation. One takes two DWs separated by a distance ℓ and computes the Euclidean action. More precisely, we will treat this here as follows. Put the system in a box by restricting the $t \in [-L, L]$. In such a case, a DW located at t_0 has on-shell action

$$e^{A(t-t_0)} = \alpha(2\cosh 3a(t - t_0))^{\frac{1}{3}} \Rightarrow I^L_{\text{on-shell}} = \left(6\hat{Vol}_3\right)\frac{4}{3}a\alpha^3 e^{3a(L-t_0)}. \tag{12.94}$$

We then consider a piecewise solution of two DWs located at $t = \pm\frac{L}{2}$ This is not a solution of the e.o.m. because solutions of non-linear equations cannot be simply superimposed i.e. it fails at the midpoint between the DWs. However, if we simply evaluate the Euclidean action, we find

$$I_{\text{on-shell}}^{\pm(L/2)} = \left(\frac{4}{3}a^2\hat{Vol}_3\right)4a\alpha^3 \sinh\frac{3aL}{2}. \tag{12.95}$$

Note that this is positive, so one might naively conclude that the DWs repel each other. However, recall that the DW profile exists not in flat space–time, but in the metric given by (12.72), which rises asymptotically. As a result, as we move the DWs further apart, there is a corresponding rise in the metric between the DWs. So, we should directly evaluate the force at a point $t = \ell$ as

$$F = -\frac{dI^{\pm(L/2)}}{d\ell} \propto \cosh(3a\ell/2) < 0. \tag{12.96}$$

Thus we conclude that the DWs in fact *attract* each other. In the superconducting analogue we would conclude that we have a *Type I superconductor* where the vortices clump together forming (potentially) finite regions of normal phase. In such a superconductor, the number of vortices is determined by the total magnetic flux.

We now describe the analogous situation in our gravitational system. We have noted that the constant $\hat{H}$ plays the role of the external magnetic induction, while H is the magnetic field, varying within the DW, with $\Delta\Theta = \int *_4 H$. Following the superconducting analogue, if we put the system in a box of size $2L$ (that is we impose a cutoff on each AdS asymptotic) the flux conservation equation is of the form

$$\Delta\Theta = 2L\hat{H} \tag{12.97}$$

The DWs carry the flux in the superconductor, and so it is natural to ask what is the lowest energy configuration satisfying (12.97)? To analyze this, consider an array of n DWs in a region of size L_0. We take the DWs to be equally spaced, as one can show that deviating from such a configuration causes a rise in energy. For such a configuration, the flux quantization condition (12.97) gives a relation between n, L_0 and $\hat{H}$. Such a representative curve is shown in Fig. 12.3.

If we solve this equation for L_0 as a function of n and $\hat{H}$, we can then compute the energy as a function of n. One obtains a curve as in Fig. 12.4.

One notes that the energy is minimized for large n, and in that case, the size L_0 asymptotes to a fixed value, which is found to be

$$L_0 = \frac{\hat{H}}{6a} \cdot 2L = \alpha^3 \cdot 2L \tag{12.98}$$

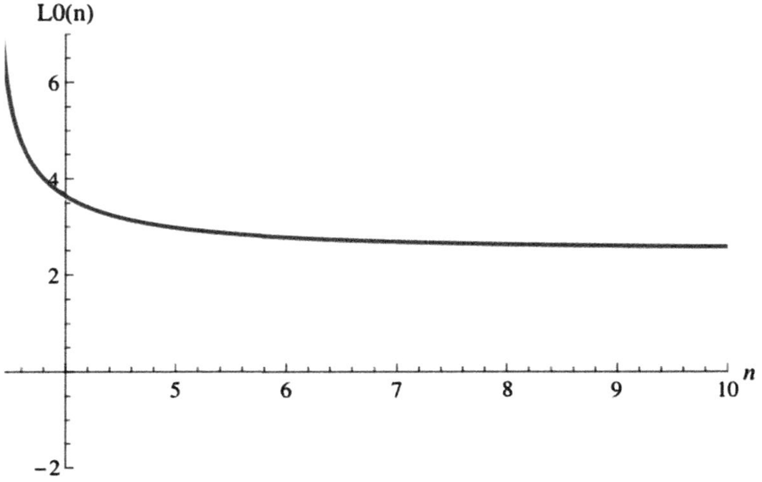

Fig. 12.3 Size of normal state droplet versus n for multi-domain walls

Fig. 12.4 A sketch of the phase diagram for a Kalb–Ramond superconductor

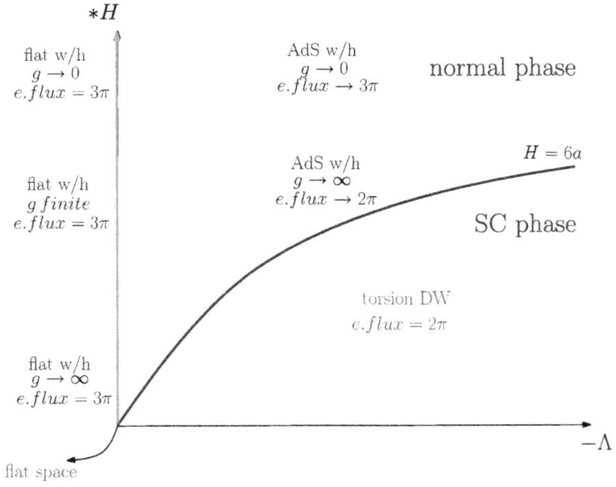

We conclude that the preferred configuration, given a fixed external flux, is a continuum of DWs arrayed over a finite size region. Within this droplet, the system is in the normal phase. We have noted that the DW core is spatially flat, and so we surmise that within the droplet, the space–time is flat. The asymptotic value of energy in Fig. 12.3 is precisely minus that contributed by the cosmological constant. Again, the size of the droplet is set by the value of the external H-flux, and the boundary conditions are AdS. Note that for a fixed cutoff, there is a critical field (given by $\hat{H} = 6a$) for which the entire spacetime is flat.

The result above motivate us to take a further step and suggest rather appealing physical picture that we may call *Kalb-Ramond superconductivity*. Here we present a qualitative description of Kalb–Ramond superconductivity. A detailed description will appear shortly in [43]. As discussed above, *minus* the cosmological constant may be interpreted as $T - T_c$. This, together with the interpretation of H as a magnetic induction and $\hat{H} = 6a$ as a critical magnetic field above which the spacetime is flat due to DW condensation, motivates to draw a $-\Lambda, *H$ graph analogously with the $T - T_c$, *magnetic field* graph in superconductivity. To do so, we need to consider all known solutions of an Einstein-axion system in $d = 4$ with and without cosmological constant. For $\Lambda = 0$ we recall the axionic DW solutions of [35]. For $\Lambda < 0$ DW wormhole-like solutions were found in [36].

These solutions have the following properties (to be discussed in more detail in [43]).

The $\Lambda = 0$ wormhole solution [35] is asymptotically flat. Its magnetic flux is proportional to an arbitrary constant g which was speculated to be quantized via a string theory embedding of the Einstein-axion system in [35]. Its electric flux is $\mathbb{Z}_2$-quantized to $\pm 3\pi$, but the electric field varies along the wormhole.

The $\Lambda < 0$ wormhole solution [36] is asymptotically AdS. Its magnetic flux is also proportional to the arbitrary constant g, and hence possibly quantized in a string theory embedding. Quite intriguingly, the electric flux on the wormhole solution interpolates between the value $\pm 3\pi$ for $g \to 0$ and $\pm 2\pi$ as $g \to \infty$. Moreover, there is a lower bound for magnetic field of the wormhole solution. Remarkably, this lower bound is the value $6a$ and coincides with the maximal magnetic field of the torsion DW.

In Fig. 5 we sketch the possible phase diagram of the Kalb–Ramond superconductor. Our torsion DW solution seems to play the role of the superconducting phase, while the wormhole solutions of [35] and [36] appear to correspond to the normal phase.

12.8 Conclusions

In this work we have presented in detail a simple toy model, the Nieh–Yan model, where torsion enters through the spacetime dependence of the coupling constant of the Nieh–Yan topological invariant. Although we have discussed the model directly in terms of torsion, it can classically be put into equivalent forms as either a massless pseudoscalar or a Kalb–Ramond field coupled to gravity. The model has an interesting and non-trivial holographic interpretation. In particular, we have shown that it possesses an exact bulk solution in Euclidean signature, termed the torsion DW, having two asymptotically AdS_4 regimes, while the pseudoscalar acquires a kink profile. We have argued then that the holographic interpretation of this torsion DW is a three-dimensional CFT with two distinct parity breaking vacua. Moreover, our bulk solution may imply that the deformation by a classically marginal pseudoscalar with a fixed coupling constant induces a transition between the two parity breaking vacua separated by a domain wall, which would be at infinity in the boundary components [43]. Remarkably, this qualitative behaviour is seen already in the three-dimensional Gross–Neveu model coupled to $U(1)$ gauge fields. The economy of our bulk model does not allow a detailed identification of the bulk and boundary theories, nevertheless we believe that our results provide a strong base where an exact bulk/boundary dictionary for AdS_4/CFT_3 can be based. A further rather intriguing property of the torsion DW is that it can be mapped into the standard Abrikosov vortex of superconductivity. Such a map identifies flat spacetime with a superconductor's normal phase, while AdS is identified with a superconducting phase. The cosmological constant would then measure the deviation from the 'critical temperature'. A phenomenon of DW condensation is found, similar to the analogous case in type I superconductors. Finally, we have briefly discussed a picture of "Kalb–Ramond superconductivity" that emerges if we view in a unified way all known four-dimensional Euclidean solutions of the Einstein-axion system. This picture will be further analysed in [43].

Our results indicate that the torsional degrees of freedom of four dimensional gravity can provide holographic descriptions for a number of interesting properties of 3D critical systems. It would be interesting to extend our analysis to more elaborate models where more torsional degrees of freedom become dynamical. It is also of interest to discuss whether our simple model can be embedded into M-theory.

Acknowledgments I thank the organizers and in particular Eleftherios Papantonopoulos for the organization of a top quality workshop and the invitation to present this talk. This work is partially supported by the FP7-REGPOT-2008-1 "CreteHEPCosmo" No 228644 and also by the University of Crete ELKE grant with KA 2745. I wish to thank R. G. Leigh and N. N. Hoang for the very fruitful and pleasant collaboration that has led to the results presented in this report.

Appendix

A.1 Parity Breaking in Three Dimensions

Consider the 3D Gross–Neveu model coupled to abelian gauge fields. The Euclidean action is[6]

$$I = -\int d^3x \left[\bar{\psi}^a (\not{\partial} - ie\not{A}) \psi^a + \frac{G}{2N} (\bar{\psi}^a \psi^a)^2 + \frac{1}{4M} F_{\mu\nu} F_{\mu\nu} \right]. \tag{A.1}$$

M is an UV mass scale. Introducing the usual Lagrange multiplier field σ, whose equation of motion is $\sigma = \frac{-2G}{N} \bar{\psi}^a \psi^a$ we can make the action quadratic in the fermions

[6] We use $\bar{\psi}^i, \psi^i (a = 1, 2, \ldots, N)$ two-component Dirac fermions. The γ-matrices are defined in terms of the usual Pauli matrices as $\gamma^i = \sigma^i \; i = 1, 2, 3$.

$$I = -\int d^3x \left[\bar{\psi}^a (\partial\!\!\!/ + \sigma - \mathrm{i}eA\!\!\!/) \psi^a - \frac{N}{2G}\sigma^2 - \frac{1}{4M} F^{\mu\nu} F_{\mu\nu} \right]. \tag{A.2}$$

The model possesses two parity breaking vacua distinguished by the value of the pseudoscalar order parameter $\langle\sigma\rangle$. This is seen as follows: switching off the gauge fields momentarily one integrates over the fermions to produce a large-N effective action as

$$\mathcal{Z} = \int (\mathcal{D}\sigma) e^{N\left[\mathrm{Tr}\log(\partial\!\!\!/ + \sigma) - \frac{1}{2G}\int d^3x\sigma^2\right]}. \tag{A.3}$$

The path integral has a non-zero large-N extremum σ_* found by setting $\sigma = \sigma_* + \frac{1}{\sqrt{N}}\lambda$

$$\mathcal{Z} = \int (\mathcal{D}\lambda) e^{N\left[\mathrm{Tr}\log(\partial\!\!\!/ + \sigma_*) - \frac{1}{2G}\int d^3x\sigma_* + \frac{1}{\sqrt{N}}\left\{\mathrm{Tr}\frac{\lambda}{\partial\!\!\!/ + \sigma_*} - \frac{\sigma_*}{G}\int d^3x\lambda\right\} + O(1/N)\right]} \tag{A.4}$$

The term in the curly brackets is the gap equation. To study it one considers a uniform momentum cutoff Λ to obtain

$$\frac{1}{G} = \int^{\Lambda} \frac{d^3p}{(2\pi)^3} \frac{2}{p^2 + \sigma_*^2} = (\mathrm{Tr}1)\left[\frac{\Lambda}{\pi^2} - \frac{|\sigma_*|}{\pi^2}\arctan\frac{\Lambda}{|\sigma_*|}\right]. \tag{A.5}$$

Defining the critical coupling as

$$\frac{1}{G_*} = \frac{\Lambda}{\pi^2}, \tag{A.6}$$

(A.5) possesses a non-zero solution for σ_* when $G > G_*$ given by

$$|\sigma_*| = \frac{2\pi}{G}\left(\frac{G}{G_*} - 1\right) \equiv m. \tag{A.7}$$

The two distinct parity breaking vacua then have

$$\sigma_* = -\frac{2G}{N}\langle\bar{\psi}^a\psi^a\rangle = \pm m. \tag{A.8}$$

Going back to (A.2) one can tune $G > G_*$ and start in any of the two parity breaking vacua. Suppose we start from $\sigma_* = +m$. To leading order in N we have

$$\begin{aligned} \mathcal{Z} &= \int (\mathcal{D}A_\mu)(\mathcal{D}\bar{\psi}^a)(\mathcal{D}\psi^a)\exp[\mathcal{S}] \\ \mathcal{S} &= \int d^3x \left[\bar{\psi}^a(\partial\!\!\!/ + m - \mathrm{i}eA\!\!\!/)\psi^a - \frac{N}{2G}m^2 + O(1/\sqrt{N}) - \frac{1}{4M}F^{\mu\nu}F_{\mu\nu}\right] \end{aligned} \tag{A.9}$$

As is well known [45, 46] for N fermions the path integral (A.9) yields an effective action for the gauge fields which for low momenta is dominated by the Chern–Simons term i.e.

$$\mathcal{Z} \approx \int e^{S_{\mathrm{CS}}}, \tag{A.10}$$

with

$$S_{\mathrm{CS}} = \mathrm{i}\frac{ke^2}{4\pi}\int d^3x \epsilon^{\mu\nu\rho} A_\mu \partial_\nu A_\rho, \quad k = \frac{N}{2}. \tag{A.11}$$

Had we started from the $\sigma_* = -m$ vacuum, we would have found again (A.10–A.11), however with $k = -\frac{N}{2}$, i.e. the vacuum with $\sigma_* = -m$ yields an effective Chern–Simons action with $k = -\frac{N}{2}$.

Consider now *deforming* the action (A.9) by the Chern–Simons term with a fixed coefficient as

$$\begin{aligned} \mathcal{Z}_{\mathrm{deform}} &= \int (\mathcal{D}A_\mu)(\mathcal{D}\bar{\psi}^i)(\mathcal{D}\psi^i) \exp[\mathcal{S}_{def}] \\ \mathcal{S}_{\mathrm{def}} &= \mathcal{S} - \mathrm{i}q \int d^3x \epsilon^{\mu\nu\rho} A_\mu \partial_\nu A_\rho \end{aligned} \tag{A.12}$$

If q is fixed to

$$q = \frac{Ne^2}{4\pi}, \tag{A.13}$$

the effective action for the gauge fields resulting from the fermionic path integrals in (A.12) is going to be *equal* to the one obtained had we started at the $\sigma_* = -m$ vacuum. In other words, deforming the $\sigma_* = +m$ vacuum with a Chern–Simons term with a fixed coefficient is equivalent to being in the $\sigma_* = -m$ vacuum. This is exactly analogous to the holographic interpretation of our torsion DW.

References

1. Herzog, C.P., Kovtun, P., Sachdev, S., Son, D.T.: Quantum critical transport, duality, and M-theory. Phys. Rev. **D75**, 085020 (2007). arXiv:hep-th/0701036
2. Hartnoll, S.A., Kovtun, P.K., Muller, M., Sachdev, S.: Theory of the Nernst effect near quantum phase transitions in condensed matter, and in dyonic black holes. Phys. Rev. **B76**, 144502 (2007). arXiv:0706.3215
3. Hartnoll, S.A., Kovtun, P.: Hall conductivity from dyonic black holes. Phys. Rev. **D76**, 066001 (2007). arXiv:0704.1160
4. Hartnoll, S.A., Herzog, C.P.: Ohm's Law at strong coupling: S duality and the cyclotron resonance. Phys. Rev. **D76**:106012 (2007). arXiv:0706.3228
5. Keski-Vakkuri, E., Kraus, P.: Quantum hall effect in AdS/CFT. arXiv:0805.4643

6. Davis, J.L., Kraus, P., Shah, A.: Gravity dual of a quantum hall plateau transition. arXiv:0809.1876
7. Hartnoll, S.A., Herzog, C.P., Horowitz, G.T.: Building a holographic superconductor. Phys. Rev. Lett. **101**, 031601 (2008). arXiv:0803.3295
8. Minic, D., Heremans, J.J.: High temperature superconductivity and effective gravity. arXiv:0804.2880
9. Nakano, E., Wen, W.-Y.: Critical magnetic field in a holographic superconductor. Phys. Rev. **D78**, 046004 (2008). arXiv:0804.3180
10. Albash, T., Johnson, C.V.: A holographic superconductor in an external magnetic field. arXiv:0804.3466
11. Gubser, S.S., Pufu, S.S.: The gravity dual of a p-wave superconductor. arXiv:0805.2960
12. Herzog, C.P., Kovtun, P.K., Son, D.T.: Holographic model of superfluidity. arXiv:0809.4870
13. Basu, P., Mukherjee, A., Shieh, H.H.: Supercurrent: vector hair for an AdS black hole. arXiv:0809.4494
14. Gubser, S.S.: Breaking an Abelian gauge symmetry near a black hole horizon. arXiv:0801.2977
15. Gubser, S.S.: Colorful horizons with charge in anti-de Sitter space. arXiv:0803.3483
16. Gubser, S.S., Rocha, F.D.: The gravity dual to a quantum critical point with spontaneous symmetry breaking. arXiv:0807.1737
17. Leigh, R.G., Petkou, A.C.: Gravitational duality transformations on (A)dS4. JHEP **11**, 079 (2007). arXiv:0704.0531.
18. de Haro, S., Petkou, A.C.: Holographic aspects of electric–magnetic dualities. J. Phys. Conf. Ser. **110**, 102003 (2008). arXiv:0710.0965
19. Mansi, D.S., Petkou, A.C., Tagliabue, G.: Gravity in the 3 + 1-split formalism I: holography as an initial value problem. arXiv:0808.1212
20. Mansi, D.S., Petkou, A.C., Tagliabue, G.: Gravity in the 3 + 1-split formalism II: self-duality and the emergence of the gravitational Chern–Simons in the boundary. arXiv:0808.1213
21. de Haro, S.: Dual gravitons in AdS4/CFT3 and the holographic cotton tensor. arXiv:0808.2054
22. Giombi, S., Yin, X.: Dual gravitons in AdS4/CFT3 and the holographic cotton tensor. arXiv:0912.3462
23. Bagger, J., Lambert, N.: Modeling multiple M2's. Phys. Rev. **D75**, 045020 (2007). arXiv:hep-th/0611108
24. Bagger, J., Lambert, N.: Gauge symmetry and supersymmetry of multiple M2-branes. Phys. Rev. **D77**, 065008 (2008). arXiv:0711.0955
25. Bagger, J., Lambert, N.: Comments on multiple M2-branes. JHEP **02**, 105 (2008). arXiv:0712.3738
26. Gustavsson, A.: Algebraic structures on parallel M2-branes. arXiv:0709.1260
27. Hosomichi, K., Lee, K.M., Lee, S., Lee, S., Park, J.: N = 4 superconformal Chern–Simons theories with hyper and twisted hyper multiplets. JHEP **0807**, 091 (2008). arXiv:0805.3662
28. Aharony, O., Bergman, O., Jafferis, D.L., Maldacena, J.: N = 6 superconformal Chern–Simons-matter theories, M2-branes and their gravity duals. arXiv:0806.1218
29. Shapiro, I.L.: Physical aspects of the space–time torsion. Phys. Rept. **357**, 113 (2002). arXiv:hep-th/0103093
30. Chandia, O., Zanelli, J.: Torsional topological invariants (and their relevance for real life). arXiv:hep-th/9708138
31. Freidel, L., Minic, D., Takeuchi, T.: Quantum gravity, torsion, parity violation and all that. Phys. Rev. D**72**, 104002 (2005). arXiv:hep-th/0507253
32. Mercuri, S.: From the Einstein–Cartan to the Ashtekar–Barbero canonical constraints, passing through the Nieh–Yan functional. Phys. Rev. **D77**, 024036 (2008). arXiv:0708.0037 [hep-th]
33. Canfora, F.: Some solutions with torsion in Chern–Simons gravity and observable effects. arXiv:0706.3538 [hep-th]
34. Leigh, R.G., Hoang, N.N., Petkou, A.C.: JHEP 0903:033 (2009). arXiv:0809.5258 [hep-th]

35. Giddings, S.B., Strominger, A.: Axion induced topology change in quantum gravity and string theory. Nucl. Phys. **B306**, 890 (1988)
36. Gutperle, M., Sabra, W.: Instantons and wormholes in Minkowski and (A)dS spaces. Nucl. Phys. **B647**:344–356 (2002)
37. D'Auria, R., Regge, T.: Gravity theories with asymptotically flat instantons. Nucl. Phys. **B195**:308 (1982)
38. Leigh, R.G., Petkou, A.C.: SL(2,Z) action on three-dimensional CFTs and holography. JHEP **12**, 020 (2003). arXiv:hep-th/0309177
39. Gibbons, G.W., Green, M.B., Perry, M.J.: Instantons and Seven–Branes in Type IIB superstring theory. Phys. Lett. **B370**, 37–44 (1996). arXiv:hep-th/9511080
40. Balasubramanian, V., Kraus, P.: A stress tensor for anti-de Sitter gravity. Commun. Math. Phys. **208**, 413–428 (1999). arXiv:hep-th/9902121
41. Kraus, P., Larsen, F., Siebelink, R.: The gravitational action in asymptotically AdS and flat spacetimes. Nucl. Phys. **B563**:259–278 (1999). arXiv:hep-th/9906127
42. Witten, E.: SL(2,Z) action on three-dimensional conformal field theories with Abelian symmetry. arXiv:hep-th/0307041
43. Leigh R.G., Nguyen-Hoang N., Petkou, A.C.: JHEP 0903, 033 (2009). arXiv:0809.5258 [hep-th]
44. Tinkham, M.: Introduction to Superconductivity, 2nd ed. Dover Publications, New York (1996)
45. Niemi, A.J., Semenoff, G.W.: Axial anomaly induced Fermion fractionization and effective gauge theory actions in odd dimensional space-times. Phys. Rev. Lett. **51**, 2077 (1983)
46. Redlich, A.N.: Parity violation and gauge noninvariance of the effective gauge field action in three-dimensions. Phys. Rev. **D29**:2366–2374 (1984)

Index

GPSR Compliance

The European Union's (EU) General Product Safety Regulation (GPSR) is a set of rules that requires consumer products to be safe and our obligations to ensure this.

If you have any concerns about our products, you can contact us on ProductSafety@springernature.com

In case Publisher is established outside the EU, the EU authorized representative is:

Springer Nature Customer Service Center GmbH
Europaplatz 3
69115 Heidelberg, Germany

Batch number: 10838015

Printed by Printforce, the Netherlands